CAMS FOR INDUSTRY

Cams for Industry

A handbook for designers of special-purpose machines

John Reeve

Mechanical Engineering Publications Limited, London

First published 1995

ISBN 0 85298 960 1

A CIP catalogue record for this book is available from the British Library.

Typeset by Paston Press Ltd, Loddon, Norfolk
Printed and Bound by Antony Rowe Ltd, Chippenham, Wiltshire

CONTENTS

AUTHOR'S PROFILE

John Reeve is a chartered engineer with a lifetime's experience in mechanical engineering, mostly concerned with the design of automatic machinery.

After a grammar school education he embarked on a full-time apprenticeship with a medium-sized firm of general mechanical engineers whose main product line was bottling machinery, dairy and brewery plant, and equipment. Four years of practical shop floor training was followed by a year in the drawing office, culminating in two further years as a designer/draughtsman.

During this period he gained a Higher National Certificate in Mechanical Engineering at evening classes, winning an IMechE National Certificate prize in the final year. After further part-time study he was elected a Graduate of both the Institutions of Mechanical and Production Engineers and later was elected a full Member of the Institution of Mechanical Engineers.

He spent seven years as the development engineer for a company that manufactured cardboard boxes, designing special purpose automatic cutting, folding, printing, and gluing machines. This was followed by a brief period as chief designer in a food processing machinery firm and another seven years as chief designer for a medium-sized general engineering company. The range of machine design experience in the latter company was very wide indeed, including gearing, power transmission, automatic bottle filling machines, bottle washing machines, labelling machines, mechanical handling equipment, etc., and other special-purpose machines for various industries.

The final phase of his career started as chief designer and later chief engineer for Manifold Machinery, a subsidiary of the (then) Thorn Electrical Industries group, who designed and manufactured machinery for the mass production of electric lamps. Lamp-making machines contain many complex component manipulating mechanisms, which are invariably operated by cams. The versatility and reliability of cams make them ideal for these machines and for automatic machines in other mass production industries. Most of the machines use intermittent motion ('indexing') to move a heavy table with work-holding fixtures from work station to work station. Cam-driven indexing mechanisms are traditionally used for this because of their smooth motion, accuracy, and reliability.

After several years an independent company, Manifold Industries Limited, was formed to design and manufacture cam-operated indexing mechanisms and other cam systems for industry in general, of which the author was technical director until his retirement. In that role he was responsible for the design of the Company's products and for advising on the best use of them in industrial machinery applications worldwide. As such he was involved for a period of more than twenty years in the design of high-performance cam mechanisms for every kind of application, ranging from automotive production to packaging machinery.

During that time he was privileged to serve for several years on the Mechanisms Committee of the Applied Mechanics Group of the Institution of Mechanical Engineers and, by virtue of his specialized cam design knowledge, also served on the Mechanisms Committee of the Engineering Sciences Data Unit during the compilation of the ESDU items on cam design.

Preface

This book contains basic principles, laws, and equations for the design and application of cams and cam-operated mechanisms, with particular reference to industrial machinery. Much of the material is also applicable to mechanisms that do not involve cams, and to motion systems in general.

The author has collected this material during the twenty-five years of his career in machine design, in which he specialized in the design, manufacture, and application of cam-operated mechanisms for industrial special-purpose machines, with particular emphasis in later years on cam-driven indexing mechanisms.

Much of the information is available from other, more authoritative sources, where it may be treated in greater detail, but other material, not available elsewhere, has been developed by the author, notably in Chapters 3–7, and 12–14. It is intended here to bring together a wide variety of topics that apply to cam and motion design, and present them in a concise form. An attempt has been made to give a 'common sense' explanation of the subject along with the mathematical treatment. The ability to evaluate formulae by substituting numerical values for the algebraic symbols is all that is needed to make practical use of the design equations. The aim is for *Cams for Industry* to be of use to both novice and expert. For readers who are well-experienced in the design of cams and motions some of the material is elementary and may seem to be self-evident, but because it is so fundamental to a thorough understanding of the subject it is hoped that for such readers the time taken to refresh one's memory will not be wasted.

Most industrial mechanisms involve constant velocity motion which is well-understood by design engineers, but linkages and cam systems essentially use non-constant velocity to achieve their purpose. The effects of acceleration and deceleration on machine parts and on the loads carried by them ('payloads') must be equally well understood for the effective design of machinery which uses reciprocating, oscillating or indexing mechanisms. It is hoped that this book will fulfil that need, and allow machine designers to feel as comfortable with such devices as they are with, say, toothed gearing.

In the early days most cam profiles were composed of circular arcs and straight lines to suit the production methods available. Nowadays profiles of *any* shape can be readily produced on numerically-controlled machines. In fact it is as easy on such machine tools to create mathematical curves as it is to produce arcs and lines properly blended – perhaps easier. It is assumed that the cam designer will take full advantage of the freedom now available to use *any* desired shape, and that profiles will be chosen to give the best mechanism performance. No space has, therefore, been given specifically to arc and line profiles.

The design philosophy used is to start with the 'best' *motion* design on the assumption that the appropriate cam profile shape can be generated from it economically. Cam geometry, dynamic properties, and mechanism performance are all developed from the motion characteristics. Those characteristics are expressed in 'normalized' format, and since it is possible to normalize the motion produced by *any* cam profile, the equations and procedures described are not restricted to standard cam laws: they can also be defined for normalized special motions and even for arc and line profiles.

Computer programs accompany the book, which evaluate the most important equations and produce a drawing of the cam profile on a personal computer screen together with certain critical values. Warning is given if those values are likely to be unacceptable, so that design parameters may be changed for the better and immediately checked. This enables very rapid optimization of a cam design, and visual assessment of it. At the same time two ASCII files are produced: one contains profile and cutter pitch curve co-ordinates, pressure angles and radii of curvature, which the user can further process to control a cam profile cutting machine: the other is a .DXF file that can be used to import the profile drawing into a CAD drawing system where it can be incorporated into the detail design of the cam.

A short list of recommended further reading is given in Appendix H.

LIST OF FIGURES

to Pearl, whose support and forbearance made this possible

CHAPTER 1

Cam and Follower Systems

A cam and its follower are components of a mechanism which interact to make some part of the mechanism move in a manner which is dependent primarily on the *shape* of the cam profile. 'Profile' is the name given to the active part of the cam which is responsible for its function. Usually the cam is driven with a simple motion, probably constant velocity, and the follower, which is in contact with the cam profile, has a more complex motion. In this case the motion of the cam is said to be the *input motion* and that of the follower and attached parts the *output motion*. Power nominally flows from the input to the output. Occasionally mechanisms are found in which this system is reversed and the 'follower' becomes the input and the cam is the output: this is known as an *inverse* cam mechanism. In all cases the component whose profile determines the motion is known as the cam and the component which transfers motion by contact with the cam profile is known as the cam follower, even though it may not, strictly speaking, follow.

The most widely used type of cam and follower system in industrial machinery has a cam which rotates in contact with a cylindrical roller follower. However a cam may also reciprocate (translate) with straight line motion, oscillate with a swinging motion or remain stationary. Similarly the motion of the follower may be rotating, translating, or stationary. Followers may have other shapes, such as conical, flat-faced, knife-edge, or ball, and some mechanisms even have one cam profile driving another cam profile: the best known example of this is a pair of gear teeth, although they are designed to impart constant velocity motion, unlike most cams.

The follower itself is attached to, and imparts motion to, other mechanism components, finally to produce the motion of the 'tool' which performs the function for which the mechanism is designed. The whole system consists of a driver (perhaps an electric motor), an input transmission between the driver and the cam, the cam, the follower, the output transmission between the follower and the tool, and the tool itself. In this context a 'tool' may be a conventional tool, such as a drill or punch, or it may be a rotary table, a linear conveyor, a pusher, a workpiece carrier, or similar.

The cam cannot be designed in isolation. The design of the other components of the mechanism profoundly affects the performance of the cam. This applies particularly to the input and output transmissions, which usually consist of a variety of conventional machine elements, such as shafts, gears, couplings, sliders, and linkages, connected together in series. The inertia and rigidity of the transmissions are of particular importance and are dealt with in detail in Chapter 11. The cam profile itself can have an infinite variety of shapes, but its design must take into consideration the design of the other system components to produce the desired tool motion, economically, and with good dynamic performance.

By and large, the chapters on cam motion are concerned with the motion of the tool, rather than the follower, and it is assumed that the cam profile shape can be 'adjusted' to suit the geometry of the output transmission if necessary. The distortion of the tool motion when the cam motion is *not* adjusted is not significant if the output transmission is suitably designed, and many, perhaps most, of the cam-driven mechanisms in industrial machinery do not have adjusted cam motions.

A special kind of motion adjustment, useful for high speed operation, is known as 'tuned' motion. In designing a tuned cam an estimate is made, usually by mathematical analysis, of the elastic deflections in the whole system throughout the motion cycle. Those deflections are deliberately incorporated into the shape of the cam profile so that the follower departs slightly from the desired motion pattern, but the final 'tool' motion is as specified, and smooth. This can only work properly at one speed of operation – the design is 'tuned in' to that speed. Nevertheless, tuned systems can still give smoother performance than untuned ones at speeds close to, but not exactly at, the design speed. At speeds *very* different from the design speed, however, they can give a worse performance.

By definition, tuned cams require special design effort for each case. This is rarely justified in industrial machinery because the vibration generated by elastic deformation can and should be minimized by sound design of structures and transmissions, and by the use of 'good' cam laws. Standard cam laws are used to good effect with untuned systems in a wide range of machines, some of which run at very high speeds – over 1000 cycles per minute.

CAM AND FOLLOWER CLASSIFICATION

There is no universal standard for the classification or description of cam and follower systems. Consequently confusion often arises when systems are described without the aid of a drawing. The number of different types of system is huge, but each of the several different aspects of a system has only a few possibilities. This allows for a simple, but comprehensive, means of classifying cam and follower systems, as follows.

(1) *The basic shape of the surface*: generated by the interaction of cam and follower motions, in which the profile is traced. In the case of a rotating cam this can be a disc, a cylinder, a cone, a globoid or a 'concave globoid'. In the case of a translating cam it can be a plane or a cylinder.

(2) *The basic mode of cam motion constraint*: rotating, translating or stationary. Rotating includes swinging back and forth, sometimes called oscillating. Translating, which means moving in a straight line, includes reciprocating back and forth.

(3) *The basic mode of follower motion constraint*: rotating, translating or stationary. Rotating includes swinging, sometimes called oscillating. Translating, which means moving in a straight line, includes reciprocating.

(4) *The type of follower*: cylindrical roller, crowned roller, tapered roller (conical), spherical roller, knife-edge, flat-faced or curved (non-roller).

(5) *The type of track*: open, where the follower can only have one contact with the cam in any one position of the system, or enclosed, where the follower can have two contacts at any one position (usually these are on opposite sides of a roller follower). Open track disc cams usually have external profiles, but can have internal profiles. Enclosed tracks are also called grooves or 'boxed' tracks, and cams with such tracks are also known as groove cams or box cams.

(6) *The type of cycle repetition*: non-progressive, where the follower returns to the same position at the end of each cycle by reversing its direction of motion at some point in the cycle; or progressive, where the follower moves progressively to a new postion at the end of each cycle.

The term *Indexing* is generally used for progressive motions, which usually contain a dwell period resulting in an intermittent motion. In most cases the follower is replaced by another follower at the end of a cycle, and only returns to its original position after a number of cycle repetitions. Most indexing systems use several roller followers equally spaced around a follower wheel or 'turret'.

The following are common names for cam types.

- Disc cam, also called a plate cam, an edge cam, or a radial cam.
- Cylindrical cam, also called a drum cam or a barrel cam.
- Face cam, which is usually a disc cam with a groove track, often called a box cam.
- End cam, which is a cylindrical cam with an open track cut at one end.
- Plane cam, also called a wedge cam, a fat plate cam, or a translating cam.
- Blade cam, which is a cam with two open tracks facing in opposite directions with a fairly narrow blade in between. The blade runs between two roller followers which are fixed to the same moving support.
- Globoidal (convex) cam, sometimes called a barrel cam.
- Concave globoidal cam. This is the more common than the convex globoidal cam and tends to be called just globoidal.

Cams are often named after their approximate shape, such as the cardioid cam, which is a disc cam with a heart-shaped profile. They are sometimes named after the function they perform, e.g., valve cam, feed cam, gripper cam, etc. Names can be ambiguous and it is best to use an accurate description, better still a drawing, when specifying cam and follower systems.

Combining the many features listed above leads to a very large number of feasibile configurations.

Some of the more common types of cam and follower used in industrial machinery are shown in Figs 1.1 and 1.2, with suitable descriptions. In those descriptions the cam is assumed to be a rotating open track type unless otherwise stated. Many other combinations of cam and follower systems can be inferred from these diagrams. The variety of possibilities is too great to be illustrated here in its entirety.

The diagrams referred to below are those shown in Figs 1.1 and 1.2.

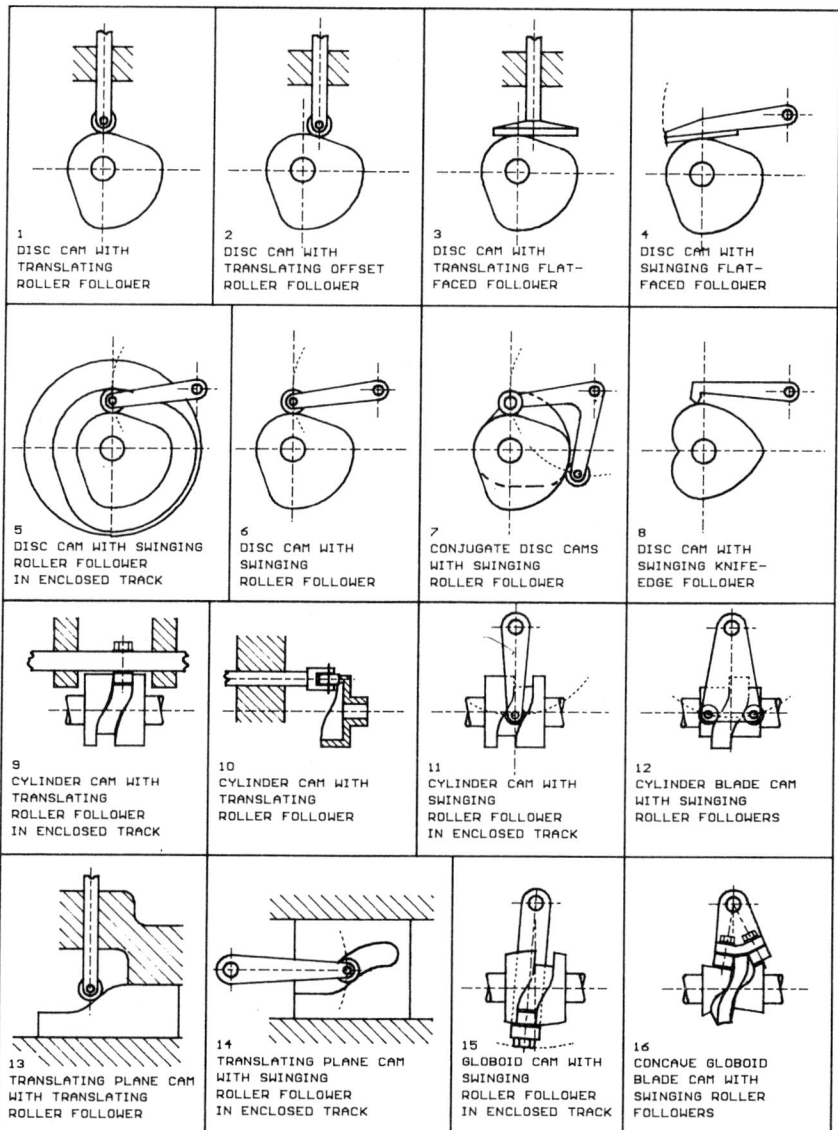

Fig. 1.1   Cam types

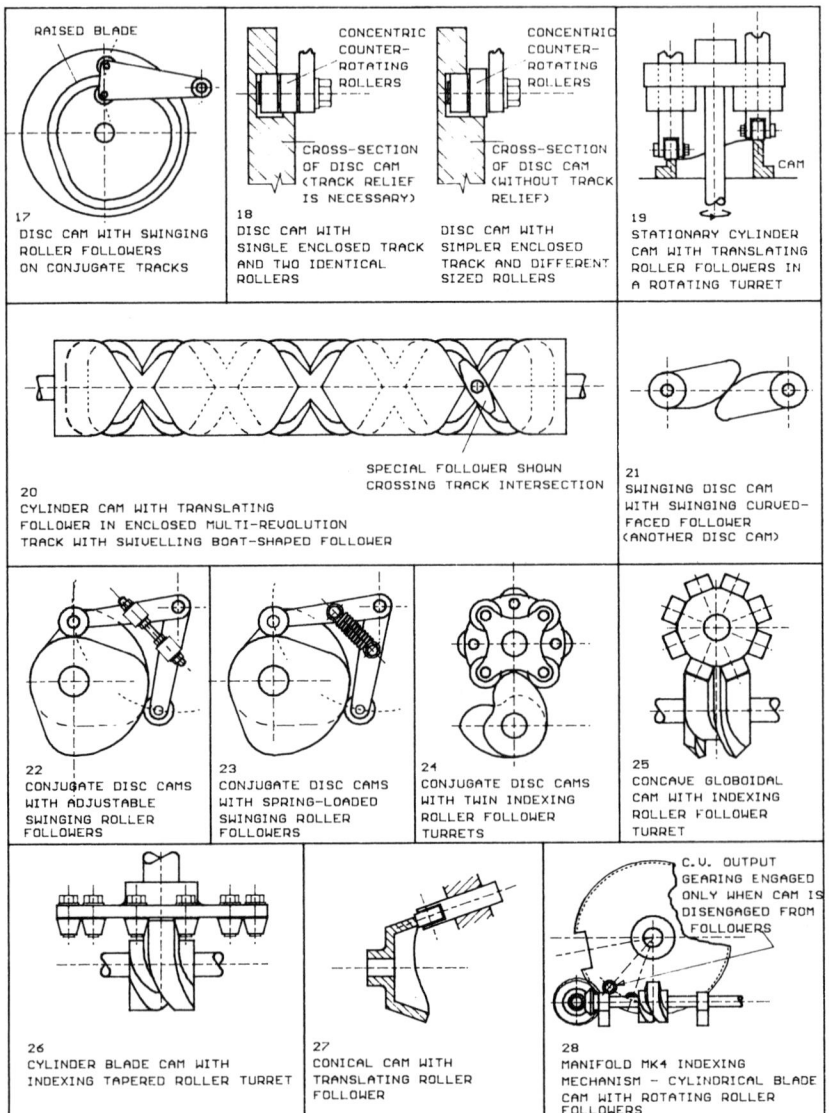

Fig. 1.2   More cam types

POSITIVE ACTION CAMS

Open track cams (diagrams 1–4) need an external force applied to the follower system to maintain cam/follower contact. This is usually provided by a spring. Enclosed track cams however (diagrams 5, 9, 11, 14, and 15) require no external force because they have a 'positive action' in both directions, that is, the cam can drive the follower either way. When positive action is required the enclosed track cam is satisfactory for most applications, but it has two drawbacks, which may be significant.

(1) To allow the follower roller to rotate the track must be larger than the roller diameter, albeit only slightly. This clearance, like any other backlash* in the system, can cause unacceptable impact forces (knocking) at high running speeds or heavy loads. Inevitably the backlash increases as the cam track and roller follower wear in service, aggravating the effects.
(2) When roller contact changes from one track wall to the other the roller must reverse its direction of rotation, which it cannot do without some degree of skidding. This can cause premature wear on the roller or the track at high running speeds or heavy loads.

Both of these drawbacks can be overcome by using a pair of *conjugate* cams (diagram 7), or conjugate tracks on one cam (diagrams 12 and 17), to provide positive action. Conjugate cams have different profiles engaging separate followers which are constrained to move in a fixed relationship to each other. The two profiles have to be manufactured very accurately and are therefore more expensive than two ordinary cam profiles or one enclosed track, but their better performance justifies the cost in some applications.

A special design of positive action cam has a single enclosed track (diagram 18) with two counter-rotating rollers on one axis, the cross-section of the track being such as to contact one roller on one side and the other roller on the opposite side. The shape of the track profile has to be no more accurately made than a conventional enclosed track, but it avoids the roller skid problem at high speed. The backlash problem

---

*Strictly speaking, 'backlash' is the shock effect of a reversal of forces in a system which has significant clearance between adjacent members that are transmitting the force. After many decades of usage it has come to mean the clearance itself, and that is how it is used throughout this book.

cannot easily be eliminated with this track design, but it is usually held within bounds if the track width and roller diameters are made to good manufacturing tolerances.

A method often employed to overcome the backlash problem is to use one of the forms of conjugate cam system, but with a means of adjusting the positions of the followers relative to each other. Diagram 22 shows one design for this on a conjugate disc cam – swinging roller follower system. The adjustment is locked when the backlash has been eliminated and, perhaps, a pre-load applied, but further adjustment may become necessary as the mechanism wears. This snag can be dealt with by replacing the lockable adjustment by a spring-loaded one (diagram 23): the strength of the spring loading must be greater than the maximum cam/ follower separating force, and this imposes extra load on both followers throughout the cycle, increasing the wear rate. As with most machine design problems, a judgement must be made to achieve an acceptable compromise.

Tapered roller followers, although mostly used to achieve near-perfect rolling with cylindrical cams, are convenient for eliminating backlash in a blade track system by moving the whole follower system closer to the cam. The cross-section of the blade is wedge-shaped to match the taper of the rollers.

The concave globoidal cam (diagram 16) inherently has a wedge-section blade when used with *cylindrical* roller followers and this too can be pre-loaded by adjusting the shaft centres distance. This is a popular design, especially for indexing mechanisms, because cylindrical rollers are generally more practical than tapered: the cam track can be produced with a cylindrical cutting tool, and high load capacity, cylindrical roller followers are commercially available.

CYCLES PER REVOLUTION

Most industrial cam systems have a reciprocating or oscillating output with a motion cycle which occupies one full revolution of the (rotating) cam. Most cam-operated indexing mechanisms also have one output motion cycle (consisting of a motion period and a dwell period) occupying one full revolution of the cam. There are usually several identical output cycles in one full revolution of the indexing turret before the mechanism returns to exactly the same position as where it started. The number of output cycles in one revolution of the *turret* is generally known as the *number of stops* of the indexing mechanism (see Chapter 8).

However, mechanisms may have two or more identical output motion cycles in one full revolution of the cam, particularly with indexing mechanisms (diagram 25). Two-cycle mechanisms are sometimes known in the cam business as *type* 2, three-cycle as *type* 3, and so on. A normal one-cycle mechanism is a type 1, but this is understood, unless otherwise stated. This terminology is *not recommended* because 'type' commonly has a wider meaning, and it can lead to confusion. It is better to specify the number of motion cycles per revolution when it is other than one, e.g., 2-cycle, 3-cycle, etc.

It is also possible to have a cam system in which the output motion cycle occupies more than one full revolution of the cam, but this is rare. The best known examples are the reel-winding mechanism cam and the Manifold Mk 4 indexer.

The reel-winding cam (diagram 20), which is used to guide wire or thread uniformly back and forth along a rotating reel, is a long cylindrical cam with both left-hand and right-hand multi-turn helical tracks connected at each end by short acceleration, dwell and deceleration zones making the complete track endless. The helices cross each other in several places, interrupting the continuity of the track walls. To safely negotiate these intersections the follower is in the form of a boat-shaped slider (diagram 20).

The diagram shows a fairly short cam with high helix lead angles for clarity of illustration, but there is almost no practical limit to the length of such a cam. It can be used to provide a very compact mechanism for quite wide reel-winding machines. A convenient way of producing a very long stroke cam is to make it in three sections all threaded onto one keyed shaft. The end sections have the deceleration, dwell and acceleration zones, produced on a cam cutting machine, while the centre section has only the helical tracks, which can be produced on a lathe or a universal milling machine.

The well-known pump-action screwdriver is an example of an inverse multi-revolution cam similar to the reel-winding cam, but without deceleration/dwell/acceleration zones, in which the 'follower' is reciprocated by hand and the 'cam' rotates to drive the tool.

The patented Manifold Mk 4 indexer uses a conventional form of indexing cam for deceleration, dwell and acceleration, but disengages the follower from the cam during the middle part of the motion cycle while the output is driven by gearing at constant velocity. During the CV period the cam does one or more idle revolutions. This mechanism comes into its own when long strokes and high inertias are involved. It is commonly used in a globoidal cam, one-stop version with reduction gearing directly attached to its output to drive a multi-stop rotary worktable. This again makes for a very compact mechanism to drive quite large indexing machines, and the relatively small cam is the only part that has to be produced on a special cam cutting machine tool, all other components being comparatively inexpensive 'conventional' machine parts.

ELECTRONIC MOTION CONTROL

So-called 'electronic cams' are frequently used nowadays. These are not really cams, but are motion control systems comprising an electronic control system and an electric motor which drives the payload, often through a gear train. Electronic motion control is outside the scope of this book, but a very brief discussion is given here for comparison with conventional mechanical cams. There are two main categories of system in general use: stepping motor systems and servo motor systems.

The stepping motor rotates by a small angular step in response to a pulse from the control system. A rapid stream of pulses causes the payload to move fairly smoothly according a predetermined pattern, which can be readily modified according to the design of the control system. This is an open loop system with no feedback of payload position so that if the pull-out torque of the motor is exceeded one or more steps will not take place and the system permanently loses positional synchronism until it is reset. Despite this drawback stepper motors are very successfully used for high speed, low power applications. More powerful motors are constantly being developed, but their limitations preclude their use in many industrial applications.

The servo motor and its control essentially form a closed loop system with positional feedback and, in the more sophisticated systems, velocity and other feedbacks. Output commands from the control system (digital or analogue) are constantly compared with the payload feedback signals and the motor power is modified to try to reduce the positional error to zero. Servo systems naturally work with a small error all the time, although the error is usually so small as to have an insignificant effect on performance. The payload lags or leads the command by a small variable amount throughout the motion, but permanent loss of synchronism is not a problem, provided that there is not a sustained overload.

Both systems usually have programmable motion (of various degrees of complexity and ease of use) provided by a micro-processor. The output motions may be simple functions, such as constant velocity or velocity 'ramps', i.e. constant acceleration periods, or they may emulate the cam laws described later for mechanical cams. Timings and strokes may also be programmable, whereas all these features are usually fixed in mechanical systems. This programmability makes the electronic 'cam' useful in the design and development stage of a machine, when it can save time and therefore cost: an attractive proposition for a one-off machine.

Generally electronic motion systems are more expensive than mechanical cam and follower systems, especially for series production, and less compact. Because of their relative complexity they are also less reliable in some circumstances. They have the particular disadvantage of an inherent power limitation which precludes their use in many high speed, high inertia applications. The types of motion for which cams are used involve the transmission of short bursts of energy into and out of the payload, requiring instant availability of power. With a conventional cam system there is a reservoir of kinetic energy at the input of the cam which, by design, can be more than adequate for that purpose (see Chapter 13). The rotor of the drive motor of an electronic motion control system and all parts connected to it have to be accelerated and decelerated along with the payload: the higher the inertia of those parts the higher must be the instantaneous power required, and the higher the power the greater is the inertia of the rotor and power transmission components. Special motors have been developed with a high power/inertia ratio but there is an upper limit imposed by available materials of construction.

CHAPTER 2

Dynamics, Units and Symbols

DYNAMICS – ACCELERATING FORCE AND TORQUE

The fundamental principle which governs the change of state of motion of machines is that a force is required to bring about that change. The force is proportional to mass and to acceleration. Mass is a measure of the amount of matter in a body. Other forces are also present to support weight, overcome friction, apply pressure when needed, etc., but the dynamic force, attributed to Newton's laws of motion, is dealt with here.

According to those laws, to change the state of motion of a body a force must be applied to it which is proportional to the rate of change of *momentum* of the body. Momentum is the product of mass and velocity. Therefore, if the mass of a body remains constant (which is true of machine parts and 'payload'), the rate of change of momentum is the product of its mass and its rate of change of velocity. Thus, *force is proportional to the product of mass and acceleration.*

This law for linear motion can be adapted for rotary motion, whereby a *torque* must be applied to change the state of rotary motion of a body, which is proportional to the product of its *moment of inertia* and its *angular acceleration.*

'Torque' is the common name for a turning moment and is the product of a tangential force and the radius at which it acts.

'Inertia' in common language means a tendency to remain stationary. In physics it is defined as the reluctance of a body to change its state of rest or its *state of motion*. It is measured by mass for linear motion and by moment of inertia for rotary motion. Generally in this book the term 'mass' is used for inertia in a linear motion context and the term 'inertia' is used as a short form of 'moment of inertia' in a rotary motion context.

DYNAMICS – WORK, ENERGY AND POWER

Work is done whenever an object moves under the influence of a force. This is quantified as 'energy'. *Energy* takes many forms: heat, electrical energy, various forms of mechanical energy, etc. In its simplest terms, when a force acts on a body to move it, the work done is the force multiplied by the distance moved.

In the relatively slow speed world of mechanisms, energy is strictly speaking neither destroyed nor created, it is merely converted from one form to another. Work done to overcome friction, for example, is converted to heat which is then dissipated to warm up (very slightly) the environment. The friction energy is lost by the machine, but not lost altogether. Most of the electrical energy supplied to the electric drive motor is converted to mechanical energy which is gained by the machine: the remainder is converted to heat energy by friction and electrical resistance in the windings.

A machine in motion has a store of energy which, in many cases, flows from one part of the machine to another and back again, but is still retained within the machine.

An important form of mechanical energy is *kinetic energy*, the 'energy of motion'. Assuming that the mass of a moving body does not change when its motion changes, which is the case in virtually all industrial machinery, its kinetic energy is proportional to its mass and to the square of its velocity.

Power is the rate of doing work or the rate at which energy is used: the more powerful a motor is, the more work it can do *in a given time*. Since work is force × distance, therefore power is force × distance ÷ time. In other words power is force multiplied by velocity.

COHERENT UNITS OF MEASURE

Physical quantities, such as mass, force, length, time, speed, etc. are expressed in various systems of units, of which the most common are Metric (metres, kilogrammes, etc.) and Imperial (feet, inches, pounds, etc.). Within each system a number of different unit sizes can be used to describe the same quantity, e.g. 1.5 feet or 18 inches (imperial), 2 tonnes or 2000 kg (metric), etc. Any equations used in dynamics must contain special conversion factors, depending on the particular units used, *unless those units are consistent*. A system of consistent units is often called a 'coherent' system.

In a coherent system the equation expressing Newton's laws (mentioned earlier) for linear motion is:

Force = Mass × Linear Acceleration   Eq. (2.1)

and the equation for rotary motion is:

Torque = Inertia × Angular Acceleration   Eq. (2.2)

Using coherent units, the kinetic energy equation for linear motion is:

Kinetic energy = Mass × (velocity)^2 ÷ 2   Eq. (2.3)

and for rotary motion:

Kinetic energy = Inertia × (angular velocity)^2 ÷ 2   Eq. (2.4)

Using coherent units, the power equation for linear motion is:

Power = Force × velocity   Eq. (2.5)

and for rotary motion:

Power = Torque × angular velocity   Eq. (2.6)

Units in common use for engineering in the metric and imperial systems in the past were not coherent, and the same name was used for units of mass and force; the imperial pound or the metric kilogramme for example were commonly used for either mass or force. This difficulty was overcome by introducing special units such as the imperial 'poundal' for force or 'slug' for mass, but they were not convenient when dealing with other technical concepts such as power, stress, etc. Nowadays the Systeme Internationale (SI) system introduced by the International Standards Organisation (ISO) has been almost universally adopted. Its use is strongly recommended to avoid the risk of expensive mistakes arising out of the confusion caused by the old technical units ('engineer's' units).

Refer to page 16 for abbreviations of units of measure.

Table 2.1 gives coherent units for use in equations (2.1), (2.3) and (2.5), above for linear motion.

Table 2.2 gives coherent units for use in equations (2.2), (2.4) and (2.6), above, for rotary motion:

For safety it is recommended that only SI units be used for all dynamics calculations, as indicated in Table 2.3 (last column).

Table 2.1 Coherent units for linear motion

Power	Force	Mass	Velocity	Acceleration
W	N	kg	m/s	m/s^2
kgf.m/s	kgf or kp	kg	m/s	g_n (normal gravity)
	kgf or kp	$kg.s^2/m$*		cm/s^2
10^{-7} W	dyn	g (gramme)	cm/s	cm/s^2
lbf ft/s	lbf	lb	ft/s	g_n (normal gravity)
lbf ft/s	lbf	slug*	ft/s	ft/s^2
	pdl	lb		ft/s^2

Table 2.2 Coherent units for rotary motion

Power	Torque	Moment of inertia	Angular velocity	Angular acceleration
W	Nm	$kg.m^2$	rad/s	rad/s^2
10^{-1}	dyn.cm	$g.cm^2$	rad/s	rad/s^2
kgf m/s	kgf.m	$kg\,m\,s^2$*	rad/s	rad/s^2
	lbf in	$lb\,in\,s^2$*	rad/s	rad/s^2
lbf ft/s	lbf ft	$lb\,ft\,s^2$*	rad/s	rad/s^2
	pdl ft	$lb\,ft^2$	rad/s	rad/s^2

*The units of mass or moment of inertia in these cases use the concept of 'mass' being the weight of a body (kg or lb) divided by the gravitational acceleration constant in appropriate units. This frequently gives rise to confusion, since 'mass' and 'weight' are interchangeable colloquially, but have quite different technical definitions. Mass is the amount of matter in a body, irrespective of its environment, but weight is the *force* of gravity on the body (a body in free fall is said to be weightless although it still has mass). Weight or force in the old technical (engineer's) systems were numerically equal, and the same unit name was used for both, e.g. kilogramme or pound. Later, force units were distinguished from mass units by the suffix '-force' (kilogramme-force, kgf and pound-force, lbf), but were still numerically equal and the suffix was not always used in practice.

The specially named unit of force, the *Newton*, was created in the SI system to avoid the use of conversion factors in dynamic equations, and to avoid confusion between the force and mass units.

Table 2.3   Abbreviations and equivalents of units of measure

Quantity	Name of unit	Abbreviation	Equivalent	
Length	metre	m		SI
	kilometre	km	1000 m	
	decimetre	dm	0.1 m	
	centimetre	cm	0.01 m	
	millimetre	mm	0.001 m	
	yard	yd	3 ft	
	foot	ft	12 in	
	inch	in	25.4 mm	
Mass	kilogramme	kg	1000 g	SI
	gramme	g		
	tonne	t	1000 kg	
	pound	lb	0.45359 kg	
	ton (UK)		2240 lb	
	ton (US)		2000 lb	
	slug		32.18 lb	
Time	second	s		SI
	minute	min	60 s	
	hour	h	60 min	
Force	Newton	N		SI
	decaNewton	dN	10 N	
	kilogramme-force (or kilopond)	kgf (or kp)	9.807 N	
	dyne	dyn	0.00001 N	
	pound-force	lbf	4.448 N	
	poundal	pdl	0.0311 lbf	
Torque	Newton metre	N.m		SI
	decaNewton metre	daN.m	10 N.m	
	kilogramme-force.metre	kgf.m	9.807 N.m	
	pound-force inch	lbf.in	0.1129 N.m	
Angle	radian	rad	57.296°	SI
	degree	°	0.017453 rad	
Linear velocity	metre/second	m/s		SI
	kilometre/hour	km/h	0.2778 m/s	
	inch/second	in/s	0.0254 m/s	
	foot/second	ft/s (fps)	0.3048 m/s	
	mile/hour	mile/h (mph)	0.4470 m/s	
Linear acceleration	metre/second^2	m/s^2		SI
	inch/second^2	in/s^2	0.0254 m/s^2	
	foot/second^2	ft/s^2	0.3048 m/s^2	

Table 2.3 (continued)

Quantity	Name of unit	Abbreviation	Equivalent	
Angular velocity	radian/second	rad/s		SI
	degree/second	°/s	0.01745 rad/s	
	revolution/minute	rev/min (rpm)	0.10472 rad/s	
Angular acceleration	radian/second^2	rad/s^2		SI
	degree/second^2	°/s^2	0.01745 rad/s^2	
Moment of inertia	kilogramme metre^2	kg.m^2	3417 lb in^2	SI
	pound inch^2	lb.in^2	0.0002959 kg.m^2	
	pound foot^2	lb.ft^2	0.004213 kg.m^2	
	pound foot second^2	lb.ft.s^2	see notes below	
	kilogramme metre second^2	kg.m.s^2	see notes below	
Energy, work	joule	J	1 Nm	SI
	kilowatt hour	kW.h	3600 kJ	
See also the	calorie	cal	4.1868 J	
units for	foot pound-force	ft.lbf	1.356 J	
torque	horsepower hour	hp.h	2.685 MJ	
Power	watt	W	1 J/s	SI
	kilowatt	kW	1000 W	
	metric horsepower (cheval	CV or PS	735.499 W	
	vapeur or Pferdestarke)		(75 kgf.m/s)	
	horsepower	hp	745.7 W (550 ft.lbf/s)	

MOMENT OF INERTIA (see Appendix D)

The moment of inertia of a *point* rotating about an axis is its mass multiplied by the square of its distance from the axis. The moment of inertia of a *body* rotating about an axis is the sum of the moments of inertia of all the points which make up that body. This can be expressed as the mass of the whole body multiplied by the square of its 'radius of gyration'. The radius of gyration is a property of the size and shape of the body.

It follows that the proper compound unit for moment of inertia is (mass unit) × (length unit)^2 such as kg m^2 or lb in^2. However some engineers express moment of inertia in kgf m s^2 (or kp m s^2), kgf cm s^2, lbf in s^2 or lbf ft s^2 units. This is because in non-coherent systems of units it is necessary to divide the true mass (kg or lb) by the gravitational acceleration constant to be able to use the old technical units of force (kgf or lbf) in dynamics calculations. Usually, therefore, when such inertia units are encountered it is necessary to multiply the values by the appropriate gravity constant to convert them into proper inertia units, for example:

$$15 \text{ kgf m s}^2 \times 9.807 \text{ m/s}^2 = 147.105 \text{ kg m}^2 \text{ (SI units)}$$

$$4 \text{ lbf ft s}^2 \times 32.18 \text{ ft/s}^2 = 128.68 \text{ lb ft}^2 = 18530 \text{ lb in}^2$$

It is not safe to assume that such conversions are correct without referring back to the source of the inertia information to verify its derivation.

A further danger in using *metric* inertia values without knowing their derivation is caused by the practice in some European countries, notably Germany, of quoting a moment of inertia as a 'GD^2' value in kp.m^2 units, sometimes loosely dimensioned as kg.m^2. GD^2 stands for weight $\times$ diameter^2, (Gewicht $\times$ Durchmesser^2) where the diameter is twice the radius of gyration. A GD^2 value is therefore 4 times higher than the true moment of inertia in SI units.

SIGNS AND SYMBOLS

Many symbols are in common use to represent the physical quantities described above, and they differ somewhat according to the context in which they are used. The list below gives the symbols that are used in this book.

Table 2.4   Notation

Symbol	Quantity
α	Instantaneous angular displacement of a cam
β	Instantaneous angular displacement of a swinging follower arm or turret
β'	Geometric angular velocity $= B/A \cdot w' = \mathrm{d}\beta/\mathrm{d}\alpha$
β''	Geometric angular acceleration $= B/A^2 \cdot w'' = \mathrm{d}^2\beta/\mathrm{d}\alpha^2$
δ	Deflection
ν	Poisson's ratio
ν_1	Poisson's ratio of body 1 in contact
ν_2	Poisson's ratio of body 2 in contact
θ	Angle of a polar co-ordinate system
a	SCCA cam law parameter, first zone; semi-major axis of contact ellipse

Table 2.4 **(continued)**

Symbol	Quantity
a_i	Input acceleration
a_o	Output acceleration
A	Angular input stroke of a rotating cam; area, e.g. of a cross-section
A_0, A_1, etc.	Coefficients of polynominal function
AF	Asymmetry factor
b	SCCA cam law parameter, second zone; semi-minor axis of contact ellipse
b_k	Normalised (dimensionless) backlash
B	Linear output stroke of a translating follower; angular output stroke of a swinging follower or turret
B_k	Real cumulative backlash referred, to the cam output
B_{10}	Criterion for 10% fatigue failure rate
B_{50}	Criterion for 50% fatigue failure rate
BM	Bending moment
c	SCCA cam law parameter, third zone
C	Distance between cam centre and swinging follower pivot
C_a	Normalised coefficient of acceleration
C_c	Input torque coefficient
C_d	Normalised coefficient of deceleration
C_m	Load mix coefficient
C_t	Torsion factor
C_v	Normalised coefficient of velocity
d	SCCA cam law parameter, fourth zone
d_n	Normalised (dimensionless) natural deceleration
D	Initial angular displacement of swinging follower from common centre-line; diameter
D_n	Real natural deceleration of the payload
e	Exponent of load-life equation
E	Young's modulus (modulus of elasticity); follower offset; efficiency
E_e	Equivalent elasticity of two materials in contact
E_1	Young's modulus of body 1
E_2	Young's modulus of body 2
E_c	Efficiency of cam and follower system
E_g	Efficiency of gear unit
f	Normal stress; natural frequency
f_{mean}	Mean compressive (Hertzian) stress over the contact area
f_{max}	Maximum compressive (Hertzian) stress, at the centre of the contact area
F	Length of swinging follower arm; contact force
F_1, F_2, etc.	Contact forces during periods 1, 2, etc.
F_d	Decelerating force
F_e	Equivalent force for all periods
F_f	Maximum non-inertia force
F_i	Inertia force
F_p	Minimum non-inertia force (pre-load)
F_s	Impact (shock) force
G	Modulus of rigidity (shear modulus); gear ratio
H	Initial height of translating follower; hours life
h	Linear displacement of translating follower
h'	Geometric velocity of translating follower

Table 2.4   (continued)

Symbol	Quantity
h''	Geometric acceleration of translating follower
Hp	Horsepower
H_v	Vickers (diamond pyramid) hardness number, D.P.N.
I	Moment of inertia
I_c	Input inertia (referred to the cam)
I_e	Equivalent moment of inertia
J	Second moment of area of the cross-section of a beam
K	Pre-load ratio
k	Radius of gyration; parameter used in point contact stress theory
L	Length; distance; life; length of rectangular contact area
M, m	Moment, torque
M_d	Decelerating torque
M_f	Maximum non-inertia torque
M_i	Inertia torque
M_p	Minimum non-inertia torque (pre-load)
M_s	Impact (shock) torque; stop/start torque
M_t	Total output torque
n	Period ratio
N	Number of shaft revolutions per minute
N_1, N_2, etc.	Number of stress cycles during periods 1, 2, etc.
p	General purpose parameter; constant in torsion factor equation
P	Power
P_{max}	Peak power used by the cam
q	Shear stress; constant in torsion factor equation
q_{max}	Maximum value of contact shear stress
q_{yield}	Yield shear stress
r	$f(\theta)$, radius of a polar co-ordinate system; constant in torsion factor equation; radius at which a force acts
Q	Inertia ratio
R	Rigidity (of transmission)
R_{11}	Radius of curvature of body 1 in the plane of rolling
R_{12}	Radius of curvature of body 1 in the perpendicular plane
R_{21}	Radius of curvature of body 2 in the plane of rolling
R_{22}	Radius of curvature of body 2 in the perpendicular plane
R_c	Radius of curvature
R_e	Equivalent radius of curvature of two bodies in contact
R_{e1}	Equivalent radius of curvature in the plane of rolling
R_{e2}	Equivalent radius of curvature in the perpendicular plane
s_{lc}	Surface stress factor for line contact
s_{pc}	Surface stress factor for point contact
S	Slope; linear stiffness = force/deflection
t	Time elapsed from the start of motion; contact ellipse aspect ratio parameter, a function of the radii of curvature R_{e1} and R_{e2}
T	Time taken for total displacement (time period)
TM	Turning moment, torque
u	Normalised input displacement; function of the contact ellipse aspect ratio parameter t

Table 2.4 (continued)

Symbol	Quantity
v	Velocity; function of the contact ellipse aspect ratio parameter t; normalised (dimensionless) impact velocity
v_i	Input velocity
v_o	Output velocity
V	Real impact velocity, linear or angular
w	Normalised output displacement $f(u)$, the cam law function
w'	Normalised output velocity $f'(u)$, first derivative of w
w''	Normalised output acceleration $f''(u)$, second derivative of w
W	Mass
W_e	Equivalent mass
x	Real input displacement; abscissa of a rectilinear co-ordinate system
X	Input stroke
X_f	Forward motion period of a D–R–R–D motion
X_{fr}	Whole motion period of a D–R–R–D motion
X_r	Return motion period of a D–R–R–D motion
y	Real output displacement $f(x)$; ordinate of a rectilinear co-ordinate system
Y	Output stroke
z	Blend factor
Z	Number of gear teeth

CHAPTER 3

Motion Laws: Cam Laws – Definition and General Use

MOTION LAWS – CAM LAWS

The mathematical function which defines the motion of a mechanism is generally known as its *motion law* or, in the case of a cam-operated mechanism, its *cam law*. The position of the mechanism at every point in its motion is determined by the motion law, and consequently its velocity, acceleration and other derivatives are known at every point, depending on the state of motion of the mechanism input (e.g. the speed of the camshaft).

For many mechanisms the motion law is determined solely by their geometry, e.g. the number and length of members of a linkage, but in cam-operated mechanisms the shape of the cam profile determines the motion law, and this is a free choice of the designer. Cam laws therefore tend to be specified as mathematical functions, chosen to suit particular applications. This is the great advantage of cam-operated mechanisms, since the choice of a suitable motion nearly always results in a higher performance (e.g. higher speed) than is possible with non-cam mechanisms. This can outweigh the small extra cost that it usually entails, and it is very important for the machine designer to understand fully the properties of cam laws.

A number of industry standard cam laws have now evolved, which are dealt with in Chapter 4 and Appendices A and B. Appendix B has factor tables for standard cam laws to make mathematical manipulation unnecessary.

In all rigid mechanisms, including cam-operated ones, the position of the output member is uniquely determined by the position of the input member: there is a mechanical connection between them. In practice, the elasticity of the mechanism components produces minute variations in the position of the output member which often take the form of a vibration of fairly high frquency (see Chapter 12). Such deviations from the position of a truly rigid mechanism are ignored by the motion law.

Cam profiles are most often made to impart the designed motion law to the cam follower. In many cases the cam follower drives a mechanism that may have an output displacement which is not proportional to the follower displacement at all points in the motion. This results in a 'distorted' motion: often the distortion is insufficient seriously to affect the dynamic performance of the mechanism. However, if the distortion is not acceptable, the profile can be designed to produce a favourable motion law at the output of the mechanism: the profile itself is 'distorted'. In that case the displacement, velocity, etc. at all points in the *mechanism output* motion will obey the chosen cam law.

Mechanisms of the same type, with the same kind of motion, can be constructed in different sizes and can be operated at different speeds. However, motion laws, and cam laws in particular, are usually designed to be independent of scale and speed, for easy use over a wide range of applications.

To make the law independent of speed, the output position is defined as a function of input position (position is usually described as the displacement from a starting point). The derivatives (velocity, acceleration, etc.) are therefore *geometric* and must be converted to *real* values, i.e. time-based values, by taking the input speed, etc., into account.

To make them independent of scale, cam laws are *normalised*. Normalisation assumes a mechanism size where the total output displacement of the motion is one unit (of any dimension) and the corresponding input displacement is one unit (of any dimension). Real displacement at any point in the motion is thus the normalised displacement multiplied by the real *total* displacement.

GENERAL FORM OF CAM LAW – APPLICATION TO A PARTICULAR MECHANISM

The use of mathematics, and differential calculus in particular, is unavoidable when analysing cam motions. However, the derivation of the main equations can be skipped (at the expense of a more thorough understanding of the principles of motion) and where possible a simply worded form of those equations has been provided. The reader is urged to grasp the physical meaning of the symbols used in the equations: the word form should help to this end. Central to the definition of a cam law function is the concept of normalisation (see Fig. 3.1).

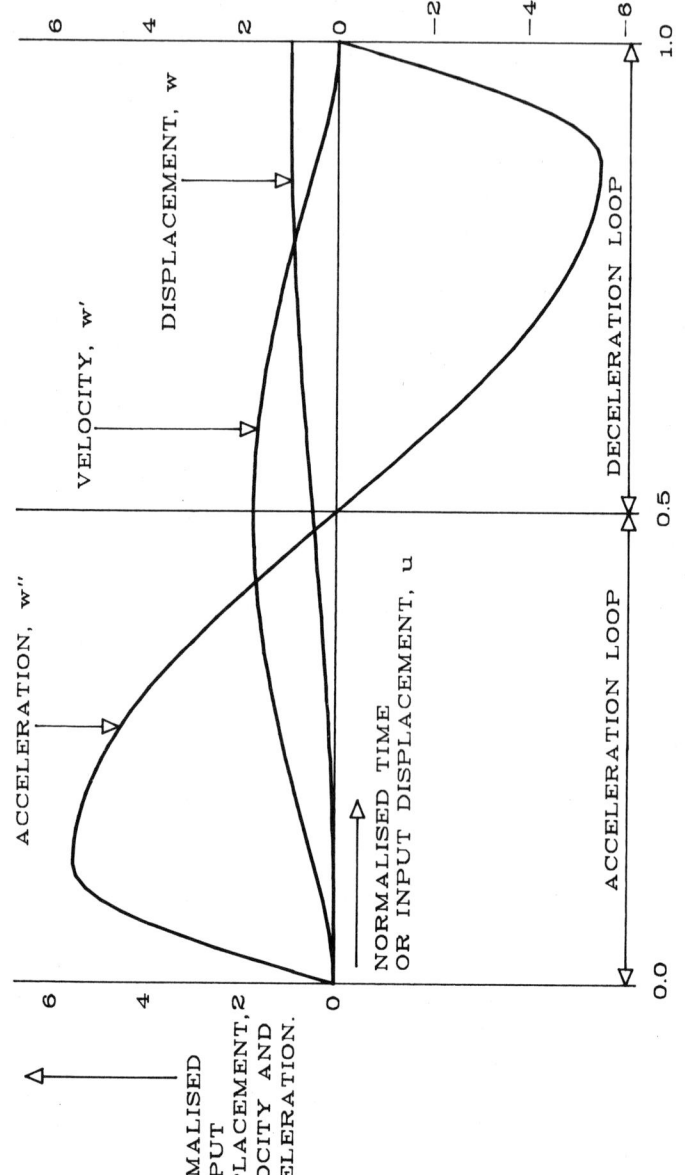

Fig. 3.1 Normalised cam law curves

The general mathematical form of a normalised cam law function is:

$w = f(u)$ where: u = normalised input displacement
and w = normalised output displacement.

This mathematical form is stated as 'w equals a function of u'.

The normalised *geometric* velocity, acceleration and jerk* are the direct derivatives of this function:

$$w' = f'(u), \ w'' = f''(u) \quad \text{and} \quad w''' = f'''(u) \qquad\qquad \text{Eq. (3.1)}$$

In the above notation:

w' means $\dfrac{dw}{du}$ i.e. the rate of change of w *with respect to u*.

w'' means $\dfrac{d^2w}{du^2}$ i.e. the rate of change of w' *with respect to u*.

w''' means $\dfrac{d^3w}{du^3}$ i.e. the rate of change of w'' *with respect to u*.

In the particular mechanism to which these normalised** functions are applied let the real values be:

a_i = input acceleration = d^2x/dt^2
a_0 = output acceleration = d^2y/dt^2
t = time elapsed from the start of motion
T = time taken for total displacement, or 'time period'
v_i = input velocity = dx/dt
v_0 = output velocity = dy/dt
x = instantaneous input displacement at time t
X = total input displacement in time T, or 'input stroke'
y = instantaneous output displacement at time t
Y = total output displacement in time T, or 'output stroke'

At this point it is necessary to explain some of the terminology frequently used in cam technology and used throughout this book.

---

*'Jerk' is the name given to the rate of change of acceleration.
**The velocity acceleration and jerk functions are not strictly normalised in the sense that they do not range from 0 to 1. They are derivatives of the normalised displacement function and for convenience they are referred to as normalised throughout this book.

The term 'stroke' is most often used to describe the total output displacement of a mechanism from the beginning to the end of a specified motion. It can, however, describe the displacement that takes place during a well defined 'period' (see below) or zone of a motion, which is not necessarily the full motion. 'Stroke' always means *distance* moved (linear or angular) from beginning to end of a motion entity. Unless otherwise specified the term is taken to mean *output* stroke, but it can also be used as *input* stroke, for example the rotation angle of a cam, in which case its meaning is made clear in context.

The term 'period' is often used instead of 'input stroke', although its most common use is to describe the *time* taken to perform a complete motion or a well defined phase of motion. Thus an indexing period could be said to be, for example, 2.5 seconds or 270° of cam rotation. We can also talk of an acceleration period or a dwell period. In all cases the term applies to the change of state of the *input* of a mechanism from beginning to end of a motion entity.

By definition: Real output displacement $= y = Y \cdot w$ Eq. (3.2)

and   Real input displacement $= x = X \cdot u$ Eq. (3.3)

To obtain the real output velocity differentiate the output displacement *with respect to time*:

$$\text{output velocity} = v_0 = \frac{dy}{dt} = \frac{dy}{dx} \cdot \frac{dx}{dt} = \frac{Y}{X} \cdot \frac{dw}{du} \cdot v_i = \frac{Y}{X} \cdot w' \cdot v_i$$

Eq. (3.4)

This is the *velocity transmission equation*, which can be expressed as

---

real output velocity = (geometric output velocity)

$\times$ (real input velocity)

If the input velocity is constant then $v_i = X/T$

and $v_0 = \dfrac{Y}{T} \cdot w'$ Eq. (3.5)

w' is the instantaneous 'Coefficient of Velocity' of the cam law. Its maximum value is designated C_v.

To obtain the real output acceleration differentiate the velocity *with respect to time*:

therefore output acceleration $= a_0 = \dfrac{\mathrm{d}^2 y}{\mathrm{d}t^2} = \dfrac{\mathrm{d}}{\mathrm{d}t}\left[\dfrac{\mathrm{d}y}{\mathrm{d}x} \cdot \dfrac{\mathrm{d}x}{\mathrm{d}t}\right]$ Eq. (3.6)

$$a_0 = \dfrac{\mathrm{d}^2 y}{\mathrm{d}x^2}\left[\dfrac{\mathrm{d}x}{\mathrm{d}t}\right]^2 + \dfrac{\mathrm{d}y}{\mathrm{d}x} \cdot \dfrac{\mathrm{d}^2 x}{\mathrm{d}t^2} = \dfrac{\mathrm{d}^2 y}{\mathrm{d}x^2} \cdot v_i^2 + \dfrac{\mathrm{d}y}{\mathrm{d}x} \cdot a_i \qquad \text{Eq. (3.7)}$$

This is the *acceleration transmission equation* and can be expressed in everyday language as:

---

real output acceleration =
   (geometric output acceleration) × (real input velocity)²
   + (geometric output velocity) × (real input acceleration)

Eq. (3.8)

This is a very important equation for all mechanisms. It shows that the output acceleration has two components. The first, which is usually the most important, is proportional to the input velocity squared. The second component is proportional to input acceleration. Input acceleration and, therefore, the second component, are zero if the input motion is perfectly smooth with constant velocity, but often this is far from the truth in practice when the input transmission is flexible or has backlash.

The transmission equations, above, apply to all mechanisms that have a unique output displacement for every possible input displacement (i.e. a 'motion law'): cam systems, linkages, gears, etc. Any motion law can be normalised, by choosing a particular input displacement as the 'input stroke' X, and the corresponding output displacement as the 'output stroke' Y. If $y = f(x)$ is the real displacement function, then $w = y/Y$ and $u = x/X$ so that the normalised motion law is $w = f(uX)/Y$.

In terms of the normalised motion law the acceleration equation becomes:

$$a_0 = \dfrac{Y}{X^2} \cdot w'' \cdot v_i^2 + \dfrac{Y}{X} \cdot w' \cdot a_i \qquad \text{Eq. (3.9)}$$

and if the input velocity is constant then $v_i = X/T$ and $a_i = 0$

output acceleration $= a_0 = \dfrac{Y}{T^2} \cdot w''$ Eq. (3.10)

w'' is the instantaneous 'Coefficient of Acceleration' and its maximum (positive) value is designated C_a.

Camm laws normally have a deceleration phase (negative acceleration) and the most negative value of w'' is designated C_d: the coefficient of maximum deceleration. If the motion is symmetrical then $C_d = C_a$ and the C_a value is used for both.

It follows that the maximum output velocity and maximum output acceleration of a mechanism with constant input velocity are:

$$v_{o\,max} = C_v \cdot \dfrac{Y}{T} \quad \text{and} \quad a_{o\,max} = C_a \dfrac{Y}{T^2} \qquad \text{Eq. (3.11)}$$

or

Maximum output velocity

$$= \dfrac{(C_v \text{ of cam law}) \times (\text{output stroke})}{(\text{stroke time})} \qquad \text{Eq. (3.12)}$$

Maximum output acceleration

$$= \dfrac{(C_a \text{ of cam law}) \times (\text{output stroke})}{(\text{stroke time})^2} \qquad \text{Eq. (3.13)}$$

These are the nominal values used in simple design calculations.

The use of these nominal values is illustrated in the following Examples 3.1 to 3.3 below.

Example 3.1

A cam-operated mechanism has an output stroke of 150 mm (0.15 m) which takes place in 2.4 seconds, and uses a Modified Sine (MS) cam law. Find the maximum output velocity and acceleration.

From the table on page 41 we see that for MS motion,
  $C_v = 1.759$ and $C_a = 5.528$

Nominal maximum output velocity $= v_{0\,max}$
  $= 1.759 \times 0.15 \div 2.4 = 0.1099$ m/s

Nominal maximum output acceleration = $a_{0\,max}$
  = $5.528 \times 0.15 \div 2.4^2 = 0.144$ m/s^2

Example 3.2

A conventional 4-station Geneva mechanism (q.v.) has an indexing period of 0.75 seconds. Find the maximum output velocity and acceleration.

A 4-station Geneva mechanism has an output stroke of 90° ($\pi/2$ rad) and from the table on page 42 we see that $C_v = 2.414$ and $C_a = 8.493$

Nominal maximum output velocity = $v_{0\,max}$
  = $2.414 \times \pi/2 \div 0.75 = 5.056$ rad/s

Nominal maximum output acceleration = $a_{0\,max}$
  = $8.493 \times \pi/2 \div 0.75^2 = 23.72$ rad/s^2

In both of these examples please note that the results are in the same units as the data used in the calculation: metres and seconds were used in the linear example to give m/s and m/s^2 results, and radians and seconds were used in the rotary example to get rad/s and rad/s^2 results.

An important principle is revealed by Equations 3.11 and 3.13. The peak nominal output acceleration (and consequently the maximum force or torque in the mechanism and its transmission) is inversely proportional to the motion time period *squared*. Once the cam angle for a mechanism has been fixed the acceleration forces are inversely proportional to the *cycle* time squared, that is, they are proportional to *operating speed squared*.

In the overall machine design procedure a decision must be made as to how much of the machine cycle time is to be used for moving the various masses and how much mechanism dwell time is needed for product processing. This usually translates into a camshaft 'timing diagram' showing the amount of camshaft rotation devoted to each mechanical function.

The time squared effect becomes a cam angle squared effect, so that for example increasing the cam angle for a motion by a factor of 1.5 (50% increase) reduces the forces by a factor of 2.25. There is also a substantial improvement in the amplitude of dynamic response vibrations with such an improvement in cam angle (see Chapter 12). There are, of course, conflicting demands on the allocation of cycle time, but in general the best compromise is achieved by allotting more time to the high inertia mechanisms and less to the low inertia ones, attempting to balance the inertia forces.

An important technique that is used to increase the potential operating speed of a machine, in addition to 'balancing' the timing diagram as described, is to overlap the timing of sequential motions. This can usually be done to a far greater extent than is at first apparent. Reference to the standard cam law factor tables (Appendix B) will reveal that the output displacements at the beginning and end of the motion are very gradual. In most sequential operations one motion does not have to come to rest *completely* before the next motion starts, even though it is convenient to think of it that way in the conceptual stage of machine design. Some small displacement of the second motion is permissible before the first motion has stopped, and a small output displacement at that point corresponds to a fairly large input displacement (cam rotation).

Example 3.3

To illustrate the point, imagine a machine in which a plunger has to be inserted into a close fitting hole in a rotary table after the table has been indexed into position. Both the plunger movement and the table indexing motion are cam-operated. During table movement there is a clearance of 5 mm between the end of the plunger and the table surface. The plunger stroke is 50 mm with a Mod Sine motion occupying 120° of camshaft rotation. Its return stroke is similar and the table indexing motion occupies the remaining 120° (see Fig. 3.2).

A plunger displacement of 5 mm from the end of the stroke represents a normalised value of

$$w = 5 \div 50 = 0.1$$

From the Mod Sine factor table we see that the normalised input displacement for $w = 0.09952$ (when the plunger almost meets the table surface) is $u = 0.233$ (step 28 in the 120 step list).

This is a camshaft rotation of $0.233 \times 120 = 28°$

Since the plunger withdrawal motion is the same, then a similar situation exists at the other end of the index period and we could theoretically overlap the timing by 28° each end, a total of no less than 56°. In practice, we could safely overlap the timing by, say, 20° each end, giving a plunger clearance of about 2.8 mm when the table stops moving.

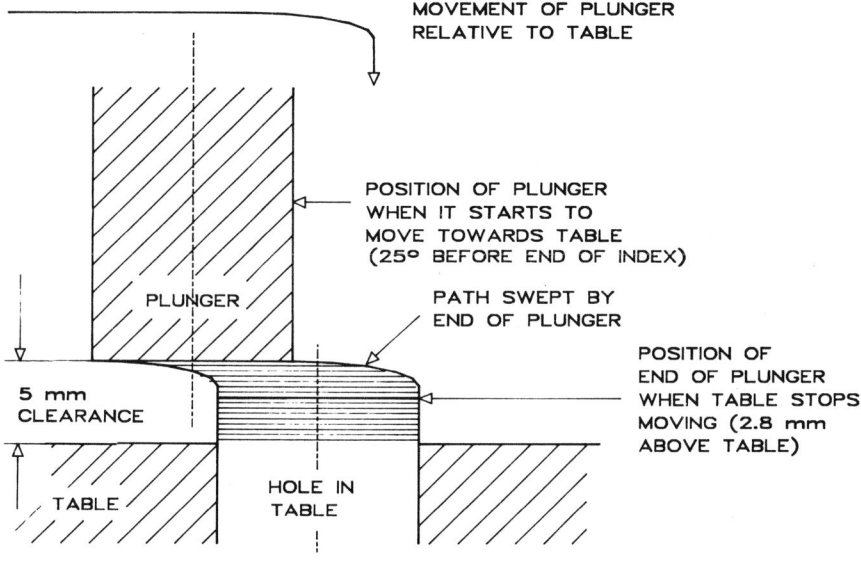

Fig. 3.2   Timing overlap

This allows the indexing motion to have a 160° cam angle (120 + 20 + 20) with timing overlap instead of 120° without. If the operating speed of the machine is limited by the dynamic performance of the fairly heavy indexing table, a potential 33% increase in operating speed is available 'for free'. The effect of the interaction of the two overlapped motions is illustrated in Fig. 3.2.

It is not unusual deliberately to increase the stroke of a mechanism (in this case the plunger mechanism) in order to get the speed benefit of a significant timing overlap. Had this machine been designed with 10 mm clearance between plunger and table instead of 5 mm, even more overlap would have been possible.

CHAPTER 4

Standard Cam Laws: Composites and Combinations

STANDARD CAM LAWS

As described in the previous chapter, every mechanism has a motion law determined by the mechanism design, and in the case of a cam-operated mechanism this is known as a cam law. Over several decades various cam laws have been developed for both specific and general use. The special purpose machine industry has, by common usage rather than any formal arrangement, standardised on a number of these. There is no single cam law that is best for all applications, each being more or less suitable for a particular mechanical system or environment.

Virtually all standard cam laws are symmetrical, but asymmetrical motions may be derived from them as described later. Symmetry in this context means that velocity increases from zero at the beginning to a maximum value at the mid-point of the motion in a certain pattern, and then decreases from that value to zero velocity at the end of the motion in a mirror image of that pattern.

In consequence of this symmetry the acceleration pattern of the second half of the motion is the same as the pattern of the first half, but reversed and negative. The motion therefore consists of an 'acceleration loop' followed by a similar 'deceleration loop' and the cam law can be fully described by specifying the acceleration loop alone. Later it will be seen that this fact simplifies the construction of composite motions and cam law 'blending'.

A distinction must be made between the nominal maximum values of force, acceleration, etc., and the real maximum values. All mechanisms are made of elastic materials: there is no such thing as a *perfectly rigid* construction. Consequently, the deformations produced by the varying forces during the motion of the mechanism tend to generate a vibration superimposed on the nominal displacement pattern. The amplitude of this vibration is added to the nominal values, resulting in peak real values of acceleration, force, torque, etc. which are always greater than nominal. Tuned cams (q.v.) theoretically eliminate this phenomenon, but they have economic limitations, as previously mentioned. In this book we deal only with untuned cam motions and examine the vibration effect in Chapter 12.

Probably the first cam law to be used as such was Simple Harmonic Motion (SH). This has an acceleration pattern which is half a cosine curve, and is exactly the motion generated by a Scotch Yoke mechanism or a simple eccentric cam and flat-faced follower mechanism. If, as in most cam applications, the motion is preceded or followed by a dwell period, there is a sudden change (a step) from zero to maximum acceleration or from maximum deceleration to zero at the transition point(s). It is an excellent motion for fairly slow speed, rigid mechanisms or those which have no dwell periods, but is not recommended for most high speed applications.

This was followed by Constant Acceleration/Constant Deceleration, now known as Parabolic (PAR) because its displacement curve is made up of two parabolas. Its acceleration pattern is rectangular with steps at the beginning, middle and end of the motion period. The middle step is a sudden change from maximum acceleration to maximum deceleration. It has the lowest possible peak acceleration and consequently produces the lowest possible *nominal* dynamic forces in the machine, which is the main reason for its popularity. However, it is the *worst* of the basic standard cam laws for generating vibrations, and this outweighs its advantage in all but the slowest or most rigid mechanisms and machines.

To overcome the drawbacks of these two popular motions, modifications were introduced to replace the sudden acceleration steps with gradual changes. Quarter sine and cosine curves were used for this purpose and the resulting cam laws are known as Modified Sine Acceleration (MS), 'Mod Sine' for short (which replaced Simple Harmonic) and Modified Trapezoidal Acceleration (MT), 'Mod Trap' for short (which replaced Parabolic).

A third cam law, which was already well known, became very popular when there was a better understanding of the dynamic response and vibration of mechanisms. This is Sine Acceleration, better known as 'Cycloidal' (CYC) because its displacement function is the same as the forward displacement of a point on the circumference of a rolling wheel which traces out a cycloidal curve. Its acceleration pattern is a full sine curve. There are no steps in it and the vibration generated is the lowest of all cam laws. However, it has a relatively high peak acceleration and relatively high peak velocity. Its advantage over Mod Sine, for example, is only apparent in very flexible or high speed mechanisms with good input transmission.

Mod Sine is probably the most useful general purpose cam law.

In addition to these cam laws, other standard laws have been developed for a variety of reasons, the most well known being the algebraic polynomials and the trigonometric types: trigonometric polynomials and Fourier series.

Probably the most common reason for using algebraic or trigonometric polynomials instead of MS or MT (or their offspring) is ease of calculation. MS and MT are not mathematically continuous throughout the motion: the motion period is divided into zones, each of which has a different equation, whereas a single equation covers the whole period of a polynomial. However, this causes very little difficulty nowadays with the universal use of computers.

Most standard cam laws are symmetrical and of the Dwell-Rise-Dwell type: D-R-D for short. This kind of motion is designed to be preceded by a dwell period, move smoothly in one direction and then be followed by a dwell period. Its acceleration gradually increases from zero at the start to a peak value, gradually decreases through zero to a peak deceleration, and then gradually changes to zero again. Such a pattern avoids abrupt changes of acceleration, making for a smooth transition from and to the dwell periods. It is the most common type of cam motion used in industrial machinery.

COMPOSITES

Any cam law can be manipulated to produce various offspring for particular purposes. Two common examples are the *asymmetrical* ones and the *composite* ones. Asymmetry is dealt with later, for it can be applied to composite laws as well as to standard laws.

A composite motion is defined here as one having a number of zones connected *in series*, each zone having its own continuous displacement function (cam law) and being joined to its neighbouring zone with continuity of displacement, velocity and acceleration (this must not be confused with a combination, described below, which has two component motions running in parallel). By this definition Mod Sine and Mod Trap are composite motions, but they are so often used that they have become standard cam laws in their own right and need not be thought of as composites. Any composite motion can be manipulated in exactly the same way as a pure continuous function.

Apart from Mod Sine and Mod Trap, the composites of particular interest consist of a standard cam law acceleration loop, followed by a period of constant velocity and followed again by a standard cam law deceleration loop. The acceleration and deceleration loops need not be of the same cam law, but often are. Such motions are special cases of cam law blending (q.v.).

It has often been found that a standard cam law, while more than adequate for producing a smooth motion, has an undesirably high peak velocity. Either the velocity is too high for the process performed by the mechanism or it forces the use of a cam size that is larger than convenient or economical. In such cases the introduction of a period of constant velocity into the middle of an otherwise standard motion usually overcomes the problem. Thus a number of *secondary standard* motions which are composites have come into common use. They are designated according to the standard cam law used for the acceleration and deceleration loops and the percentage of the whole period that is taken by the constant velocity zone.

For example, MSC.50 is composed of:

25% Mod Sine acceleration
50% constant velocity
25% Mod Sine deceleration

Common composites are included in the tables of standard cam laws.

COMBINATIONS

Whilst most cam applications are only concerned with moving a mass from one position to another in a specified period as smoothly as possible, there are some in which the pattern of movement is subject to constraints dictated by the machine's purpose. For example, the mechanism output may have to reach a certain point by a specified time and then proceed at a certain velocity for some distance before decelerating. There are very many other examples.

One convenient way of dealing with this is to design a special polynomial cam law (as mentioned above), but if the constraints are complex a smooth solution may be impracticable or difficult to find. A fairly simple solution is often available, even for complex special motion problems, by blending standard laws with periods of constant velocity and with *combination cam laws* (see Chapter 5 for cam law blending).

A *combination* is the superimposition of one standard motion on another or, more usually, on constant velocity (CV). With this kind of motion the output displacement at any point is the sum of two components. One component is derived from one cam law and the other from a different cam law or from a constant velocity function. The total output stroke is the sum of two partial strokes and each intermediate displacement component is the product of the appropriate normalised displacement factor and the partial stroke. In effect one entire partial motion is added to another entire partial motion.

The importance of this concept is that the velocity, acceleration and other derivatives are also additive. Calculation of these is therefore quite

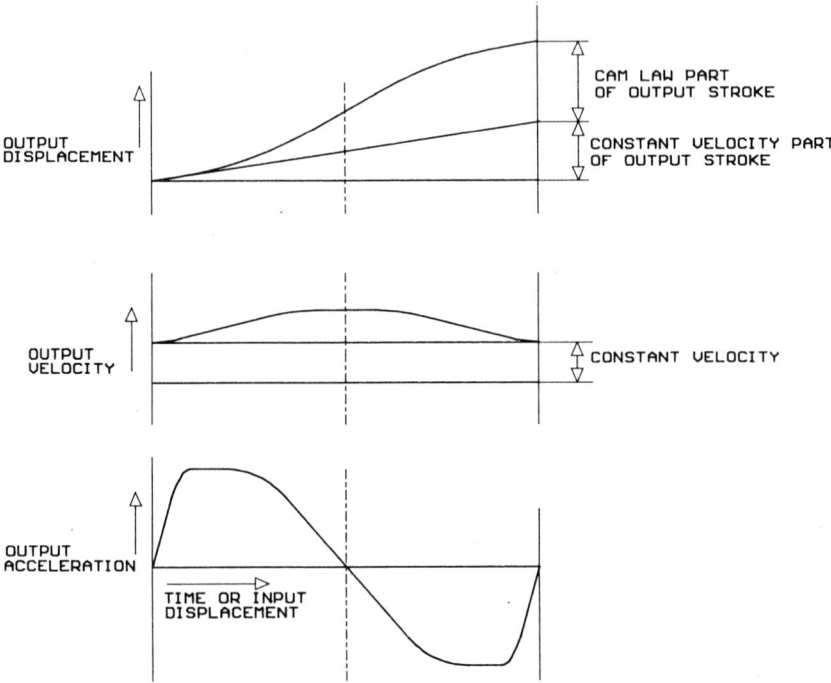

Fig. 4.1   Combination of SCCA cam law and constant velocity

straightforward, and the smoothness and other dynamic properties of the motion are easily and directly related to the component cam laws used.

Example 4.1

A motion occupies a 3 seconds time period with a stroke of 400 mm (0.4 m) and it starts and finishes with a constant velocity of 50 mm/s (0.05 m/s). A Mod Sine acceleration pattern is to be used.

Find the output displacement, velocity and acceleration after 0.9 second has elapsed.

This is a combination of MS and constant velocity (CV) motions. The velocity of MS at the beginning and end is zero, so that the CV velocity must be

$$0.05 - 0 = 0.05 \text{ m/s}$$

The CV output stroke is therefore

$$0.05 \times 3 = 0.15 \text{ m}$$

Total stroke is 0.4 m therefore the MS stroke is

$$0.4 - 0.15 = 0.25 \text{ m}$$

After 0.9 second out of a 3 second period the normalised input displacement is

$$u = 0.9 \div 3 = 0.3$$

From the table in Appendix B we find the normalised MS values:

$$w = 0.17789 \qquad w' = 1.3230 \qquad w'' = 4.1081$$

The corresponding real values for (partial) displacement, velocity and acceleration are:

$$y = Y.w = 0.25 \times 0.17789 = 0.044472 \text{ m} = 44.472 \text{ mm}$$

$$v = Y.w'/T = 0.25 \times 1.3230 \div 3 = 0.11025 \text{ m/s}$$

$$a = Y.w''/T^2 = 0.25 \times 4.1081 \div 3^2 = 0.1141 \text{ m/s}^2$$

The CV displacement after 0.9 second is $0.05 \times 0.9 = 0.045$ m

The CV velocity is 0.05 m/s

and its acceleration is zero.

Combining the two partial motions by adding the respective components we get:

$$\text{output displacement} = 0.044472 + 0.045 = 0.089472 \text{ m} = 89.472 \text{ mm}$$

$$\text{output velocity} = 0.11025 + 0.05 = 0.16025 \text{ m/s}$$

$$\text{output acceleration} = 0.1141 + 0 = 0.1141 \text{ m/s}^2$$

POLYNOMIAL CAM LAWS

A 'polynomial' cam law usually means an algebraic series of the form:

$$w = A_0 + A_1.u + A_2.u^2 + A_3.u^3 + A_4.u^4 + \cdots \qquad \text{Eq. (4.1)}$$

The coefficients A_0, A_1, etc. can be chosen to fit various conditions that apply to the particular application, but for a *standard* cam law certain general conditions have to be met, such as zero displacement, zero velocity and zero acceleration at the beginning of the motion ($u = 0$, $w = 0$, $w' = 0$, $w'' = 0$), and zero velocity and zero acceleration at the end of the motion ($u = 1$, $w' = 0$, $w'' = 0$), as well as that imposed by normalisation, i.e. unity displacement at the end of the motion ($u = 1$, $w = 1$). The first three terms of the above series disappear with these conditions, so that the lowest term of a standard cam law polynomial is the u^3 one. Also the sum of all coefficients of a non-reflective polynomial (see below) is one.

The usefulness of a polynomial may be extended if the function applies only up to the mid-point of the motion, ignoring the equation after that point. The second half of the motion is the reversed reflection of the first half (as with most other standard cam laws). This could be termed a 'semi-polynomial' or 'reflective polynomial' as opposed to a full or non-reflective polynomial. Some full polynomials are in fact symmetrical in this way and could be treated as reflective if preferred, for instance the standard '3-4-5' cam law mentioned below. A reflective polynomial has to fulfil the conditions of zero displacement zero velocity and zero acceleration at the beginning; and 0.5 displacement and zero acceleration at the mid-point ($u = 0.5$, $w = 0.5$, $w'' = 0$). Reflective polynomials offer much more variety of acceleration loop *shapes* than is possible with full ones, making them easier to design for specific dynamic characteristics.

The only way fully to specify a polynomial is to give the coefficients of *all* the terms. This is a rather cumbersome means of naming standard cam laws, so it has become popular to name them according to the powers of the terms in the series. The best example is the cam law:

$$w = 10.u^3 - 15.u^4 + 6.u^5 \hspace{3cm} \text{Eq. (4.2)}$$

which is known as a '3-4-5' polynomial. While it is possible to have a different set of coefficients for the three terms, these particular values fulfil all the requirements of a standard full polynomial and incidentally give a symmetrical motion. It is reasonable, therefore, to restrict the name '3-4-5' to this equation. This is a useful motion with $C_a = 5.7735$ and $C_v = 1.875$, and a performance between Mod Sine and Cycloidal (see below).

If more terms are included, say a u^6 term, then a large number of different sets of coefficients would give practical cam laws with varying degrees of asymmetry and other properties. Designations such as '3-4-5-6' are not really adequate. There is, however a useful *reflective* polynomial:

$$w = (40/3).u^3 - 40.u^4 + 64.u^5 - (128/3).u^6 \hspace{2cm} \text{Eq. (4.3)}$$

Which is designated here as a '3-4-5-6-R' polynomial. This motion has $C_a = 5$ and $C_v = 2$, and has a similar performance to the Mod Trap cam law (see below).

SCCA CAM LAWS

Many of the popular standard cam laws and their composites can be considered as belonging to a family of laws known as SCCA (Sine-Constant-Cosine-Acceleration). This is very convenient, particularly for computer programs, because one set of general equations can be used for all members of the family by specifying only three parameters. Equations and diagrams of these laws can be found in Appendix A. Although the equations are more complex than those needed for some of the motions (e.g. CYC and PAR), they are no more complex than, for instance, the specific equations for MS and MT.

The general form of the SCCA *acceleration loop* is divided into four zones:

(1) A quarter sine curve with period parameter *a*
(2) Constant acceleration with period parameter *b*

(3) A quarter cosine curve with period parameter c

(4) Zero acceleration with period parameter d

The deceleration loop is a reversed negative copy of the acceleration loop.

The paramaters a, b, c and d are scaled so that:

$$a + b + c + d = 1$$ Eq. (4.4)

Any parameter may be zero, but not all of them, and none may be negative. To describe a particular SCCA motion it is only necessary to specify a, b and c since d must then satisfy Equation 4.4.

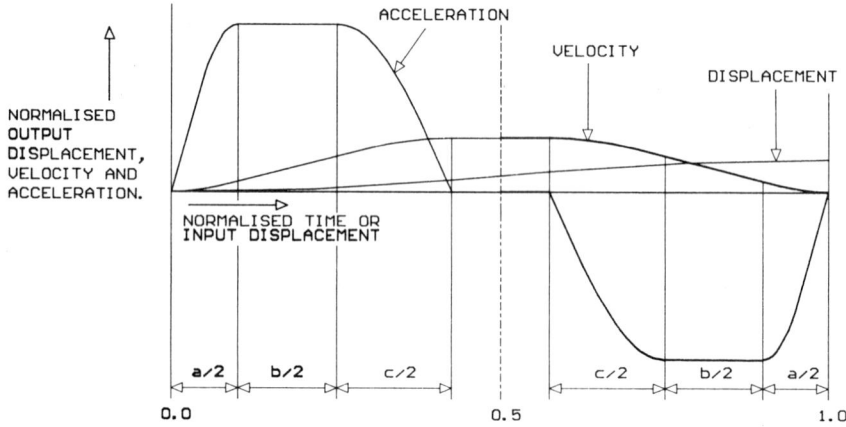

Fig. 4.2   General form for the SCCA cam laws

The *shape* of the acceleration loop is one of the most important characteristics of a cam law, and most of the best shapes are available as SCCA laws with appropriate parameters.

Table 4.1 lists SCCA parameters for popular standard laws, and their C_v and C_a values.

Interesting points to note

For non-composite laws ($a + b + c = 1$);

when $a = c$ (i.e. the acceleration loop is symmetrical),

then $C_v = 2$,

and when a is less than c, then C_v is less than 2.

Table 4.1   Popular SCCA cam law parameters

Popular cam law name	Designation	SCCA parameters			Coefficients	
		a	*b*	*c*	C_v	C_a
Parabolic	PAR	0	1	0	2	4
Simple Harmonic	SH	0	0	1	1.571	4.935
Mod Trap	MT	0.25	0.5	0.25	2	4.888
Mod Sine	MS	0.25	0	0.75	1.759	5.528
Mod Sine 20%CV	MSC.20*	0.2	0	0.6	1.528	5.999
Mod Sine 33%CV	MSC.33*	0.16667	0	0.5	1.404	6.616
Mod Sine 50%CV	MSC.50*	0.125	0	0.375	1.275	8.013
Mod Sine 66%CV	MSC.66*	0.08333	0	0.25	1.168	11.009
Cycloidal	CYC	0.5	0	0.5	2	6.283
Cycloidal 50%CV	CYCC.50	0.25	0	0.25	1.3333	8.378

*The number represents the percentage of the whole motion period that is occupied by a central constant velocity zone. The figures 33 and 66 are used in designations instead of 33.33333 and 66.6666 respectively, for brevity.

For composite laws $(a + b + c < 1)$:

C_v is always lower and C_a higher than the basic non-composite version.

There are, of course, an infinite number of possible sets of a, b, c parameters, and it is sometimes useful to vary them a little from one of the standard sets to achieve a particular value of C_v, especially when using cam law blending (q.v.) to match a specified motion pattern. The above table gives a guide to the best starting point for modifying SCCA parameters.

Other standard motion laws of interest are given in Table 4.2.

The equations for standard cam laws are given in Appendix A, together with displacement, velocity and acceleration diagrams. When you compare different cam laws you will find little difference in their displacement diagrams, but rather more difference in their velocity diagrams. There is a distinct difference in their acceleration diagrams, however, which makes the shape of its acceleration loop the most characteristic feature of a cam law.

It is true to say that any cam law that has an acceleration loop of similar shape to that of a standard cam law will perform in a very similar manner, no matter what mathematical equation(s) are used to produce that shape. Standard SCCA cam laws can thus be simulated with different (often simpler) equations, with predictable results. However, the generality of the SCCA cam law equations, given modern high-speed computing, is a great advantage and makes the need for simulation questionable.

Table 4.2   Popular non-SCCA cam law characteristics

Popular cam law name	Characteristics	Coefficients	
		C_v	C_a
Double Harmonic	Trigonometric D-R-R-D asymmetrical cam law	2.041	−9.870 (C_d)
3-4-5 Polynomial	Algebraic D-R-D symmetrical cam law	1.875	5.7735
3-4-5-6-R Polynomial	Algebraic D-R-D symmetrical cam law	2	5
3 stations Geneva mechanism	Indexing mechanism with a symmetrical motion dictated by its geometry	3.232	16.437
4 stations Geneva mechanism		2.414	8.494
6 stations Geneva mechanism	The motion pattern is particular to each number of stations. Abrupt acceleration step at each end of the motion	2.000	5.653
8 stations Geneva mechanism		1.860	4.946

The *normalised* values calculated from the cam law equations are generally known as cam law *factors*: displacement factor, velocity factor, etc. The factors for any point in the motion can be calculated from the appropriate equations and then used to calculate the real displacement, etc. using Equations 3.2, 3.5 and 3.10 as shown in the example below. This is an ideal method for a computer program, but the evaluation of the cam law equations is rather time-consuming otherwise. For convenience, therefore, standard *cam law factor tables* have been included in Appendix B, which give normalised displacement, velocity, acceleration and jerk factors for various standard laws.

The tables have factors for 120 steps (equal intervals of normalised input displacement). They also include 'input torque factors' for use as described in Chapter 13. Intermediate factors can be obtained from the tables by interpolation if the input displacement falls between two of the steps listed. The factors are calculated from the cam law equations and can be used directly to calculate real values as follows:

Example 4.2

An oscillating mechanism has an output stroke of 55° and is driven by a cam with an input stroke (period) of 60°. The camshaft runs at 40 rev/min.

Find the output displacement, velocity and acceleration at 40° of cam rotation after the start of the motion period. The cam law is Mod Sine.

The Mod Sine factor table in Appendix B has 120 steps, so for a 60° cam period each step is 0.5° of rotation. 40° after the start is therefore step number 80. The factor table gives:

$u = 0.666$ $w = 0.77583$ $w'' = 1.45085$ $w'' = -3.5533$

From Equation 3.2 output displacement $= y = 55 \times 0.77583 = 42.671°$

The output stroke is $Y = 55 \times \pi \div 180 = 0.95993$ rad so the output displacement at $u = 0.666$ could be expressed as:

$y = 0.95993 \times 0.77583 = 0.7747$ rad $(=42.671°)$

One revolution of the camshaft takes $60 \div 40 = 1.5$ s, therefore the time period of the motion (60° cam period) is

$T = 60 \div 360 \times 1.5 = 0.25$ s

From equation 3.5 output velocity

$= 0.95993 \times 1.45085 \div 0.25$

$= 5.5709$ rad/s

From equation 3.10 output acceleration

$= 0.95993 \times (-3.5533) \div 0.25^2$

$= -54.575$ rad/s^2

(The negative sign shows that this is in fact a *deceleration*)

CAM LAWS COMPARED

Table 4.3 gives merit ratings for common standard cam laws to enable a choice to be made in the initial stage of machine design. Merit for each characteristic ranges from 1 (relatively bad) to 5 (excellent).

Explanatory notes

(1) Peak acceleration is the nominal maximum output acceleration during the motion period, calculated by the cam law equation.

Table 4.3   Merit ratings of SCCA cam laws

Cam law designation	Peak acceleration (1)	Output vibration (2)	Peak velocity (3)	Impact (4)	Input torque (5)	Input vibration (6)	Residual vibration (7)
PAR	5	1	2	1	1	1	1
SH	3	1	4	4	5	2	1
MT	3	3	2	2	2	3	3
MS	2	4	3	3	4	4	4
CYC	1	5	2	3	3	4	5

(2) A vibration is superimposed on the nominal output acceleration increasing the nominal peak value. Its severity depends on the elasticity and operating speed of the mechanism and the merit rating shown applies to mechanisms of average rigidity running at fairly high speed.

(3) Peak velocity is the nominal maximum output velocity during the motion period, calculated by the cam law equation. Its value is also increased by superimposed vibration.

(4) Impact forces occur at the locations of backlash in the mechanism when the changeover from acceleration to deceleration occurs. The severity of the impact depends on how gradually the changeover takes place, i.e. on how low the jerk is at the point of impact. Strictly speaking it is the changeover from positive to negative force or torque that matters, but in most high speed systems that almost coincides with the acceleration changeover.

(5) The nominal input torque of a mechanism varies throughout the motion period and is a function of the output load profile and the velocity pattern. The peak acceleration and the peak velocity do not coincide and neither of them coincides with the peak input torque. Cam laws with good, that is low, acceleration do not necessarily have good input torque.

(6) The elasticity and backlash of the input transmission can cause serious *overrun*. This is when the sudden reversal of input torque at the changeover causes the cam to jump forwards before it can transmit a decelerating force to the output. The more gradually that the nominal input torque changes over, the less severe is the overrun and its consequences (see Transmission design, Chapter 11).

(7) Residual vibration takes place in the dwell period immediately following the motion period in high speed or very elastic systems. Its

amplitude depends on the vibration generated during the motion period (see note 2.) and the degree of damping present in the output transmission. It is very difficult to add sufficient damping to high speed mechanisms to eliminate residual vibration, so the choice of a suitable cam law is vital in some cases.

On scanning the merit ratings above, it can be seen that, of the laws listed, Mod Sine is the best for general purposes. Its particular merit is that it is very tolerant of a 'bad' input drive and transmission (elasticity, backlash, wear, low inertia). It is nowadays the first choice of cam designers and is almost universally used by commercial manufacturers of cam-operated indexing and oscillating mechanisms.

GENEVA MECHANISM

This is the name given to a very popular type of indexing mechanism, sometimes called a *Maltese Cross* mechanism, which is illustrated in its various forms in Fig. 4.3. This is not usually thought of as a cam mechanism because there is no specially curved profile on any of the components: all shapes are made of straight lines and circular arcs. However, it can be thought of as an inverse cam mechanism. The slots in the output 'starwheel' are usually straight as shown, but they could be curved to produce a non-standard motion if necessary.

A fundamental characteristic of this mechanism is that for each number of stops (number of stations) there is a particular motion period/dwell period and a particular motion law, whereas with a conventional *cam* mechanism there is virtually a free choice of both motion period and motion law. In the cases of the 3-stop and 4-stop Geneva mechanisms in particular, the dynamic characteristic of the motion law is inferior to a good standard cam law.

Unless otherwise specified, a Geneva mechanism is assumed to be an external one, but the internal Geneva is useful for producing a shorter dwell period for a given number of stops without using a conventional cam. It also has an acceleration pattern close to that of Simple Harmonic motion.

The locking crescents engage automatically during the dwell period, but because of the need to have a sliding clearance the locking effect is not very positive: the crescents also tend to wear rapidly, especially on the tips. Many ingenious devices have been developed to provide positive

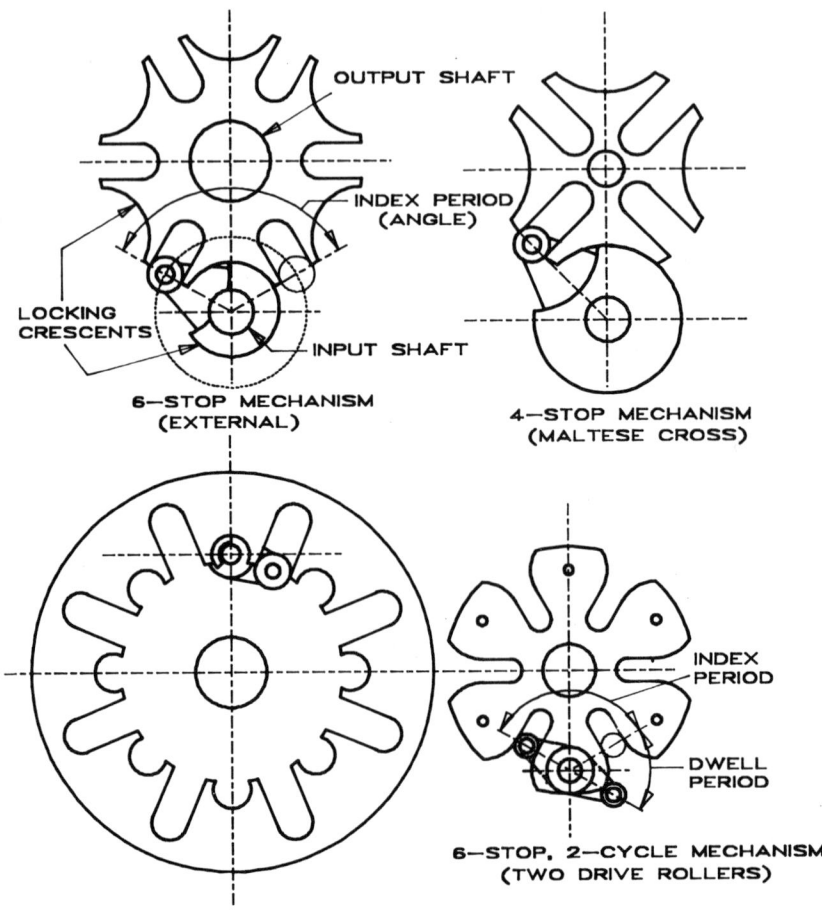

OUTPUT SHAFT

INDEX PERIOD
(ANGLE)

LOCKING
CRESCENTS

INPUT SHAFT

6—STOP MECHANISM
(EXTERNAL)

4—STOP MECHANISM
(MALTESE CROSS)

INDEX
PERIOD

DWELL
PERIOD

6—STOP, 2—CYCLE MECHANISM
(TWO DRIVE ROLLERS)

8—STOP INTERNAL MECHANISM

Fig. 4.3 Geneva mechanisms

locking for high performance Geneva mechanisms, but it is doubtful
whether there are many occasions when the expense and complication
are worth while in view of the other shortcomings of the Geneva
compared with cam operated indexing mechanisms.

An interesting version of the Geneva is the two-cycle one shown here
as a 6-stop mechanism, but it can be any other number of stops that is

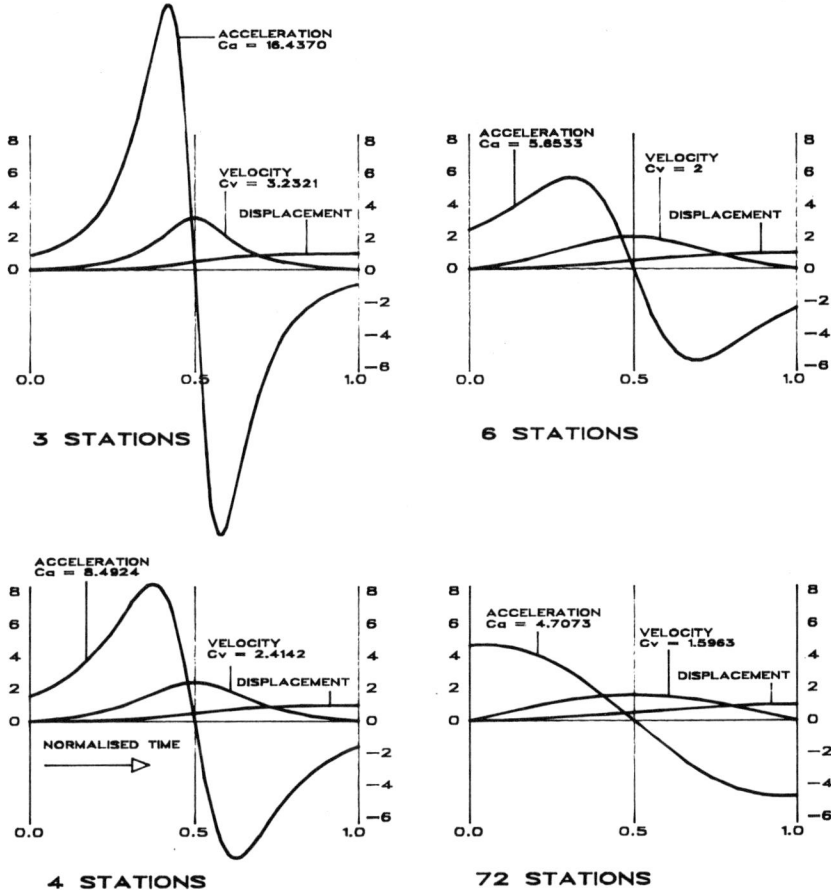

Fig. 4.4   Motion diagrams of Geneva mechanisms

practical for an external Geneva. Two complete index and dwell cycles are performed for each revolution of the input shaft. The dwell period is shorter than the index period, unlike a conventional external Geneva, and locking during the dwell period is effected by the two drive rollers straddling a 'lobe' on the starwheel, possibly with some pre-load.

This is a more positive locking arrangement than that of a normal Geneva mechanism and is subject to less wear.

Details of the 'cam law' – motion law – of some Geneva mechanisms are included in Appendix B for comparison with standard cam laws, and some Coefficients of Velocity and Acceleration are given in Table 4.4. A few diagrams of the normalised acceleration, etc. are also shown in Fig. 4.4. It can be seen that there is high peak acceleration with 3-stop and 4-stop mechanisms, and it is interesting to see that for a very large number of stations (72) the motion is almost the same as Simple Harmonic.

Table 4.4   **Geneva mechanism motion characteristics**

Motion coefficients of external Geneva mechanisms

No. of stations	C_v	C_a
3	3.232	16.437
4	2.414	8.493
5	2.139	6.499
6	2.000	5.653
8	1.860	4.946
10	1.789	4.673
12	1.746	4.551
16	1.697	4.470

CHAPTER 5

Cam Law Blending, Special Motion Design

If the purpose of a cam motion is simply to move the output of the mechanism from one position to another in a certain cam rotation period, then the criterion for choosing the motion is usually just to get a smooth action with reasonable cost. Very low cost designs that are not very smooth may be acceptable for some applications, but for best performance, e.g. high speed, the best economical choice is one that uses a standard cam law, probably Mod Sine. If, however, there is an additional special requirement, e.g. to pass through a particular intermediate point (known as a precision point) at a particular time in the motion, or, to avoid loss of contact with a product during the deceleration phase of a pusher mechanism, or to mimic the motion of some other mechanism in the machine, etc., then a special motion may have to be designed.

There are many ways of designing a special cam motion to fulfil a special requirement, some of which will not necessarily result in a smooth motion with a good dynamic performance. It is self-evident that if the special motion is composed of the acceleration and deceleration loops of standard cam laws, a dynamic performance similar to that of the standard laws may be expected. Such a motion may be easily designed using the *cam law blending* technique described here. The term 'blending' is used to describe the use of contiguous periods of motion with smooth transition from one type of motion to an adjacent different type of motion. Incidentally, some of the standard cam laws referred to in this book are themselves blended motions, but they are used just as if they were single, continuous mathematical functions.

Smoothness requires that there be, in order of importance, *continuity of displacement, continuity of velocity and continuity of acceleration*. Continuity means without a break or step, but not necessarily without a kink (see Fig. 5.1).

In general, any special motion must consist of periods of acceleration, periods of deceleration and possibly periods of constant velocity in a particular sequence. For good dynamic performance it is necessary to

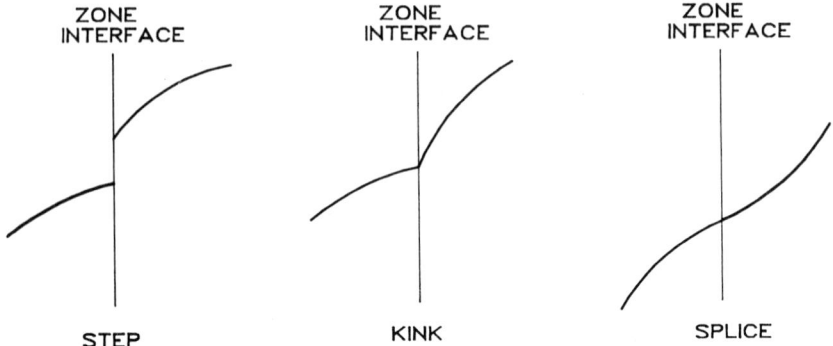

Fig. 5.1   Cam law blending

avoid steps in the acceleration pattern, i.e. discontinuities. However, the jerk pattern does not necessarily have to be continuous: a step in the jerk function might be the result of a 'kink', rather than a step, in the acceleration function, which does not in itself cause adverse dynamic effects. A kink in a function is a point on its graph where the slope suddenly changes, but without a break.

As we have seen in Chapters 3 and 4, the so-called standard cam laws are symmetrical, with an acceleration loop occupying the first half of the motion and a similar deceleration loop occupying the second half. The dynamically 'good' laws have loops which start and finish with zero acceleration. Constant velocity periods, of course, also start and finish with zero acceleration, and so do loops of combined acceleration and constant velocity.

When such loops and pure CV periods are strung together, therefore, there will only be acceleration kinks, not steps, at the interfaces of adjacent zones. Furthermore, those kinks occur at points of zero acceleration, where, by definition, the nominal dynamic forces in the system are zero, so that the nominal strain energy in the system is minimal and adverse effects are least likely.

Choosing complete standard acceleration or deceleration loops, then, satisfies one important requirement for smooth transition between zones of motion. To satisfy the other requirements, the stroke (overall displacement) and period (cam rotation) for *each* zone must be chosen to give continuous displacement and velocity at all zone interfaces. At first sight this is a daunting problem, but the solution can be found using the tangent intersect method which involves simple calculations of similar triangles constructed on the *displacement diagram*. A displacement diagram is usually drawn as a graph of real output displacement, *y*, against real input displacement, *x*. Input displacement – cam rotation – is proportional to time if the cam rotates at constant velocity, so plotting *y* against time, *t*, is equally valid and is sometimes more convenient.

Rather than have two unknowns for each zone (stroke and period) it has been found in practice that with most blending problems it is easy to decide the period of each zone by considering the special conditions imposed by the application. Occasionally it may be necessary to revise one's first choice of these, but the first trial results will usually indicate a successful second choice. This simplifies the design process considerably.

BLEND POINTS, INTERSECT POINTS AND THE TANGENT INTERSECT METHOD OF CAM LAW BLENDING

Continuity of displacement simply requires that each zone of motion starts with the same displacement as that with which the preceding zone finishes. This condition needs no special calculations and is quite obvious on inspection of the displacement diagram. There is a single point on the displacement curve at the interface between two zones of motion, which we shall call the *blend point* (BP) for those zones. Having decided on the periods for each motion zone the *x* co-ordinates of the blend points are fixed, but not the *y* co-ordinates. However, given that the periods and cam laws of adjacent zones are known, the value of the displacement at the blend point can be calculated to give the same velocity at the end of one zone as at the beginning of the next, i.e. continuity of velocity. Velocity is the slope of the displacement curve, and continuity of slope means smoothness without any kinks. Suitable blend point displacements can be found using the *tangent intersect method*.

A straight line drawn through the blend point at the same slope as the displacement curve is a tangent to the curve (see Fig. 5.2). If the slopes (velocities) at the start and finish of a zone are *not* the same, the tangents from those two blend points will intersect at some point, not on the curve, which we call the tangent intersect point or just the *intersect point* (IP). The important property of an intersect point is that its input displacement (*x* co-ordinate) is fixed by a 'blend factor' which is *only* dependent on the cam law of that zone.

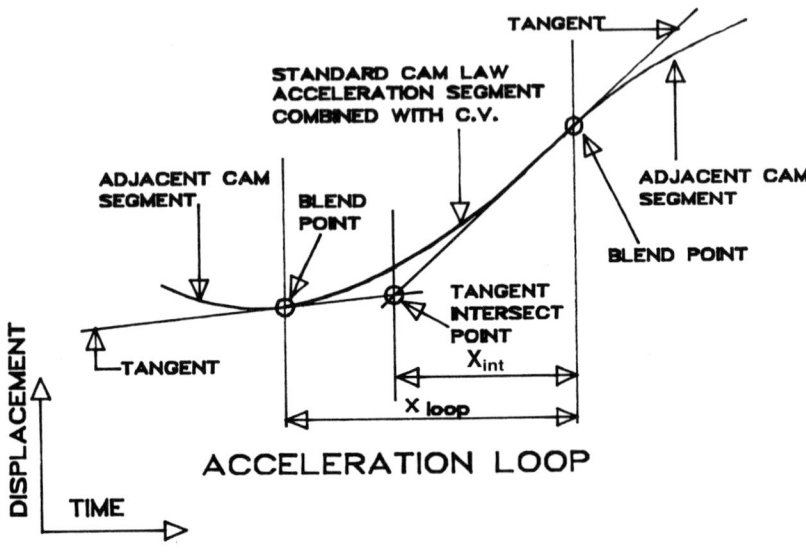

ACCELERATION LOOP

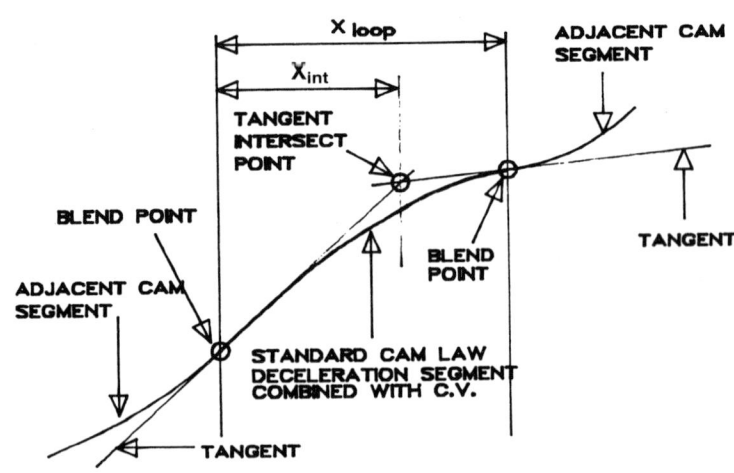

DECELERATION LOOP

Fig. 5.2   Tangent intersect method of blending

The blend factor for any point in a cam law is derived from its normalised displacement and velocity factors at that point:

$$z = u \cdot w'/w \qquad \text{Eq. (5.1)}$$

where:

  z = blend factor
  u = input displacement factor
  w' = output velocity factor
  w = output displacement factor

The blend factor at the maximum velocity end of a standard cam law acceleration loop is its C_V value (coefficient of velocity).

Once the x co-ordinates of the intersect points are known their y co-ordinates and those of the blend points can be found by triangulation and simple arithmetic.

In general, the zones of a blended motion consist of loops of cam law acceleration (or deceleration) and constant velocity combined (see Chapter 4). Either component may be missing to give the special cases of a pure cam law or pure constant velocity. Intersect points do not occur in pure CV zones.

A standard cam law acceleration loop has a velocity of zero at the start and a normalised velocity of C_V at the finish (mid-point of the full motion). Conversely the deceleration loop velocities are C_V at the start (mid-point of the full motion) and zero at the finish.

The intersect point (IP) of a zone falls within the zone and is displaced along the x-axis by the amount (zone period $\div C_V$) from the end at which C_V applies. For an acceleration loop the IP displacement is *back* from the finish of the zone, and for a deceleration loop it is *forward* from the start of the zone.

$$x_{\text{int}} = x_{\text{loop}}/z \qquad \text{Eq. (5.2)}$$

where

  x_{int} = distance along the x-axis from the intersect point to the end of the acceleration loop.

or

  x_{int} = distance along the x-axis from the start of the acceleration loop to the intersect point.
  x_{loop} = length along the x-axis of the acceleration or deceleration loop.

For example, a MS *acceleration* + CV zone of 30° period has a C_V value of 1.7596, so its IP occurs 30 ÷ 1.7596 = 17.049° *before the end* of the zone, i.e. 12.951° after the start. On the other hand a MS *deceleration* + CV zone of 40° period has its IP occurring at 40 ÷ 1.7596 = 22.732° *after the start* of the zone.

It is worth noting that if the cam law used has a C_V value of 2, such as Cycloidal or Mod Trap, then the IP occurs exactly in the centre of the zone, whether it is an acceleration or a deceleration one.

CAM LAW BLENDING

Example 5.1

In this example a low velocity *ramp* (CV zone) is required at the start of the motion to minimise the impact when taking up clearance between a pusher and a product, where the clearance is variable up to 5 mm. It has been estimated that the impact velocity will be acceptable if the ramp has a slope of 1 mm per 6 degrees of camshaft rotation and therefore occupies 30° of rotation to cover the 5 mm maximum clearance. A constant velocity zone is required in the middle of the motion to match the velocity of another machine component: this works out to be a geometric velocity of 0.4 mm per degree of camshaft rotation, and it can continue for 60°. It is estimated that 45° cam rotation can be allowed for accelerating from the starting ramp to the middle CV zone and 60° for decelerating *to rest* at the end. There is no constraint on the total output stroke, but a smooth motion is desirable and it is considered that the speed and rigidity of the machine allow Mod Trap acceleration and deceleration to be used.

Designate the zones, intersect points (IPs), blend points (BPs) and tangent slopes (Ss), in sequence, thus:

Zone 1.	CV ramp as specified.	
BP1.	Blend point at end of zone 1.	Slope of tangent = S1
Zone 2.	Acceleration zone.	Contains IP2.
BP2.	Blend point at end of zone 2.	Slope of tangent = S2
Zone 3.	Middle CV zone as specified.	
BP3.	Blend point at end of zone 3.	Slope of tangent = S3
Zone 4.	Deceleration zone as specified.	Contains IP4.

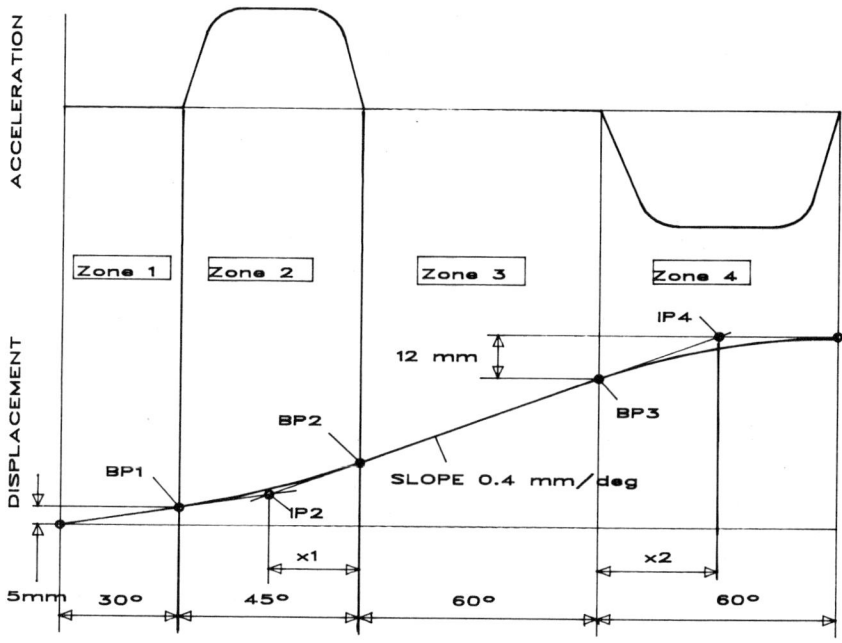

Fig. 5.3   Blending, Example 5.1

Referring to Fig. 5.3:

$x1 = 45 \div 2 = 22.5°$ $x2 = 60 \div 2 = 30°$
$S1 = 5 \div 30 = 0.16667$ mm/deg, $S2 = S3 = 0.4$ mm/deg (given)
y-value of IP1 $= (30 + 22.5) \times 0.16667 = 8.75$ mm
y-value of BP2 $= 8.75 + 22.5 \times 0.4 = 17.75$ mm
y-value of BP3 $= 17.75 + 60 \times 0.4 = 41.75$ mm
y-value of IP4 $= 41.75 + 30 \times 0.4 = 53.75$ mm (total stroke)

Zone 1 is CV, 30° input × 5 mm output.
The output displacement of Zone 2 is 17.75 − 5 = 12.75 mm. Of this 45 ×
0.16667 = 7.5 mm is the CV part, leaving 5.25 mm for the MT acceleration part, therefore:

Zone 2 is *half of* a standard MT motion with 90° input × 10.5 mm output,
superimposed on CV of 45° input × 7.5 mm output.

A short designation of this motion is 45° × (5.25 mm MTA + 7.5 mm CV). Here MTA means Mod Trap Acceleration loop.

The output displacement of Zone 3 is 60 × 0.4 = 24 mm.

Zone 3 is CV with 60° input × 24 mm output.

The output displacemnent of Zone 4 is 30 × 0.4 = 12 mm, therefore:

Zone 4 is *half of* a standard MT motion with 120° input × 24 mm output with *no* superimposed CV.

The short designation is 60° × 12 mm MTD. Here MTD means Mod Trap Deceleration.

The short form designation that could be used for the *whole* of this blended motion is:

30° × 5 mm CV: 45° × (5.25 mm MTA + 7.5 mm CV):
60° × 24 mm CV: 60° × 12 mm MTD.

The input stroke of the full motion (motion period) is 30 + 45 + 60 + 60 = 195° and the total output stroke is 5 + 5.25 + 7.5 + 24 + 12 = 53.75 mm.

This motion is a smooth one in itself, which conforms to the specification, but there is an unsatisfactory condition at the very beginning of the motion which was overlooked in the original specification. Assuming that the mechanism is at rest at the start (i.e. the specified motion is preceded by a dwell period – zero velocity) then there is a kink in the displacement curve at that point, which will produce shock forces and excessive vibration if the output inertia is significant.

A simple cure for this is to make the first zone a Mod Trap acceleration loop instead of a ramp. The final slope of the zone (at BP1) can be the same as the original ramp, which gives an acceptable maximum pusher impact velocity, and of course the preceding velocities will be lower than this maximum. The output displacement and velocity at BP1 must remain 5 mm and 1/6 mm/deg as specified. The IP for the modified zone is the original start point of the motion and the new start point is 30° earlier.

The new Zone 1 is a 60° × 5 mm MTA loop. The total period of the motion is increased by 30°, but this is the price of a shock-free movement. The modification is shown in Fig. 5.4.

It must be said that the modified initial MTA period can be reduced at the expense of a higher peak acceleration if Zone 1 is split into two zones – MTA followed by CV – but some extension to the overall motion period cannot be avoided if the original kink is to be removed.

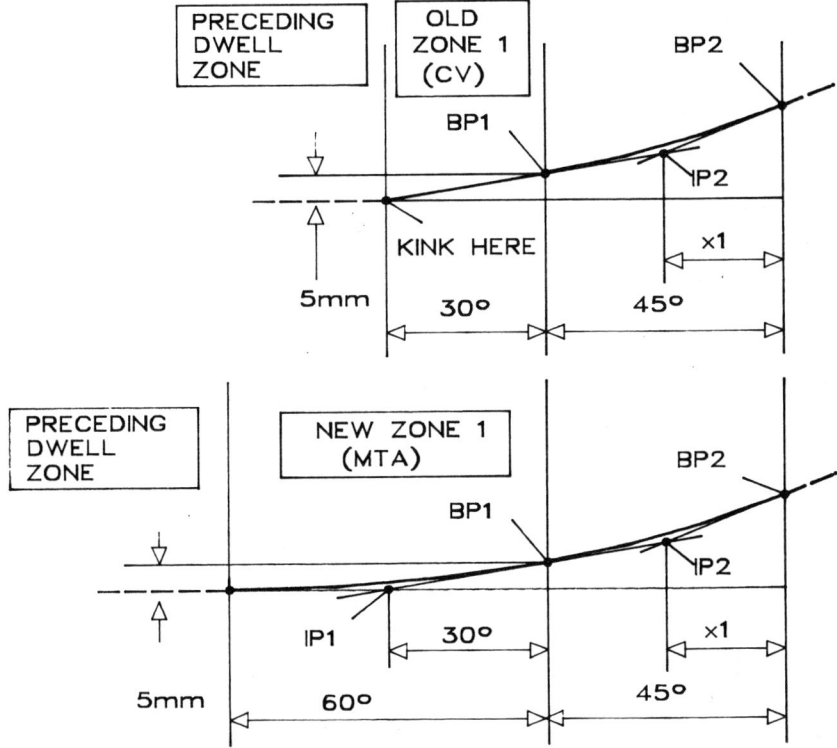

Fig. 5.4   Blending, Example 5.1 – modification

For manufacturing purposes the displacements at incremental rotations within each zone can be calculated from the equations in Appendix A or from the factor tables in Appendix B. See Chapter 4 for the
treatment of combined motions.

ASYMMETRICAL CAM LAWS: A SPECIAL CASE OF BLENDING

It is quite common, for a variety of reasons, to use asymmetrical motions
based on standard cam laws. One may want a slow start and a fast finish
(low acceleration followed by high deceleration) or vice versa. This can

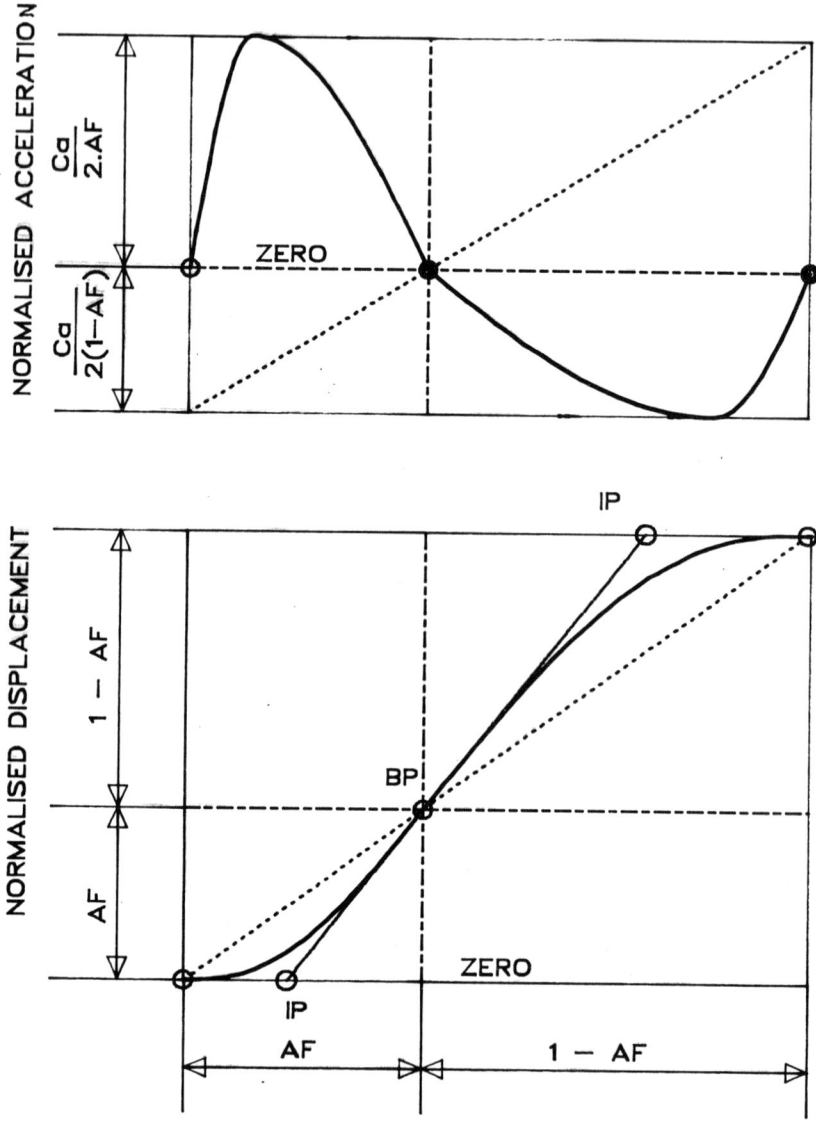

Fig. 5.5   Asymmetrical motion

best be achieved by blending a standard acceleration loop with a standard deceleration loop. These loops need not be of the same cam law, but they usually are (see Fig. 5.5).

A useful property of asymmetrical motions, with the same cam law for both loops, is that the intermediate blend point always lies on the straight line joining the start to the finish of the motion (the displacement diagram diagonal). This simplifies the calculations. Such motions are designated simply as asymmetrical cam laws with a specified *asymmetry factor*, *AF*. For example, Asymmetrical Mod Sine (AF = 0.35).

The asymmetry factor is the fraction of the whole motion period that is occupied by the acceleration zone. A motion with 100° cam angle and AF = 0.4 has 40° of acceleration and 60° of deceleration. Also, because of the linear disposition of the blend point, the output displacements of the two zones are also proportioned by the asymmetry factor. A motion with a 100 mm stroke and AF = 0.4 accelerates for 40 mm of that stroke and decelerates for the remaining 60 mm.

When an asymmetrical motion has a constant velocity zone in the middle (e.g. MSC.50) then AF is the fraction of the symmetrical acceleration and deceleration periods together that is occupied by its acceleration period.

The maximum normalised velocity of the asymmetrical motion is the same as that of the symmetrical law, namely C_V. The maximum normalised acceleration and deceleration are simple functions of the C_a value of the symmetrical law:

$$\text{Maximum acceleration} = \frac{C_a}{2 \cdot AF} \qquad \text{Eq. (5.3)}$$

$$\text{Maximum deceleration} = \frac{C_a}{2 \cdot (1 - AF)} \qquad \text{Eq. (5.4)}$$

With AF = 0.35 for example, Max. Accel. = $C_a / 0.7$
and Max. Decel. = $C_a . / 1.3$

The usefulness of the general purpose SCCA family of cam laws (see Chapter 4) is further extended by including the AF value as an extra parameter. There are now only 4 SCCA parameters required – a, b, c and AF – fully to specify a very large number of cam laws, symmetrical and asymmetrical, with and without constant velocity zones, to cover the vast majority of high performance industrial cam applications. For *symmetrical* laws, of course, AF = 0.5.

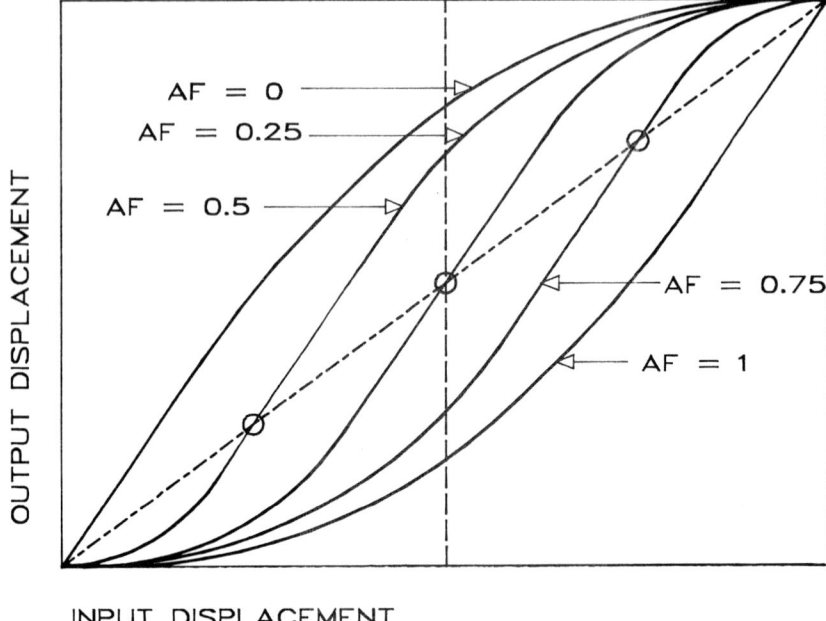

AF = 0

AF = 0.25

AF = 0.5

AF = 0.75

AF = 1

OUTPUT DISPLACEMENT

INPUT DISPLACEMENT

Fig. 5.6   Asymmetrical motion domain

Asymmetrical motions are useful for a number of purposes. To reduce the peak acceleration at the expense of the peak deceleration, *or vice versa*, is quite common. This may be in order to reduce the force or torque in one part of the system, e.g. the pressure exerted on a delicate payload, or to minimise spillage in an open liquid container being transported by the mechanism. It may be in order to increase the radius of curvature at one part of the cam profile to avoid undercutting. Yet again it may be to make a part of the mechanism pass through a particular point in space (a 'precision point') or to avoid collision with an obstacle. Whatever the purpose, it must be appreciated that there are practical limits to the amount of asymmetry that can be used and a motion more complex than simple asymmetry may be necessary to achieve one's end.

The displacement diagram in Fig. 5.6 shows the domain covered by all asymmetrical Cycloidal motions. No point outside that domain can possibly be reached by a simple asymmetrical Cycloidal law. Obviously the limiting values, AF = 0 and AF = 1, are not practical where the motion is preceded or followed, respectively, by a dwell zone.

Similar diagrams can be drawn for any asymmetrical standard law. If a precision point or collision point is specified outside the domain, a composite motion must be designed with more than two zones, properly blended of course. This is illustrated later, in Example 5.3.

CAM LAW BLENDING

Example 5.2

An oscillating mechanism has an output stroke of 60° during a cam rotation period (input stroke) of 90°. The motion is preceded and followed by dwell periods. It is required that the output displacement be 9° at 30° cam rotation after the start. Cycloidal acceleration and deceleration is preferred (see Fig. 5.7).

The precision point (30, 9) falls within the asymmetrical Cycloidal domain, so a simple asymmetrical motion will do. Referring to the diagram below let X_a and Y_a be the co-ordinates of the blend point, and

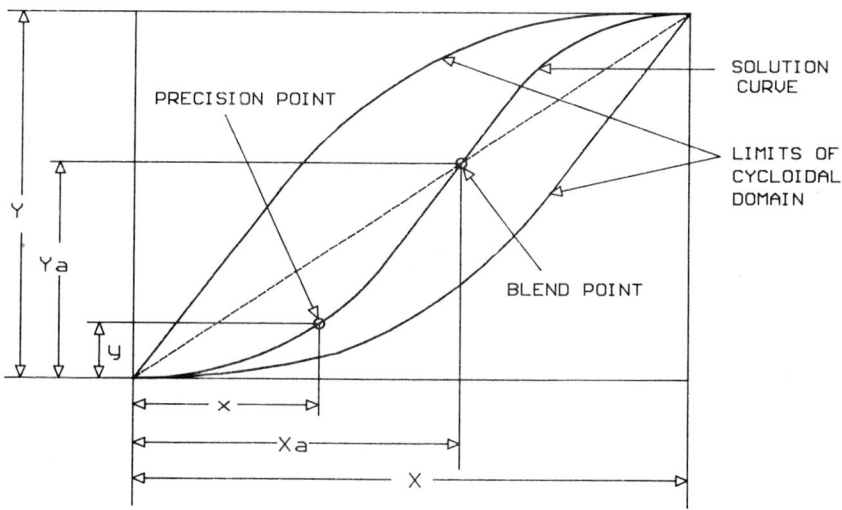

Fig. 5.7   Blending, Example 5.2

let x and y be the co-ordinates of the precision point. We know that the blend point must lie on the diagonal of the diagram, so $Y_a/X_a = Y/X$.

By definition, the normalised displacements of the precision point are:

$$u = \frac{x}{2X_a} \quad \text{and} \quad w = \frac{y}{2Y_a}$$

Therefore: $\quad w/u = \dfrac{y \cdot X_a}{x \cdot Y_a} \quad$ and by substitution $\quad w/u = \dfrac{y \cdot X}{x \cdot Y}$

Inserting values we get: $\quad w/u = \dfrac{9 \times 90}{30 \times 60} = 0.45$

Searching the Cycloidal factor list in Appendix B we find that at step 34:

$$u = 0.283 \qquad w = 0.12765 \qquad w/u = 0.4505$$

This is probably close enough, but if we needed a more accurate solution we could interpolate from the table to find a value of u where $w/u = 0.45$ exactly.

Since $\quad u = \dfrac{x}{2 \cdot X_a} \quad$ then $\quad X_a = \dfrac{x}{2 \cdot u} = \dfrac{30}{0.566} = 53°$

Now the asymmetry factor, $AF = 53 \div 90 = 0.58889$

Therefore $\quad y_a = AF \cdot Y = 0.5889 \times 60 = 35.333°$

The solution is an asymmetrical Cycloidal motion ($AF = 0.58889$)

The actual y value of the calculated precision point is

$\quad 0.12765 \times 35.333 = 9.0206°$

which could be improved as described above.

Figure 5.8 shows a practical design of cam at the beginning of the motion found by this calculation. The cam segment looks very steep, but the maximum pressure angle was found to be 46.66° which, though fairly high, is acceptable for this kind of mechanism. A point of interest is that the drawing shows that the maximum cam radius is so large that it interferes with the follower pivot shaft. This forces the machine designer to make the follower lever cantilevered on its shaft and the follower roller also cantilever mounted: a good example of the need for cam design to be an integral part of the machine design process.

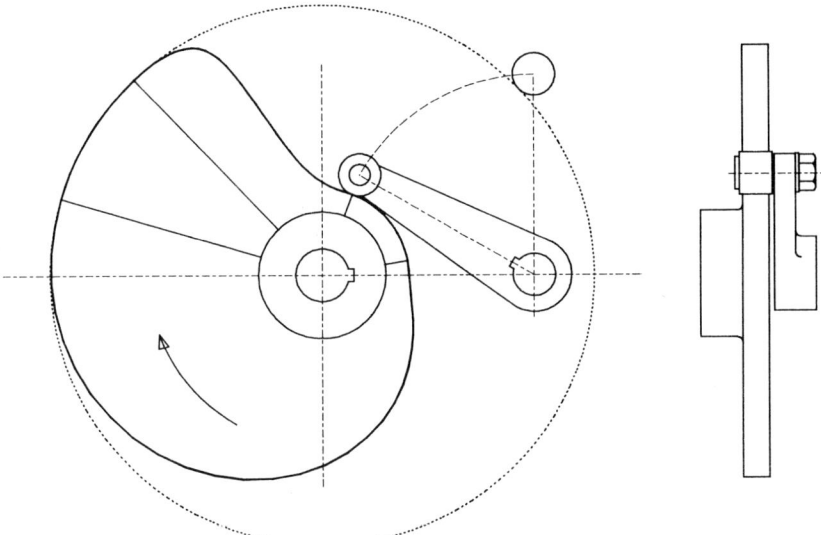

Fig. 5.8 Practical design, Example 5.2

CAM LAW BLENDING

Example 5.3

A lifting mechanism has an output stroke of 80 mm during a cam rotation period of 105°. The motion is preceded and followed by dwell periods. It is necessary to reach a lift of 65 mm when the cam has rotated 42° (see Fig. 5.9).

In the absence of any special cam law requirement use Mod Sine. The precision point (42, 65) lies outside the asymmetrical Mod Sine domain, so it is not possible to satisfy the specification with a simple asymmetrical motion.

It is obvious from Fig. 5.9 that the acceleration required to reach the precision point must be much higher than the deceleration from there to the end of the motion. The lowest peak deceleration would be achieved if the whole of the 63° period were a full deceleration loop, and this would have a tangent through the precision point to cut the final dwell tangent at IP as shown.

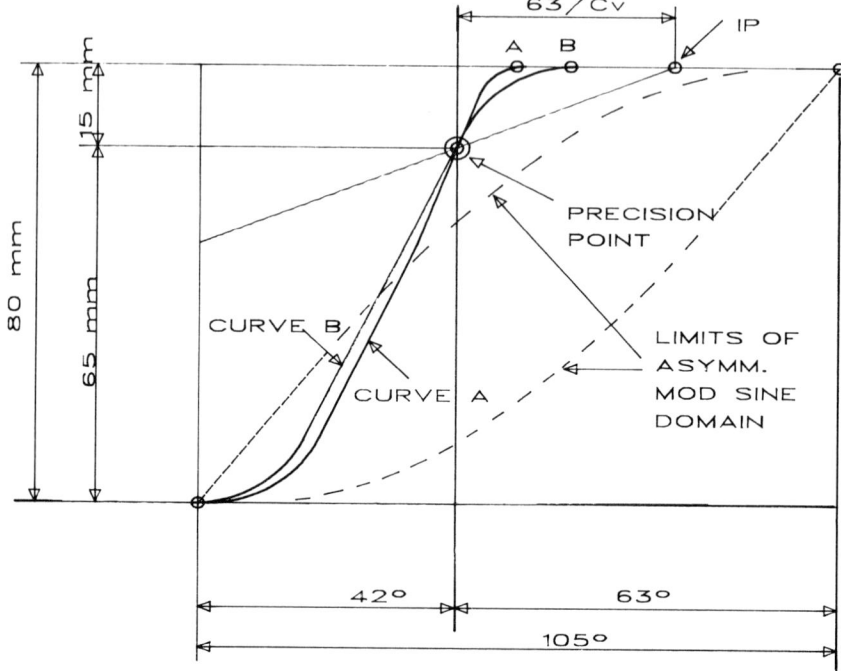

Fig. 5.9 Blending, Example 5.3

When this tangent is extended backwards we find that it does not intersect the initial dwell tangent (diagram base line) within the motion period, which means that the 43° zone would need to be made up of 2 or more composite zones to be able to have a true blend at the precision point. Inevitably this would result in a higher peak acceleration than if the 42° zone were a full Mod Sine acceleration loop.

Curve A is such a curve and it can be seen to have quite a high slope (velocity) at the precision point, which is also the blend point. The simplest way to blend with this is to have a short deceleration period followed by an extra dwell period. The motion is then an asymmetrical Mod Sine ending at point A on the diagram. The remaining dwell period simply increases the following dwell. By proportion, bearing in mind that the blend point of an asymmetrical motion lies on the diagonal, the total motion period is 80 × 42 ÷ 65 = 51.693°.

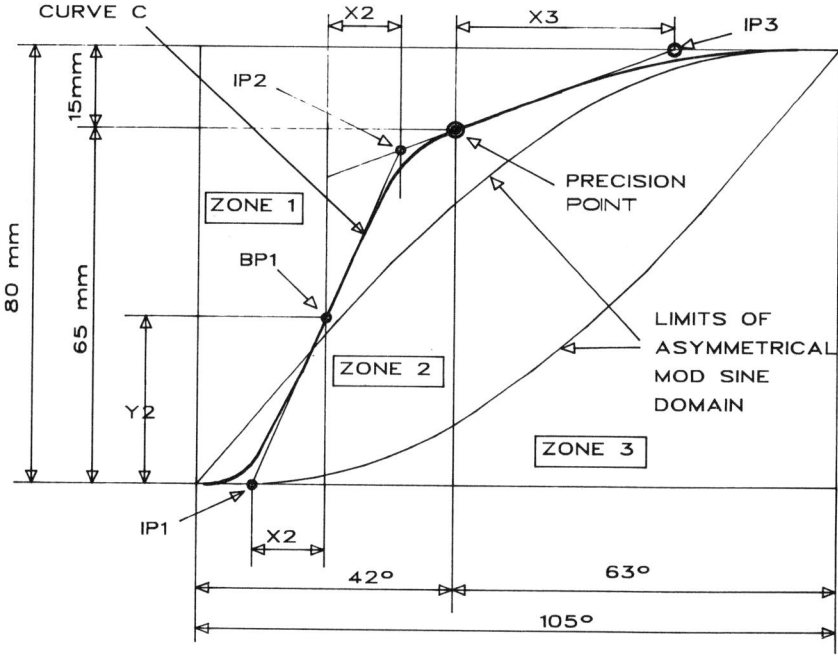

Fig. 5.10 Blending, Example 5.3 – modification

Unfortunately curve A now has quite a high peak deceleration, but we can reduce this considerably, at the expense of a slightly higher peak acceleration, by using standard symmetrical Mod Sine motion, curve B, with its period adjusted to make it pass through the precision point. The period adjustment is calculated as follows:

Normalised output displacement at the PP = w = 65 ÷ 80 = 0.8125

From the Mod Sine factor table we find by interpolation that this corresponds to *normalised* input displacement $u = 0.6928$.

The *real* input displacement at this point is $x = 42°$, therefore the input stroke is $X = x/u = 42 ÷ 0.6928 = 60.623°$. This is shown as point B on the diagram.

The diagram, curve C, Fig. 5.10, shows a solution using the whole of the final 63° period for deceleration as already mentioned.

For the Mod Sine motion $C_V = 1.7596$ and $C_a = 5.528$

The position of IP3 is given by

$\quad$ X3 = 63 ÷ 1.7596 = 35.803°

The slope of the tangent through the precision point is therefore

$\quad$ 15 ÷ 35.803 = 0.41895 mm/deg

There is an arbitrary choice of the x co-ordinate of the blend point, BP1, but if we make it halfway in the 42° period the acceleration will be higher than the deceleration because of the non-zero velocity at the precision point. This means that the peak acceleration would be slightly higher than necessary.

Let BP1 have an x-value of 23°, leaving 19° for the subsequent deceleration period.

The position of IP2 is given by
$\quad$ X2 = 19 ÷ 1.7596 = 10.798°

The y-value of IP2 is
$\quad$ 65 − 0.41895 (19 − X2) = 61.564 mm

The x-position of IP1 is given by
$\quad$ X1 = 23 ÷ 1.7596 = 13.071°

The y-value of blend point BP1 is
$\quad$ Y2 = 61.564 × X1 ÷ (X1 + X2) = 33.713 mm

Curve C consists of:
$\quad$ Zone 1: $\quad$ 23° × 33.713 mm MSA
$\quad$ Zone 2: $\quad$ 19° × (23.327 mm MSD + 7.960 mm CV)
$\quad$ Zone 3: $\quad$ 63° × 15 mm MS

Compare peak geometric accelerations, decelerations and velocities of the three solutions (see Figs 5.9 and 5.10):

	Curve A	Curve B	Curve C
Peak acceleration:	0.1018	0.1203	0.1761 mm/deg^2
Peak deceleration:	0.4414	0.1203	0.1786 mm/deg^2
Peak velocity:	2.7232	2.3220	2.5792 mm/deg

Curve B (symmetrical Mod Sine) is the best of the three.

CAM LAW BLENDING

Example 5.4

There is a fairly common requirement to engage workpieces that are equally spaced on a constantly moving conveyor with a tool (a gripper for example) in such a way that the tool motion matches that of the conveyor during the period of engagement. The conveyor may be a linear motion chain or belt conveyor, or a rotating table; the problem is exactly the same for both. To have enough time to perform its function the tool must travel with the conveyor, at the same velocity, for a considerable portion of the cycle time before disengaging and returning to engage the next, approaching, workpiece. When the motion of one mechanism exactly copies the motion of another it is known as 'motion tracking'.

The tool motion, or at least the component of it that is co-directional with the conveyor, is a reciprocating or oscillating motion which, for a smooth and accurate performance is best designed as a blend of standard cam law and constant velocity zones.

The velocity of the CV zone must of course exactly match that of the conveyor during the engagement period. The rest of the cycle consists of deceleration, reverse acceleration, reverse deceleration and forward acceleration to blend again with the CV zone. If these four modes of acceleration and deceleration are designed as individual zones a complicated motion may result without achieving the best dynamics. A typical layout and motion diagrams are shown in Fig. 5.11.

However, provided that no restrictions are imposed on the overall output stroke of the motion, there need only be two zones, the CV and a combination zone (cam law superimposed on CV). The keystone of such a motion design is the conveyor pitch, that is the spacing of the workpieces on the conveyor.

For our example the conveyor has workpieces equally spaced at 200 mm intervals and two-thirds of the cycle time is required for tool engagement. We are concerned here only with the horizontal component of the tool motion. The vertical component is separately designed to suit the tool function, and the shape of the motion path may vary accordingly, not necessarily being the same as that shown in the drawing, Fig. 5.11.

The conveyor pitch is 200 mm and its velocity can therefore be expressed as 200/360 = 5/9 mm per degree of camshaft rotation, assuming

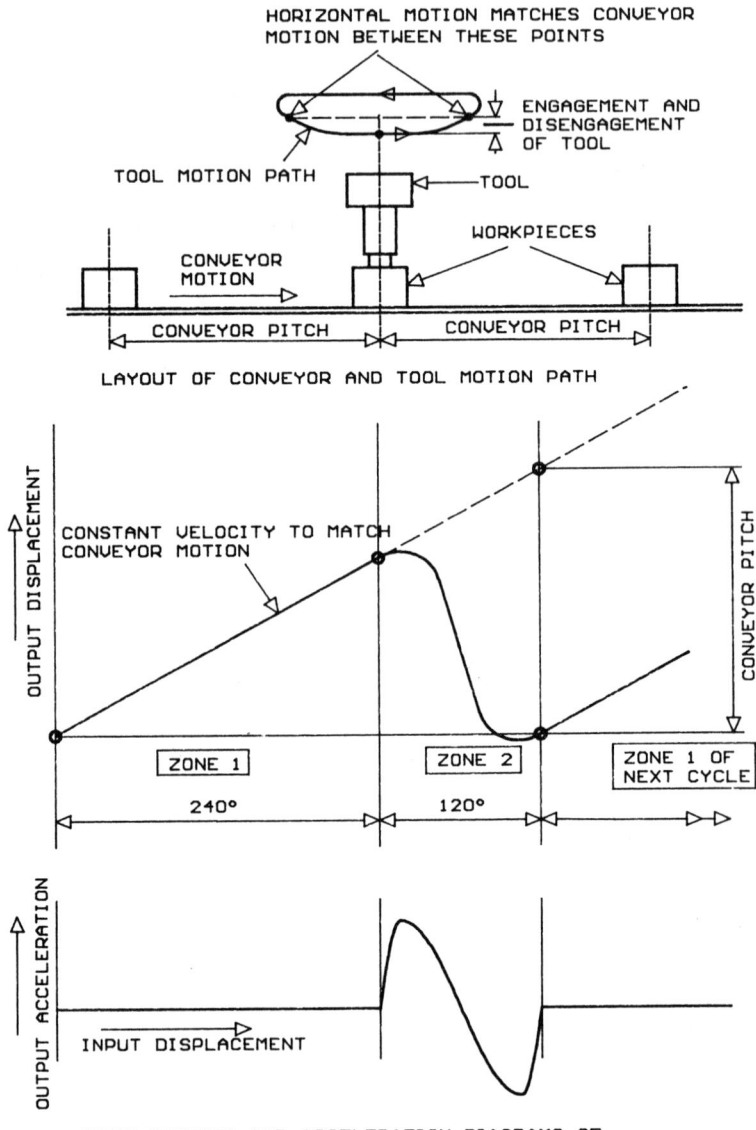

Fig. 5.11   Blending, Example 5.4 (motion tracking)

one camshaft revolution per operating cycle. This is the geometric velocity, which is the same as the slope of the displacement curve plotted as mm of output against degrees of input.

The tool motion's CV zone has the same geometric velocity and an input stroke (period) of 2/3 of 360° = 240°. The output stroke of this zone is therefore 5/9 × 240 = 133.333 mm.

The combination zone has an input stroke of 120° and consists of a CV component with the same slope as before (5/9 mm per degree) from which is *subtracted* a *full* cam law motion, i.e. both the acceleration and deceleration loops. The portion of the output stroke attributed to CV is 5/9 × 120 = 66.667 mm.

When the straight line of the CV zone is projected to overlap the CV zone of the *next* (*repeat*) cycle it is displaced from it by exactly the conveyor pitch. Hence the output stroke of the cam law component of the combination zone is equal to the conveyor pitch, in this case 200 mm.

Using a Mod Sine cam law we have as a complete specification:

Zone 1:   240° × 133.333 mm CV
Zone 2:   120° × (66.667 mm CV − 200 mm MS)

None of the output stroke components calculated so far represents the perceived output stroke of the whole motion. That stroke is the difference between the farthest forward and the farthest backward displacement. These points occur somewhere in the combination zone (zone 2) where the geometric output velocity is zero. Since the velocity is composed of a cam law component subtracted from a CV component of 5/9 mm/degree, it follows that the geometric velocity component of the cam law must also be 5/9 mm/degree at the extremities of the motion where the combined velocity is zero.

$$\text{Thus:} \quad v = \frac{w' \times 200}{120} = \frac{5}{9} \text{ therefore } w' = \frac{5 \times 120}{9 \times 200} = 0.333333$$

From the Mod Sine factor table we find the two extremities to be:

$u = 0.1055$ $w = 0.01249$ $w' = 0.3333$ (interpolated)
and $u = 0.8945$ $w = 0.98751$ $w' = 0.3333$ (interpolated)

The real displacement components of the cam law are:

$x = 0.1055 \times 120 = 12.660°$, $y = 0.01249 \times 200 = 2.498$ mm
and $x = 0.8945 \times 120 = 107.340°$, $y = 0.98751 \times 200 = 197.502$ mm

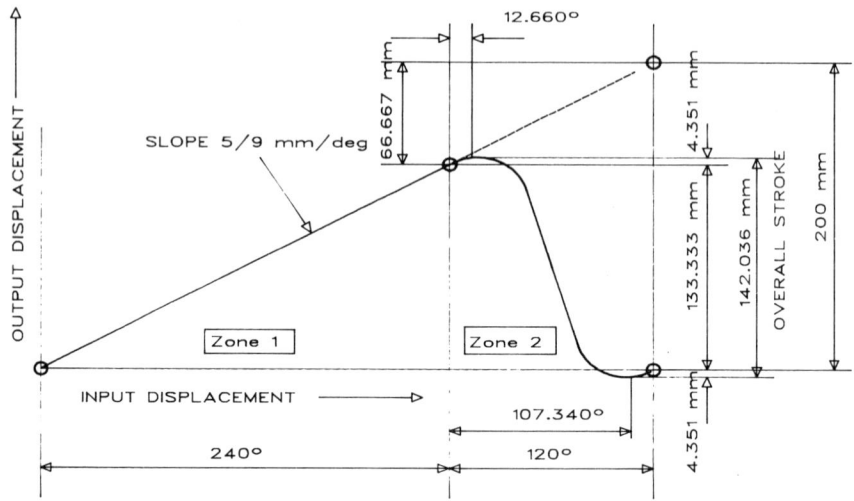

Fig. 5.12   Blending, Example 5.4

Combining these with the CV components we get the actual output displacements of the extremities:

$$5/9 \times 12.660 - 2.498 = 4.535 \text{ mm}$$
$$\text{and } 5/9 \times 107.340 - 197.502 = -137.869 \text{ mm}$$

These are the displacements from the beginning of the combination zone and their difference is:

$$4.535 - (-137.869) = 142.404 \text{ mm}$$

which is the actual output stroke of the motion. It transpires that the overall stroke is only a little greater than the distance travelled at the matching velocity. Figure 5.12 shows the full result.

It is noteworthy that in this calculation no mention was made of time or real velocity. By using the conveyor pitch as the basic design parameter a result has been achieved which is accurate and quite independent of the operating speed of the machine (cycles per minute). The geometric velocities and accelerations can be converted to real values by introducing the input velocity according to the equations in Chapter 3.

A disc cam is shown in Fig. 5.13 which produces the motion calculated

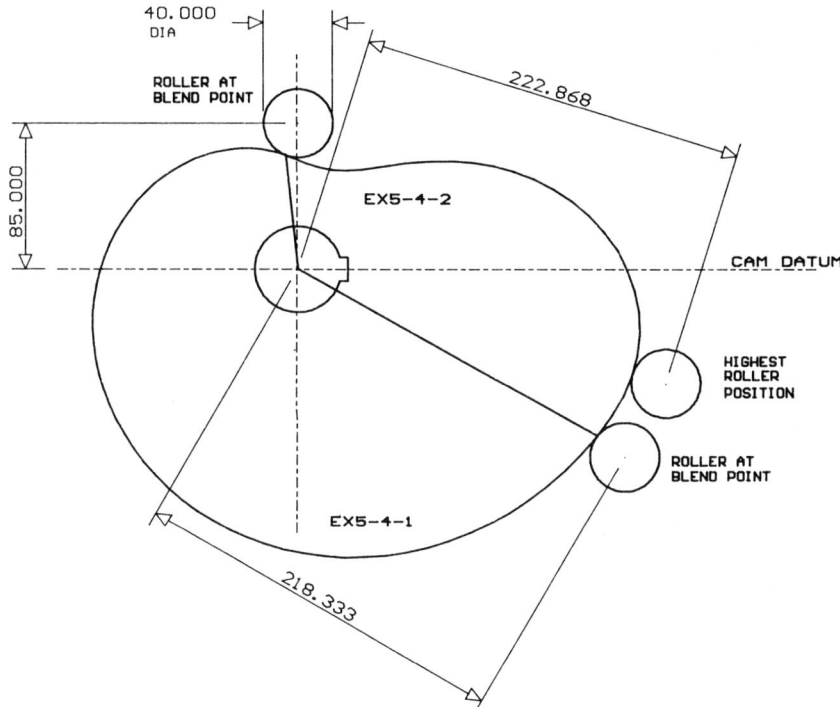

Fig. 5.13   Cam for horizontal motion: Example 5.4

above, using a roller diameter of 40 mm at a starting height of 85 mm. The cam was drawn with the aid of the CAM-DISK computer program.

An interesting principle is illustrated by this example. The 'obvious' way to start designing the tool motion is to decide on the overall output stroke, which enables the mechanical design to proceed immediately. This decision is somewhat arbitrary, however, without considering the constraints that it imposes on the dynamic performance of the motion. It is, of course, quite easy in that event, especially with the tangent intersect method of blending, to establish the input strokes (periods) of the forward acceleration and deceleration zones, and of the return motion, using, say, Mod Sine cam laws. That would be an effective way of achieving smooth performance, but not necessarily the *optimum* performance.

There are two reasons for this. Firstly, if the chosen stroke is too short the forward acceleration and deceleration will be higher than necessary, and if it is too long the return acceleration and deceleration will be higher than necessary. Secondly, even if the stroke can be chosen to balance these values (peak forward acceleration = peak return acceleration), they will still be higher than those calculated above, because the acceleration at each extremity will return to zero unnecessarily between peak forward deceleration and peak reverse acceleration: unnecessary because forward deceleration is the same thing as reverse acceleration.

Before leaving this example, at the risk of belabouring it, let us consider a convenient way of reducing the overall stroke, if required, at the expense of higher peak acceleration. The calculation procedure is exactly the same, but a composite cam law is chosen, which has a higher C_a value, such as MSC.50. This gives a central period of CV in the return stroke, which is strictly unnecessary, but it is a convenient method of getting a faster, and therefore shorter, turn round at the ends of the stroke. In practice it is seldom necessary to do this.

The purpose of this discussion has been to emphasise the need to understand motion design well enough to make it part of the total design process of cam-operated mechanisms, particularly in the early stages.

CHAPTER 6

Dwell-Rise-Return-Dwell Motion

A dwell-rise-return-dwell cam, usually just called D-R-R-D, is a reciprocating or oscillating cam which only has a dwell period at one end of its stroke. The more common cam is the dwell-rise-dwell type, D-R-D, which has dwell periods at both ends of its stroke: if it is reciprocating it has a D-R-D forward motion and a D-R-D return motion.

Obviously a cam can be designed with a forward D-R-D motion followed immediately by a return D-R-D motion with no dwell period in between. This, however, is not as good dynamically as a true D-R-R-D motion because at the return point the deceleration, having just declined from a peak value to zero, immediately increases to a peak value *in the same sense*, i.e. reverse acceleration. Such a motion works well for most applications, but to get the best performance a D-R-R-D motion should be used, which has a continuous acceleration pattern through the return point. Example 5.4 in the previous chapter illustrates the principle of continuous acceleration through the return point, although in that case there is no dwell at either end of the stroke.

Any standard cam law which finishes with peak deceleration (not reducing to zero) may be used for the forward stroke, with a mirror image of it used for a return stroke of the same period, to achieve continuous acceleration at the return point. However, this only works if the overall action is symmetrical, which may not be convenient. The two obvious standard laws for the purpose also have drawbacks:

- *Simple Harmonic* has a high step at the beginning of the acceleration curve where it joins the adjacent dwell zone.
- *Double Harmonic* has a high peak acceleration at the return point.

D-R-R-D motions are notoriously difficult to design, particularly if fairly strict constraints are imposed on output stroke and forward and return periods. The *Tangent Intersect* method of cam law blending described in the previous chapter, however, can be used effectively for the purpose.

There are two broad categories of D-R-R-D motion: *Symmetrical* and *Asymmetrical*. These will be dealt with separately. In both categories we shall use three zones, each consisting of a standard cam law acceleration loop, the middle one containing the return point.

SYMMETRICAL D-R-R-D MOTION

In Fig. 6.1 the total output stroke, H, and the three zone periods, $T1$, $T2$ and $T3$, are known. All calculations are carried out with reference to the displacement diagram. The input displacements are regarded as x co-ordinates and the output displacements as y co-ordinates and the Blend Points (BP) and Intersect Points (IP) are referred to as having x-values and y-values accordingly.

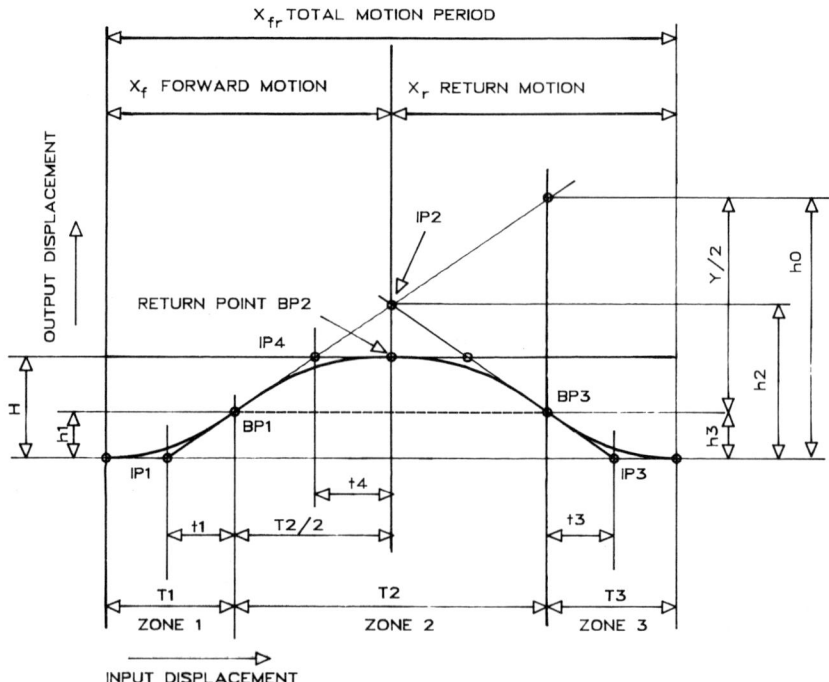

Fig. 6.1   Symmetrical D-R-R-D motion

For the motion to be truly symmetrical the middle zone must be a symmetrical acceleration loop, typically Mod Trap or Cycloidal, with a C_v value of 2, and the last zone must be a mirror image of the first zone. In zones 1 and 3 the y-values of IP1 and IP3 are zero because the motion is preceded and succeeded by dwells. The x-values are easily found because by definition (see Chapter 5)

$$t1 = t3 = T1/C_v1 \qquad\qquad \text{Eq. (6.1)}$$

where C_v1 is the blend factor at the end and beginning of the zones respectively. By virtue of the symmetry of zone $\underline{2}$, $\overline{\text{IP2}}$ occurs at mid-period.

The value of the input displacement factor, u, at BP2 is 0.25 (bearing in mind that the acceleration loop is *half* of a full cam law), so from the factor table for the zone 2 cam law we can find the factors w, w', w'' and the blend factor z.

The x-value of IP4 is:

$$t4 = \frac{T2}{2 \cdot z} \text{ and its } y\text{-value is } H \qquad\qquad \text{Eq. (6.2)}$$

The slope of the tangent through BP1 is found from the positions of IP1 and IP4 to be:

$$S1 = \frac{H}{t1 + T2/2 - t4} \qquad\qquad \text{Eq. (6.3)}$$

The y-values of BP1, BP3 and IP2 are then easily found by similar triangles to be:

$$h1 = h3 = S1 \cdot t1 \qquad\qquad \text{Eq. (6.4)}$$

$$h2 = S1 \cdot (t1 + T2/2) \qquad\qquad \text{Eq. (6.5)}$$

also

$$h0 = S1 \cdot (t1 + T2) \qquad\qquad \text{Eq. (6.6)}$$

The cam law part of the zone 2 displacement is the same as the CV part (because $h1 = h3$):

$$Y/2 = h0 - h3 = S1 \cdot T2 \qquad\qquad \text{Eq. (6.7)}$$

D-R-R-D MOTION

Example 6.1

A symmetrical D-R-R-D cam has an output stroke of 40 mm and 90°
forward period, 90° return period and 180° dwell. The forward acceleration period is 50° Mod Sine and the central, return zone has a cycloidal
deceleration/acceleration pattern. Use Fig. 6.1.

We are given:

$$T1 = 50° \quad T2 = 80° \quad T3 = 50° \quad H = 40 \text{ mm}$$

$$t1 = t3 = 50 \div 1.7596 = 28.415° \qquad \text{Eq. (6.1)}$$

From the Cycloidal factor table, at $u = 0.25$,

$$w = 0.09085 \quad w' = 1.0 \quad w'' = 6.28319 \quad w''' = 0$$

and the blend factor, $z = 2.75194$

$$t4 = \frac{T2}{2 \cdot z} = \frac{40}{2.75194} = 14.535° \qquad \text{Eq. (6.2)}$$

$$S1 = \frac{H}{t1 + T2/2 - t4} = \frac{40}{(28.415 + 40 - 14.535)} = 0.74239 \text{ mm/deg}$$

$$\text{Eq. (6.3)}$$

$$h1 = h3 = S1 \cdot t1 = 0.74239 \times 28.415 = 21.095° \qquad \text{Eq. (6.4)}$$

$$h2 = S1 \cdot (t1 + T2/2) = 0.74239 \times 68.415 = 50.791 \text{ mm} \quad \text{Eq. (6.5)}$$

$$h0 = S1 \cdot (t1 + T2) = 0.74239 \times 108.415 = 80.486 \text{ mm} \quad \text{Eq. (6.6)}$$

$$Y/2 = h0 - h3 = 80.486 - 21.095 = 59.391 \text{ mm} \qquad \text{Eq. (6.7)}$$

The complete motion specification is:

50° × 21.095 mm MSA

+80° × (59.391 mm CV − 59.391 mm CYCD)

+50° × 21.095 mm MSD

ASYMMETRICAL D-R-R-D MOTION, STROKE AND ALL PERIODS KNOWN

In Fig. 6.2 the total output stroke, *H*, and the three zone periods, *T*1, *T*2 and *T*3, are known. This is the simplest type of asymmetrical D-R-R-D motion to calculate. The periods would be chosen, probably by trial and error, to produce forward and return periods approximately as required.

The extreme output displacement occurs at the *return point*, which is the blend point BP2 somewhere in zone 2, its timing being as yet unknown. It is essential for the design of this category of motion that *H*, *T*1, *T*2 and *T*3 are all fixed, and the *timing* of the return point be determined by the design process. A method for coping with a fixed return point timing is dealt with later.

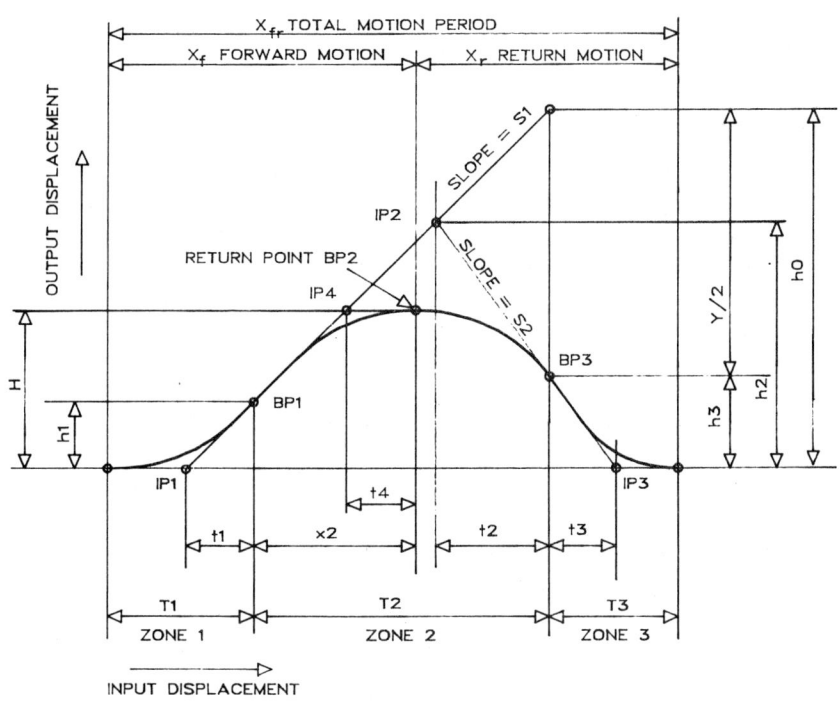

Fig. 6.2   Asymmetrical D-R-R-D motion

All calculations are carried out with reference to the displacement diagram. The input displacements are regarded as x co-ordinates and the output displacements as y co-ordinates and the Blend Points (BP) and Intersect Points (IP) are referred to as having x-values and y-values accordingly.

In zones 1 and 2 the y-values of IP1 and IP2 are zero because the motion is preceded and succeeded by dwells. The x-values are easily found because by definition (see Chapter 5)

$$t1 = T1/C_v1 \quad t2 = T2/C_v2 \quad \text{and} \quad t3 = T3/C_v3 \qquad \text{Eq. (6.8)}$$

where C_v1, C_v2 and C_v3 are the blend factors at the ends and beginning of the zones respectively.

As yet the y-value of IP2 is unknown, but IP1 and IP3 have been totally defined.

Let the slope of the common tangent through BP1 $= S1$

and the slope of the common tangent through BP3 $= S3$

Consider zone 2

This is a cam law acceleration loop subtracted from the constant velocity line of slope $S1$. The loop is half of a full cam law whose input stroke is X and output strike is Y, say. The return point of the motion is at BP2, where the tangent is horizontal (slope $= 0$).

The geometric velocity of the cam law component at BP2 is $w' . Y/X$ therefore, for zero slope at that point:

$$S1 = w' . Y/X \qquad \text{Eq. (6.9)}$$

The geometric velocity of the cam law component at BP3 is $C_v2 . Y/X$ therefore the slope at BP3 is:

$$S3 = S1 - C_v2 . Y/X$$

Combining these two equations we get:

$$S3 = (w' - C_v2) . Y/X \qquad \text{Eq. (6.10)}$$

Now that we know the slope $S3$ we can calculate the y-value of IP2:

$$h2 = -S3 . (t2 + t3) = (C_v2 - w') . (t2 + t3) . Y/X$$

Also

$$h2 = S1.(t1 + T2 - t2) = w'.(t1 + T2 - t2).Y/X$$

The right-hand sides of these equations are equal, from which:

$$w' = C_v2 . \frac{(t2 + t3)}{(t1 + T2 + t3)} \qquad \text{Eq. (6.11)}$$

From the appropriate normalised factor table for zone 2 (see Appendix B), the values of u, w and z can be found which correspond to the w' value just calculated. These are the normalised factors of zone 2 cam law component at BP2. Equation 6.11 is an important one: it defines the return point in any 3-zone D-R-R-D motion.

We already know the y-value of the return point, BP2: it is the specified output stroke, H. Now that we also know u, the normalised input displacement factor of BP2, we can fix its x-value also. Zone 2 is half of a full cam law of input stroke X. Therefore $X = 2.T2$ and by definition $x2 = u.X$, therefore

$$x2 = 2.u.T2 \qquad \text{Eq. (6.12)}$$

However, we cannot calculate the y-values of the remaining BPs and IPs until we know the zone 2 cam law component stroke, Y. First we use the blend factor, z, at BP2 to find the x-value of IP4:

By definition,

$$t4 = x2/z \qquad \text{Eq. (6.13)}$$

and the y-value of IP4 is H. From the diagram we can see that

$$S1 = H/(t1 + x2 - t4) \qquad \text{Eq. (6.14)}$$

and we already know that

$$S1 = w'.Y/X$$

therefore

$$Y = S1.X/w' \quad \text{or} \quad Y = 2.S1.T2/w' \qquad \text{Eq. (6.15)}$$

The y-value of BP1 is

$$h1 = S1.t1 \qquad \text{Eq. (6.16)}$$

and

$$h0 = S1.(t1 + T2) \qquad \text{Eq. (6.17)}$$

so the *y*-value of BP3 is $h3 = h0 - Y/2$, or

$$h3 = S1.(t1 + T2) - Y \div 2 \qquad \text{Eq. (6.18)}$$

and

$$h2 = h0 - S1.t2 \qquad \text{Eq. (6.19)}$$

The calculation is now complete and the method is repeated in the following example.

Example 6.2

A reciprocating motion is required with a slow forward and fast return followed by a 60° dwell. The output stroke is 150 mm. The precise period of the forward motion is not important, but is required to be approximately 50% more than the return period.

Using the symbols of the previous diagram, the forward and return periods occupy 300° of cam rotation, the remaining 60° being the specified dwell period,

therefore $T1 + T2 + T3 = 300°$

No particular cam law has been specified, so we shall choose Mod Sine for the acceleration and deceleration zones and for the middle zone a symmetrical acceleration pattern, say Cycloidal. We now have to estimate the individual values of $T1$, $T2$, and $T3$ that will apportion the forward and return periods in a 1.5:1 ratio approximately. If we were using two D-R-D motions, one for each direction, then we should have 180° of forward motion and 120° of return motion, comprising 90° forward acceleration, 90° forward deceleration, 60° return acceleration and 60° return deceleration. Since we expect to have lower peak accelerations with the D-R-R-D motion the zone 3 period should be longer than 60°, and zone 2 should be longer still. A good approximation for the three zones of the D-R-R-D motion is likely to be:

$$T1 = 140° \text{ (MSA):} \quad T2 = 90° \text{ (CV-CYCA):} \quad T3 = 70° \text{ (MSD)}$$

We are given that

$$H = 150 \text{ mm}$$

and from the factor tables we find

$$C_v1 = C_v3 = 1.7596 \quad \text{and} \quad C_v2 = 2$$

Consider zone 2

Because the cam law component is a complete acceleration loop, i.e. half a complete cam law motion we know that the input stroke of the full law is

$$X = 2 . T2 = 2 \times 90 = 180°$$

$$t1 = T1/C_v1 = 140 \div 1.7596 = 79.564° \qquad \text{Eq. (6.8)}$$

$$t2 = T2/C_v2 = 90 \div 2 = 45°$$

$$t3 = T3/C_v3 = 70 \div 1.7596 = 39.782°$$

$$w' = C_v2 . \frac{(t2 + t3)}{(t1 + T2 + t3)} = 2 \times \frac{45 + 39.\backslash 2}{79.564 + 90 + 39.782} \qquad \text{Eq. (6.11)}$$

$$w' = 0.80997$$

From the Cycloidal factor table (Appendix B) we find that this value of w' occurs between steps 26 and 27, so by interpolation we estimate that the factors at the BP2 position are:

$$u = 0.21956 \quad w = 0.063355 \quad w' = 0.80997 \quad w'' = 6.16670$$

and the blend factor $z = u . w'/w = 2.8070$

$$x2 = 2 . u . T2 = 2 \times 0.21956 \times 90 = 39.521° \qquad \text{Eq. (6.12)}$$

The return point, BP2, occurs at $140 + 39.521 = 179.521°$ after the start: this is the forward period. This is close enough to the suggested 180° value to be acceptable so we shall continue with the calculation. The closeness of the calculated return point to the target is due to a very good first guess for $T1$, $T2$ and $T3$. The inexperienced designer is unlikely to get so close with the first attempt, and intelligent adjustment of the T values might be necessary at this stage.

$$t4 = x2/z = 39.521 \div 2.8070 = 14.079° \qquad \text{Eq. (6.13)}$$

$$S1 = H/(t1 + x2 - t4) = 150/(79.564 + 39.521 - 14.079)$$

$$= 1.42849 \text{ mm/deg} \qquad \text{Eq. (6.14)}$$

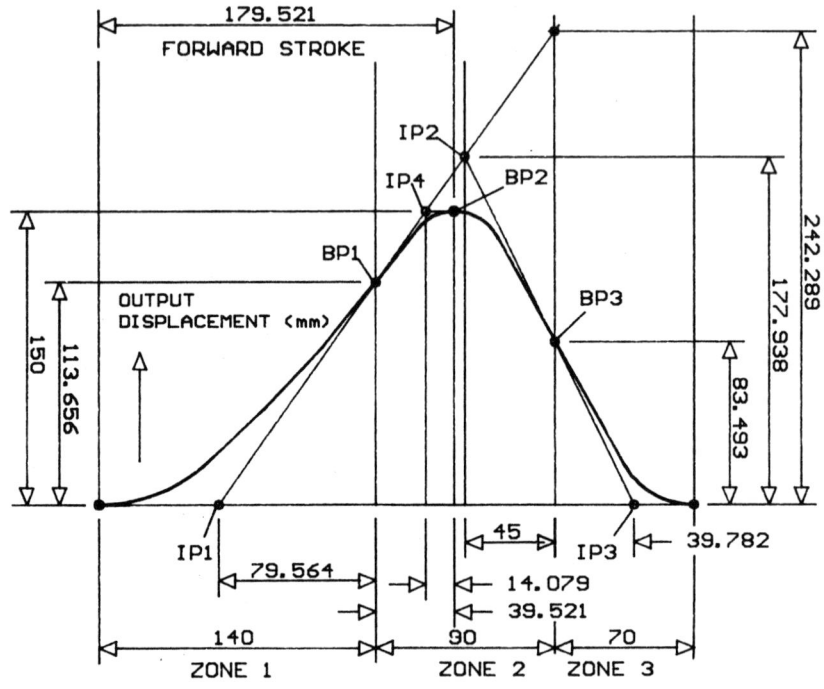

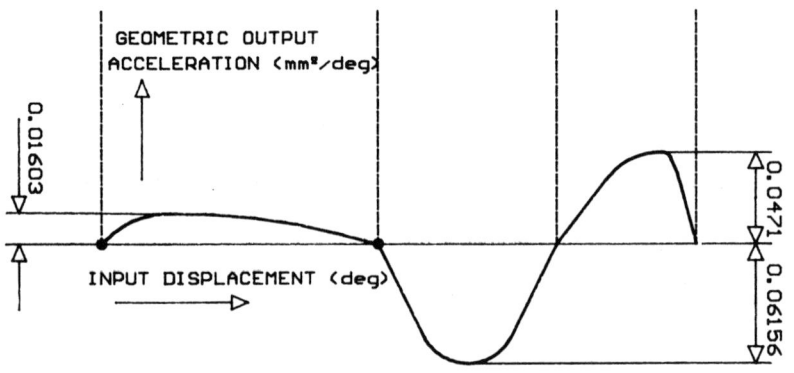

Fig. 6.3   D-R-R-D motion, Examples 6.2 and 6.3

$$Y = 2 . S1 . T2/w' = 2 \times 1.42849 \times 90 \div 0.80997 \qquad \text{Eq. (6.15)}$$

$$= 317.454°$$

$$S3 = (w' - C_v2) . Y/X \qquad \text{Eq. (6.10)}$$

$$= (0.80997 - 2) \times 317.454 \div 180 = -2.09878 \text{ mm/deg}$$

$$h1 = S1 . t1 = 1.42849 \times 79.564 = 113.656 \text{ mm} \qquad \text{Eq. (6.16)}$$

$$h0 = S1 . (t1 + T2) = 1.42849 \times 169.564 = 242.220 \text{ mm} \quad \text{Eq. (6.17)}$$

$$h3 = h0 - Y/2 = 242.220 - 317.454/2 = 83.493 \text{ mm} \qquad \text{Eq. (6.18)}$$

$$h2 = h0 - S1 . t2 = 242.220 - 1.42849 \times 45 \qquad \text{Eq. (6.19)}$$

$$= 177.938 \text{ mm}$$

The CV part of the zone 2 output stroke is $h0 - h1$

$$= 242.220 - 113.656 = 128.564 \text{ mm}$$

and the cam law part is

$$Y/2 = 158.727 \text{ mm}$$

The entire motion can now be specified as:

$$140° \times 113.656 \text{ mm MSA}$$

$$+ 90° \times (128.564 \text{ mm CV} - 158.727 \text{ mm CYCD})$$

$$+70° \times 83.493 \text{ mm MSD}$$

This is shown in Fig. 6.3 together with the (geometric) output acceleration. The peak acceleration and peak deceleration almost balance, and a slight adjustment of the zone 2 and zone 3 periods could be made to improve his, but in practice it is probably not worth the extra design time. However, in the following section a method is described which, with the aid of special tables, avoids the guesswork when choosing zone periods (T values), and enables one immediately to choose optimum periods whereby the peak acceleration and peak deceleration balance.

ASYMMETRICAL D-R-R-D MOTION WITH BOTH STROKE AND TIMING OF RETURN POINT KNOWN

This is similar to the previous case, but the duration of the forward and return movements are known, as well as the output stroke itself. Such is usually the case in practice, when the cam timing must fit in with the timing of other machine functions.

The choice of period for each of the three zones is very limited, particularly when optimum peak acceleration is required. The position of BP2 is known and there is *not* a unique solution for the positions of BP1 and BP3, although one is dependent on the other. As can be seen from the previous example, it is possible to make an educated guess of the zone periods, calculate the return point timing and then by trial and error adjust the timings of BP1 and BP3 to get the correct return point timing, but this will not necessarily give the best solution.

The following method, although more complex in its derivation, gives a solution much more quickly, particularly if the constraint of using the cam laws incorporated in the special tables can be accepted. The method provides the means of constructing tables of various solutions for an exact return point, including the optimum solutions. Some such tables are included in Appendix F. If preferred the derivation can be skipped and the calculation method described in the example, which uses simple arithmetic on factors found in the tables, can be followed instead.

Refer to the general diagram, Fig. 6.2. Considering zone 2, knowing the values of X_f and X_{fr}, it is necessary to find values of u, w, w' etc. at the return point BP2 that allow a valid combination of $T1$, $T2$ and $T3$ to be chosen. By definition:

$$X_f = T1 + t4 = T1 + 2 . u . T2$$

therefore

$$T2 = \frac{X_f - T1}{2 . u} \qquad \text{Eq. (6.20)}$$

and

$$X_{fr} = T1 + T2 + T3$$

therefore

$$T2 = X_{fr} - T1 - T3 \qquad \text{Eq. (6.21)}$$

Substituting the $t = T/Cv$ equivalents in Equation 6.11 and rearranging it we get:

$$w' = \frac{T2 + C_v2 . T3/C_v3}{T1/C1 + T2 + T3/C_v3}$$

from which:

$$T2 = \frac{w' \cdot T1}{(1 - w') \cdot C_v 1} - \frac{(C_v 2 - w') \cdot T3}{(1 - w') \cdot C_v 3} \qquad \text{Eq. (6.22)}$$

Eliminating $T2$ from Equations 6.20 and 6.21 we get:

$$T3 = X_{fr} + T1 \cdot \left(\frac{1}{2 \cdot u} - 1 \right) - \frac{X_f}{2 \cdot u} \qquad \text{Eq. (6.23)}$$

Eliminating $T2$ from Equations 6.20 and 6.22 we get:

$$T3 = \frac{C_v 3}{(C_v 2 - w')} \cdot \left\{ \left[\frac{w'}{C_v 1} + \frac{(1 - w')}{2 \cdot u} \right] \cdot T1 - \frac{(1 - w')}{2 \cdot u} \cdot X_f \right\}$$
$$\text{Eq. (6.24)}$$

Now eliminating $T3$ from Equations 6.23 and 6.24 gives us:

$$\frac{T1}{X_f} = \frac{(C_v 2 - w') \cdot (2 \cdot u \cdot X_{fr}/X_f - 1) + (1 - w') \cdot C_v 3}{(C_v 2 - w') \cdot (2 \cdot u - 1) + (1 - w') \cdot C_v 3 + 2 \cdot u \cdot w' \cdot C_v 3/C_v 1}$$
$$\text{Eq. (6.25)}$$

Although the cam laws, and therefore the C_v values are known, this equation does not give a unique value of $T1$, but expresses it as a function of u and its dependent variable w'. The equation is too complex to invert (i.e. to express u as a function of $T1$), but it could be used to find a solution by first choosing a value of u, looking up the corresponding value of w', and then calculating the value of $T1$ that corresponds to the chosen value of u. Equation 6.20 could then be used to find $T2$ and the rest of the solution would follow the procedures of the previous example.

Unfortunately, it turns out that for any particular value of X_f/X_{fr} there is a very narrow range of u that produces practical values of $T1/X_f$, and moreover the value of u is very sensitive: a small change of u results in a large change of $T1/X_f$.

It is therefore difficult to estimate a good value of u and it is better to choose a practical value of $T1$ and find the value of u that satisfies Equation 6.25. This is best done by numerical methods. If we restrict ourselves to a few of the standard cam laws, and use the same law for zone 1 and zone 2, we can simplify the problem and enable practical values of $T1/X_f$ and X_f/X_{fr} to be listed in a few simple tables for each law.

Furthermore, Equation 6.25 is considerably simplified by letting all three zones have a cam law with a C_v value of 2, such as Cycloidal or Mod Trap. This is acceptable for many applications, Cycloidal being a good cam law for high speed anti-vibration, and Mod Trap being good for moderate speed applications where a low peak acceleration is required.

Equation 6.25 then reduces to:

$$\frac{T1}{X_f} = \frac{2.u.(2 - w').X_{fr}X_f - w'}{4.u - w'} \qquad \text{Eq. (6.26)}$$

The tables in Appendix F have been constructed by using u and its dependent variable w' as parameters for Equations 6.25 or 6.26, 6.23 and 6.21. Normalised output displacement values at the blend points ($h1/H$ etc.) are also listed.

The *optimum* values of $T1/X_f$ are also indicated, which produce zones wherein the peak nominal acceleration is equal to the peak nominal deceleration – a dynamically favourable balance. The optimum values should always be used, unless there is a compelling reason not to do so. The following example illustrates the use of the tables.

Example 6.3

This is a repeat of the previous example, but this time we shall use the special D-R-R-D tables to solve for an exact return point. The motion required is 150 mm output stroke with a 180° forward period and 120° return period. Forward acceleration and return deceleration are both Mod Sine, and the middle zone containing the return point is Cycloidal.

We are given that

$$H = 150 \text{ mm}, \qquad X_f = 180° \quad \text{and} \quad X_r = 120°$$

therefore

$$X_{fr} = X_f + X_r = 180 + 120 = 300°,$$

$$X_f/X_{fr} = 180/300 = 0.6$$

and

$$X_r/X_{fr} = 120/300 = 0.4$$

From the tables (Appendix F) for $X_f/X_{fr} = 0.6$ we find the optimum values:

$T1/X_f = 0.74748$ $h1/H = 0.72314$

$T2/X_{fr} = 0.34555$ $Y/2/H = 1.21349$ $u = 0.21923$

$T3/X_r = 0.51491$ $h3/H = 0.49003$

From these we easily calculate:

$$T1 = 0.74748 \times X_f = 0.74748 \times 180 = 134.546°$$

$$h1 = 0.72314 \times H = 0.72314 \times 150 = 108.471 \text{ mm}$$

$$T2 = 0.34555 \times X_{fr} = 0.34555 \times 300 \div 103.665°$$

$$Y/2 = 1.21349 \times H = 1.21349 \times 150 = 182.024 \text{ mm}$$

$$T3 = 0.51491 \times X_r = 0.51491 \times 120 = 61.789°$$

$$h3 = 0.49003 \times H = 0.49003 \times 150 = 73.505 \text{ mm}$$

The slope at BP1 is the slope of the constant velocity component of the zone 2 motion, and since the C_v of Mod Sine is 1.759603 we find:

$$S1 = C_v . h1/T1 = 1.759603 \times 108.471 \div 134.546$$

$$= 1.41859 \text{ mm/degree}$$

The CV component of the zone 2 displacement is therefore

$$S1 \times T2 = 1.41859 \times 103.665 = 147.058 \text{ mm}$$

This is the complete optimum solution and can be specified as:

134.546° × 108.471 mm MSA

+103.665° × (147.058 mm CV − 182.024 mm CYCA)

+61.789° × (−73.505 mm MSD)

To prove the arithmetic:

$$X_{fr} = 134.546 + 103.665 + 61.789 = 300°$$

The displacement at the end of the full motion is:

$$108.471 + 147.058 - 182.024 - 73.505 = 0°$$

The timing of the return point is:

$$134.546 + 0.21923 \times 103.665 \times 2 = 179.999°$$

This should be 180°. Rounding errors account for the difference.

From the Cycloidal factors table we find by linear interpolation that for $u = 0.21923$, $w = 0.06308$, so that the displacement at the return point is:

$$108.471 + 1.41859 \times (180 - 134.546) - 0.06308 \times 2 \times 182.024$$

$$= 149.986 \text{ mm}$$

This should be 150 mm: rounding and interpolation errors account for the small difference. The more accurate value of $w = 0.06304$ is found from the cam law equation instead of by linear interpolation from the factors table. Using this value of w the return point displacement is calculated to be 150.002 mm.

Using the special D-R-R-D tables is evidently easier and quicker than the other method, but unfortunately it inevitably produces fractional values of the zone periods. Depending on the equipment available for producing the cam profile, this may not be convenient. However, the periods calculated from the tables can be rounded off to convenient values which will always be workable and dynamically sound. With the revised periods the displacements must be calculated as described in the previous example, and there will be a slight discrepancy in the timing of the return point.

Of course, sets of factors other than the optimum set can be chosen from the tables, but the further away from the optimum then the worse will be the imbalance of peak accelerations, for example:

In the optimum example, above, the peak (geometric) accelerations for the three zones are respectively:

0.016562, -0.053212 and 0.053214 mm/deg^2

For the previous example, which is in fact fairly close to the optimum, they are:

0.016028, -0.061562 and 0.047096 mm/deg^2

The peak deceleration in the middle zone is seen to be about 15% higher than the optimum.

It is interesting also to compare the peak accelerations of a motion with the same output stroke (150 mm), same forward period (180°) and same reverse period (120°), but using symmetrical Mod Sine D-R-D motions for both forward and return strokes. The peak accelerations are:

± 0.025593 ± 0.057583 mm/deg^2

The highest peak acceleration is only 8.2% higher than that of the optimum D-R-R-D motion. This questions the value of using the more complicated D-R-R-D solution for such a case, and shows perhaps that, if it *is* used, one must aim for the *optimum* solution. However, it can be shown that one complete D-R-R-D motion generates less vibration than two separate D-R-D motions for forward and return strokes (see Chapter 12).

CHAPTER 7

Cam Geometry, Curvature, Interference, Pressure Angle: Disc Cams

The phrase 'cam geometry' here is intended to embrace only the properties of the cam profile, the part of the cam follower that interacts with that profile, and the relationship between these. It does not include the properties of other parts of the cam or follower such as bearings, hubs, etc.

Apart from the obvious need to be able to accommodate the size and shape of the cam and follower within the space available in the machine, the most important geometric properties of the cam and follower are the specification of the cam profile shape, the curvature of that profile and the pressure angle at the point of contact. Curvature is crucial to the ability of the contact surfaces to withstand the contact force. Pressure angle is a vital ingredient of that force.

PRESSURE ANGLE

When a cam moves a follower it does so by exerting a force along a 'line of pressure' at the point of contact. This motive force does not necessarily act in the direction of the follower motion. The angle between the line of pressure and the line of motion is the *pressure angle*. When there is no lateral friction (traction) at the cam surface the line of pressure is normal to the surface: this is almost true if a roller follower with an anti-friction bearing is used. It is conventional, therefore, to specify the pressure angle between the line of motion and the normal to the profile: this may be termed the 'nominal' pressure angle and is always assumed unless otherwise specified.

In most industrial machine cams the pressure angle varies throughout the motion period. When the pressure angle is not zero the motive force, also known as the contact force, has two components: a 'useful' one acting in the direction of follower motion, and a 'useless' one at right-angles to it. The useful component does useful work, i.e. it transmits energy to the mechanism output, but the useless component merely exerts lateral force on the follower, which contributes to the wasted friction energy and unwanted elastic deflections in the system without doing any useful work.

It is desirable to have a low pressure angle, but a very low angle can only be achieved with a large cam. A compromise is always necessary, and some designers assert that the pressure angle should never exceed some arbitrary figure, such as 30° or 45°. Certainly it is useful as a guide in the initial stage of design to use such a 'rule', and also as a rough check on the practicability of the final design. But it is a rule that may be justifiably broken when the adverse effects of a high pressure angle can be tolerated for the sake of some desirable feature, such as squeezing a cam into a limited available space. The consequences of using a high pressure angle, over 45° say, must always be considered very carefully.

CURVATURE

The curvature of a line is the degree to which it deviates from a straight line. A straight line, by definition, has no curvature and a circle has a constant curvature. Most cam profiles, however, have a varying curvature, each point on the profile having its own particular curvature, which might not be the same as at any other point.

Curvature is quantified usually by stating the *radius of curvature*. For a pure circle this is its radius, and so for a line of varying curvature the radius of curvature at any point is the radius of a small arc of a circle that exactly matches the line at that point. The greater the curvature of a line the smaller is its radius of curvature, and vice versa. The curvature as such is therefore defined as the reciprocal of the radius of curvature. A sharp point or kink in a line has infinite curvature and therefore zero radius of curvature. At the other extreme a straight line has no curvature, so its radius of curvature is infinity.

The curvature of any point on any line can thus be given a numerical value. However it is also convenient to give it a sign (plus or minus) for some purposes. Curvature changes sign when its direction changes (see Fig. 7.1). The choice of positive or negative for a particular direction is arbitrary, but once chosen the designation must be consistent throughout

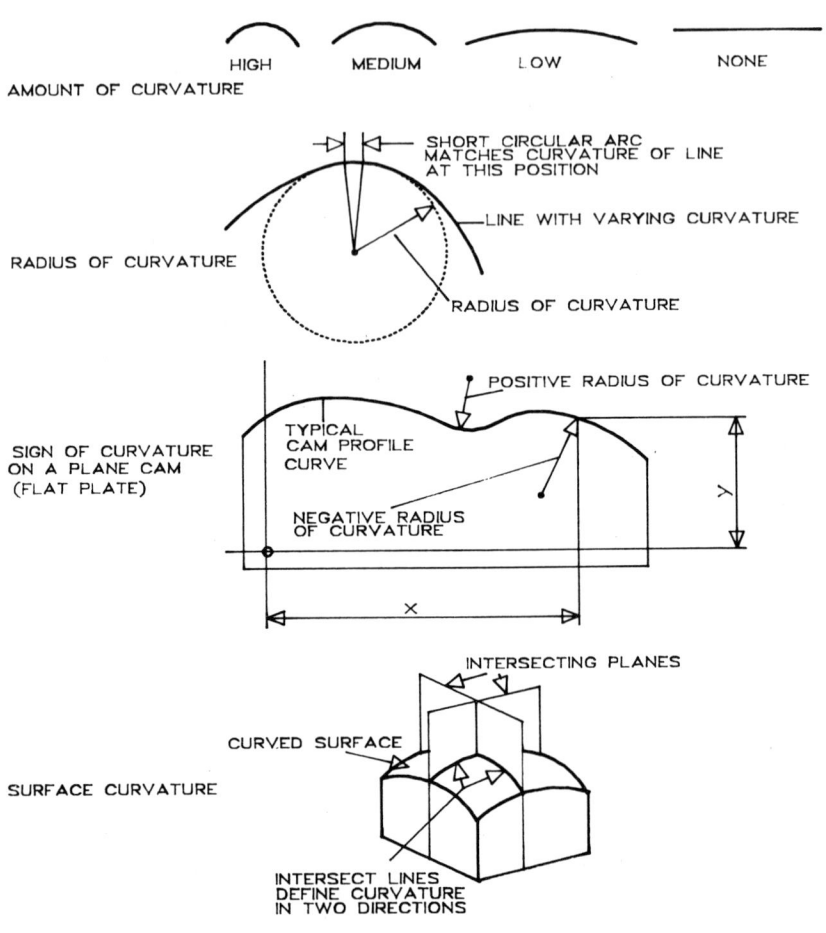

Fig. 7.1 Curvature definitions

the curve. For example, if on the third diagram of Fig. 7.1 the parts of the curve which are convex in relation to the area below the line are called positive then all concave parts of the curve must be negative. Attention to sign is important when calculating curvature, particularly if the value is to be used in a formula for estimating surface stress at the point of contact.

In cam systems, of course, we deal with curved *surfaces*, as opposed to just lines, but the curvature of a surface can be defined in terms of the curvature of a line at the intersection of the surface with a plane. This is done for two planes which are perpendicular to each other and normal to the curved surface at the point under consideration, see Fig. 7.1. The curvatures of the intersection lines give us the curvature of the surface in two directions at right angles. Very often in industrial cams the curvature of the contact surface is zero (straight) in one direction, e.g. across the edge of a disc cam. When a surface is curved in both directions it is said to have a compound curvature.

Usually cam track surfaces are machined using tools (milling cutters or grinding wheels) which are of cylindrical form, so that the curvature across the track is zero as mentioned above. In the case of a disc cam, although the curvature around the cam profile varies, it is constant throughout the track width at any one point on the profile. This is not necessarily so for other types of cam: the curvature of the track surface of a cylinder or barrel cam (see Fig. 1.1.) increases towards the centre of the cam.

Cams can be specially machined to have cambered tracks, but if so the amount of curvature of the camber is small enough to be ignored in this discussion. The purpose of a camber is usually to allow for small misalignments in the cam/follower assembly which might otherwise cause high stress concentration and premature failure of the surface. A more economical way of dealing with this problem is to camber the follower: cambered roller followers are commercially available. This subject is discussed further in Chapter 9.

Unfortunately, in general, the curvature of a line cannot be quantified accurately without recourse to differential calculus. Nevertheless, an approximate value of the radius of curvature of a disc cam can be calculated from Equation 7.5 without using calculus, by substituting values of $f(\theta)$, $f'(\theta)$ and $f''(\theta)$ from Equations 7.7, 7.8 and 7.9 using simple, if somewhat tedious arithmetical calculations.

To be able to use most of the following equations, however, it is necessary to understand the symbolism of calculus. Here we use the forms:

$y = f(x)$ for "y is a function of x",

$f'(x)$ or y' for the first derivative of $f(x)$, i.e. the rate of change of y with respect to x,

$f''(x)$ or y'' for the second derivative of $f(x)$, i.e. the rate of change of $f(x)$ with respect to x,

and so on.

The alternative fully written forms of the derivatives,

$$\frac{dy}{dx}, \frac{d^2y}{dx^2}, \frac{dy}{dp}, \frac{d^2y}{dp^2} \text{ etc.}$$ are used to avoid possible misunderstanding

when more than two variables are involved.

PLANE CAM PROFILE: CURVATURE

The curvature of a plane cam profile can be calculated exactly if the orthogonal co-ordinates of the curve, expressed as an algebraic function, $y = f(x)$, and its derivative functions $y' = f'(x)$ and $y'' = f''(x)$ are known. The x and y axes in this system have their origin anywhere in the plane: see Fig. 7.1. In this case the radius of curvature is conventionally given as:

$$R_c = \frac{[1 + (y')^2]^{1.5}}{y''} \qquad \text{Eq. (7.1)}$$

In many cases the profile or pitch curve cannot easily be expressed in the form $y = f(x)$, but sometimes the co-ordinates x and y, and their derivatives, are easily expressed as separate functions of some independent parameter. For example, let the parameter be p so that for every value of p both x and y (and their derivatives) can be calculated, then the radius of curvature is given by:

$$R_c = \frac{\left[\left(\dfrac{dy}{dp}\right)^2 + \left(\dfrac{dx}{dp}\right)^2\right]^{1.5}}{\dfrac{d^2y}{dp^2} \cdot \dfrac{dx}{dp} - \dfrac{dy}{dp} \cdot \dfrac{d^2x}{dp^2}} \qquad \text{Eq. (7.2)}$$

The parameter p in this equation could be the cam movement from the start of the motion of the elapsed time or any other convenient variable.

PRESSURE ANGLE

This is the simplest of all the pressure angle calculations:

$$\phi = \arctan y' \qquad \text{Eq. (7.3)}$$

or in terms of an independent parameter, p:

$$\phi = \arctan \frac{dy/dp}{dx/dp} \qquad\qquad \text{Eq. (7.4)}$$

DISC CAM PROFILE: CURVATURE

The profile of a disc cam lies in a plane and therefore Equation 7.1 can be used. The x and y axes in this case have their origin at the centre of rotation of the cam and are, as it were, scribed on the cam and rotate with it. Unfortunately, this equation has two limitations which make it inconvenient for use with a disc cam profile.

As previously mentioned, the sign of the radius of curvature is important, and in particular it should indicate whether the profile is convex or concave. Equation 7.1 gives a positive value for curves which are 'u' shaped and negative for 'n' shaped (looked at with the x, y axes in the conventional orientation) wherever they occur on the cam. This means that on a disc cam the equation makes a convex curve negative above the x-axis, but positive below the x-axis.

The other limitation is that at points on the profile where its tangent is parallel to the y-axis, and there must be at least two such points on a complete disc profile, the value of y' and y'' are infinity, making it impossible to determine the value of the equation.

Equation 7.2, which yields a positive value of convex curvature at any point on the profile could be used to overcome the first limitation, but the second limitation may still apply (in this case dx/dp, dy/dp, etc. may be infinite at some point), depending on what variable is chosen for the parameter p. For these reasons it is more reliable to use another form of the equation.

The profile curve is very often expressed as polar co-ordinates, $r = f(\theta)$ (with derivatives $f'(\theta)$ and $f''(\theta)$). This form of specifying the shape is the most convenient if the workpiece is to be mounted on a rotary table for profile machining and the co-ordinates can be used directly as input data. The origin of this system is also the centre of rotation of the cam, and the x-axis ($\theta = 0$) is virtually scribed on the cam and rotates with it. In this case the radius of curvature is:

$$R_c = \frac{\{r^2 + [f'(\theta)]^2\}^{1.5}}{r^2 + 2 \cdot [f'(\theta)]^2 - r \cdot f''(\theta)} \qquad\qquad \text{Eq. (7.5)}$$

This equation yields a positive value of convex curvature at any point on the profile.

As with Cartesian co-ordinates it is also possible to express the radius of curvature in terms of the polar co-ordinates and an independent parameter, p, thus:

$$R_c = \frac{\left[\left(r \cdot \frac{d\theta}{dp}\right)^2 + \left(\frac{dr}{dp}\right)^2\right]}{\frac{d\theta}{dp} \cdot \left[\left(r \cdot \frac{d\theta}{dp}\right)^2 + 2 \cdot \left(\frac{dr}{dp}\right)^2\right] - r \cdot \left[\frac{d^2r}{dp^2} \cdot \frac{d\theta}{dp} - \frac{d^2\theta}{dp^2} \cdot \frac{dr}{dp}\right]} \qquad \text{Eq. (7.6)}$$

As before, the parameter p can be cam rotation, elapsed time or any other convenient independent variable.

Throughout this book we have encouraged the design of cams to produce a motion according to a known cam law. The equation of the normalised cam law, $w = f(u)$ is then known (see Chapters 3 and 4) but the equation of the actual cam profile, which depends on the type of cam and follower system and its geometry, usually is not. Later, equations for the radius of curvature in terms of the cam law function will be given for certain cam types.

However, using a computer, it is relatively easy to calculate pairs of polar co-ordinates of the profile at regular time increments. Indeed this is the usual procedure for producing the input data for N.C. (Numerical Control) or C.N.C. (Computerised Numerical Control) profile machining with a rotating workpiece.

Approximate values of $f'(\theta)$ and $f''(\theta)$ can be found from three adjacent pairs of co-ordinates *which must be very close together*:

Let the co-ordinates be, in sequence: $r_1, \theta_1; r_2, \theta_2; r_3, \theta_3$ then

$r = r_2$ Eq. (7.7)

$$f'(\theta) \approx \frac{r_3 - r_1}{\theta_3 - \theta_1} \qquad \text{Eq. (7.8)}$$

$$f''(\theta) \approx \frac{r_3 - 2 \cdot r_2 + r_1}{(\theta_3 - \theta_2) \cdot (\theta_2 - \theta_1)} \qquad \text{Eq. (7.9)}$$

Note that all θ values in the above equations are in radians.

Substituting these values in Equation 7.5 gives an approximate value of the radius of curvature at the position r_2, θ_2. This is a tedious process if

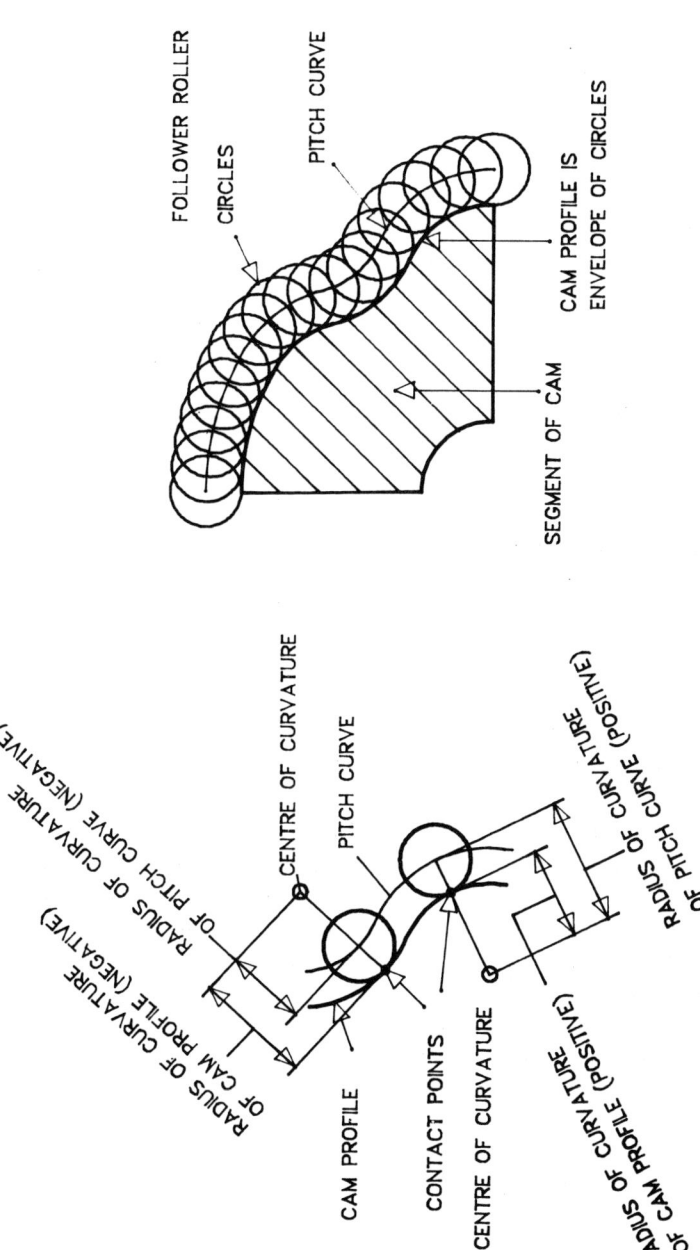

FOLLOWER ROLLER
CIRCLES

PITCH CURVE

CAM PROFILE IS
ENVELOPE OF CIRCLES

SEGMENT OF CAM

CENTRE OF CURVATURE

PITCH CURVE

RADIUS OF CURVATURE
OF PITCH CURVE (NEGATIVE)

RADIUS OF CURVATURE (POSITIVE)
OF PITCH CURVE

RADIUS OF CURVATURE
OF CAM PROFILE (NEGATIVE)

CAM PROFILE

CONTACT POINTS

CENTRE OF CURVATURE

RADIUS OF CURVATURE (POSITIVE)
OF CAM PROFILE

Fig. 7.2   Curvature of cam profile and pitch curve

carried out manually, but it can be easily written into a computer program for analysing any disc cam profile that can be expressed as a series of polar co-ordinates, $r = f(\theta)$.

The most common type of follower used with disc cams in industry is the cylindrical roller. The path traced by the centre of the roller is known as the *pitch curve or pitch profile* of the cam. The pitch curve is directly derived from the required output motion of the mechanism and the actual cam track profile is derived from that. Choosing convex curvature to be positive in relation to the cam centre and concave to be negative, the radius of curvature of the disc cam track surface is the radius of curvature of the pitch curve minus the radius of the roller, see Fig. 7.2. This is self evident from the fact that the radius of curvature and the radius of the roller are both normal to the curve at the point of contact, and are therefore on the same line. The cam profile is the envelope curve of all the roller circles distributed along the pitch profile, as shown in Fig. 7.2.

Note that in absolute magnitude (ignoring sign) the radius of curvature of a convex cam profile is less than its pitch curve radius (positive minus positive), whereas that of a concave cam profile is greater (negative minus negative).

DISC CAM WITH TRANSLATING FOLLOWER

Figure 7.3 shows a typical disc cam with translating roller follower, with and without offset. The direction of the offset is known as leading or trailing relative to the direction of cam rotation. The centre of an in-line roller follower travels along a line which passes through the centre of rotation of the cam (the cam centre-line). An offset roller follower moves on a line parallel to such a centre-line. Travelling in the direction of cam rotation, a leading follower is encountered before the centre-line and a trailing one is encountered after the centre-line.

The sector of the cam we shall consider has a rise or *lift* (output stroke) of B with a *cam rotation* (input stroke) of A. Note that without any follower offset the cam angle occupied by the sector is identical to the input stroke, but that is not true for an offset follower system.

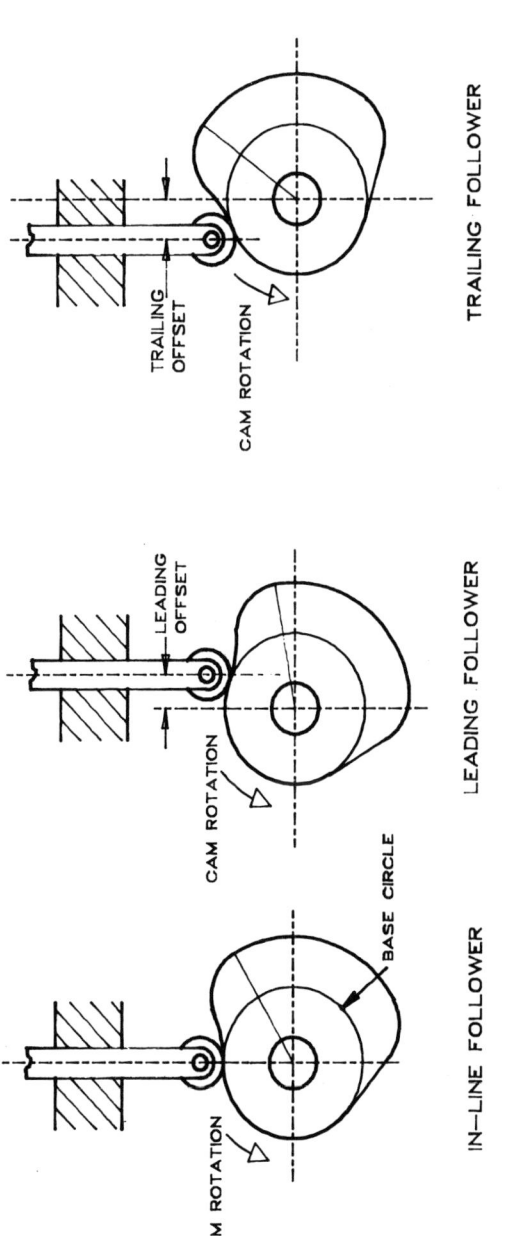

IN—LINE FOLLOWER　　　　LEADING FOLLOWER　　　　TRAILING FOLLOWER

DISC CAMS WITH TRANSLATING FOLLOWERS
ALL SHOWN AT THE START OF THE RISE SEGMENT

Fig. 7.3　Follower offset

RADIUS OF CURVATURE FROM KNOWN CAM LAW

Referring to Fig. 7.4, we consider a system in which the linear displacement of the follower obeys a known cam law function of the cam rotation: this is the most common requirement of an industrial cam system. We assume a leading follower (positive offset), but the equations for this can be adapted for in-line or trailing offset followers as shown later.

At any instant the input displacement (cam rotation from the start of the motion) is

$$a = A \cdot u$$ Eq. (7.10)

the output displacement (height of roller above cam centre) is

$$h = H + B \cdot w$$ Eq. (7.11)

and the geometric velocity and acceleration are respectively

$$h' = \frac{dh}{da} = \frac{B}{A} w'$$ Eq. (7.12)

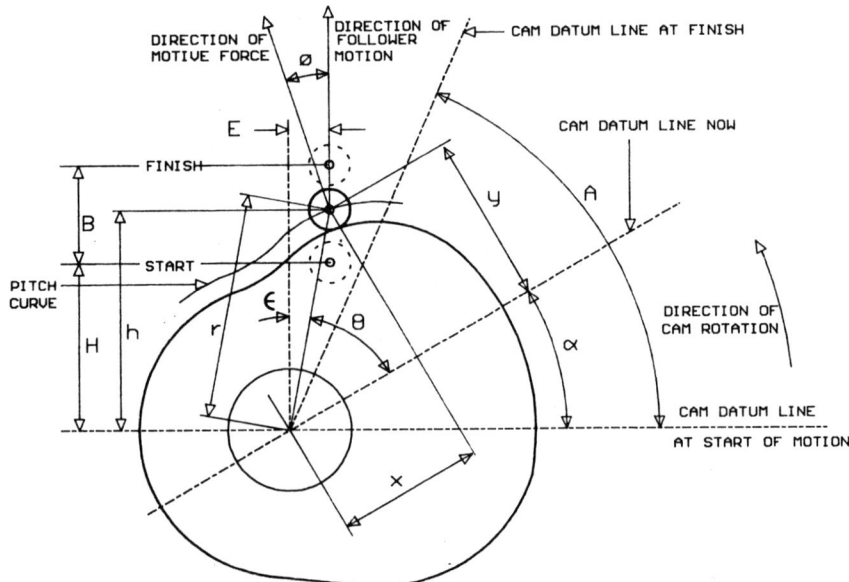

Fig. 7.4 Geometry of disc cam with translating follower

and

$$h'' = \frac{d^2h}{d\alpha^2} = \frac{B}{A^2} w''$$ Eq. (7.13)

where
   A = input stroke = cam rotation angle for the whole motion, (must
      be in radians for use in the radius of curvature equations),
   B = output stroke (linear dimension in this case),
   H = initial height of the roller above the cam centre,
and
   $w = f(u)$ is the normalised cam law function, w' and w'' are its
      derivatives (see Chapters 3 and 4 and Appendix A).

The values of linear displacement of the follower, the Cartesian co-ordinates, x and y, and the polar co-ordinates, r and θ, of the pitch curve are derived from the cam law displacement factor by the next five equations (see Fig. 7.4):

$h = H + B.w$ Eq. (7.14)

$r = (h^2 + E^2)^{0.5}$ Eq. (7.15)

$\varepsilon = \arctan(E/h)$ Eq. (7.16)

$\theta = \pi/2 - \varepsilon - \alpha$ Eq. (7.17)

$x = r.\cos\theta$ Eq. (7.18)

$y = r.\sin\theta$ Eq. (7.19)

Using Equations 7.11, 7.12 and 7.13 we can find the radius of curvature of the *pitch curve* at any instant from the following equation

$$R_c = \frac{[h^2 + (h')^2 + (E^2 - 2.E.h')]^{1.5}}{h^2 + 2.(h')^2 - h.h'' + (E^2 - 3.E.h')}$$ Eq. (7.20)

This equation yields a positive value of convex curvature at any point on the profile.

When there is no offset then $E = 0$, $h = r$, $\alpha = \pi/2 - \theta$, $h' = -f'(\theta)$, $h'' = f(\theta)$ and the equation becomes:

$$R_c = \frac{\{r^2 + [f'(\theta)]^2\}^{1.5}}{r^2 + 2.[f'(\theta)]^2 - r.f''(\theta)}$$ Eq. (7.21)

which is the same as Equation 7.5.

Note that all A, α and θ values in the above equations are in radians.

The offset value, E, is taken as positive if it is leading and negative if it is trailing (see Fig. 7.3). Equation 7.20 can be used for either case by using the appropriate sign for E, and can be used for a system with no offset by using $E = 0$. The follower is leading or trailing (positive or negative E) in relation to the direction of rotation of the cam. Figure 7.4 is for an anti-clockwise cam rotation: for clockwise rotation its mirror image would apply.

The equation applies to a rising motion, but can be used for a falling motion by using a negative value of B and taking H as the height of the highest position of the roller (the starting point of the falling motion) instead of the lowest position (the starting point of a rising motion). If the motion is symmetrical a simpler alternative is to reverse the direction of rotation of the cam and treat it as a rising motion, *but only if the motion is symmetrical*.

Since the cam law describes the motion of the centre of the follower roller, the equations give the radius of curvature of the *pitch curve*. The radius of curvature of the actual cam profile is found by adding or subtracting as appropriate the roller radius. Although the pitch curve co-ordinates are generally used for profile machining, the *profile* curvature is important for contact stress analysis and to detect the presence of interference and/or undercutting.

PRESSURE ANGLE FROM KNOWN CAM LAW

The pressure angle is dependent on the geometric velocity, which is derived from the velocity factor of the normalised cam law and is given by:

$$\phi = \left[\arctan\left(\frac{h' - E}{h}\right) \right] \qquad \text{Eq. (7.22)}$$

It should be noted that always *in the dwell periods $h' = 0$* by definition, and therefore $\phi = -\arctan(E/h) = -\varepsilon$.
Also when there is *no offset $E = 0$*, and $\phi = \arctan(h'/h)$.

Example 7.1

The various methods of calculating the radius of curvature described above are illustrated by this example.

A sector of a disc cam drives a translating roller follower vertically away from it a distance of 25 mm, with a mod sine motion, during 90° of cam rotation. The 20 mm diameter roller is offset (leading) by 15 mm and it starts at 60 mm above the centre of the camshaft. Find the radius of curvature of the cam profile and pressure angle after the first 54° of rotation.

We are given:

$A = 90° = 1.570796$ radians
$B = 25$ mm
$H = 60$ mm
$E = 15$ mm
Cam law is Mod Sine.
$\alpha = 54° = 0.942478$ radians.

To calculate the displacement, velocity and acceleration of the roller we shall use the Mod Sine factor table in Appendix B. The normalised input is $u = 54°/90° = 0.6$, which is found at step number 72 in the table (the step can be found directly as $54 \times 120/90 = 72$). Reading from the table the normalised output displacement, velocity and acceleration are:

$$w = 0.67213, \qquad w' = 1.64551, \qquad w'' = -2.24842$$

Therefore:

$$h = 60 + 25 \times 0.67213 = 76.80325 \text{ mm} \qquad \text{Eq. (7.14)}$$

$$r = (76.80325^2 + 15^2)^{0.5} = 78.25432 \text{ mm} \qquad \text{Eq. (7.15)}$$

$$\theta = \arctan(76.80325/15) - 54 = 24.9490° = 0.435442 \text{ radians} \qquad \text{Eq. (7.16)}$$

$$x = 70.95191 \text{ mm} \qquad \text{Eq. (7.18)}$$

$$y = 33.00856 \text{ mm} \qquad \text{Eq. (7.19)}$$

To use the approximate method we must first find values of r and θ for near points either side of the required one. It is easiest to take the adjacent steps in the factor table and carry out the same calculations as above. The results, including the above, are shown in the following small table:

Table 7.1 Calculations for approx. radius of curvature

Step	u	α	w	h	r	θ
71	0.59167	0.929393	0.65835	76.45875	77.91624	0.447679
72	0.6	0.942478	0.67213	76.80325	78.25432	0.435442
73	0.60833	0.955563	0.68577	77.14425	78.58903	0.423188

Therefore;

$$r = 78.25432 \qquad\qquad \text{Eq. (7.7)}$$

and approximately:

$$f'(\theta) = \frac{78.58903 - 77.91624}{0.423188 - 0.447679} = \frac{0.67279}{-0.024491} = -27.47091 \text{ mm/rad}$$

$$\text{Eq. (7.8)}$$

and:

$$f''(\theta) = \frac{78.58903 - 2 \times 78.25432 + 77.91624}{(0.423189 - 0.435442) \times (0.435442 - 0.447679)}$$

$$= \frac{-0.00337}{0.00014995} = -22.474 \text{ mm/rad}^2 \qquad \text{Eq. (7.9)}$$

Substituting these values in Equation 7.5 we get:

$$R_c = \frac{(78.25432^2 + 27.47091^2)^{1.5}}{78.25432^2 + 2 \times 27.47091^2 + 78.25432 \times 22.474}$$

$$= \frac{570466}{9392} = 60.74 \text{ mm} \qquad\qquad \text{Eq. (7.5)}$$

This result is positive, so the curvature is convex. The calculation was for the pitch curve and, since the *radius* of the roller is 10 mm, the radius of curvature of the cam profile at the specified position is $60.74 - 10 = 50.74$ mm approximately.

In this example the *exact* radius of curvature can be found from Equation 7.20. We have already calculated $h = 76.80325$ mm, but now we must evaluate the derivatives from Equations 7.12 and 7.13.

$$h' = \frac{25}{1.570796} \times 1.64551 = 26.1891 \text{ mm/rad} \qquad \text{Eq. (7.12)}$$

$$h'' = \frac{25}{1.570796^2} \times (-2.24842) = -22.7813 \text{ mm/rad}^2 \qquad \text{Eq. (7.13)}$$

Substituting these values in Equation 7.20 we get:

$$R_c = \frac{[76.80325^2 + 26.1891^2 + (15^2 - 2 \times 15 \times 26.1891)]^{1.5}}{76.80325^2 + 2 \times 26.1891^2 + 76.80325}$$
$$\times 22.7813 + (15^2 - 3 \times 15 \times 26.1891)$$

$$= \frac{467541}{8066.5} = 57.961 \text{ mm}$$

This is the pitch curve radius of curvature. It is positive and therefore convex. The cam profile radius of curvature, found by subtracting the 10 mm roller radius, is 47.961 mm.

Note that this result is about 6% less than the approximate one calculated from Equations 7.5, 7.7, 7.8 and 7.9, showing that the approximate method is reasonably satisfactory in this case. However, if the pitch circle radius of curvature is nearly the same as the roller radius there may be a significant error in the approximate cam profile radius of curvature that could give a false impression as to the occurrence of interference (see below). If there is any doubt about interference it is safest to use the exact method of Equation 7.20 to investigate the problem.

For the pressure angle we have already calculated h' and h for 50° of cam rotation, so now we can calculate

$$\phi = \arctan \left[(26.1891 - 15)/76.80325 \right] = 8.288° \qquad \text{Eq. (7.22)}$$

The pressure angle in the dwell periods are

$$-\varepsilon = -\arctan (15/60) = -4.036° \quad \text{and}$$

$$-\varepsilon = -\arctan (15/85) = -10.008° \qquad \text{Eq. (7.16)}$$

INTERFERENCE AND UNDERCUTTING

These phenomena occur in all kinds of cams (disc, cylinder, etc.) whose pitch curves have either high curvature or *looping*.

Figure 7.5 shows an example of interference due to high curvature (small radius of curvature). The cam profile shape must be such as to allow the unimpeded passage of the follower roller as its centre travels along the pitch curve. In fact, if the profile is machined with a cutter of the same diameter as the roller travelling along the specified pitch curve (a common manufacturing procedure) then a clear passage of the roller *must* ensue. However, this sometimes entails the *interference* of one part of the profile with an adjacent part of it and the removal of metal that would otherwise contact the roller, as shown in Fig. 7.5.

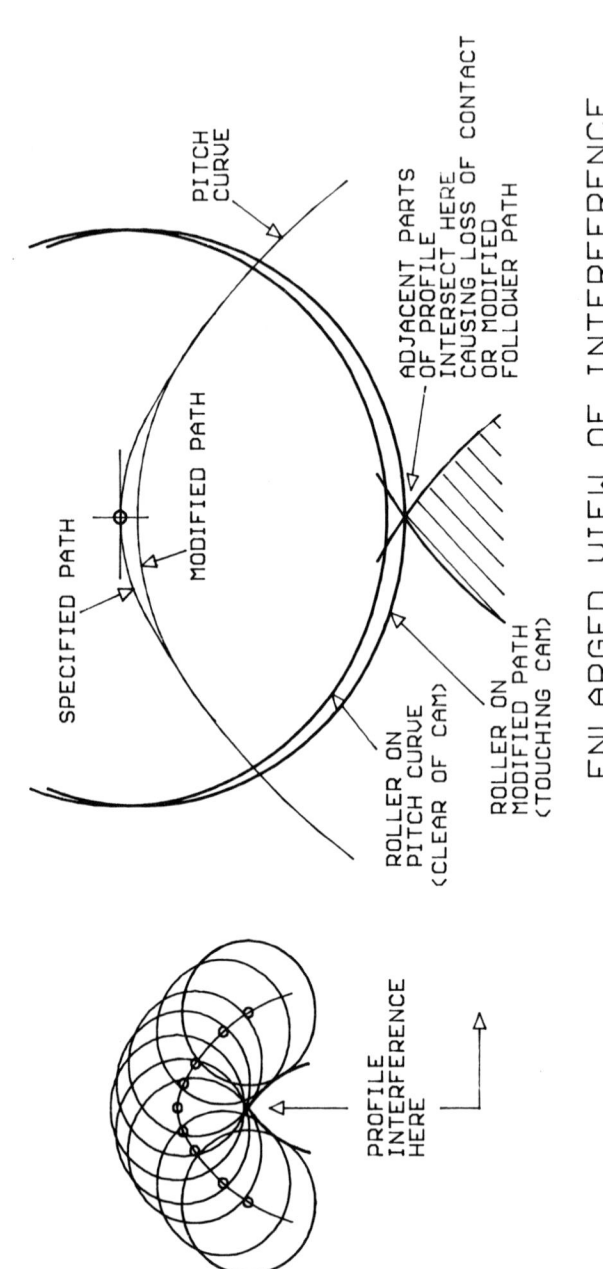

PITCH CURVE

SPECIFIED PATH

MODIFIED PATH

ADJACENT PARTS
OF PROFILE
INTERSECT HERE
CAUSING LOSS OF CONTACT
OR MODIFIED
FOLLOWER PATH

ROLLER ON
PITCH CURVE
(CLEAR OF CAM)

ROLLER ON
MODIFIED PATH
(TOUCHING CAM)

ENLARGED VIEW OF INTERFERENCE

PROFILE
INTERFERENCE
HERE

Fig. 7.5   Cam profile interference

If the system is such as to produce a *separating* force rather than a *closing* force at this point, possible in a conjugate or box cam system, then there will be loss of contact and the follower will not be totally controlled by the cam. If, on the other hand, there is a closing force, e.g. in a spring-loaded open track system, the roller will remain in contact at this point, but will not follow the prescribed pitch curve: the motion will be modified.

Interference of this kind happens when the radius of curvature of the pitch curve is less than the radius of the follower roller. It can be avoided either by reducing the size of the roller or altering the cam law, making it asymmetrical for example, or by increasing the size of the cam. In some cases changing the follower offset will increase the radius of curvature sufficiently to avoid interference. Any of these remedies, however, must be taken with due regard to other important design criteria to reach an acceptable compromise. Trial and error tactics usually have to be employed at this stage.

In general, loss of contact is very detrimental to the performance of the mechanism, although there are some special cases where it can be tolerated. If possible avoid the condition by modifying the design of the cam or follower as suggested.

Figure 7.6 shows an example of undercutting due to a pitch curve *loop*. Such loops seldom occur with translating roller followers, but can often occur with swinging roller followers and especially with the conjugate disc cam indexing mechanism (see Fig. 1.2., diagram 24). Pitch curve loops produce two points of interference which give rise to the undercutting,

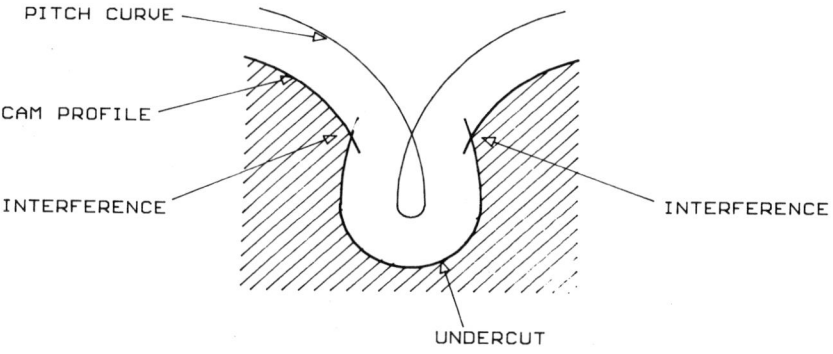

Fig. 7.6   Cam profile undercutting

irrespective of roller radius. This kind of interference cannot be cured by changing the size of the roller – only by eliminating the loop. Changing the cam law, the cam size, the output stroke or the motion period are the obvious remedies to try.

Where there is interference or undercutting there is inevitably a surface stress and rapid wear of the cam.

DISC CAM WITH SWINGING ROLLER FOLLOWER

In this class of mechanisms the follower consists of a roller, usually at the end of an arm, which is constrained to oscillate (swing) bodily around a pivot under the control of a disc cam. It includes indexing mechanisms, where several rollers are pitched around a circle on a follower disc (usually called a follower wheel or turret).

The concept of follower offset is less meaningful than with a translating follower because the direction of motion of the roller changes as it swings round its arcuate path, so that it may be trailing at one time and leading at another with respect to a cam radius parallel to the direction of motion. Therefore the terms trailing and leading are applied to the follower roller with respect to the line that passes through both the centre of rotation of the cam and the pivot of the follower arm – the *common centre-line*. Travelling in the direction of cam rotation, a leading follower roller is encountered before the common centre-line and a trailing one is encountered after the common centre-line (see Fig. 7.7).

If the shape of the pitch curve can be expressed directly in terms of orthogonal co-ordinates, x and y, or in terms of the polar co-ordinates, r and θ, with appropriate derivative functions, then Equations 7.1 to 7.9 inclusive can be used to find the radius of curvature at any point: the procedure is the same as for the translating follower case. The cam datum, which is used as a reference line for measuring the cam profile and cam rotation, is taken as the x-axis, which is virtually scribed on the cam and rotates with it. The cam datum may be different from the common centre-line by some arbitrary angle if preferred, but for convenience here they coincide at the start of the motion.

However, the reference line for measuring the displacement of the follower arm *must* be the common centre-line for the equations below to be valid.

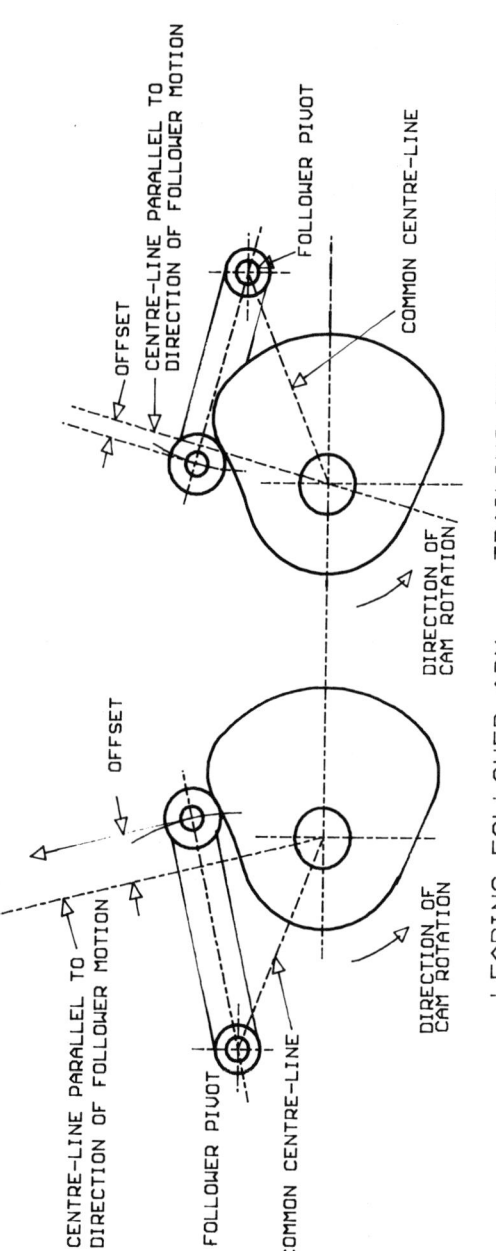

Fig. 7.7 Disc cams with swinging followers

RADIUS OF CURVATURE AND PRESSURE ANGLE FROM KNOWN CAM LAW

When the angular motion of the follower arm obeys a known cam law, as it often does in industrial applications, the co-ordinates can be calculated from the following equations (see Fig. 7.7).

$$\alpha = A \cdot u \qquad\qquad\qquad \text{Eq. (7.23)}$$

$$\beta = D + B \cdot w \qquad\qquad\qquad \text{Eq. (7.24)}$$

$$r = [C^2 + F^2 - 2 \cdot C \cdot F \cdot \cos(\beta)]^{0.5} \qquad\qquad \text{Eq. (7.25)}$$

$$\theta = \pi - \arcsin[F \cdot \sin(\beta)/r] - \alpha \qquad\qquad \text{Eq. (7.26)}$$

$$x = r \cdot \cos \theta \qquad\qquad\qquad \text{Eq. (7.27)}$$

$$y = r \cdot \sin \theta \qquad\qquad\qquad \text{Eq. (7.28)}$$

where

A = angular input stroke of the cam

B = angular output stroke of the swinging follower

D = initial angular displacement of swinging follower from common centre-line

C = the distance from the centre of rotation of the cam to the follower pivot

F = the length of the follower arm, pivot to roller centre

Note that $w = f(u)$ is the normalised cam law function as before, but B, β, D, α and A are now angular dimensions in radians.

The radius of curvature can be calculated directly from the geometric velocity and acceleration derived from the cam law factors thus:

$$R_c = \frac{F \cdot \left[\left(\frac{C}{F}\right)^2 + (1 - \beta')^2 - 2 \cdot \frac{C}{F} \cdot (1 - \beta') \cdot \cos(\beta)\right]^{1.5}}{\left(\frac{C}{F}\right)^2 + (1 - \beta')^3 - \frac{C}{F} \cdot (1 - \beta')(2 - \beta') \cdot \cos(\beta) - \frac{C}{F} \cdot \beta'' \cdot \sin(\beta)}$$

$$\text{Eq. (7.29)}$$

and the pressure angle is found from:

$$\tan \phi = \frac{\cos \beta - \frac{F}{C}(1 - \beta')}{\sin \beta} \qquad\qquad \text{Eq. (7.30)}$$

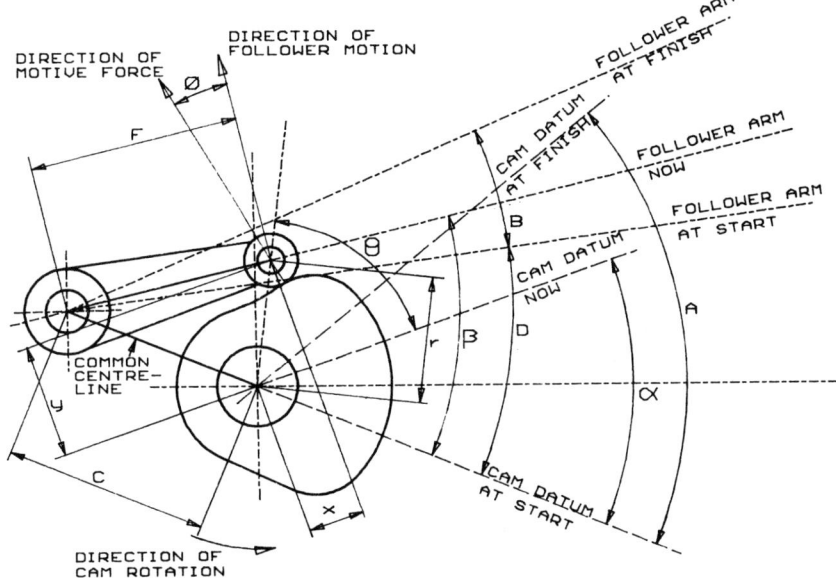

Fig. 7.8 Swinging arm geometry

where β' = geometric velocity $= \dfrac{B}{A} \cdot w' = \dfrac{d\beta}{d\alpha}$
Eq. (7.31)

and β'' = geometric acceleration $= \dfrac{B}{A^2} \cdot w'' = \dfrac{d^2\beta}{d\alpha^2}$
Eq. (7.32)

B, A, β and α must be in radians.

This equation yields a positive value of convex curvature at any point on the profile. It is for a mechanism with a *leading* follower arm and a rising motion as shown in Fig. 7.8.

For a *trailing* follower and rising motion Fig. 7.9 can be used. In this case Equations 7.29 and 7.30 must be modified by changing the sign of the β' terms thus:

$$R_c = \frac{F \cdot \left[\left(\dfrac{C}{F}\right)^2 + (1 + \beta')^2 - 2 \cdot \dfrac{C}{F} \cdot (1 + \beta') \cdot \cos(\beta)\right]^{1.5}}{\left(\dfrac{C}{F}\right)^2 + (1 + \beta')^3 - \dfrac{C}{F} \cdot (1 + \beta')(2 + \beta') \cdot \cos(\beta) - \dfrac{C}{F} \cdot \beta'' \cdot \sin(\beta)}$$

Eq. (7.33)

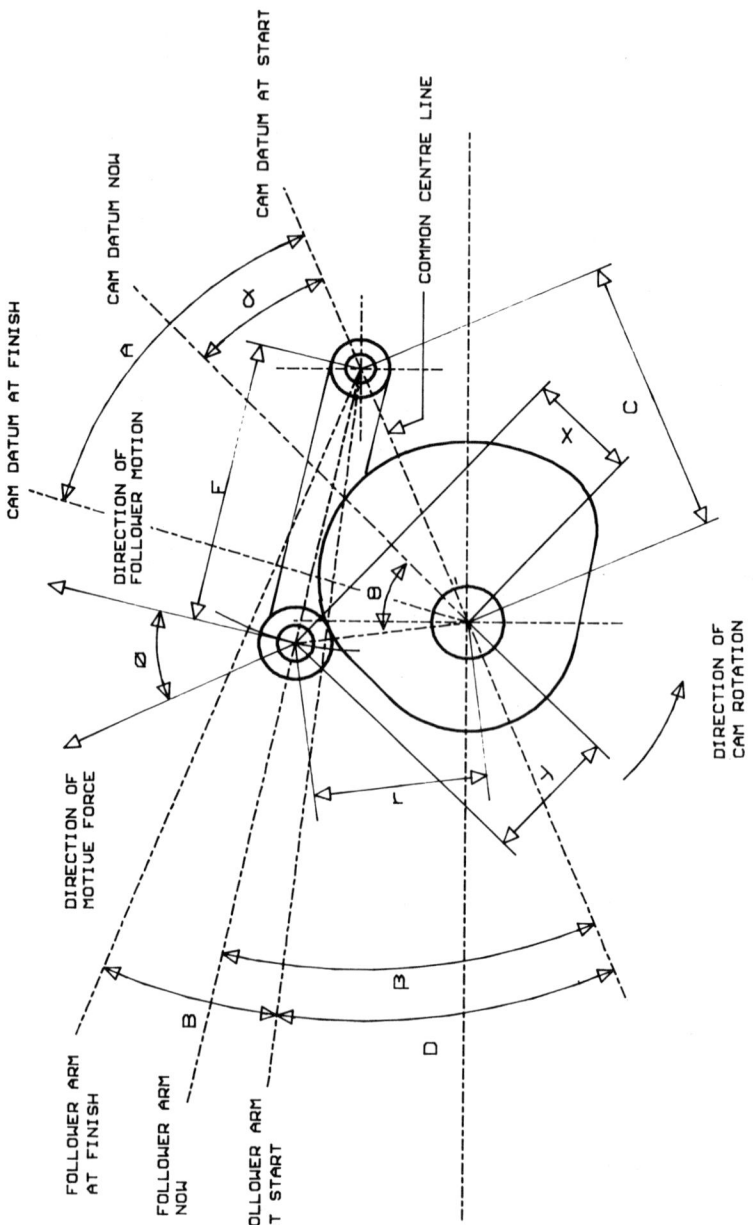

CAM DATUM AT FINISH

CAM DATUM NOW

CAM DATUM AT START

COMMON CENTRE LINE

DIRECTION OF
FOLLOWER MOTION

DIRECTION OF
MOTIVE FORCE

DIRECTION OF
CAM ROTATION

FOLLOWER ARM
AT FINISH

FOLLOWER ARM
NOW

FOLLOWER ARM
AT START

Fig. 7.9 Swinging arm geometry

$$\tan \phi = \frac{\cos \beta - \dfrac{F}{C}(1 + \beta')}{\sin \beta} \qquad \text{Eq. (7.34)}$$

Also, for a trailing follower, the value of θ must be found from Equation 7.35 instead of Equation 7.26:

$$\theta = \arcsin \left[F \cdot \frac{\sin \beta}{r} \right] - \alpha \qquad \text{Eq. (7.35)}$$

For a *falling* motion Equation 7.29 or 7.34 can be used for leading or trailing follower arm, respectively, by taking D as the angle from the cam datum to the highest position of the follower arm (start of motion) and *using a negative value of the stroke B.*

Example 7.2

An oscillating mechanism is driven by a swinging arm with a 25 mm diameter roller follower running in a groove in the face of a disc cam. The arm is 150 mm long and swings between horizontal and 20° above horizontal. Its pivot is 155 mm to the right and 60 mm above the centre of rotation of the cam. The Mod Sine motion takes place in 60° of anti-clockwise cam rotation.

Find the pressure angle and radius of curvature of the cam track at 5° of cam rotation after the start. Referring to Fig. 7.10 we get:

A = input stroke = 60° = 1.04720 rad

B = output stroke = 20° = 0.34907 rad

C = cam centre to pivot centre distance = $\sqrt{(155^2 + 60^2)}$ = 166.208 mm

D = initial follower displacement = arctan (60/155) = 21.161° = 0.369329 rad

F = length of follower arm = 150 mm

This is a mechanism with a trailing follower arm and a rising motion, so we must use Equation 7.34 to calculate the radius of curvature. First we must calculate the normalised input displacement, from which we can find the normalised output factors and then calculate geometric values, α, β, β' and β'', for substitution in the equation.

$\alpha = 5° = 0.0872665$ rad   and therefore $u = 5 \div 60 = 0.0833333$

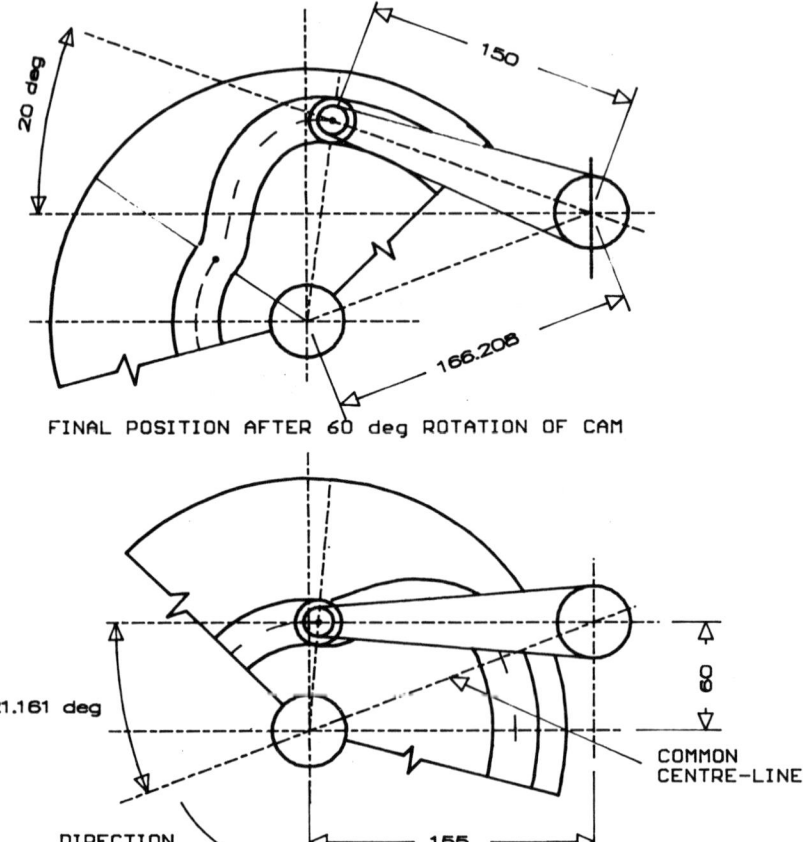

FINAL POSITION AFTER 60 deg ROTATION OF CAM

INITIAL POSITION OF CAM AND FOLLOWER

Fig. 7.10   Example 7.2

From the Mod Sine factors table in Appendix B we find that this corresponds to step number 10 (out of 120 total) and we read off:

$$w = 0.00634, \quad w' = 0.21995, \quad w'' = 4.78735$$

Now

$$\beta = D + B \cdot w = 0.34907 \times 0.00634 = 0.371542 \text{ rad } (=21.288°)$$

$$\beta' = \frac{B}{A} \cdot w' = \frac{0.34907}{1.04720} \times 0.21995 = 0.073317$$

$$\beta'' = \frac{B}{A^2} \cdot w'' = \frac{0.34907}{1.04720^2} \times 4.78735 = 1.523872 \text{ rad}^{-1}$$

$$\sin \beta = 0.363056$$

$$\cos \beta = 0.931767$$

$$\frac{C}{F} = \frac{166.208}{150} = 1.108053$$

Substituting these values in Equation 7.34 gives:

$$R_c = \frac{150 \times [1.108053^2 + 1.073317^2 - 2 \times 1.108053 \times 1.073317 \times 0.931767]^{1.5}}{1.108053^2 + 1.073317^3 - 1.108053 \times 1.073317 \times 2.073317 \times 0.931767 - 1.108053 \times 1.523872 \times 0.363056}$$

$$R_c = \frac{9.917128}{-0.446331} = -22.220 \text{ mm}$$

This is the radius of curvature of the pitch curve at the specified position (5° of cam rotation after the start). It turns out to be negative, which means that the pitch curve is concave at this point. Its value is greater than the radius of the roller and therefore there is no interference of the outer wall of the track.

The radius of curvature of the outer and inner track surfaces at this point are respectively:

$$-22.220 + 12.5 = -9.720 \text{ mm}$$

(concave relative to the cam centre, but convex in practice)

and

$$-22.220 - 12.5 = -34.720 \text{ mm (concave)}$$

The pressure angle is found from Equation 7.34 by substituting values already calculated:

$$\tan \phi = \frac{\cos 21.288° - \dfrac{150}{166.208}(1 + 0.073317)}{\sin 21.288°} = -0.10159$$

so

$$\phi = \arctan(-0.10159) = -5.801°$$

For comparison we can calculate the approximate radius of curvature of the pitch curve using Equations 7.23 to 7.26, 7.31, 7.32, 7.5, 7.7 and 7.8 (see Table 7.2).

In this instance the approximate result is remarkably close to the 'exact' one, but there could be a significant difference if the radius of curvature were much smaller.

This point on the cam profile was chosen for this example because from an accurately drawn plot it was seen to be near to the point of maximum curvature (minimum *radius* of curvature). As such it could be a critical position for surface stress and load capacity.

However, if the force on the cam at this point is downwards (upwards force on follower) then the small convex radius of curvature of the outer wall of the track is not detrimental, since nominally it carries no load. This is likely to be the case if most of the payload is inertial, because the cam is imparting an upwards acceleration to the follower at this point. Had there been interference, though, the loss of control of the output motion through loss of contact between cam and follower might have been unacceptable.

In most cases it is important to find the point of minimum radius of curvature on a profile. The equations are too complex to derive a simple

Table 7.2   Calculations for approx. and exact radius of curvature

QUANTITY	VALUE			SOURCE
STEP	9	10	11	Mod Sine table
u	0.075	0.08333	0.09167	Mod Sine table
w	0.00467	0.00634	0.00834	Mod Sine table
w'	0.18133	0.21995	0.26098	Mod Sine table
w''	4.47221	4.78735	5.05004	Mod Sine table
α	0.078540	0.087266	0.095993	Eq. 7.23
β	0.370960	0.371543	0.372241	Eq. 7.24
β'	0.060444	0.073317	0.086992	Eq. 7.31
β''	1.423549	1.523861	1.607478	Eq. 7.32
exact R_c	-24.00319	-22.21988	-21.14703	Eq. 7.29, 7.34
r	60.45122	60.53839	60.64289	Eq. 7.25
θ	1.040109	1.031499	1.022911	Eq. 7.26
approx $f'(\theta)$		-11.14488		Eq. 7.8
approx $f''(\theta)$		234.4158		Eq. 7.9
approx R_c		-22.69358		Eq. 7.5

equation for the position of minimum radius of curvature, so several calculations must be carried out at adjacent positions to find the minimum by trial and error. This is very time-consuming if done manually, but a computer program can be written fairly easily to yield an accurate solution quickly by numerical methods.

It is recommended that in all cases of cam profile analysis an accurate profile drawing should be plotted, preferably by computer, for visual assessment. This is invaluable for making quick design decisions as to best roller size, overall cam size, best follower arm length, etc. and whether the minimum radius of curvature is likely to be critical.

CAM GEOMETRY: FLAT-FACED FOLLOWERS

The only kinds of cam in general use that can be used with flat-faced followers are disc cams and translating plate cams.

DISC CAM WITH TRANSLATING FLAT-FACED FOLLOWER

When a flat-faced follower is used its contact face is usually at 90° to its direction of motion. However, it can be at some other angle, and the difference between this and 90° is its 'face angle'. The equations here are for the general case of a non-zero face angle, which of course may also be used for the normal case.

Figure 7.11 shows a typical disc cam and translating flat-faced follower (inclined), at the start of a motion period, at the end of the period and at an intermediate position. If the friction between the cam track and the follower is negligible the contact force is normal to the follower face and therefore makes an angle with the direction of motion – the nominal pressure angle – which is identical to the face angle defined above. The nominal pressure angle of a flat-faced follower and cam is equal to the face angle and *is constant throughout the motion*.

In practice the friction at the contact point may be significant because it is pure sliding. The true pressure angle is the sum or difference of the nominal angle and the friction angle, depending on the direction of cam rotation. In some mechanisms this may be an important feature in the design of the follower guide system. For example, the face angle may be chosen to equal the friction angle approximately, to give a nearly zero true pressure angle: this could remove most of the lateral forces on the follower guide.

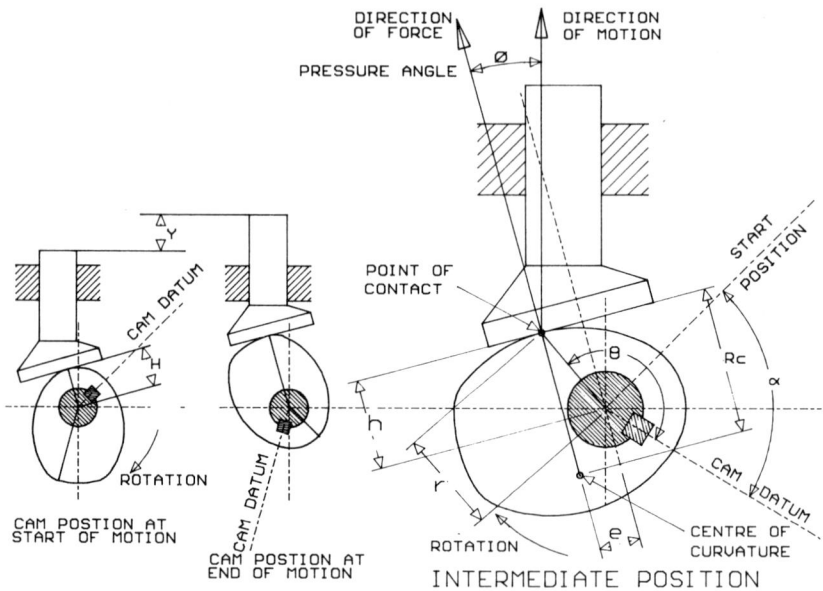

DIRECTION OF FORCE

DIRECTION OF MOTION

PRESSURE ANGLE

Ø

POINT OF CONTACT

START POSITION

θ

Rc

α

CAM DATUM

ROTATION

CAM DATUM

CENTRE OF CURVATURE

e

r

h

CAM POSTION AT START OF MOTION

ROTATION

CAM POSTION AT END OF MOTION

INTERMEDIATE POSITION

Fig. 7.11 Disc cam with translating flat-faced follower

As the cam rotates the contact point moves across the face of the follower, so that the offset – the perpendicular distance from the cam centre to the nominal line of force – varies throughout the motion. The nominal line of force is normal to the follower face and passes through the point of contact. The cam track has a curvature at the point of contact whose centre lies on that line.

There is a simple relationship between the offset and the geometric velocity of the follower, thus:

$$e = (B \cdot \cos \phi / A) \cdot w'$$ Eq. (7.36)

The distance from the cam centre to the follower face is:

$$h = H + (B \cdot \cos \phi) \cdot w = (D + B \cdot w) \cdot \cos \phi$$ Eq. (7.37)

The polar co-ordinates of the cam profile are:

$$r = \surd(h^2 + e^2) = \left(B \cdot \cos \phi \cdot \sqrt{\left[\left(w + \frac{H}{B \cos \phi}\right)^2 + (w'/A)^2\right]}\right.$$

Eq. (7.38)

$$\theta = \alpha + \arctan(e/h) = A \cdot u + \arctan\left[\frac{w'}{A \cdot \left(w + \dfrac{H}{B \cdot \cos\phi}\right)}\right]$$

Eq. (7.39)

and the radius of curvature of the profile at the point of contact is:

$$R_c = (B \cdot \cos\phi) \cdot \left[w + \frac{H}{B \cdot \cos\phi} + \frac{w''}{A^2}\right]$$

Eq. (7.40)

where:

α = instantaneous input displacement (cam rotation) from start
θ = angular co-ordinate of cam profile
ϕ = pressure angle = face angle
A = input stroke, cam rotation for whole motion period [rad]
B = output stroke
D = distance from cam centre *in the direction of motion* to the plane of the follower face (i.e. from cam centre to face intersect).
e = offset
h = instantaneous distance from follower face to cam centre
H = perpendicular distance from follower face to cam centre at start of motion (i.e. the base radius)
r = radial co-ordinate of cam profile
R_c = radius of curvature of cam profile at the point of contact
u = normalised input displacement
$w = f(u)$ = normalised output displacement
$w' = dw/du$ = normalised output velocity
$w'' = d^2w/du^2$ = normalised output acceleration

If the follower face is not inclined (face angle = 0) the $\phi = 0$ and $\cos\phi = 1$ can be substituted in the above equations.

The sign of R_c is dependent on the sign and value of the acceleration factor w''. In the last phase of the motion w'' is negative (deceleration), so it is theoretically possible for the radius of curvature to be zero or negative, i.e. concave. A zero radius of curvature is a sharp point, which is undesirable because of rapid wear. A concave curvature is not practicable because of interference (see Fig. 7.12). The minimum radius of curvature occurs at the position of peak deceleration (negative acceleration), which is when $w'' = -C_d$ by definition.

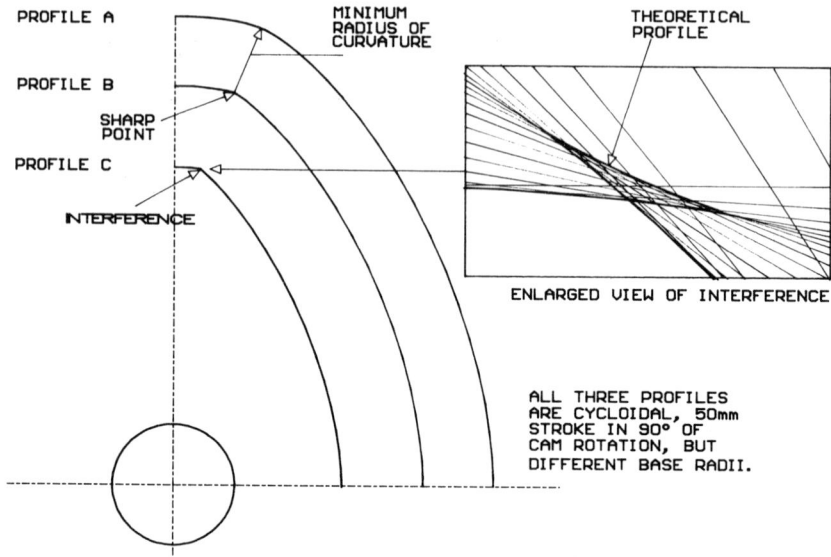

PROFILE A

MINIMUM
RADIUS OF
CURVATURE

THEORETICAL
PROFILE

PROFILE B

SHARP
POINT

PROFILE C

INTERFERENCE

ENLARGED VIEW OF INTERFERENCE

ALL THREE PROFILES
ARE CYCLOIDAL, 50mm
STROKE IN 90° OF
CAM ROTATION, BUT
DIFFERENT BASE RADII.

Fig. 7.12   Interference of flat-faced follower

For the curvature always to be positive we have the limiting condition:

$$h > (B . \cos \phi) . C_d/A^2 \qquad\qquad \text{Eq. (7.41)}$$

or

$$H > (B . \cos \phi) . (C_d/A^2 - w) \qquad\qquad \text{Eq. (7.42)}$$

where h and w correspond to the position of maximum deceleration.

The latter equation sets a lower limit to the value of the base radius H for a cam with known output stroke, input period and cam law. A larger base radius than this is necessary to have a practical minimum radius of curvature.

The importance of Equation 7.42 is illustrated in Fig. 7.12 which shows three versions of a disc cam and flat-faced follower with a cycloidal motion of 50 mm output stroke and 90° input stroke (cam rotation). Profile A is a satisfactory design with a base radius of $H = 105$ mm, which is greater than the critical lower limit. Profile B has $H = 81.866$ mm, which is exactly at the limit set by Equation 7.42, and is unsatisfactory because the minimum radius of curvature is zero: it is a sharp point.

Profile C is worst of all because $H = 55$ mm, well below the critical value, and severe interference results in a profile that not only has a sharp point but also cannot produce the specified motion: this is illustrated in the enlarged inset.

It is important to realise that if the motion does not start and finish with zero velocity then the angle subtended by the cam profile is different from the input stroke (angle of rotation of the cam). This is because the value of arctan(e/h) in Equation 7.39 can only be the same at both ends of the stroke if the offset e is zero in both places.

DISC CAM WITH SWINGING FLAT-FACED FOLLOWER

In this case the *angular* displacement of the follower conforms to a cam law and the follower pivot may lie in the plane of the face or be offset from it by a distance E, as shown in Fig. 7.13 for a *leading* follower and Fig. 7.14 for a trailing one.

The instantaneous position of the follower is given by the angle:

$$D + \beta = D + B \cdot w \qquad \text{Eq. (7.43)}$$

The distances from the point of contact to the centre of rotation of the cam, taken parallel and perpendicular to the follower face, are e and h respectively. These co-ordinates are similar to the previous case, but their orientation varies throughout the motion as the follower rotates about its pivot.

Leading follower

$$f = C \cdot \left[\frac{\cos (D + \beta)}{1 - \beta'} \right] = C \cdot A \cdot \left[\frac{\cos [D + \beta)}{A - B \cdot w'} \right] \qquad \text{Eq. (7.44)}$$

$$e = C \cdot \beta' \cdot \left[\frac{\cos (D + \beta)}{1 - \beta'} \right] = C \cdot A \cdot \beta' \cdot \left[\frac{\cos (D + \beta)}{A - B \cdot w'} \right] \qquad \text{Eq. (7.45)}$$

$$h = C \cdot \sin (D + \beta) - E \qquad \text{Eq. (7.46)}$$

The pressure angle depends on the offset E. If $E = 0$ then the pressure angle is constantly zero. Otherwise it varies throughout the motion according to:

$$\phi = \arctan (E/f) \qquad \text{Eq. (7.47)}$$

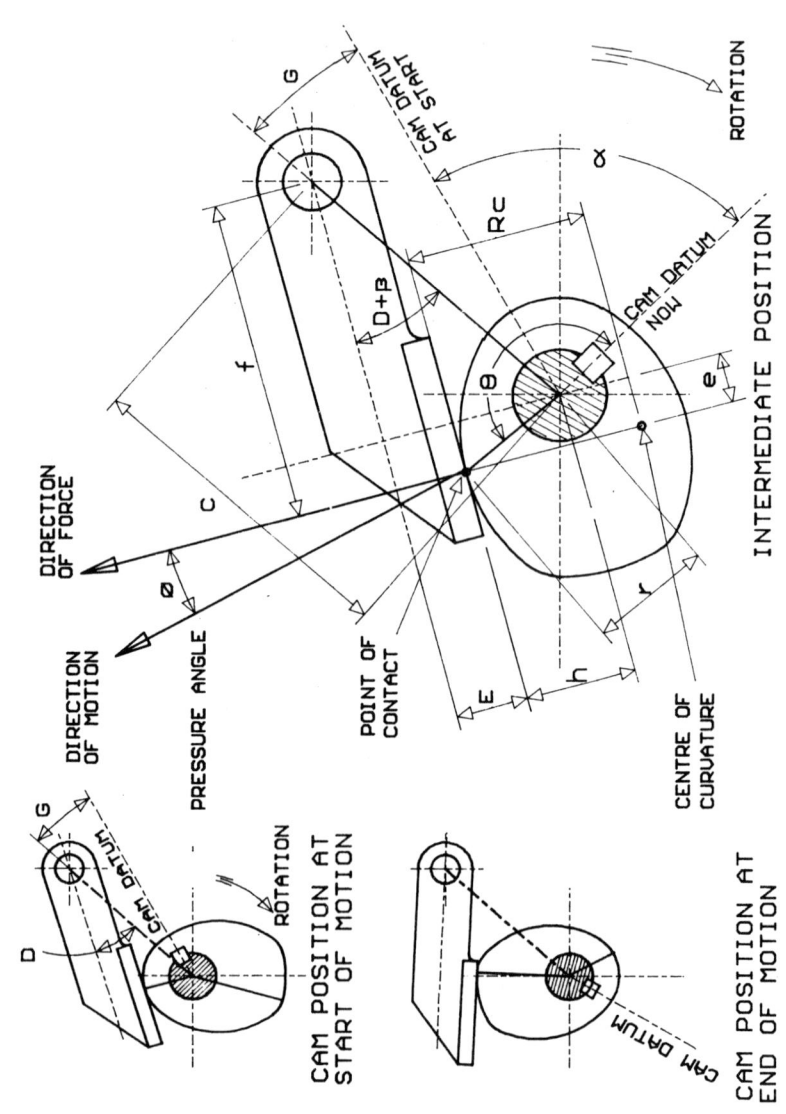

Fig. 7.13   Disc cam with swinging flat-faced follower (leading)

The polar co-ordinates of the cam profile relative to the cam datum are:

$$r = \sqrt{(h^2 + e^2)}$$ Eq. (7.48)

and

$$\theta = \alpha + \arctan(e/h) + \pi/2 + G - D - \beta \, [\text{rad}]$$ Eq. (7.49)

The radius of curvature at the point of contact is:

$$R_c = C \cdot \frac{\beta'' \cos(D + \beta) + (1 - \beta') \cdot (1 - 2\beta') \cdot \sin(D + \beta)}{(1 - \beta')^3} - E$$

   Eq. (7.50)

or

$$R_c = A \cdot C \cdot \frac{B \cdot w'' \cos(D + \beta) + (A - B \cdot w') \cdot (A - 2B \cdot w') \cdot \sin(D + \beta)}{(A - B \cdot w')^3} - E$$

   Eq. (7.51)

where:

B = output stroke of follower [rad].
C = distance from the centre of rotation of the cam to the follower pivot.
D = initial angular displacement of follower from common centre-line.
E = offset of follower face from follower pivot.
f = distance from point of contact to follower pivot, parallel to follower face.
G = the initial angle between the common centre-line and the cam datum.
β = angular displacement of follower from the start = $B \cdot w$
β' = geometric angular velocity of follower = $B \cdot w'/A = d\beta/d\alpha$.
β'' = geometric angular acceleration of follower = $B \cdot w''/A^2 = d^2\beta/d\alpha^2$.

and all other symbols are as before.

Similar considerations of undercutting apply to this as to the translating follower case dealt with earlier, namely that R_c cannot be negative. Also, it is not practicable for the offset e to be negative during a rising motion, so we get the two conditions:

$$A \geq B \cdot C_v$$ Eq. (7.52)

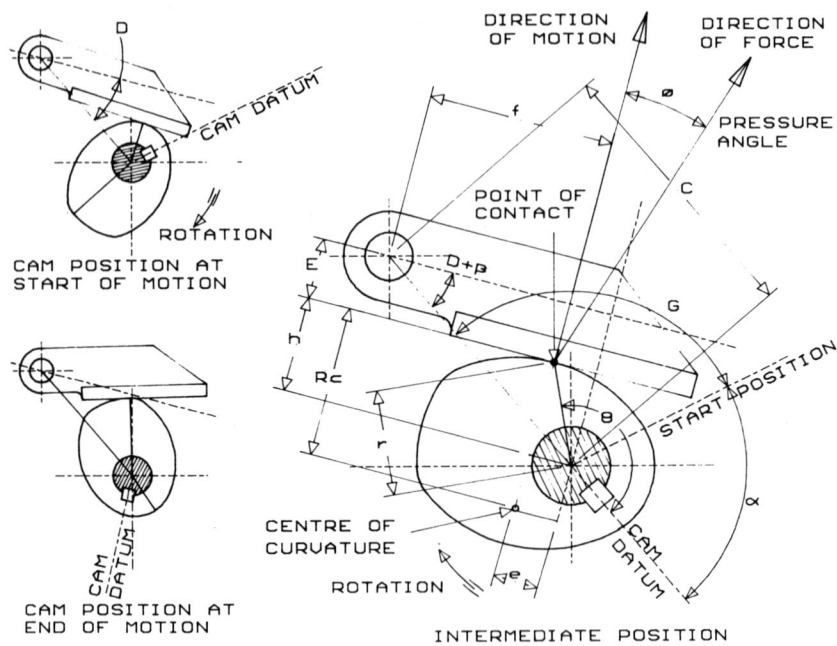

Fig. 7.14   Disc cam with swinging flat-faced follower (trailing)

where C_v = Coefficient of velocity of the cam law. Therefore $(1 - \beta')$ is always positive and the smallest value of R_c occurs when β'' is negative, near the position of maximum deceleration, when $\beta'' \approx -C_d$. Therefore approximately:

$$C_d < (1 - \beta') \cdot (1 - 2\beta') \cdot \tan(D + \beta) - (1 - \beta')^3 \cdot \sec(D + \beta) \cdot E/C$$
$$\text{Eq. (7.53)}$$

where C_d = Coefficient of deceleration of the cam law and β and β' apply at the position of maximum deceleration. The last term of this equation is usually small, so as a rough check for practicability it may be ignored.

For a given output stroke B it may not be possible to use the required input stroke A or cam law and still satisfy Equation 7.53 by adjusting D and/or E. Sometimes an asymmetrical cam law with a low value of C_d can be used effectively. In fact this type of mechanism is only capable of producing fairly small output strokes.

The above equations apply to rising motion and leading follower as shown in Fig. 7.13. For a falling motion the same equations can be used by substituting $-B$ for B. However, a slightly different set of equations must be used for a trailing follower system as shown in Fig. 7.14.

Trailing follower

$$f = C \cdot \left[\frac{\cos (D + \beta)}{1 + \beta'} \right] = C \cdot A \cdot \left[\frac{\cos (D + \beta)}{A + B \cdot w'} \right] \qquad \text{Eq. (7.54)}$$

$$e = C \cdot \beta' \cdot \left[\frac{\cos (D + \beta)}{1 + \beta'} \right] = C \cdot A \cdot \beta' \cdot \left[\frac{\cos (D + \beta)}{A + B \cdot w'} \right] \qquad \text{Eq. (7.55)}$$

$$h = C \cdot \sin (D + \beta) - E \qquad \text{Eq. (7.56)}$$

The polar co-ordinates relative to the cam datum are:

$$r = \sqrt{(h^2 + e^2)} \qquad \text{Eq. (7.57)}$$

and

$$\theta = \alpha + \arctan (e/h) - \pi/2 + G + D + \beta \, [\text{rad}] \qquad \text{Eq. (7.58)}$$

The radius of curvature at the point of contact is:

$$R_c = C \cdot \frac{\beta'' \cos (D + \beta) + (1 + \beta') \cdot (1 + 2\beta') \cdot \sin (D + \beta)}{(1 + \beta')^3} - E$$

$$\text{Eq. (7.59)}$$

or

$$R_c = A \cdot C \cdot \frac{B \cdot w'' \cos (D + \beta) + (A + B \cdot w') \cdot (A + 2B \cdot w') \cdot \sin (D + \beta)}{(A + B \cdot w')^3} - E$$

$$\text{Eq. (7.60)}$$

The equations are very similar to those of the leading follower case: this is not surprising, because a rising motion with a leading follower becomes a falling motion with a trailing follower when the direction of rotation of the cam is reversed.

The same arguments apply as regards interference and minimum radius of curvature, leading to the following condition:

$$C_d < (1 + \beta') \cdot (1 + 2\beta') \cdot \tan (D + \beta) - (1 + \beta')^3 \sec (D + \beta) \cdot E/C$$

$$\text{Eq. (7.61)}$$

SYMMETRY

It is important to realise that with a swinging flat-faced follower the angle
subtended by the cam profile is different from the input stroke (angle of
rotation of the cam). This is because the values of both β and $\arctan(e/h)$
in Equations 7.49 and 7.58 are different at each end of the stroke. Figure
7.15 shows a cam with a totally symmetrical *motion*, but the actual cam
profile clearly is not symmetrical. The rise segment is a different shape
from the fall segment, and it occupies a smaller angle on the cam. In this
case the rise occupies $120° - 15° = 105°$ and the fall occupies $120° + 15° =$
$135°$, while the dwells are unchanged at $60°$ each.

An important consequence of this is that even though the cam be
correctly manufactured it is essential that it be fitted into the machine the
correct way round relative to the follower pivot and direction of cam
rotation. Otherwise the motion would be severly distorted: this applies
also to asymmetrical motions (the diagram uses a symmetrical motion to
make the asymmetry of the profile obvious).

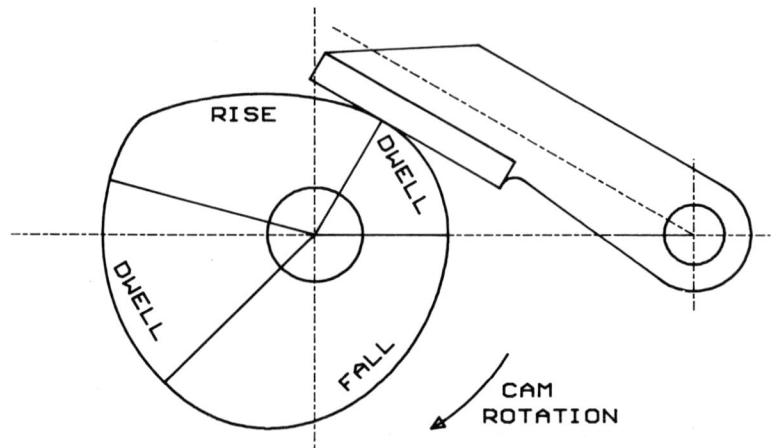

MOTION SPECIFICATION
120° × 15° CYCLOIDAL RISE
60° DWELL
120° × 15° CYCLOIDAL FALL
60° DWELL

Fig. 7.15   Asymmetrical profile produces symmetrical motion

CHAPTER 8

Cam Geometry, Cylindrical Cams and Globoidal Cams

CYLINDRICAL CAM WITH TRANSLATING ROLLER FOLLOWER

In this type of mechanism the cam track is wrapped around the surface of a cylinder, and is best visualised by developing it (unwrapping it) into a flat plane. It is feasible for the linear path of the follower roller not to be parallel to the cam axis, but this is so rare that it is not dealt with here. However, the case of a parallel, but offset, follower path is included.

Figure 8.1 shows, on the left, the more common case of a tapered roller

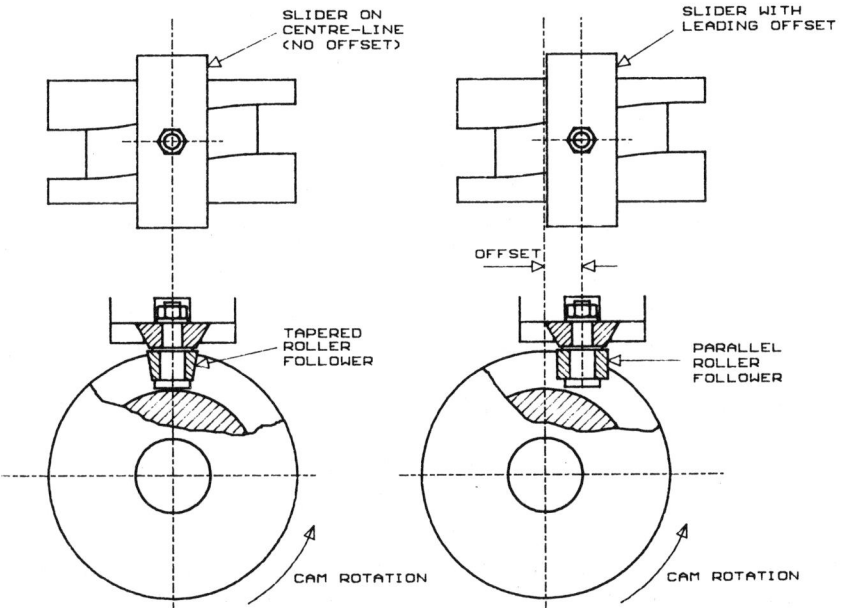

Fig. 8.1   Cylindrical cam with translating follower

follower with no offset, and on the right an offset cylindrical roller follower. A tapered (i.e. conical) roller is often used with a cylindrical cam to minimise the surface slip (skidding) due to differential track wall speed. The part of the track wall near the outside of the cam must move at a higher speed than the part near the bottom of the track, being at a greater distance from the axis. With a cylindrical roller all parts of its surface must move at the same speed and cannot match the cam track surface speed at all points. Pure rolling is not possible, therefore, with a parallel roller. If a tapered roller is used with its cone apex at the cam axis, then pure rolling can occur during the dwell periods, and nearly pure rolling during the motion periods.

For this reason tapered rollers are preferred in this class of mechanism, and they have the additional practical advantage that clearance (or preload) between cam track and follower can be adjusted easily by moving the roller nearer to or further from the cam axis. However, it must be said that it is economically advantageous to use commercial cylindrical roller followers, particularly those with high load capacity needle bearings, if the track lubrication is adequate to prevent undue wear through surface skidding.

The offset follower shown on the right of Fig. 8.1 is an unusual case and is exaggerated to illustrate its disadvantages. The roller could, of course, be conical to minimise skidding as described above, but skidding occurs anyway because the direction of motion of the cam track surface is not normal to the roller axis: pure rolling is impossible. Another effect of the skewed surface motion is to apply a force on the roller in an axial direction imposing more friction and wear on the mechanism. The other major disadvantage is that the roller does not have a full line of contact unless the track groove is made deeper and the roller put further into mesh than shown, and then the outside of the cam would be in danger of fouling the slide mechanism. Nevertheless, for the sake of completeness the offset follower will be dealt with, and the in-line follower taken as a special case (with zero offset).

As before, the radius of curvature of the pitch curve is calculated and the radius of curvature of the track surface derived from it by adding or subtracting the roller radius. There is, however, a different pitch curve for every position along the axis of the roller. Unlike a disc cam, then, the radius of curvature varies across the track face, being smaller nearer the cam axis. The cam can be regarded as a number of concentric cylinders,

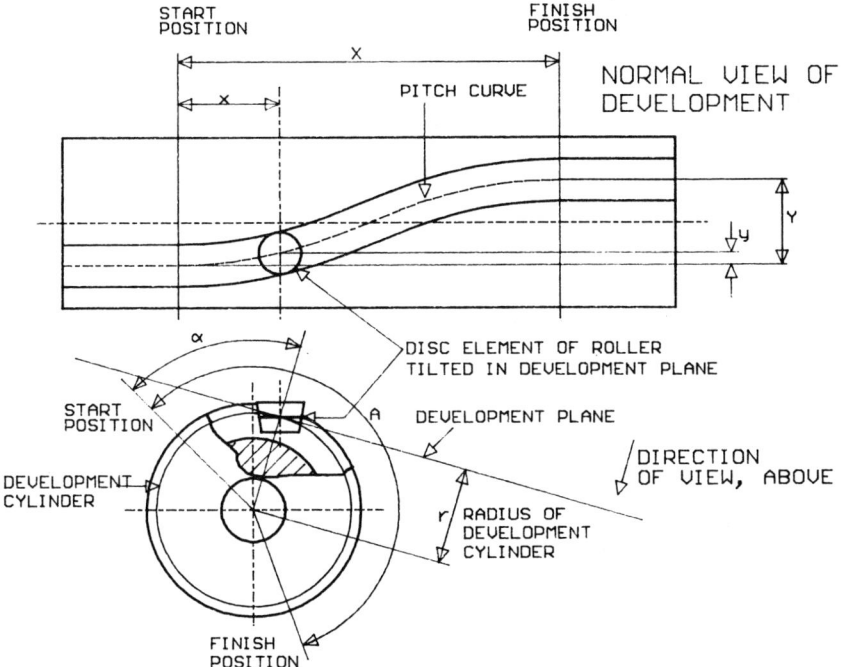

Fig. 8.2   Development of cylindrical cam with translating follower

each of which can be unwrapped to produce a flat development containing its own pitch curve as shown in Fig. 8.2.

The developed profile is the same as a translating plane cam (flat plate) of appropriate scale, so the radius of curvature and pressure angle of the pitch curve can be found from Equations 7.1 and 7.2 repeated here:

$$R_c = \frac{[1 + (y')^2]^{1.5}}{y''}$$ Eq. (8.1)

$$\phi = \arctan y'$$ Eq. (8.2)

where:

$$x = \text{linear input displacement} = r.\alpha$$ Eq. (8.3)

$$X = \text{linear input stroke} = r.A$$ Eq. (8.4)

α = cam rotation from the start of the motion = $A.u$ Eq. (8.5)

A = input stroke (cam rotation) in radians

y = output displacement = $Y.w$ Eq. (8.6)

$$y' = f(x) = \frac{Y}{X} \cdot w'$$ Eq. (8.7)

$$y'' = f''(x) = \frac{Y}{X^2} \cdot w''$$ Eq. (8.8)

Y = linear output stroke

$w = f(u)$ = normalised cam law displacement

$w' = f'(u)$ = normalised cam law velocity

$w'' = f''(u)$ = normalised cam law acceleration

A positive value of R_c represents a 'u'-shaped curve in the illustration, so the radius of curvature of the lower track wall is found by adding the roller radius, and that of the upper wall by subtracting the roller radius.

This simple analysis is not strictly true. The elemental disc of an offset roller is tilted in the development plane and is therefore seen as an ellipse, not a circle. Rather less than the roller radius should therefore be added or subtracted to get the radius of curvature of the track walls. Since the offset is likely to be small, the error is not significant for all practical purposes. There is no such error in the case of zero offset.

Except in the dwell periods, the slope of the curve causes the point of contact at the track wall to be offset from the centre-line of the follower path and consequently on an elemental disc that is somewhat nearer the cam axis and tilted relative to the developed plane because of the curvature of the surface of the development cylinder. This tilt has the same effect as already described, but if the roller is a tapered one the elemental disc also has a reduced radius. The effect is not very significant and can be ignored if the roller diameter is small compared with the cam diameter.

The cone angle of a tapered roller gives rise to a further aberration in that the curved *surface* of the track wall is not normal to the development plane (this is most obvious in the dwell period when the track wall is part of a flat disc if the roller is cylindrical, but is part of a cone if the roller is tapered).

The calculation of a true surface curvature for the track wall is thus very complex, but in practice it is found that the simple treatment given above is adequate for most purposes. However, the radius of curvature of the *pitch curve* is not affected by either the tilt of an offset follower or the development cylinder curvature or the cone angle of the roller and the simple equations can be regarded as exact.

It is apparent that the amount of offset does not enter into the above equations, but is only used in the derivation of the development radius, *r*.

Example 8.1

A cylindrical cam, shown in Fig. 8.3, has an outside diameter of 100 mm and drives a reciprocating slider backwards and forwards a distance of 50 mm with a Cycloidal motion. The timing is 150° forward, 30° dwell and

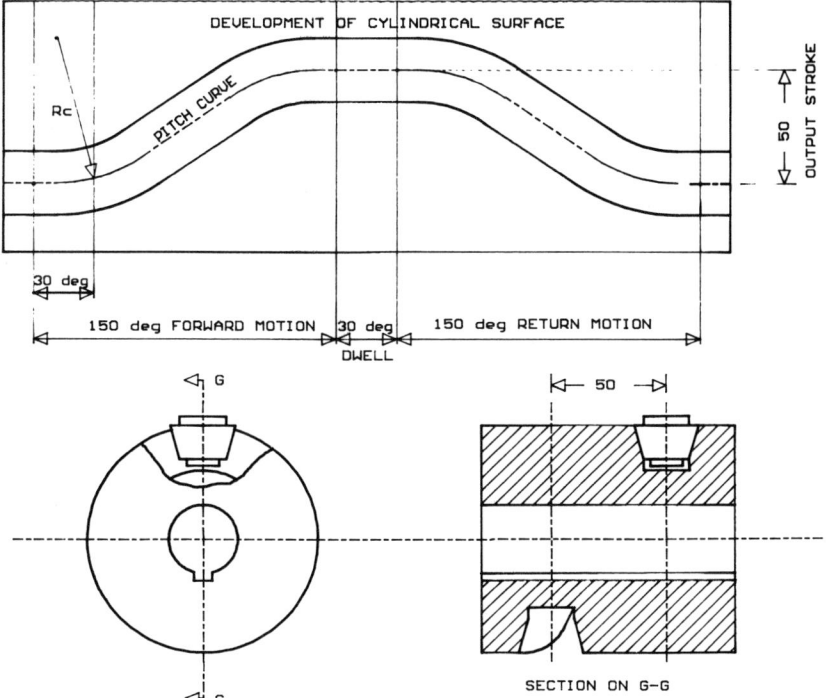

Fig. 8.3   Example 8.1

150° return. The slider uses a tapered roller follower engaged full depth in an enclosed track. The large end of the roller is 28 mm diameter, the small end is 20 mm diameter and it is 15 mm long. Find the radius of curvature of the track walls and the pressure angle *at both ends of the roller* 30° after the start of the motion.

From the information given we can see that

$A = 150° = 2.617994$ rad

$\alpha = 30° = 0.523599$ rad

$Y = 50$ mm

First calculate the radius of curvature at the large end of the roller, which is given to be flush with the outside surface of the cylinder, so we get a development radius of

$r = 50$ mm and therefore

$x = 50 \times 0.523599 = 26.180$ mm   Eq. (8.3)

$X = 50 \times 2.617994 = 130.8997$ mm   Eq. (8.4)

The normalised input displacement $= u = \alpha/A$ Eq. (8.5)
$u = 30°/150° = 0.2$

From the Cycloidal factors table in Appendix B we read at Step 24 (out of 120):

$u = 0.20000 \qquad w = 0.04863 \qquad w' = 0.69098 \qquad w'' = 5.97566$

Therefore

$y = 50 \times 0.04863 = 2.4315$ mm   Eq. (8.6)

$y' = \dfrac{50}{130.899} \times 0.69098 = 0.263936$ Eq. (8.7)

$y'' = \dfrac{50}{130.899^2} \times 5.97566 = 0.017437 \text{ mm}^{-1}$ Eq. (8.8)

and

$R_c = \dfrac{(1 + 0.26396^2)^{1.5}}{0.017437} = 63.446$ mm   Eq. (8.10)

This is the radius of curvature of the pitch curve. The track walls have radii of curvature of

$$63.446 - 14 = 49.446 \text{ mm convex}$$

and

$$63.446 + 14 = 77.446 \text{ mm concave}$$

The pressure angle at this point is

$$\phi = \arctan 0.263936 = 14.785° \qquad \text{Eq. (8.2)}$$

Now calculate the radius of curvature at the small end of the roller.

The development cylinder is 15 mm below the outer surface of the cam at a radius of

$$r = 50 - 15 = 35 \text{ mm}$$

The quantities that are affected by the change of r are the *linear* input stroke X, input displacement x, geometric velocity y' and geometric acceleration y'':

$$x = 35 \times 0.523599 = 18.326 \text{ mm}$$

$$X = 35 \times 2.617994 = 91.62979 \text{ mm}$$

$$y' = \frac{50}{91.62979} \times 0.69098 = 0.37705$$

$$y'' = \frac{50}{91.62979^2} \times 5.97566 = 0.035586 \text{ mm}^{-1}$$

$$R_c = \frac{(1 + 0.37705^2)^{1.5}}{0.035586} = 34.301 \text{ mm}$$

The roller radius is now only 10 mm, so the track radii of curvature at this position are:

$$34.301 - 10 = 24.301 \text{ mm convex}$$

and

$$34.301 + 10 = 24.301 \text{ mm concave}$$

The pressure angle here is

$$\phi = \arctan 0.37705 = 20.659°$$

It can be seen that the radius of curvature of the track wall at the 'root' of the track (i.e. at the 'tip' of the roller) is considerably smaller, and the pressure angle considerably larger, than at the outside diameter of the cam. This is invariably so, and it is therefore best to calculate for this position at an early stage in the cam design to avoid having to re-design because of interference, etc.

CYLINDRICAL CAM WITH SWINGING ROLLER FOLLOWER

This is similar to the cylindrical cam with translating follower analysed above, in that the track profile can be developed in a flat plane and the radius of curvature calculated from Equation 8.10. All the comments about roller offset equally apply to this case, and it should be noted that there is *always* a varying offset with a swinging follower arm. It is important to keep the offset to a minimum by choosing the length of the follower arm and the pivot-to-cam centres distance carefully in relation to the output stroke.

If the follower arm length (pivot to roller centre) is greater than the pivot-to-cam centres distance the cam axis plane intersects the arcuate path of the roller centre in a chord, and the maximum distance from the chord to the arc is called 'overthrow'. Overthrow is the follower arm length minus the pivot-to-cam centres distance. If possible one should choose an overthrow that approximately equalises the maximum positive and maximum negative offsets that occur during the stroke, so that neither of them is excessive.

An important datum in the geometry of this mechanism is the plane that contains the follower pivot axis and is normal to the cam axis: this is called the 'datum plane' (see Fig. 8.4). In particular, the angular displacement of the follower arm is measured from this plane when dealing with a known angular displacement function, in order to simplify the equations.

In the unusual event that the profile is specified as a Cartesian function, $y = f(x)$, in the development plane then the geometry is exactly the same as for a plane translating cam with translating follower, but note that x is *not* proportional to the angular displacement of the cam. It is far more common for the motion to be be expressed as an angular function, $\gamma = f(\alpha)$, and usually in the form of a standard cam law. The equations below

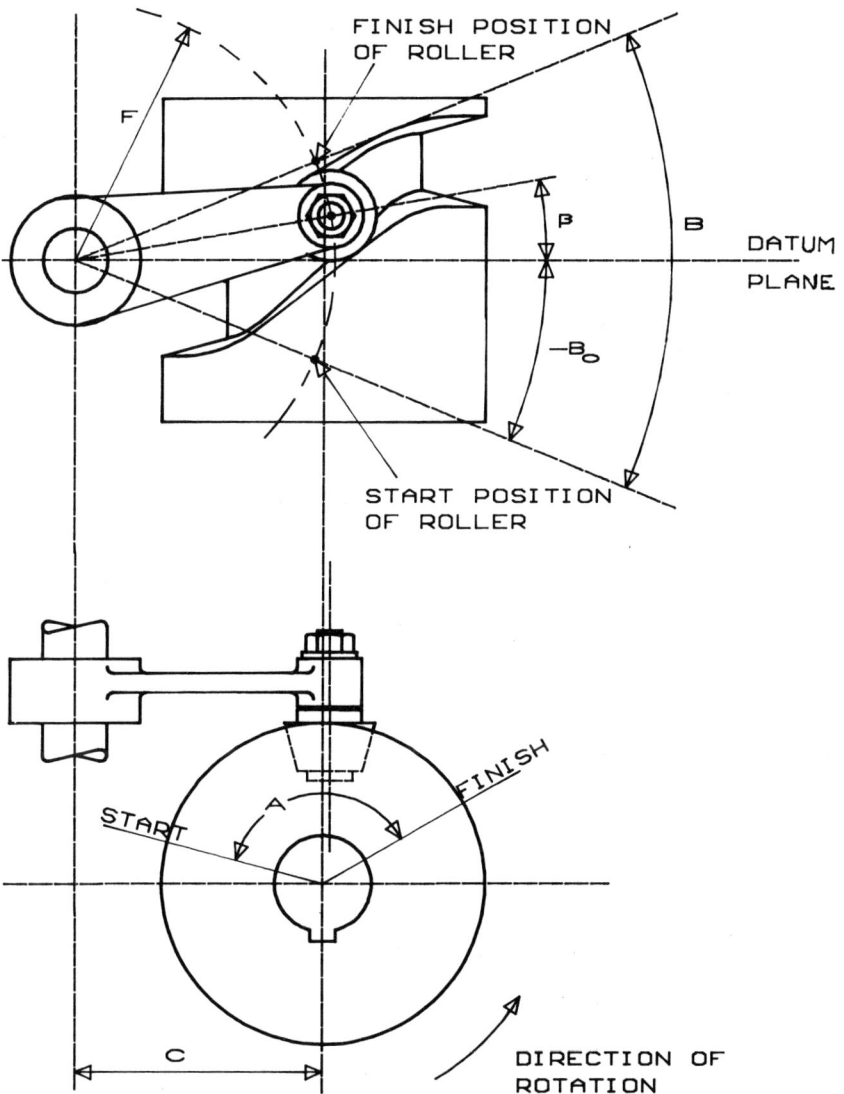

Fig. 8.4   Cylindrical cam with swinging roller follower

make use of the normalised cam law function, $w = f(u)$, so that standard cam laws can be used conveniently, but any function, $\gamma = f(\alpha)$, can be converted to a normalised form by simply substituting

$$u = \alpha/A \quad \text{and} \quad w = \gamma/B; \quad w' = A\gamma'/B; \quad w'' = A^2\gamma''/B$$

Eq. (8.9)

The radius of curvature of the pitch curve is as before:

$$R_c = \frac{[1 + (y')^2]^{1.5}}{y''} \qquad \text{Eq. (8.10)}$$

but now:

$$\alpha = A.u \qquad \text{Eq. (8.11)}$$

$$\beta = B.w + B_0 \qquad \text{Eq. (8.12)}$$

$$\beta' = \frac{d\beta}{d\alpha} = \frac{B}{A}.w' \qquad \text{Eq. (8.13)}$$

$$\beta'' = \frac{d^2\beta}{d\alpha^2} = \frac{B}{A^2}.w'' \qquad \text{Eq. (8.14)}$$

$$x = r.\alpha + F.\cos\beta - F.\cos B_0 \qquad \text{Eq. (8.15)}$$

$$y = F.\sin\beta \qquad \text{Eq. (8.16)}$$

$$y' = \frac{\beta'.\cos\beta}{(r/F) - \beta'.\sin\beta} \qquad \text{Eq. (8.17)}$$

$$y'' = \frac{(r/F).\beta''.\cos\beta - (r/F)(\beta')^2.\sin\beta + (\beta')^3}{F.[(r/F) - \beta'.\sin\beta]^3} \qquad \text{Eq. (8.18)}$$

where:

α = angular input displacement.
β = angular output displacement *from datum plane*.
A = angular input stroke.
B = angular output stroke.
B_0 = angle from datum plane to start of follower motion (measured in direction of motion). If the datum plane is encountered after the start of motion then B_0 is negative.
C = perpendicular distance from cam axis to follower pivot axis.

F = length of follower arm.

r = radius of development cylinder.

R_c = radius of curvature of pitch curve.

$w = f(u)$, the normalised cam law function.

w', w'' = the first and second derivatives of w, the normalised velocity and acceleration.

x, y = Cartesian co-ordinates of developed track.

y', y'' = first and second derivatives of y with respect to x.

Example 8.2

An oscillating mechanism is operated by a lever 80 mm long with a roller at the end engaging a groove in a cylindrical cam of 100 mm outside diameter. The lever swings 30° forwards and backwards with a MSC.50 motion symmetrically disposed about a datum plane which is normal to the cam axis. The timing of the oscillating cycle is: 120° forwards, 90° dwell, 120° backwards, 30° dwell. Find the radius of curvature of the pitch curve at the outside of the cam 105° after the start of the motion.

The mechanism is similar to that shown in Fig. 8.4 and the development of the track profile is shown in Fig. 8.5 with the follower arm in the

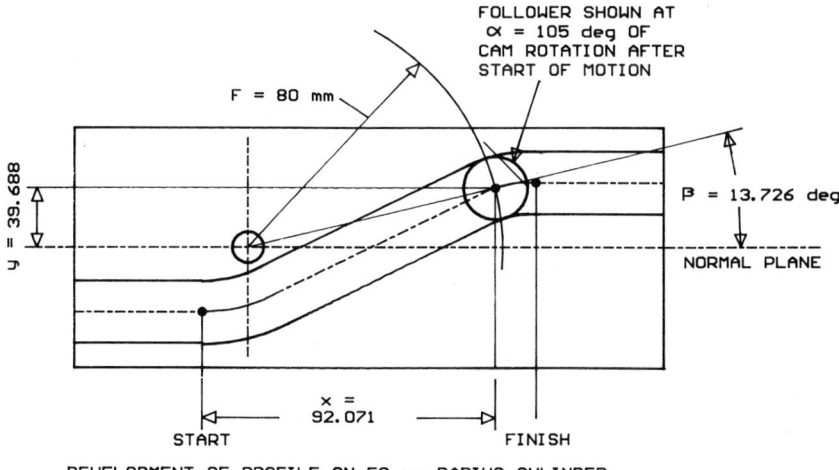

DEVELOPMENT OF PROFILE ON 50 mm RADIUS CYLINDER
A = 120 deg, B = 30 deg, Bo = -15 deg, MSC50 CAM LAW

Fig. 8.5 Example 8.2

specified position. From the specification we see that the cam law is
MSC.50 and that:

$$A = 120° = 2.094395 \text{ rad}$$

$$B = 30° = 0.523599 \text{ rad}$$

$$B_0 = -15° = 0.261799 \text{ rad}$$

$$F = 80 \text{ mm}$$

$$r = 50 \text{ mm}$$

C is not given, but to minimise offset its best value would be halfway
between 80 and 80 cos 15° = 78.637 mm, say 78.5 mm (an overthrow of
1.5 mm). First we find the normalised input displacement so that we can
use the MSC.50 factors table in Appendix B for the output displacement,
etc.:

$$u = 105/120 = 0.875 \text{ and from the table we read:}$$

$$w = 0.95754 \qquad w' = 0.79704 \qquad w'' = -6.93919$$

Therefore

$$\beta = 30° \times 0.95754 - 15° = 13.7262° = 0.23957 \text{ rad} \qquad \text{Eq. (8.12)}$$

$$\beta' = \frac{0.523599}{2.094395} \times 0.79704 = 0.19926 \qquad \text{Eq. (8.13)}$$

$$\beta'' = \frac{0.523599}{2.094395^2} \times (-6.93919) = -0.828305 \text{ rad}^{-1} \qquad \text{Eq. (8.14)}$$

$$r/F = 50/80 = 0.625$$

$$y' = \frac{0.19926 \times \cos 13.7262°}{0.625 - 0.19266 \times \sin 13.7262°} = 0.335058 \qquad \text{Eq. (8.17)}$$

$$y'' = \frac{\begin{array}{c} -0.625 \times 0.828305 \cos 13.7262° \\ -0.625 \times 0.19926^2 \sin 13.7262° + 0.19926^3 \end{array}}{80 \times (0.625 - 0.19926 \sin 13.7262°)^3}$$

$$= -0.032471 \qquad \text{Eq. (8.18)}$$

$$R_c = \frac{(1 + 0.335058^2)^{1.5}}{-0.032471} = -36.126 \text{ mm} \qquad \text{Eq. (8.1)}$$

This result is negative, indicating that the curve is 'n'-shaped at this point, as expected. As in previous examples, the radius of curvature of the track wall is found by adding or subtracting the roller radius.

Although the developed profile looks symmetrical in Fig. 8.5 it is not, even though the motion and layout *are* symmetrical. This arises from the fact that the direction of the follower roller motion is inclined to the cam axis by a varying amount throughout the stroke. The degree of asymmetry is small in this example because the offset is small, but when there is a larger offset it can be significant. In the case of a blade cam (two rollers on the follower arm straddling a blade on the cam), or an indexing cam (multiple blades) there can be a considerable difference in thickness of blade from one end to the other purely because of the asymmetry of the profile. Indeed, one constraint in the design of a cylindrical indexing cam is the need for adequate thickness of the blades at their thin ends. The blade asymmetry is clearly seen in a typical indexing mechanism shown in Fig. 8.6.

Such a mechanism is a special case of a swinging follower mechanism. The follower consists of a turret (follower wheel) with a number of rollers spaced around a pitch circle. The cam geometry conforms to the same equations, but instead of the follower rollers returning to their starting position at the end of one motion cycle they continue in the same direction and are replaced by the subsequent rollers.

Rollers pass from one side of the cam to the other successively, with each cycle imparting an indexing motion to the turret. In the simplest kind of mechanism as shown in Fig. 8.6, only one roller passes through the cam in each cycle. This is known as a one-pass or single-pass cam and it has only one track, which is open at both ends, and only one blade. Multiple pass mechanisms, however, are quite common, the number of blades being the same as the number of passes.

When the rollers are equally spaced on the turret pitch circle, the mechanism is analogous to a pair of toothed gears, the blades and rollers corresponding to the gear teeth. The number of cycles of the cam to produce one revolution of the turret is the ratio of the number of rollers to the number of passes (or blades), in the same way as a gear ratio is the ratio of the number of teeth on one gear to the number of teeth on the other.

It is also possible to have a turret with unequally spaced rollers. Then the rollers are arranged in sets of two, the space between each pair being less than the space between adjacent pairs, for example twelve rollers

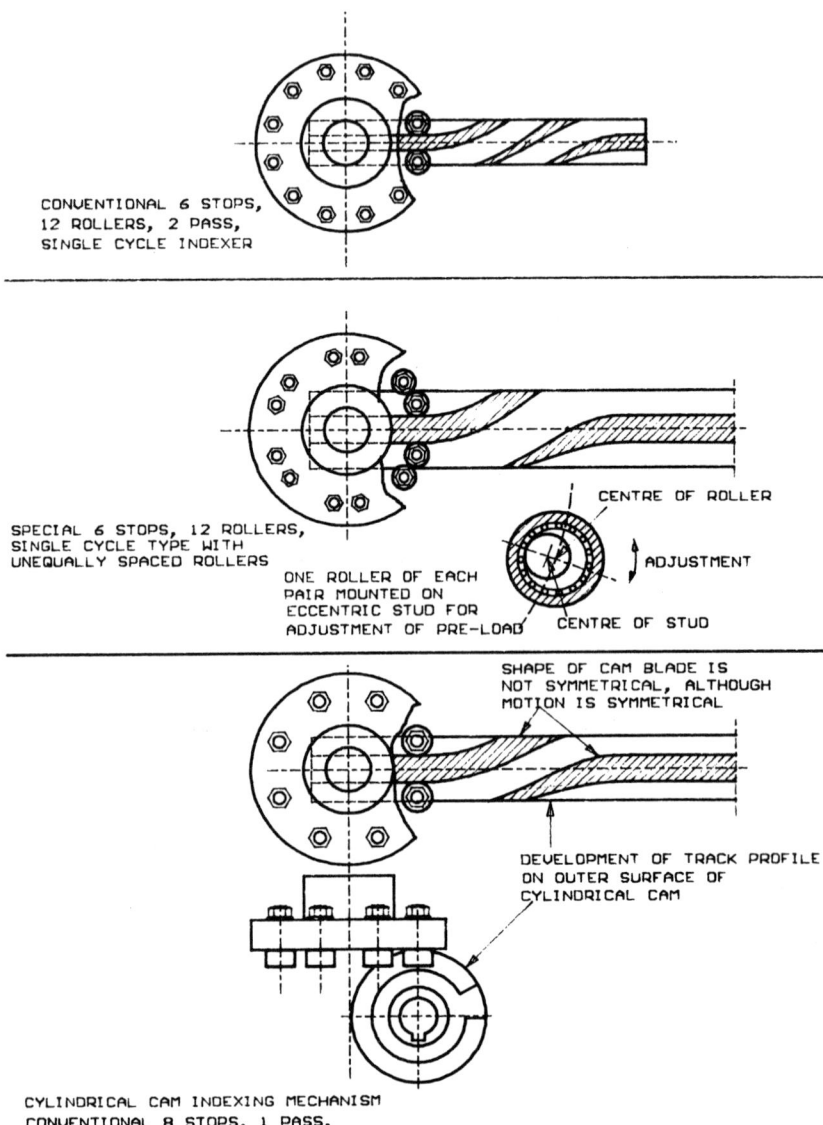

CONVENTIONAL 6 STOPS,
12 ROLLERS, 2 PASS,
SINGLE CYCLE INDEXER

SPECIAL 6 STOPS, 12 ROLLERS,
SINGLE CYCLE TYPE WITH
UNEQUALLY SPACED ROLLERS

ONE ROLLER OF EACH
PAIR MOUNTED ON
ECCENTRIC STUD FOR
ADJUSTMENT OF PRE-LOAD

CENTRE OF ROLLER

ADJUSTMENT

CENTRE OF STUD

SHAPE OF CAM BLADE IS
NOT SYMMETRICAL, ALTHOUGH
MOTION IS SYMMETRICAL

DEVELOPMENT OF TRACK PROFILE
ON OUTER SURFACE OF
CYLINDRICAL CAM

CYLINDRICAL CAM INDEXING MECHANISM
CONVENTIONAL 8 STOPS, 1 PASS,
SINGLE CYCLE TYPE

Fig. 8.6 Cylinder cam indexers

equally spaced would be at 30° pitch, but unequally spaced rollers might be at alternate pitches of 20° and 40°. This device, illustrated in the central diagram of Fig. 8.6, can be used to facilitate adjustable preloading between the rollers and the cam blade and to avoid reversal of roller rotation. In this case *a pair* of rollers engage one blade and are analogous to a single gear tooth. The single-bladed cam appears to be a single-pass cam, but it passes two rollers through in each cycle. The number of cycles of the cam to produce one revolution of the turret is *not* the same as the ratio of rollers to blades. The analogy with toothed gearing breaks down.

A further complication arises in the description of an indexing mechanism, because it is also possible to have more than one motion cycle in one revolution of the cam: a multi-cycle cam. Indexing motion cycles usually consist of repetitions of a motion period followed by a dwell period (motion – dwell – motion – dwell – and so on), so that the turret stops a number of times during its revolution. Hence the number of indexing cycles per revolution of the turret is known as the *number of stops*, S, and is used as an unambiguous designation. The output stroke, B, of the mechanism is $360/S$ degrees or $2\pi/S$ radians. For a conventional mechanism (as first described, with equally spaced rollers and a single-cycle cam) the number of stops is the number of rollers divided by the number of passes. The lower diagram of Fig. 8.6 shows a conventional 8 rollers, *8 stops, 1 pass* single cycle mechanism, and the upper diagram shows a conventional 12 rollers, 6 stops, 2 pass, single cycle mechanism.

CONCAVE GLOBOIDAL CAM WITH SWINGING FOLLOWER

A true globoidal cam, as shown in Fig. 1.1 diagram 15, is not often used because of the rather awkward shape of the follower arm. The concave globoidal cam, Fig. 1.1 diagram 16, which is popular both as an oscillating mechanism and especially as an indexing mechanism, is the version we shall deal with here. Tapered roller followers can be used in such mechanisms to minimise surface skidding, but they offer no advantage in preload adjustment, since this is achieved by adjustment of the cam-to-pivot centres distance equally well with parallel roller(s). It is far more usual, therefore, to use parallel rollers with high capacity needle bearings. Unless otherwise specified a 'globoidal cam' is taken to mean a *concave* globoidal cam, because it is so common.

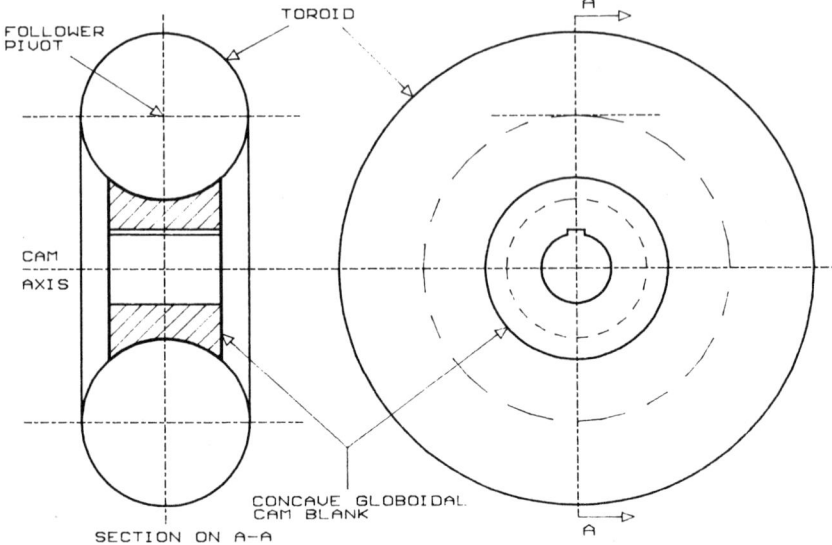

Fig. 8.7   Concave globoidal cam surface as part of a toroid

The surface in which the cam profile is traced is the inner part of a toroid (a 'ring doughnut') which is generated by the circle of revolution of the follower as its pivot rotates around the cam axis, as shown in Fig. 8.7. It is impossible to 'unwrap' the cam surface to present the profile in a flat plane with true measurements. Rather as a flat map of the world cannot be true to scale all over, so any flat representation of a globoidal cam profile must be distorted.

The angular displacement of the follower roller, β, is taken from the 'datum plane', which is the plane normal to the cam axis that contains the follower pivot axis. As with the cylindrical cam, the sign of the initial displacement from the datum plane, B_0, is important. If, when moving in the direction of follower motion, the start position is encountered before the datum plane then B_0 is negative; if encountered after the datum plane then B_0 is positive.

However, for illustration purposes a small region can be projected, and its geometry analysed as if it were a true development. For clarity Fig. 8.8. shows a single roller on the projected view, although two rollers straddle a blade. This is a common kind of oscillating mechanism, and a

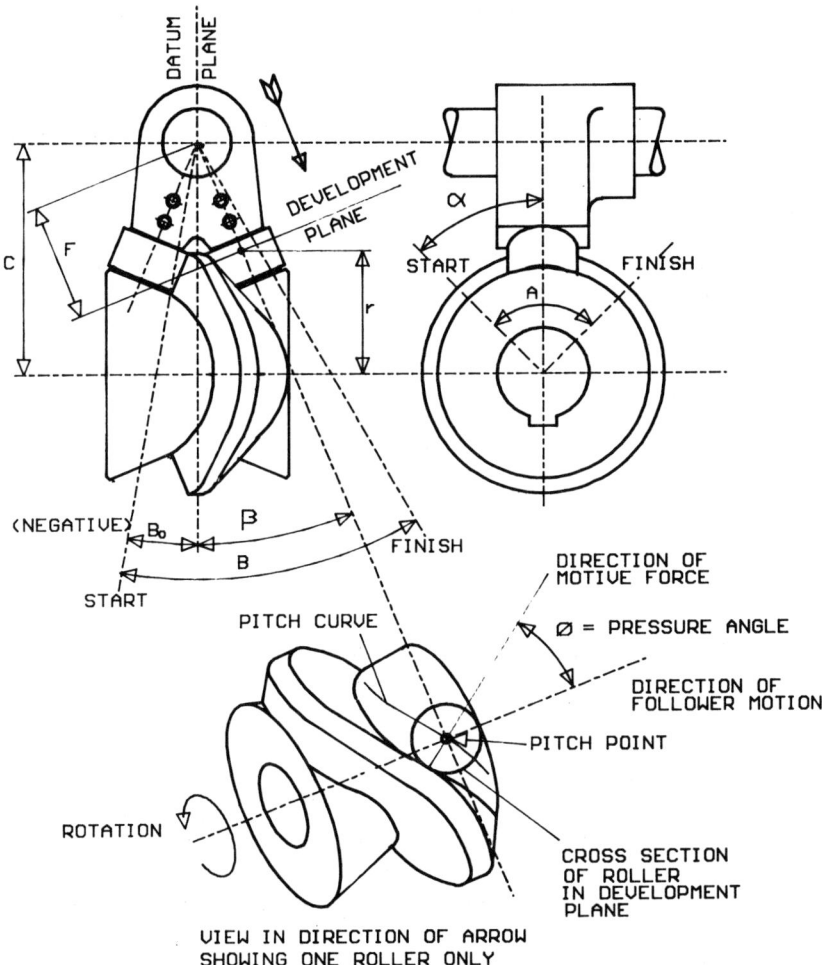

Fig. 8.8   Concave globoidal cam with swinging roller followers

similar diagram can be drawn for an indexing mechanism with a multi-roller turret.

The 'development plane' is anywhere along the length of the roller, normal to its axis. The curvature at a particular point in the motion differs according to the position of the plane along the roller axis, in the same way as it does for a cylindrical cam explained earlier. The geometry is in many ways similar to the cylindrical cam geometry, but the use of Cartesian co-ordinates (x, y) has little meaning in this case, and equations are given in terms of the angular displacement function, $\beta = f(\alpha)$, and its derivatives. Again, it is convenient to use a motion which is defined as a normalised function, but any angular displacement function can be normalised (see Equation 8.9).

It is convenient to calculate the pressure angle of the curve before calculating its radius of curvature:

$$\phi = \arctan\left(\frac{F}{r} \cdot \beta'\right) \qquad \text{Eq. (8.19)}$$

The *curvature* of the pitch curve (i.e. the reciprocal of its radius of curvature) is given by the following equation:

$$\frac{1}{R_c} = \frac{\cos\phi}{r}\left[\frac{F}{r} \cdot \beta'' \cdot \cos^2\phi - \sin\beta \cdot (1 + \sin^2\phi)\right] \qquad \text{Eq. (8.20)}$$

where:

α = angular input displacement (cam rotation).
β = angular output displacement *from datum plane.*
β' and β'' are found from Equations (8.13) and (8.14).
ϕ = pressure angle.
A = angular input stroke.
B = angular output stroke.
B_0 = angle from datum plane to start of follower motion (measured in direction of motion). If the datum plane is encountered after the start of motion then B_0 is negative.
C = perpendicular distance from cam axis to follower pivot axis.
F = distance from follower pivot to the pitch point (the centre of the roller in the development plane).
R_c = radius of curvature of pitch curve.
r = perpendicular distance from cam axis to the pitch point, found from:
$$r = C - F\cos\beta \qquad \text{Eq. (8.21)}$$

The *radius* of curvature of the pitch curve is of course the reciprocal of Equation 8.20, and the radius of curvature of the cam track profile is found by adding or subtracting the roller radius from it, as in previous cases. If the result of this equation is positive the curve is *u*-shaped and if negative it is *n*-shaped, similar to the plane curve convention.

Concave globoidal cams are used extensively for indexing mechanisms, in which the follower is in the form of a turret (follower wheel) with a number of rollers equally spaced on a pitch circle. These mechanisms can have the same variations of design as the cylindrical cam indexing mechanisms, described on pages 134–141, but of course the cam geometry is different and the questions of follower offset and 'overthrow' do not arise. The above equations apply to every roller, but it must be understood that any particular roller often engages a different part of the cam track in successive indexing cycles. For each cycle the roller would then have a different starting position, although its input and output strokes and its cam law must be the same.

Example 8.3

A globoidal cam indexing mechanism has 6 stops and a 150° index period with cycloidal motion. The centres distance between the input and output shafts is 125 mm and the turret has 6 rollers 30 mm diameter by 14 mm long equally spaced on a 62.5 mm pitch circle. The track is cut with 3 mm clearance beyond the end of the rollers. Find the pressure angles and radii of curvature of the cam track walls at the mid-point of the line of contact and at the root of the blade, after 45° of cam rotation from the start of the motion. The mechanism is illustrated in Fig. 8.9.

Since this is a 6-roller, 6-stops mechanism it has a single-pass cam. Each roller of the turret behaves in exactly the same way as every other roller and there is only one cam track, which occupies more than one revolution of the cam. However, two adjacent rollers engage different parts of the track at the same time, so we must do two sets of calculations, one set for each roller.

Designate the two rollers 1 and 2: see Figs 8.9 and 8.8. The starting position for roller 1 (at the dwell) is 30° *after* the datum plane and for roller 2 it is 30° *before* the datum plane. We are given:

$$A = 150° = 2.61799 \text{ rad} \qquad B = 60° = 1.047197 \text{ rad} \qquad C = 125 \text{ mm}$$

The displacement, velocity and acceleration of the turret after 45° of cam rotation are found using the Cycloidal factors table in Appendix B:

$$u = 45°/150° = 36/120 = 0.3$$

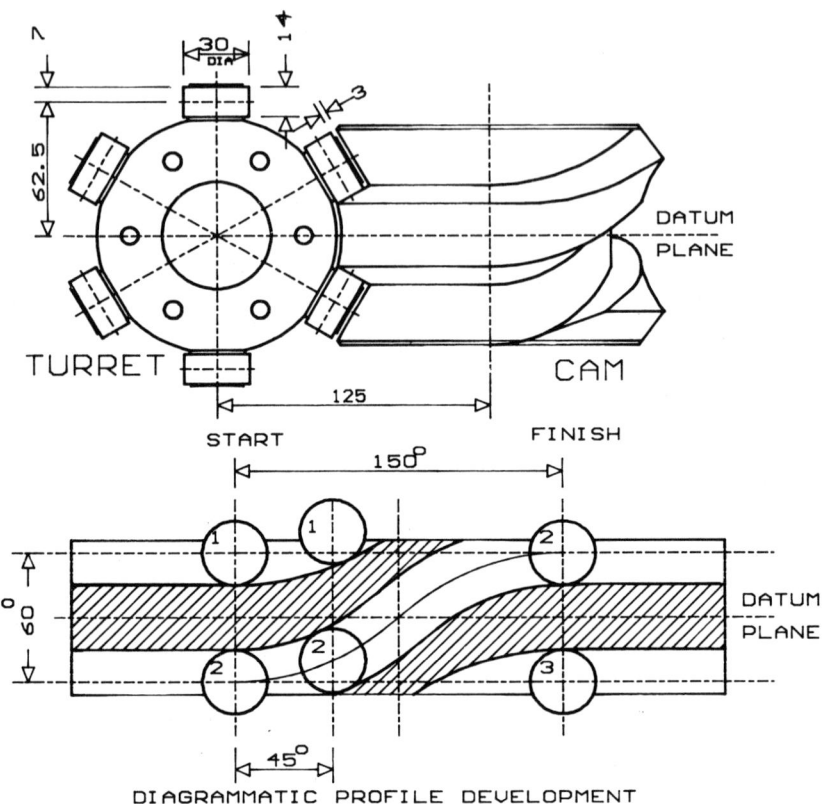

Fig. 8.9   Example 8.3: globoidal cam indexing mechanism

This is found at step 36 in the table, from which we read:

$$w = 0.14863 \qquad w' = 1.30902 \qquad w'' = 5.97566$$

Mid-point of line of contact for roller 1

$$B_0 = +30°$$

$$\beta = 30° + B.w = 30° + 60° \times 0.14863$$

$$= 38.9178° = 0.679244 \text{ rad} \qquad \text{Eq. (8.12)}$$

$$\beta' = \frac{60°}{150°} \times 1.30902 = 0.523608 \qquad \text{Eq. (8.13)}$$

(Note that in this equation the angles can both be in degrees because the result is dimensionless, but in the following equation they *must* be in radians for the result to be in the correct dimensions for subsequent use.)

$$\beta'' = \frac{B}{A^2} w'' = \frac{1.047197}{2.61799^2} \times 5.97566 = 0.913016 \ \text{rad}^{-1} \qquad \text{Eq. (8.14)}$$

First we must analyse the pitch curve before we can determine the track wall curve. The mid-point of the line of contact is in the same development plane as the pitch point at the turret pitch circle radius.

$F = 62.5$ mm, therefore

$$r = 125 - 62.5 \cos (38.9178°) = 76.372 \ \text{mm} \qquad \text{Eq. (8.21)}$$

$$\phi = \arctan \left(\frac{62.5}{76.372} \times 523608 \right) = 23.1952° = 0.404833 \ \text{rad}$$

$$\text{Eq. (8.19)}$$

$$\frac{1}{R_\text{c}} = \frac{\cos 23.1952°}{76.372} \left[\frac{62.5}{76.372} \times 0.913016 \times \cos^2 23.1952° \cdots \right.$$
$$\left. \cdots -\sin (38.9178°) \times (1 + \sin^2 23.1952°) \right]$$

$$= \frac{0.919168}{76.372} \times [0.631269 - 0.725659]$$

$$= -0.001136 \ \text{mm}^{-1} \qquad \text{Eq. (8.20)}$$

$$R_\text{c} = \frac{1}{-0.001136} = -880.3 \ \text{mm}$$

The negative result indicates that the curve is *n*-shaped in the diagram and therefore the track wall is convex. The approximate radius of curvature of the track wall here is

$880.3 - 15 = 865.3$ mm convex

Root of blade, roller 1

The only difference from the data in the previous calculation is that the pitch point is 10 mm farther from the turret axis (half the roller length, 7 mm, plus 3 mm clearance), which alters the following:

$F = 62.5 + 10 = 72.5$ mm

$r = 125 - 72.5 \cos 38.9178°$

$\phi = \arctan\left(\dfrac{72.5}{68.5915} \times 0.523608\right) = 28.9621°$

Substituting these values in the calculation gives:

$1/R_c = -0.00008558$, therefore $R_c = -11{,}685$ mm

The radius of curvature of the track wall at the blade is therefore 11,670 mm convex, which is virtually a flat surface.

Mid-point of line of contact, roller 2

Following the same procedure as for roller 1 we find:

$B_0 = -30°$

$\beta = -30° + 60° \times 0.14863 = -21.0822°$

$\quad\quad = -0.367953$ rad   Eq. (8.12)

$\beta' = 0.523608$ as before

$\beta'' = 0.913016$ rad^{-1} as before

$F = 62.5$ mm, therefore

$r = 125 - 62.5 \cos(-21.0822°) = 66.6834$ mm   Eq. (8.21)

$\phi = \arctan\left(\dfrac{62.5}{66.6834} \times 0.523608\right) = 26.1399°$

$\quad\quad = 0.456228$ rad   Eq. (8.19)

$$\frac{1}{R_c} = \frac{\cos 26.1399°}{66.6834}\left[\frac{62.5}{66.6834} \times 0.913016 \times \cos^2 26.1399° \cdots\right.$$
$$\left. \cdots -\sin(-21.0822°) \times (1 + \sin^2 26.1399°)\right]$$

$\quad\quad = 0.015067$ mm^{-1} Eq. (8.20)

$R_c = \dfrac{1}{0.015067} = 66.371$ mm

The curve is *u*-shaped and the approximate radii of curvature of the track walls here are

66.371 + 15 = 81.371 mm concave

and

66.373 − 15 = 51.371 mm convex

From the diagram we see that the first of these is at the upper surface of the right-hand part of the blade, and the second is at the lower surface of the left-hand part of the blade, opposite roller 1.

Root of blade, roller 2

As with roller 1 there is a change of the pitch point radius, F, with the consequent re-calculation of other dimensions:

$F = 72.5$ mm

$r = 57.353$ mm

$\phi = 33.5003°$

$R_c = 54.078$ mm

The track wall radii of curvature are 69.078 mm and 39.078 mm, the latter being the convex curvature at the root of the lower surface of the left-hand part of the blade, opposite roller 1.

These calculations are tedious, but as mentioned before, they can be easily built into a computer program so that all parts of the cam track surface can be quickly analysed. This is very desirable to facilitate the identification of critical factors, such as interference or undercutting, excessive pressure angle, etc.

CHAPTER 9

Cam Track Surface Stress

The ability of a cam and follower system to perform its function under load for an acceptable length of time is dependent on, among other things, the stress at the surface of the cam track at the point or line of contact with the follower. This is commonly known as the contact stress, and is a function of the contact force and the curvatures of the cam and follower surfaces. The contact force at any moment in the motion cycle can be estimated from the dynamic and static forces in the system and the pressure angle at the point of contact. Pressure angle and curvature are dealt with in Chapters 7 and 8 on cam geometry. The system forces are dealt with in following chapters, and here we shall consider the contact stress, assuming that the curvature and contact force are known.

The distribution of stress in the contact region is very complex. It is affected by the nature of the contacting materials, the kind of lubrication, if any, and the relative motion of the contacting surfaces. Accurate prediction of mechanism failure due to excessive contact stress is imposs- ible. Numerous tests of ostensibly identical mechanisms are necessary to establish a statistical probability of failure for the particular mechanism. Further tests on different mechanisms must then be done to establish the effects of design variations. The results of such tests could be used to predict the load carrying capacity of a range of mechanisms of similar design operating in similar conditions, but there would still be a consider- able margin of error. Such a procedure is clearly uneconomical for general application to cam and follower systems. The best we can do is to analyse an idealised model of the contact region to provide an estimate of contact stress that can be used as a fairly reliable guide to mechanism load capacity.

The investigations of Heinrich Hertz in 1880 produced a theory of the distribution of stress in two curved bodies in contact under pressure, which is still the basis of most of today's contact stress analysis. The maximum principal stress in the contact region according to his theories is known as the Hertzian stress, and is often used as a failure criterion.

When two curved surfaces are forced together they both compress and the surfaces are distorted so that the original point or line of contact is

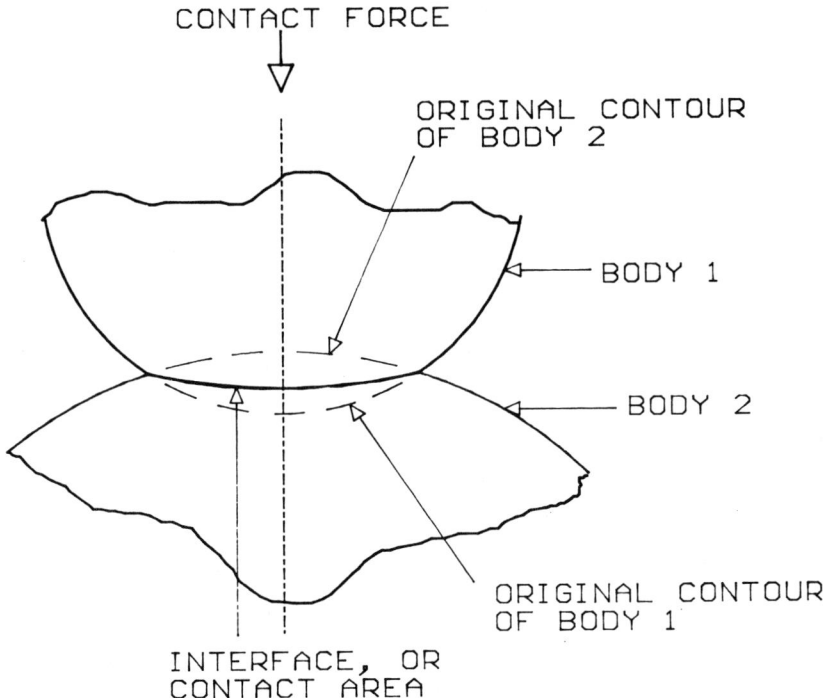

CONTACT FORCE

ORIGINAL CONTOUR
OF BODY 2

BODY 1

BODY 2

ORIGINAL CONTOUR
OF BODY 1

INTERFACE, OR
CONTACT AREA

Fig. 9.1   Surface stress contact area

expanded into an *area* of contact as shown in Fig. 9.1. At every part of the contact area the compressive stress in one body is opposed by an equal compressive stress in the other body. In other words the principal compressive stresses normal to the contact area are equal in magnitude and distribution for both bodies. That stress has a maximum value (the Hertzian stress) at the centre of the contact area and falls to zero at the edges of the area.

Although the principal contact stresses in both bodies are equal, the compressive strain (distortion) of each is dependent on the Young's modulus E (modulus of elasticity) of each material, and may be different for each body. It is assumed that the materials are homogeneous and elastic, like the metals normally used for cams.

The compressive strain in each body produces a lateral strain whose magnitude depends on the Poisson's ratio ν of the material, which may also be different for each body. For each body, then, there are three factors which affect the size and shape of the contact area and the magnitude and distribution of stress in the contact region:

curvature,
Young's modulus E
and Poisson's ratio v.

One of the bodies may have compound curvature (see Chapter 7), for example a sphere or a cambered roller. Unless the other body has a complementary curvature that matches exactly (e.g. a ball and socket), there is so-called 'point' contact. In fact, under the action of a contact force pressing the bodies together, the point expands to become an elliptical *area* of contact.

When the curvatures match exactly in one direction, but not in the other, such as in the case of a cylindrical roller on a conventional disc cam, there is so-called 'line' contact. The contact force in this case causes the line to expand to become a rectangular *area* of contact.

The effect of local distortion at the point or line of contact is illustrated in Fig. 9.1, and a diagrammatic presentation of the distribution of principal compressive stresses at the interface is shown in Fig. 9.2. These are the Hertzian stresses.

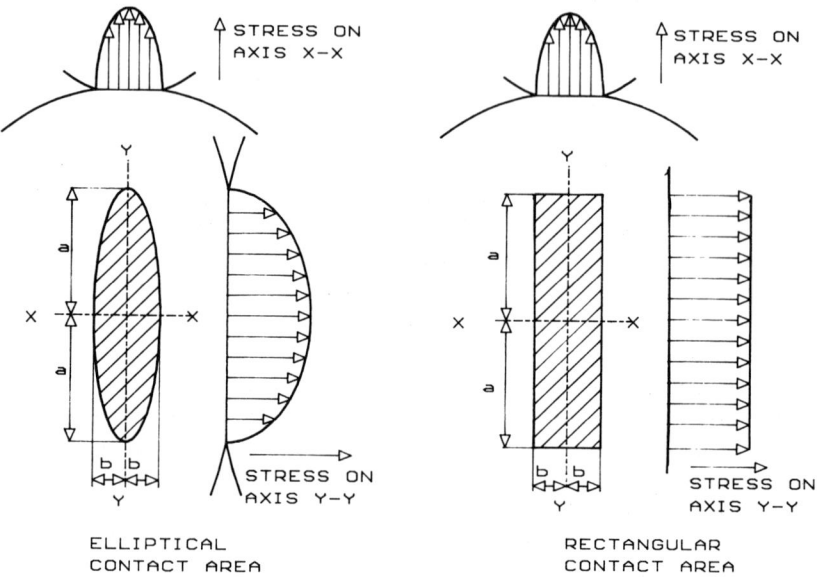

Fig. 9.2   Distribution of principal contact stresses

Hertzian theory deduces the major and minor axes of an elliptical contact area and the width of a rectangular contact area (its length is the length of the contact 'line', e.g. the width of a disc cam or the width of a cylindrical roller). These dimensions are of little practical interest in themselves, except in the case of surface-hardened materials, but they are used to derive the stress equations given below.

Of the two cases line contact (rectangular contact area) is less complex than point contact (elliptical contact area), and is probably the more common in industrial machinery cam systems. It will be dealt with first.

RECTANGULAR CONTACT AREA

The most important systems in this category are those with cylindrical roller followers in contact with non-cambered cam tracks. Such cams include disc cams, cylinder cams and globoidal cams, with open or enclosed tracks. In practice the length of the roller often does not match the width of the cam track: disc cams are often designed to be narrower than the roller, enclosed tracks are inevitably wider than the roller, and so on. The length of the contact area is the length of the 'line' of contact, allowing for overlap, corner radii on the roller, etc. (see Fig. 9.3). When

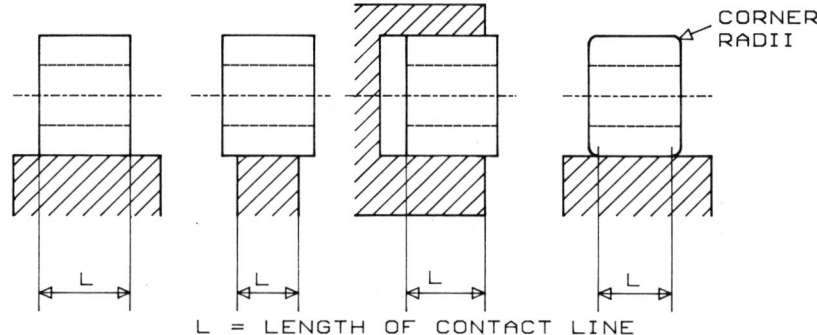

L = LENGTH OF CONTACT LINE

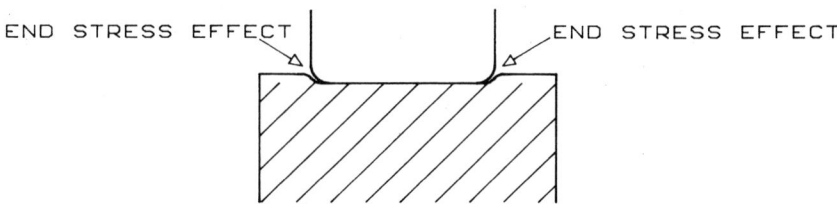

Fig. 9.3   Length of contact 'line' of roller on cam

the roller length and track width are different there is an 'edge stress' effect because the material in the overlapping body adjacent to the contact area is not free to deflect with the contact area material, as shown in Fig. 9.3. This can result in a maximum localised Hertzian stress up to 50% higher than the mean, whereas the nominal maximum is only 27% higher than the mean. To avoid this effect rollers are sometimes made with a slightly convex profile near the ends to distribute the stress more evenly.

For a given contact force, various combinations of curvature of the two bodies and various combinations of material properties can produce a contact area of the same size and shape, and with the same stress distribution, as if we substituted one curved body and one flat body, both of the same material. This enables us to use a single radius of curvature equivalent to that of the substitute curved body and a single elasticity value equivalent to that of the substitute material. These are called here the *equivalent* radius of curvature, R_e, and the *equivalent* elasticity, E_e, and are given by:

$$\frac{1}{R_e} = \frac{1}{R_{c1}} + \frac{1}{R_{c2}} \qquad\qquad \text{Eq. (9.1)}$$

$$\frac{1}{E_e} = \frac{1}{2}\left[\frac{1 - v_1^2}{E_1} + \frac{1 - v_2^2}{E_2}\right] \qquad\qquad \text{Eq. (9.2)}$$

where: R_e = equivalent radius of curvature
$\quad\quad\ R_{c1}$ = radius of curvature of body 1
$\quad\quad\ R_{c2}$ = radius of curvature of body 2
$\quad\quad\ E_e$ = equivalent elasticity
$\quad\quad\ E_1$ = Young's modulus of body 1
$\quad\quad\ E_2$ = Young's modulus of body 2
$\quad\quad\ v_1$ = Poisson's ratio of body 1
$\quad\quad\ v_2$ = Poisson's ratio of body 2

For simplicity let us consider a truly rectangular contact area, whose length is L, the length of the contact 'line', and whose width is $2b$, see Fig. 9.2. The half-width, according to Hertz is:

$$b = \left[\frac{8}{\pi} \cdot \frac{F}{L} \cdot \frac{R_e}{E_e}\right]^{1/2} \qquad\qquad \text{Eq. (9.3)}$$

where: b = half-width of contact area

F = contact force
L = length of contact area

The *mean* compressive stress over the contact area is, of course:

$$f_{mean} = \frac{F}{2.b.L}$$

Eq. (9.4)

but the maximum compressive stress, at the centre of the contact area is:

$$f_{max} = \frac{4}{\pi}.f_{mean} = \frac{2.F}{\pi.b.L}$$

Eq. (9.5)

Substituting for b and re-arranging gives us:

$$f_{max} = \left[\frac{F.E_e}{2.\pi.L.R_e}\right]^{1/2}$$

Eq. (9.6)

Because it is usual to estimate the safe contact stress level in terms of f_{max}, the maximum Hertzian stress, it is convenient to separate from Equation 9.6 two groups of variables, one relating to the contact force applied and the system geometry, and the other relating to the material properties. The equation can therefore be put in the form:

$$s_{1c} = \frac{2.\pi}{E_e}.f_{max}^2 = \frac{F}{L.R_e}$$

Eq. (9.7)

s_{1c} is called the *surface stress factor* for line contact and is very easily estimated for an application using the right-hand expression of Equation 9.7. From the central expression of Equation 9.7 the *allowable* stress factor can be calculated for a cam material if its *allowable* Hertzian stress, f_{max}, is known and the equivalent elasticity of the cam and follower materials, E_e, is known.

Equation 9.7 is both simple and important. The surface stress factor, s_{1c}, is comparable to the S_c factor originally used in British Standard 436, Spur and Helical Gearing. Because the flanks of involute gear teeth are curved surfaces in contact under pressure, similar to a cam and roller follower, the S_c values for gear materials, suitably adjusted for metal fatigue effects, could be used for cam systems when the operating conditions are similar. The main difference, however, is that gear teeth slide (skid) across each other, rather more than they roll, but at moderate speeds most cam systems have almost pure rolling. Safe s_{1c} values for rolling contact are derived specially for cams.

Example 9.1

The disc cam described in Example 7.1, Chapter 7, uses a cylindrical cam follower roller of 20 mm diameter × 13.5 mm wide. The roller has corner radii of 0.5 mm and runs on a needle bearing with negligible friction. The cam is made of high quality cast iron and the roller is made of hardened steel. The cam is 12 mm wide, so the line of contact is taken as 12 mm. The estimated force on the follower system is 1500 N in the direction of its motion at the point in the cycle for which the pressure angle and curvature have been calculated in Example 7.1. Calculate the stress factor and maximum Hertzian stress at that point.

This is a line contact case for which we know the length of the area of contact, but we must calculate the equivalent radius of curvature, the equivalent elasticity and the contact force before we can calculate the stress. It is not necessary to calculate the width of the area of contact.

Designate the cam as body No. 1 and the roller as body No. 2, then:

$R_1 = 47.961$ mm (the cam surface, not the pitch curve – see Example 7.1)

$R_2 = 10$ mm

$$\frac{1}{R_e} = \frac{1}{47.961} + \frac{1}{10} = 0.12085 \qquad \text{(from Eq. (9.1))}$$

therefore $R_e = 1 \div 0.12085 = 8.275$ mm

We know from Example 7.1 that the pressure angle is 8.735°, so the contact force must be:

$$F = 1500 \div \cos 8.735° = 1517.6 \ [\text{N}]$$

Given that $L = 12$ mm we can find the stress factor from Eq. (9.7):

$$s_{1c} = \frac{1517.6}{8.275 \times 12} = 15.283 \ [\text{N/mm}^2] = 15.283 \times 10^6 \ [\text{N/m}^2]$$

Please note that s_{1c} is a stress *factor*, not an actual stress, which is useful for comparing the load applied to the surface against the load carrying capacity of the materials. It is independent of material properties: it would have the same value whatever the cam and roller were made of. The actual stress depends also on the elasticity of the materials.

For this example we shall assume the following material properties:

Cast iron cam: $E_1 = 140 \times 10^9 \ [\text{N/m}^2]$
$\nu_1 = 0.26$

Steel roller: $E_2 = 205 \times 10^9 \, [\text{N/m}^2]$
$\nu_2 = 0.30$

$$\frac{1}{E_e} = \frac{1}{2}\left[\frac{1 - 0.26^2}{140 \times 10^9} + \frac{1 - 0.30^2}{205 \times 10^9}\right] = 5.495 \times 10^{-12} \qquad \text{(from Eq. (9.2))}$$

$E_e = 180 \times 10^9 \, [\text{N/m}^2]$

Rearranging Equation 9.7 we find the maximum Hertzian stress:

$$f_{\text{max}} = \left[\frac{s_{1c}.E_e}{2.\pi}\right]^{1/2} = \left[\frac{15.283 \times 10^6 \times 180 \times 10^9}{2 \times \pi}\right]^{1/2} = 661 \times 10^6 \, [\text{N/m}^2]$$

This stress is well within the static stress limit of both the steel and the cast iron, but failure may occur by rolling contact fatigue, so it must also be used to estimate the expected life of the mechanism, as explained in the next chapter.

ELLIPTICAL CONTACT AREA

When one or both of the bodies in contact have a compound curvature there is 'point contact', and under pressure the point of contact becomes a small elliptical *area* of contact. Typically this occurs in cam systems when a cambered roller follower is used. The object of using a cambered roller is to allow for manufacturing errors that cause the cam and roller axes to be not quite parallel. With modern high quality machining and careful assembly this should not be necessary, but if it is, then there should be sufficient camber to ensure that the point of contact is well within the width of the cam track, so that the elliptical area is not truncated at the edge of the cam.

To avoid excessive complication it is assumed that the principal curvatures of the two bodies are in planes at right angles: one in the direction of rolling and the other perpendicular to it. This is true of virtually all industrial cam systems. We now have up to four radii of curvature to deal with:

R_{11} = radius of curvature of body 1 in the plane of rolling
R_{12} = radius of curvature of body 1 in the perpendicular plane
R_{21} = radius of curvature of body 2 in the plane of rolling
R_{22} = radius of curvature of body 2 in the perpendicular plane

An equivalent radius of curvature in the plane of rolling can be defined, which is the same as for line contact, described above. In this case it is given by:

$$\frac{1}{R_{e1}} = \frac{1}{R_{11}} + \frac{1}{R_{21}}$$ Eq. (9.8)

There is also an equivalent radius of curvature in the perpendicular plane, given by:

$$\frac{1}{R_{e2}} = \frac{1}{R_{12}} + \frac{1}{R_{22}}$$ Eq. (9.9)

The combined equivalent radius of curvature of both bodies is:

$$\frac{1}{R_e} = \frac{1}{R_{e1}} + \frac{1}{R_{e2}}$$ Eq. (9.10)

It is interesting to note that if there is no curvature (infinite radius of curvature) of either body in the perpendicular plane, or if one has concave (i.e. negative) curvature exactly matching the convex curvature of the other in that plane, then Equation 9.9 yields a zero value for equivalent curvature in the perpendicular plane, and we have line contact instead of point contact. In that case R_e is the same as R_e for line contact.

The semi-axes of the contact area ellipse are given in terms of two coefficients, u and v, which are functions of an aspect ratio parameter, t, which in turn is a function of the radii of curvature R_{e1} and R_{e2}, thus:

$$a = u . \left[\frac{3 . F . R_e}{E_e} \right]^{1/3}$$ Eq. (9.11)

$$b = v . \left[\frac{3 . F . R_e}{E_e} \right]^{1/3}$$ Eq. (9.12)

$$\cos t = \left[\frac{R_{e2} - R_{e1}}{R_{e2} + R_{e1}} \right]$$ Eq. (9.13)

where a and b are the semi-major and semi-minor axes respectively of the contact ellipse, and u and v can be found from the table below, once t has been found from Equation 9.13.

If the camber is only very slight, then R_{e2} will be much larger than R_{e1}, giving a small value of t (cos t approaching 1) and, as can be seen from the

Table 9.1 Point contact (elliptical contact area) coefficients

t	cos t	u	v	k	t	cos t	u	v	k
0	1.0000	∞	0.000	∞					
1	0.9998	36.829	0.131	113.161	46	0.6947	1.895	0.610	1.546
2	0.9994	22.189	0.169	52.925	47	0.6820	1.857	0.618	1.511
3	0.9986	16.481	0.197	34.037	48	0.6691	1.821	0.626	1.478
4	0.9976	13.289	0.219	24.630	49	0.6561	1.787	0.633	1.447
5	0.9962	11.236	0.238	19.157	50	0.6428	1.754	0.641	1.419
6	0.9945	9.792	0.255	15.600	51	0.6293	1.723	0.648	1.393
7	0.9925	8.703	0.271	13.081	52	0.6157	1.693	0.656	1.368
8	0.9903	7.855	0.285	11.230	53	0.6018	1.664	0.663	1.345
9	0.9877	7.175	0.298	9.813	54	0.5878	1.636	0.671	1.322
10	0.9848	6.609	0.312	8.801	55	0.5736	1.609	0.678	1.301
11	0.9816	6.123	0.326	7.985	56	0.5592	1.583	0.686	1.281
12	0.9781	5.728	0.338	7.268	57	0.5446	1.558	0.694	1.261
13	0.9744	5.385	0.347	6.498	58	0.5299	1.533	0.701	1.243
14	0.9703	5.078	0.356	5.932	59	0.5150	1.509	0.709	1.226
15	0.9659	4.808	0.367	5.489	60	0.5000	1.486	0.717	1.210
16	0.9613	4.567	0.377	5.096	61	0.4848	1.463	0.725	1.194
17	0.9563	4.350	0.386	4.727	62	0.4695	1.441	0.733	1.179
18	0.9511	4.155	0.394	4.397	63	0.4540	1.419	0.741	1.165
19	0.9455	3.978	0.403	4.128	64	0.4384	1.398	0.750	1.152
20	0.9397	3.816	0.412	3.889	65	0.4226	1.378	0.758	1.139
21	0.9336	3.668	0.421	3.675	66	0.4067	1.358	0.766	1.127
22	0.9272	3.531	0.429	3.482	67	0.3907	1.339	0.775	1.116
23	0.9205	3.405	0.438	3.307	68	0.3746	1.320	0.783	1.105
24	0.9135	3.288	0.446	3.151	69	0.3584	1.302	0.792	1.095
25	0.9063	3.180	0.454	3.008	70	0.3420	1.284	0.800	1.086
26	0.8988	3.082	0.462	2.882	71	0.3256	1.267	0.809	1.077
27	0.8910	2.993	0.469	2.770	72	0.3090	1.251	0.818	1.069
28	0.8829	2.912	0.476	2.664	73	0.2924	1.235	0.826	1.061
29	0.8746	2.834	0.483	2.568	74	0.2756	1.219	0.835	1.054
30	0.8660	2.758	0.490	2.475	75	0.2588	1.203	0.844	1.048
31	0.8572	2.685	0.498	2.385	76	0.2419	1.188	0.853	1.042
32	0.8480	2.615	0.505	2.302	77	0.2250	1.173	0.863	1.036
33	0.8387	2.551	0.512	2.228	78	0.2079	1.158	0.872	1.031
34	0.8290	2.488	0.519	2.157	79	0.1908	1.144	0.882	1.026
35	0.8192	2.429	0.526	2.091	80	0.1736	1.130	0.891	1.021
36	0.8090	2.373	0.533	2.029	81	0.1564	1.115	0.902	1.017
37	0.7986	2.319	0.541	1.969	82	0.1392	1.101	0.912	1.013
38	0.7880	2.266	0.548	1.912	83	0.1219	1.087	0.923	1.010
39	0.7771	2.214	0.555	1.858	84	0.1045	1.074	0.934	1.007
40	0.7660	2.163	0.563	1.806	85	0.0872	1.061	0.944	1.005
41	0.7547	2.114	0.571	1.757	86	0.0698	1.049	0.955	1.003
42	0.7431	2.066	0.579	1.710	87	0.0523	1.036	0.965	1.002
43	0.7314	2.020	0.587	1.665	88	0.0349	1.024	0.977	1.001
44	0.7193	1.977	0.594	1.624	89	0.0175	1.012	0.988	1.000
45	0.7071	1.935	0.602	1.583	90	0.0000	1.000	1.000	1.000

table, u will be much larger than v and the ellipse will be long and narrow: this is when one must be careful that the whole of the ellipse is contained within the cam width.

To calculate the mean compressive contact stress we proceed as for line contact, by dividing the contact force F by the contact area πab, so

$$f_{mean} = \frac{F}{\pi . u . v . (3 . F . R_e / E_e)^{2/3}} = \frac{1}{\pi . u . v} \cdot \left[\frac{F . E_e^2}{9 . R_e^2} \right]^{1/3} \qquad \text{Eq. (9.14)}$$

and

$$f_{max} = \frac{4}{\pi} \cdot f_{mean} = \frac{4}{\pi^2 . u . v} \left[\frac{F . E_e^2}{9 . R_e^2} \right]^{1/3} \qquad \text{Eq. (9.15)}$$

Equation 9.15 can also be expressed in terms of convenient groups of variables separating the material properties from the force and dimensions:

$$s_{pc} = \frac{F}{k . R_e^2} = \frac{9 . \pi^6}{64} \cdot \frac{f_{max}^3}{E_e^2} \qquad \text{Eq. (9.16)}$$

where $k = (u . v)^3$ and is given in the table.

s_{pc} is known as the *surface stress factor* for point contact and can be fairly easily calculated from the equations above and Table 9.1.

These equations show that, unlike most other stresses, the surface stress for two curved bodies in contact is not proportional to the applied force. This is because as the force is increased the bodies distort in such a way as to increase the area supporting the force.

Thus, for example, if the contact force is doubled the line contact stress is increased by a factor of only 1.414, and the point contact stress by only 1.26: the square-root and cube-root of 2 respectively.

SHEAR STRESS

A common cause of cam surface failure is related to excessive *shear* stress just below the cam surface as a consequence of the compressive stress *at* the surface. There is a strict relationship between the two stresses, so that the stress factor, s_{1c} or s_{pc}, can be used as a criterion for this mode of failure also. A table of allowable surface stress factors, based on subsurface shear stress, for various cam and roller material combinations is given in Chapter 10.

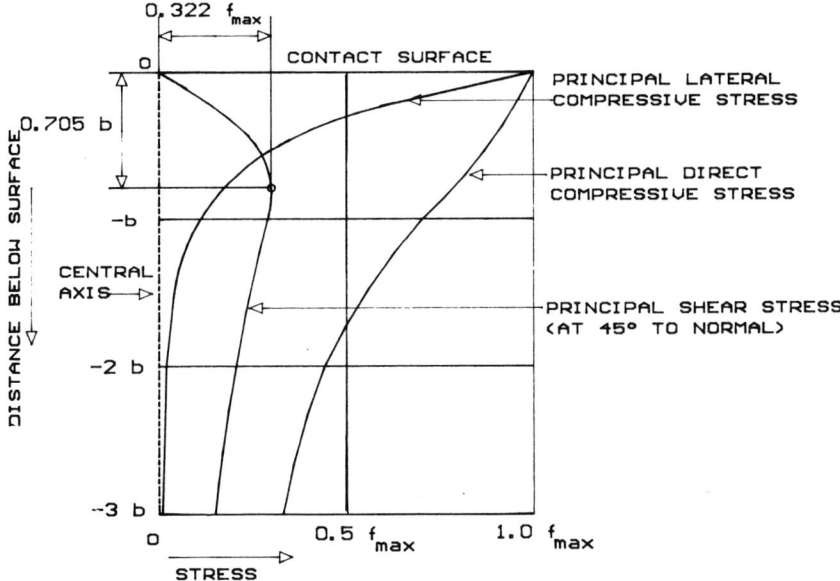

Fig. 9.4 Stresses below contact surface on central axis

Figure 9.4 shows the distribution of the stresses in the material along the axis immediately below the *centre* of the contact area. Stresses elsewhere are less than on this central axis. The stress values are plotted in terms of the maximum Hertzian stress, f_{max}, and the distance below the surface is plotted in terms of the half-width of the contact area, b. The diagram gives a simplified picture of the main stresses when the deformation is entirely elastic, i.e. when no plastic deformation or yielding has occurred and no residual stresses are present when the load is removed. This condition is rarely true for high performance cams, but it serves well as a safe basis for estimating the load-carrying capacity of a cam surface.

When the yield point of the material is exceeded during the first stress cycle, the permanent deformation of material under the surface gives rise to residual stresses which reduce the maximum values of the sub-surface stresses during subsequent cycles. The process is repeated during the next few cycles, giving successively lower maxima. This process is known as 'shakedown' and at certain stress levels it can continue until the maximum stresses are below the yield point and the situation stabilises. However, there is a 'shakedown limit' above which such stabilisation does not occur and catastrophic failure takes place. In industrial situations it is necessary to operate cams below the shakedown limit, so it is assumed that the elastic stress distribution of Figs 9.2 and 9.4 can be used to estimate *safe* cam loadings: if shakedown does occur (below the shakedown limit) it will reduce the very stresses which cause it.

The principal shear stress, q, is probably the most important sub-surface stress because one of the most common causes of cam surface failure is 'rolling contact fatigue' in which shear fractures occur just below the surface, causing small flakes or granules to break away and produce a pitted surface. This is discussed further in Chapter 10. As can be seen from the diagram, the maximum value of shear stress occurs at a depth of $0.705b$, and is:

$$q_{max} = 0.322f_{max} \qquad \qquad \text{Eq. (9.17)}$$

from which we can deduce that the maximum allowable Hertzian stress is 3.105 times the maximum allowable shear stress if rolling contact fatigue is the likely mode of failure. In that case:

$$s_{1c} = \frac{60.6}{E_e} \cdot q_{max}^2 \qquad \qquad \text{Eq. (9.18)}$$

or

$$s_{pc} = \frac{4049}{E_e^2} \cdot q_{max}^3 \qquad \qquad \text{Eq. (9.19)}$$

In a homogeneous material the depth at which q_{max} occurs is unimportant, but many cams and followers are made of surface-hardened steels, where the stress depth is significant compared to the thickness of the hard casing. A situation can arise where the hard casing is strong enough to bear the maximum shear stress, but the soft core is not. In surface-hardened cams, therefore, we must either have q_{max} low enough to suit the core, or have it occur well within the casing thickness.

The strength of the material is related to its hardness. There is a more or less gradual transition between the hardness of the casing and that of the core. Figure 9.5 shows a typical stress/strength diagram for a case-hardened cam where the shear stress is everywhere less than the allow-able shear stress of the material. If the casing were much thinner than shown, however, the core strength would become critical, even though the casing strength is more than adequate. The casing thickness is, of course, that which is left after finishing operations such as grinding.

It is generally accepted that a safe casing thickness for case-hardened

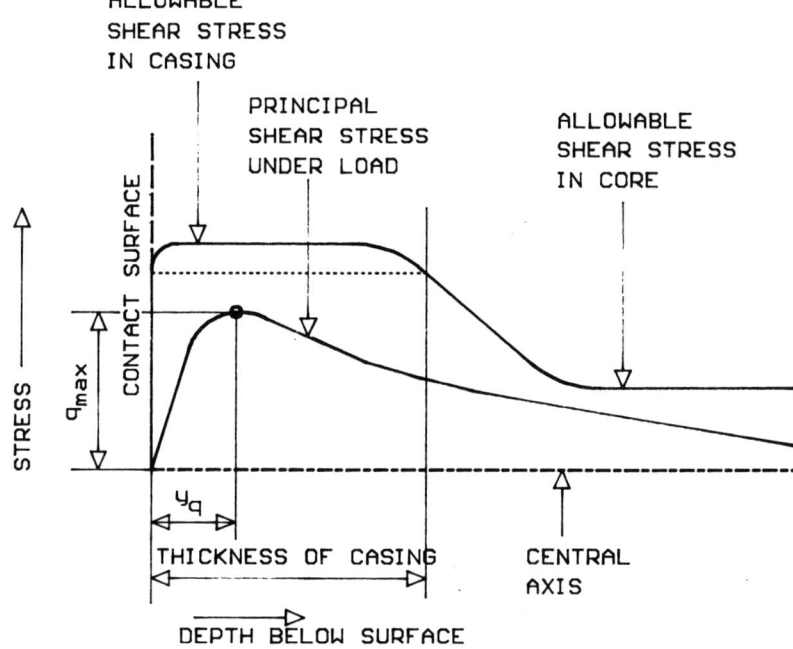

Fig. 9.5 Shear stress vs. casing strength

steel is twice the depth of the maximum shear stress, that is to say that the case thickness should be a least 1.41 times the width of the contact area *b* obtained from Equations 9.3 or 9.12.

Because of the hardness transition zone there may be some difficulty in establishing an agreed case thickness. It is standard practice in some Quality Assurance departments to take micro hardness readings on the cross-section of a test piece at various distances from the surface and assess the casing thickness as the distance within which the hardness readings exceed a specified value. For design purposes the specified hardness and minimum thickness are used in estimating the load capacity of the cam surface.

The yield shear stress of a material is directly related to the Vicker's Pyramid hardness, or diamond pyramid number D.P.N., which is expressed in kgf/mm^2 units. In the test the material yields in shear until there is sufficient contact area to support the test load without further penetration.

The relationship is:

$q_{yield} = H_v/6 \; [kgf/mm^2]$

or $1.634 \times 10^6 H_v \; [N/m^2]$ Eq. (9.20)

or $237 H_v \; [lbf/in^2]$

where H_v = the D.P.N. hardness and q_{yield} = yield shear stress.

These equations are used in Chapter 10 to estimate the safe load capacities of cam and follower systems.

CHAPTER 10

Load/Life Calculation

The most common mode of contact surface failure in a cam and roller system is that of surface pitting due to *rolling contact fatigue*. The pits are caused (almost certainly) by shearing of the metal a short distance below the surface, so that the criterion for surface failure is that the sub-surface shear stress exceeds the yield point of the material. Failure may also occur due to scuffing, when there is an insufficient film of lubricant keeping the surfaces apart, particularly when one surface skids or slides on the other to a significant extent.

Whenever possible the cam track surface should be lubricated, and because there is a high pressure on a very small area of contact it is best to use an E.P. (extreme pressure) additive in the lubricant. The E.P. additive provides chemical lubrication when the oil film breaks down. A discussion of hydrostatic and hydrodynamic lubrication is beyond the scope of this book, but it is safe to assume that any good lubrication practice for gear teeth is also good practice for cam tracks. In situations where only sparse lubrication or none at all is possible, e.g. in a hostile environment, resort to 'self-lubrication' materials such as special cast irons, and use cam and follower dimensions that keep the contact stress very low.

In rolling contact the metal suffers fatigue, that is to say that its strength reduces, with repeated stress cycles, and the relationship between allowable stress and life (the S-N curve) is well established, notably in the field of ball and roller bearings. The S-N curve is nowadays considered to continue decreasing indefinitely with increasing time: there is not an 'endurance limit' in the field of rolling contact fatigue in the same sense that it may exist in the field of structural stress, and it is unwise to assume that a cam surface will never wear out if it is sufficiently lightly loaded.

What is more, the curve cannot be extrapolated back to zero life and infinite stress, for there is a static stress limit which causes failure with a single application of stress (or at least with a small number of cycles, e.g. the 'shakedown limit' described in the previous chapter). This gives rise to two kinds of load capacity for a cam surface: the 'dynamic capacity' based on the S-N curve and the 'capacity limit' based on shakedown. Both

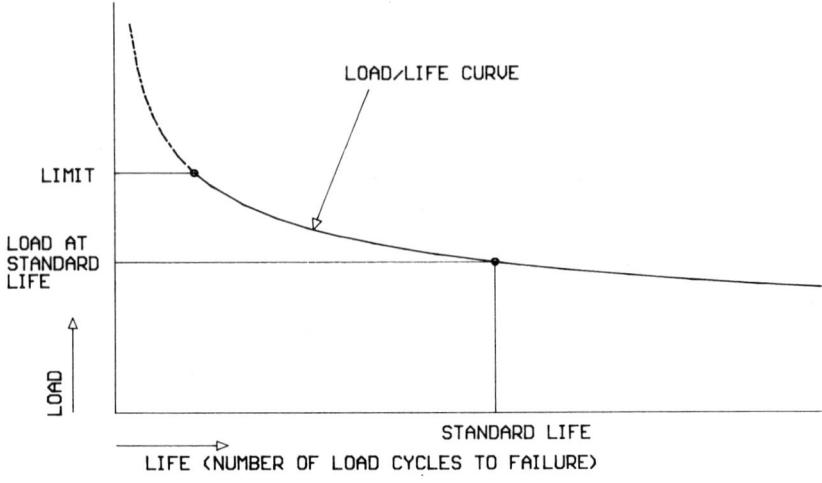

Fig. 10.1   Load life curve

kinds can be shown on a diagram such as Fig. 10.1 where load (i.e. stress, force, torque, etc.) is plotted against the total number of stress cycles to failure.

The number of stress cycles to failure is a convenient way of expressing the life of the system because it takes into account both the operating speed and the hours of operation. The *shape* of the dynamic capacity curve is independent of its magnitude, so that the whole of the curve can be deduced from one point on it. It is a common practice to choose a standard life of 1 million stress cycles for specifying a point on the curve and then to use a load/life formula for estimating allowable loads at other lives. A similar system has been adopted by some manufacturers of cam-operated mechanisms, whereby the dynamic load capacity is given at a specified speed and life (e.g. 50 cycles/min for 8,000 operating hours) with speed factors and life factors for adjusting the capacity for other speeds and lives, i.e. for other points on the load/life curve. A similar method is also used by bearing manufacturers.

It has been mentioned earlier that to establish the 'true' load capacity of any particular industrial cam system experimentally is quite impracticable. Fortunately an enormous number of tests have been carried out over many years with rolling bearings of various designs, providing a

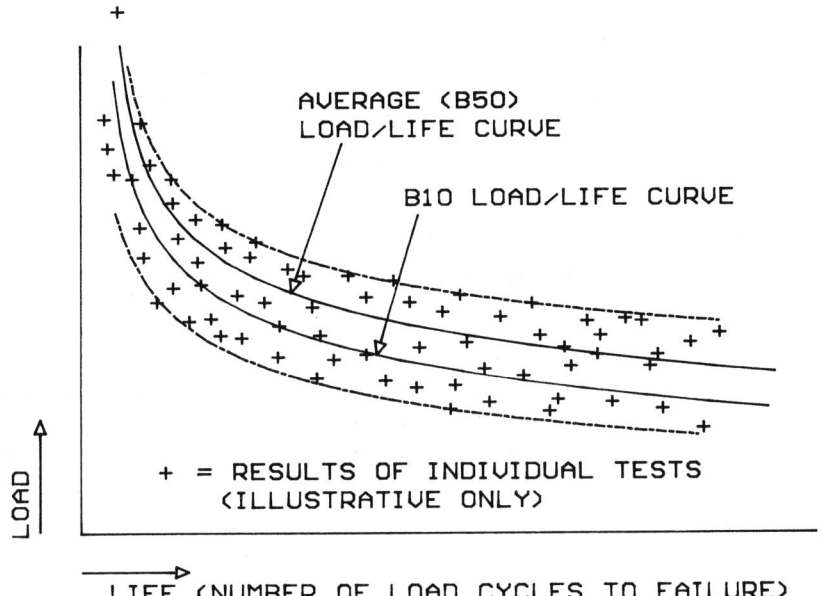

Fig. 10.2   Scatter of test results

wealth of statistical data. From these can be deduced the shapes of the S-N curves for line contact and point contact fatigue. However, there is a very wide scatter of the plotted results, as shown in Fig. 10.2 (which is illustrative only), and it is impossible to state with certainty that a mechanism will survive for a particular life under a particular load. It has therefore become standard practice to state dynamic load capacities of rolling bearings in terms of their *probability* of survival. The reader is recommended to consult specialist publications, including the technical sections of manufacturers catalogues for more information on the load capacities of rolling bearings.

The standard usually adopted for bearings is a 'B10' life of one million revolutions. This means that the bearing has a 10% probability of failing at that life when loaded to the specified capacity: in other words it has a 90% chance of survival for one million revolutions or more. In this context failure is defined as the appearance of the first perceptible sign of surface fatigue pitting. In practice a bearing (or a cam system) may carry on working satisfactorily for a long time after the first appearance of pitting, so this definition may be regarded as a safe one in most industrial environments. Occasionally a 'B50' life is used when specifying a load capacity, that means there is a 50% probability of failure (as defined above) within the specified life. B50 is of course less safe than B10 and it is therefore important to be sure that the basis of the stated dynamic capacity is known when a commercial product is used.

The relationship between B10 and B50 capacities for line contact is:

B50 life = twice the B10 life for a given load, and
B50 load = 1.23 times the B10 load for a given life.

The similarity between the modes of failure of cam tracks and rolling bearings makes it reasonable to adopt bearing technology for estimating the load/life characteristics of cams. In what follows, then, the exponent which defines the shape of the S-N curve, and the B10 probability concept, which takes into account the scatter of test results, are identical to those used for rolling bearings.

It has been found that the shape of the load/life curve conforms to the equation:

$$\frac{\text{load 1}}{\text{load 2}} = \left[\frac{\text{life 2}}{\text{life 1}}\right]^e$$

or

load ratio = (life ratio)^e $\hspace{2cm}$ Eq. (10.1)

where:

1 and 2 refer to any two points on the curve
$e = 0.3$ for line contact or
$e = 1/3$ for point contact
load = contact force
life = number of stress cycles to probable failure

For a given cam/follower geometry the equation holds true for any expression of load that is proportional to contact force, such as the surface stress factor s_{lc} or s_{pc}, and for any expression of life that is proportional to the number of stress cycles, such as operating hours at a given speed. It is the basis of 'life factors' at constant speed and 'speed factors' at constant life.

Table 10.1

Life or speed ratio	Load ratio		Life or speed ratio	Load ratio		Life or speed ratio	Load ratio	
	Line contact	Point contact		Line contact	Point contact		Line contact	Point contact
0.2	0.6170	0.5848	4.2	1.5381	1.6134	18	2.3800	2.6207
0.4	0.7597	0.7368	4.4	1.5597	1.6386	19	2.4189	2.6684
0.6	0.8579	0.8434	4.6	1.5806	1.6631	20	2.4565	2.7144
0.8	0.9362	0.9283	4.8	1.6009	1.6869	22	2.5277	2.8020
1.0	1.0000	1.0000	5.0	1.6207	1.7100	24	2.5946	2.8845
1.2	1.0562	1.0627	5.5	1.6677	1.7652	26	2.6576	2.9625
1.4	1.1062	1.1187	6.0	1.7118	1.8171	28	2.7174	3.0366
1.6	1.1514	1.1696	6.5	1.7534	1.8663	30	2.7742	3.1072
1.8	1.1928	1.2164	7.0	1.7928	1.9129	32	2.8284	3.1748
2.0	1.2311	1.2599	7.5	1.8303	1.9754	34	2.8803	3.2396
2.2	1.2669	1.3006	8.0	1.8661	2.0000	36	2.9302	3.3019
2.4	1.3004	1.3389	9.0	1.9332	2.0801	38	2.9781	3.3620
2.6	1.3320	1.3751	10	1.9953	2.1544	40	3.0242	3.4120
2.8	1.3619	1.4095	11	2.0531	2.2240	45	3.1330	3.5569
3.0	1.3904	1.4422	12	2.1074	2.2894	50	3.2336	3.6840
3.2	1.4176	1.4736	13	2.1587	2.3513	60	3.4154	3.9149
3.4	1.4436	1.5037	14	2.2072	2.4101	70	3.5771	4.1213
3.6	1.4686	1.5326	15	2.2533	2.4662	80	3.7233	4.3089
3.8	1.4926	1.5605	16	2.2974	2.5198	90	3.8572	4.4814
4.0	1.5157	1.5874	17	2.3396	2.5713	100	3.9811	4.6416

Table 10.1 gives such life factors and speed factors for both line contact and point contact. It should be noted that the life in hours at a constant speed and the speed for a constant life in hours are both proportional to the life in stress cycles. Therefore the load capacity rating must be adjusted by dividing by the appropriate factor in both cases: higher speed and longer life means lower load capacity.

It was shown in Chapter 9 how the surface stress factor is related to the Hertzian stress and how that is proportional to the shear stress. The static yield stress in shear (the failure criterion) is proportional to the diamond pyramid hardness, so it is proposed that in the absence of experimental data for a particular cam/follower design an allowable stress level can be based on the hardness and elasticity of the surfaces.

For static yield, combining Equations 9.7, 9.16 and 9.20 gives yield stress factors, for line and point contact respectively, of

$$s_{lc} = 161.8 \times 10^{12} H_v^2 E_e \ [\text{N/m}^2]$$

$$s_{pc} = 17.66 \times 10^{21} H_v^3 / E_e^2 \ [\text{N/m}^2]$$

The stress factors for a B10 life of 1 million stress cycles can be taken as half of the static yield values, giving us *allowable dynamic stress capacities* of

$$\text{Allowable } s_{lc} = 80.9 \times 10^{12} H_v^2 / E_c \ [\text{N/m}^2] \qquad \text{Eq. (10.2)}$$

$$\text{Allowable } s_{pc} = 8.83 \times 10^{21} H_v^3 / E_c^2 \ [\text{N/m}^2] \qquad \text{Eq. (10.3)}$$

The shakedown stress limit (see Chapter 9) is about 25% higher than the static yield stress, but it is recommended that for industrial machinery a factor of safety of 2 on the yield stress load is appropriate both for a static stress capacity and for a dynamic stress limit.

A single value then can be used for the *dynamic stress capacity* (B10, 10^6 cycles) and *stress limit* for a cam and follower material combination. The following table gives suggested allowable surface stress factors for various materials, based on Equations 10.2 and 10.3.

Table 10.2

MATERIAL UNDER STRESS, TO WHICH s_{lc} & s_{pc} APPLY	H_v	E 10^9 N/m²	v	Material of the mating part							
				Group A		Group B		Group C		Group D	
				s_{lc}	s_{pc}	s_{lc}	s_{pc}	s_{lc}	s_{pc}	s_{lc}	s_{pc}
							10^6 N/m²				
GROUP A											
Steel, hardened	740	203	0.29	199	72.6	224	91.3	246	110	283	146
Steel, hardened	500	·203	0.29	91.1	22.4	102	28.2	112	34.0	129	45.1
Steel, hardened	300	203	0.29	32.8	4.84	36.8	6.09	40.4	7.35	46.5	9.74
Steel	200	203	0.29	14.6	1.43	16.3	1.80	18.0	2.18	20.7	2.89
GROUP B (numbers in parentheses are British Standard grades)											
S.G.Iron (42/2)	250	180	0.28	24.3	3.20	27.1	3.97	29.6	4.74	33.9	6.19
S.G.Iron (32/7)	185	170	0.28	13.7	1.38	15.3	1.70	16.6	2.02	19.0	2.62
S.G.Iron (24/17)	155	160	0.28	10.0	0.87	11.0	1.06	12.0	1.26	13.6	1.62
Grey Cast Iron (26)	260	150	0.26	29.3	4.46	32.3	5.42	35.0	6.38	39.6	8.15
GROUP C (numbers in parentheses are British Standard grades)											
Grey Cast Iron (20)	230	145	0.26	23.4	3.21	25.7	3.89	27.9	4.56	31.5	5.81
Grey Cast Iron (17)	210	140	0.26	19.9	2.55	21.9	3.07	23.7	3.60	26.6	4.56
Grey Cast Iron (14)	190	135	0.26	16.7	1.97	18.3	2.37	19.7	2.76	22.2	3.49
Meehanite GB	210	145	0.26	19.5	2.44	21.5	2.96	23.2	3.47	26.2	4.42
GROUP D											
Al. Bronze	130	105	0.35	8.79	0.80	9.54	0.95	10.2	1.09	11.4	1.34
Ph. Bronze	90	120	0.38	3.81	0.22	4.17	0.26	4.50	0.30	5.05	0.38
Gunmetal	70	90	0.35	2.83	0.15	3.04	0.18	3.24	0.20	3.57	0.25

Table 10.2 gives allowable stress factors for some common cam and follower materials where rolling contact fatigue is expected to be the failure criterion. The values given are probably conservative, particularly for special steels where fatigue resistance is improved by minimising impurity inclusions. The factors for similar materials of different hardness can be obtained by multiplying the s_{lc} and s_{pc} values by the hardness (H_v) ratio squared or cubed respectively. For other materials use Equations 10.2 and 10.3.

Example 10.1

A disc cam with a translating roller follower, as shown in Fig. 10.3, has a minimum convex radius of curvature of the cam profile of 12.462 mm. The follower roller is 20 mm diameter × 16 mm wide, made of hardened and ground steel with a minimum surface hardness of 500 H_v. It runs on a needle bearing. The cam has a face width of 12 mm and is made of a close grained grey cast iron with a surface hardness of approximately 210 H_v. The highest contact force is 420 N and it occurs in the region of the minimum radius of curvature. Estimate the life of the cam in hours of

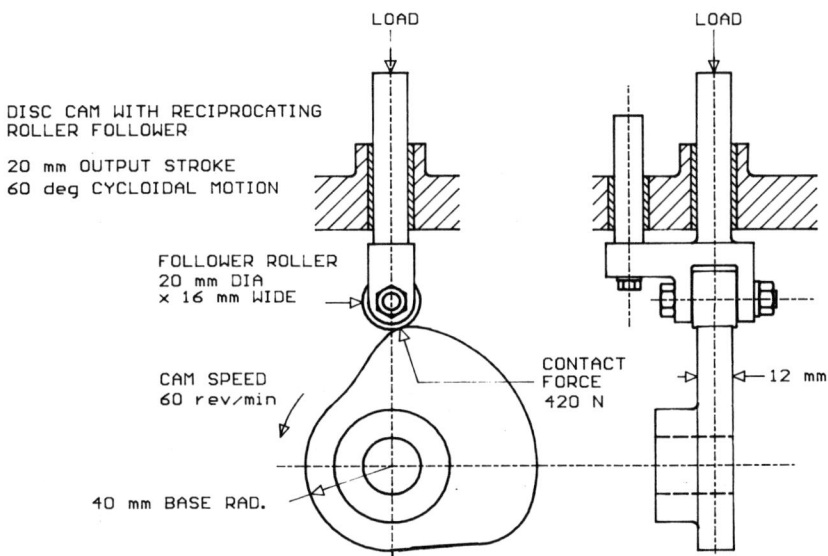

Fig. 10.3   Example 10.1

operation at a speed of 60 rev/min. Estimate also the life of the cam if a cambered (crowned) roller follower is used, with a camber radius of 500 mm.

First find the applied surface stress factor from the contact force and the cam and follower geometry.

The cylindrical roller gives a line contact which is 12 mm long, the width of the cam track. The roller radius is 10 mm, so the equivalent radius of curvature, from Equation 9.1 is:

$$1/R_c = 1/10 + 1/12.462 = 0.18024$$

$$R_c = 5.548 \text{ mm} = 0.005548 \text{ m}$$

From Equation 9.7 we find the applied surface stress factor:

$$s_{lc} = \frac{420}{0.005548 \times 0.012} = 6.309 \times 10^6 \text{ N/m}^2$$

From Table 10.2 the allowable surface stress factor for this grade of cast iron in contact with hardened steel is $19.9 \times 10^6 \text{ N/mm}^2$ for both a limiting stress and a B_{10} life of 1 million stress cycles. The applied load is well within this stress limit and we can now estimate the B_{10} life of the cam track.

The load ratio is $19.9 \div 6.309 = 3.154$. The life ratio is found by inverting Equation 10.1, thus

$$\text{Life ratio} = \text{load ratio}^{1/c} = 3.154^{1/0.3} = 46.02$$

The B_{10} life is therefore 52.75 million stress cycles. There is one stress cycle at that particular point on the cam profile in each revolution, that is 60 per minute, or 3,600 per hour. The B_{10} life in hours is therefore:

$$\frac{46.02 \times 10^6}{3,600} = 2,783 \text{ hours}$$

This is about six years on single shift, which is a fairly short life for an industrial cam.

(Note. The life could be extended considerably by changing the cam material to a higher grade of cast iron or to hardened steel. For a similar material the load ratio increases as the square of the hardness and the B_{10} life increases as hardness to the power of 6.666.)

In the case of the cambered roller we have point contact and must find the equivalent radius of curvature from Equations 9.8 to 9.10. Let the cam be body 1 and the roller body 2.

$$1/R_{c1} = 1/10 + 1/12.462 = 0.18024$$

so

$$R_{c1} = 5.548 \text{ mm in the plane of rolling, as before}$$

$$1/R_{c2} = 0 + 1/500 = 0.002$$

so

$$R_{c2} = 500 \text{ mm in the perpendicular plane}$$

$$1/R_c = 0.18024 + 0.002 = 0.18224$$

so

$$R_c = 5.487 \text{ mm} = 0.005487 \text{ m}$$

The aspect ratio parameter of the contact ellipse is found from Equation 9.13:

$$\cos t = (500 - 5.548) \div (500 + 5.548) = 0.97805$$

$$t = 12.026°$$

and from Table 9.1 we find by interpolation the coefficients

$$u = 5.719 \qquad v = 0.338 \qquad k = 7.248$$

Using this value of k in Equation 9.16 we get the point contact applied surface stress factor:

$$s_{pc} = \frac{420}{7.248 \times 0.005487^2} = 1.925 \times 10^6 \text{ N/m}^2$$

From Table 10.2 we see that the allowable value of s_{pc} is $2.55 \times 10^6 \text{ N/m}^2$ (for both stress limit and b_{10} capacity). The applied stress is therefore still within the stress limit, but the load ratio for 10^6 cycles is now

Load ratio = $2.55 \div 1.925 = 1.325$

The load/life exponent is 1/3 for point contact, so

Life ratio = $1.325^3 = 2.324$

B_{10} life = $2.324 \times 10^6 \div 3,600 = 645$ hours only.

As expected, the cambered roller imposes a higher surface stress on the cam and therefore a shorter life. In practice a harder cam material would be chosen to get an acceptable life or the cam and/or roller size would be increased.

As a matter of interest, using equations 9.3, 9.11 and 9.12, the cylindrical roller produces a rectangular contact area 12 mm $\times$ 0.104 mm $= 1.248$ mm^2, whereas the cambered roller produces an elliptical contact area of 3.866 mm $\times$ 0.228 mm, i.e. 0.692 mm^2.

There are of course possible modes of failure of the cam and follower system of the above example other than the wear of the cam track surface. Leaving aside structural failure, there are the wear of the roller surface and the wear (rolling contact fatigue) of the follower roller bearing.

The surface of the roller has the same magnitude and distribution of stress as the cam track, but the frequency of stress application and the allowable stress factor (different material) are different. The stress factor can simply be taken from the table or calculated from the equations in Chapter 9, but the frequency of application, the number of stress cycles on a particular point on the roller surface, per hour of operation is rather difficult to compute.

The peak stress condition occurs once in every revolution of the cam and always at the same point on the profile. However, the peak stress is almost certainly at a different point on the roller surface each time it occurs, and any one point on the roller (which may eventually fail) is subject to a different stress level each time it contacts the cam, i.e. each revolution of the roller. It is simple to estimate the number of roller revolutions per hour as being the ratio of the cam circumference to the roller circumference times the number of revolutions of the cam per hour.

However, it is fair to assume that the spot that fails has been subjected to every possible stress level during the life of the mechanism. It is possible to calculate the approximate cumulative duration (number of stress cycles) at each stress level and, drawing again on rolling bearing design practice, to combine them into an equivalent duration at the peak stress that would cause failure.

This procedure uses the notion that all the various levels of stress contribute to the eventual breakdown of the surface: in effect damage accumulates throughout the life of the surface.

Let N_1, N_2, etc. be the cumulative number of stress cycles at each different stress level and F_1, F_2, etc. be the relevant contact forces. The life of the surface, i.e. the total number of stress cycles to failure, is the same as for a constant contact force applied throughout the life, known as the equivalent force, F_c, given by the following equation:

$$F_c = \left[\frac{N_1.F_1^{1/c} + N_2.F_2^{1/c} + \cdots}{N_1 + N_2 + \cdots} \right]^c$$

Eq. (10.4)

The wear of the follower bearing can be dealt with in a similar way, but it is a little easier because a safe approximation is to pretend that the force on the bearing always acts in the same direction, stressing the same spot on the inner raceway. The number of stress cycles per hour is then the same as the revolutions per hour of the cam. It is still necessary, however, to take into account the variable stress level and the cumulative damage effect.

Such a complicated procedure for the design of the follower roller is only rendered practicable by the computer, and at least one commercial cam mechanism manufacturer uses specially written programs to estimate the load capacities of their products in accordance with this procedure. In one-off special purpose mechanism design, however, the load capacity of the cam can be calculated as described (see Example 10.1). A very safe load capacity for the follower roller can be estimated by assuming a constant contact force equal to the peak force, and a stress cycle frequency equal to the average roller speed. If space constraints impose the use of a smaller roller than this very safe design procedure demands, then some comfort may be taken from the fact that it is usually much easier to replace a worn out roller than a worn out cam.

CHAPTER 11

Transmission Design, Gearing

The previous chapters have dealt almost exclusively with the design of the cam profile and the part of the follower system that contacts it. Other aspects of the cam design, for example the means of attaching it to its shaft, and other parts of the follower system have been left aside, since they belong to a wider, more general machine design technology. However, some properties of the input and output power transmissions have a serious effect on the performance of the cam and follower system, and design of the transmissions must be considered as an integral part of the mechanism design.

In this context the input transmission is defined as all the power transmission components from the power source (usually an electric motor) to the cam. The output transmission is defined as all the power transmission components from the follower (usually a roller) to the payload. These components include shafts, gears, gearboxes, couplings, chain drives, belt drives, linkages, etc. There are three main aspects of transmission design that must be considered.

- Strength: the ability to withstand the necessary forces and torques without fracture or yield.
- Rigidity: the ability to transmit those forces without excessive deflection.
- Backlash*: when reversal of force or torque occurs any backlash or 'play' in transmission is detrimental.

The first of these aspects, strength, will not be dealt with here, other than to indicate in other chapters the peak force or torque imposed on the transmission due to the special nature of the cam system. The design of components to withstand the peak stresses belongs to the general field of strength of materials, not particularly to cam systems.

---

*Strictly speaking, 'backlash' is the shock effect of a reversal of forces in a system which has significant clearance between adjacent members that are transmitting the force. After many decades of usage it has come to mean the clearance itself, and that is how it is used throughout this book.

The second aspect, rigidity, is of major concern because it plays a vital role in the generation of vibration in the system, which is fully dealt with in the next chapter. All transmission components have some degree of elasticity, the reciprocal of rigidity. When a metal component is stressed within its elastic limit it suffers elastic strain, that is a distortion or deflection which is related to its size and shape, and is proportional to the load applied. When it is stressed beyond its elastic limit it suffers plastic deformation which does not recover when the stress is removed.

It may also suffer hysteresis within the elastic limit, an energy absorbing phenomenon whereby the strain produced by an increasing load is not fully reduced by a decreasing load. Hysteresis is responsible for internal damping of vibrations, but this effect is unlikely to be significant in cam systems. To simplify the analysis we shall assume that transmission components are perfectly elastic with no hysteresis.

If there are several components connected in series, as in a typical transmission, the deflections are cumulative so that the overall deflection from one end of the transmission to the other is the sum of the individual deflections. When gearing is involved different parts of the transmission may be subject to different torques and the deflection at one side of a gear pair may be transformed into a different deflection at the other side of the gear. To assess the rigidity of a transmission as a whole, therefore, it is necessary to estimate the rigidity of each component and combine them in a particular way.

Most cams used in industrial machines are of the rotating kind and are driven by rotating motors, so that most of the input transmission components are rotary, subject to torsion and with an angular deflection proportional to the applied torque (belts and chains may not be thought of as rotary components although their pulleys and sprockets are). The simplest components are shafts, but these are often constructed with sections of different diameters, which for our purpose can be considered as several different shafts connected in series.

SHAFTS

The rigidity of a rotary component is defined as the torque divided by the angular deflection (see Fig. 11.1). For a simple shaft it is given by:

$$R = \frac{\pi \cdot G \cdot (D^4 - d^4)}{32 \cdot L} \qquad \text{Eq. (11.1)}$$

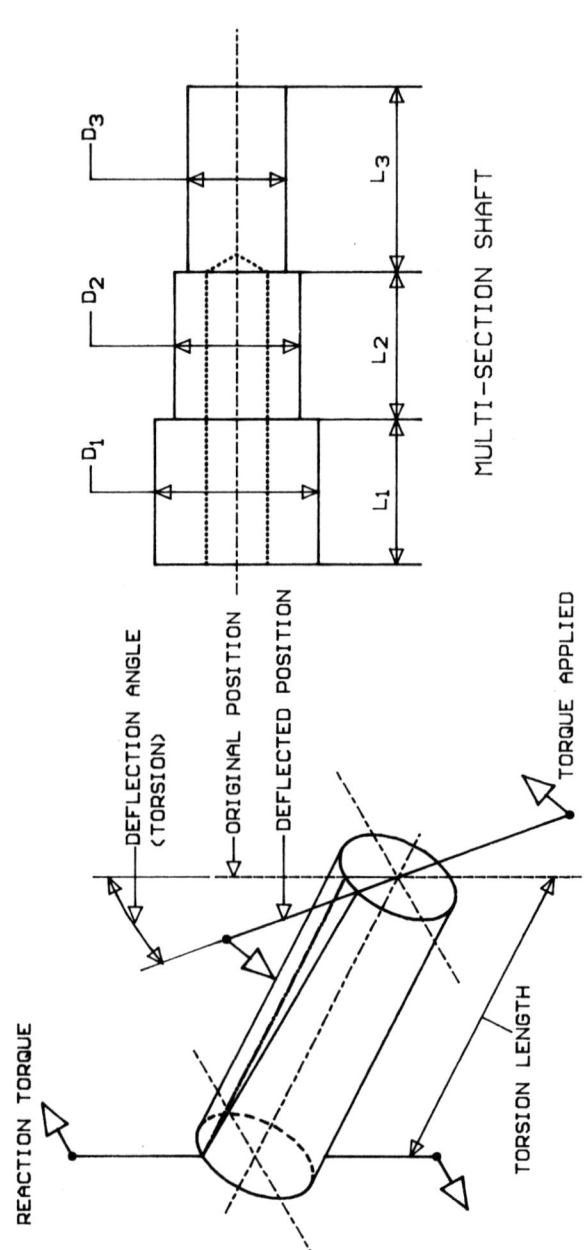

Fig. 11.1 Shaft torsion

where:

R = rigidity of shaft

G = Modulus of Rigidity of the material (shear stress ÷ shear strain)

D = outside diameter of shaft

d = inside diameter of shaft ($d = 0$ for a solid shaft)

L = length of shaft subject to torsion

For a multi-section shaft let the diameters and lengths of each section be $D_1, d_1, L_1; D_2, d_2, L_2; D_3, d_3, L_3$, etc., and the rigidity of each section calculated from Equation 11.1 be R_1, R_2, R_3, etc. then the overall rigidity is R where:

$$\frac{1}{R} = \frac{1}{R_1} + \frac{1}{R_2} + \frac{1}{R_3} \text{ etc.} \qquad\qquad \text{Eq. (11.2)}$$

This equation also applies to any mixture of diverse rotary components (gears, couplings, etc.) connected together in series.

From this it can be seen that the overall rigidity of the complete transmission is always less than the rigidity of any of its components, and that the least rigid component is the most significant. The following example illustrates this point.

Example 11.1

Calculate the rigidity of a steel shaft as shown in Fig. 11.1 which has three sections with the following dimensions:

$D_1 = 30$ mm, $\qquad d_1 = 15$ mm, $\qquad L_1 = 40$ mm

$D_2 = 25$ mm, $\qquad d_2 = 15$ mm, $\qquad L_2 = 35$ mm

$D_3 = 22$ mm, $\qquad d_3 = 0$, $\qquad\qquad L_3 = 160$ mm

Take the Modulus of Rigidity of steel as $G = 82.5 \times 10^9 \, N/m^2$

$$R_1 = \frac{\pi \times 82.5 \times 10^9 \times (0.030^4 - 0.015^4)}{32 \times 0.040} = 153{,}762 \text{ Nm/rad}$$

$$R_2 = \frac{\pi \times 82.5 \times 10^9 \times (0.025^4 - 0.015^4)}{32 \times 0.035} = 78{,}680 \text{ Nm/rad}$$

$$R_3 = \frac{\pi \times 82.5 \times 10^9 \times 0.022^4}{32 \times 0.160} = 11{,}858 \text{ Nm/rad}$$

therefore

$$\frac{1}{R} = \frac{1}{153,762} + \frac{1}{78,680} + \frac{1}{11,858}$$

$$= 0.000006503 + 0.00001271 + 0.00008433 = 0.00010354$$

therefore

$$R = 1/0.00010354 = 9,657 \text{ Nm/rad for the whole shaft.}$$

Note that for dimensional consistency the diameters and lengths were converted to metres when substituted in Equation 11.1.

This result shows that the overall rigidity is less, but not much less, than that of the least rigid part of the shaft. Had we ignored the most rigid part (R_1) the result would have been 10,305 instead of 9,657 Nm/rad, only 7% too high. In general, when it is obvious that some parts of a transmission are very much more rigid than others they may be allowed for by an estimated small reduction in the calculated overall rigidity, instead of being included in the full calculation.

GEARS

There are several types of gearing employed in industrial machines, the most common of which are spur gears, worm gears and chain drives. Generally the gear disc itself has a torsional rigidity high enough to be ignored, but the gear teeth may distort sufficiently under load to contribute significantly to the overall elasticity of the transmission, particularly on small diameter pinions. Both the driving and the driven teeth bend and also compress under Hertzian stress. These two distortions combine to give tangential linear deflection at the pitch line which varies somewhat as each tooth passes through the contact zone, but this variation can be ignored for our purposes. The *stiffness** of the tooth can be defined as the tangential force divided by the tangential deflection when the contact point is on the common centre-line, and can only be approximately

---

*The term 'stiffness' is used here to denote force divided by linear deflection in the direction of that force. The term 'rigidity' is used generally to denote the opposite of flexibility or resilience, but specifically in this book to denote torque divided by angular deflection in parts subject to torsion.

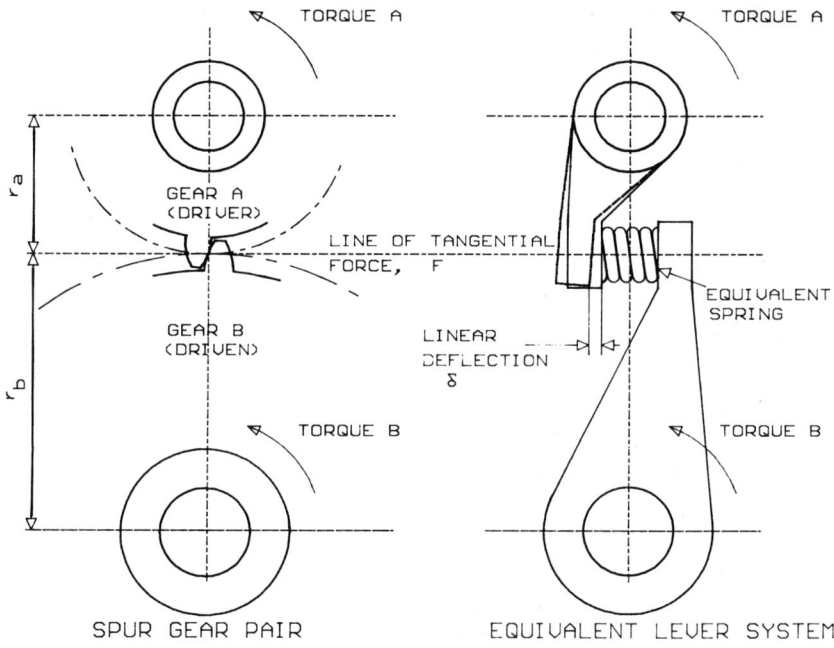

Fig. 11.2   Rigidity of spur gears

calculated from the tooth dimensions and material properties. When possible it is best to *measure* the torsional rigidity of a pair of gears rather than estimate it. However, some design guidance is derived from considering the relationship between tooth stiffness and torsional rigidity.

In Fig. 11.2 a pair of unequal spur gears is shown transmitting a torque with a consequent tangential tooth force. The deflections are analogous to those of a pair of levers whose tips are connected by a spring, as shown on the right.

The tangential tooth force *F* produces a total tangential deflection at the pitch circle of $\delta = F/S$ where S is the stiffness of both teeth in series. This causes gear A to rotate through an angle of δ/r_a radians if gear B is held stationary, or gear B to rotate through an angle of δ/r_b radians if gear A is held stationary. These rotations are, in effect, angular deflections caused by the torque applied to the relevant gear shaft.

Let M_a and M_b respectively be the torques (turning *Moments*) on gears A and B, whose pitch circle radii are r_a and r_b. The effective torsional rigidity of gear A relative to gear B is R_a ($=$ torque $\div$ angular deflection) and is found thus:

$$M_a = F \cdot r_a$$

therefore

$$R_a = \frac{M_a}{\delta / r_a} = \frac{F \cdot r_a^2}{\delta} = S \cdot r_a^2 \qquad \text{Eq. (11.3)}$$

similarly the effective torsional rigidity of gear B relative to gear A is

$$R_b = S \cdot r_b^2 \qquad \text{Eq. (11.4)}$$

In general torsional rigidity R is related to linear stiffness S acting at radius r by the equation:

$$R = S \cdot r^2 \qquad \text{Eq. (11.5)}$$

In a well designed transmission system gear tooth stiffness is seldom critical, but it is important to note that the torsional rigidity of a gear is proportional to the stiffness of the gear tooth, so that the larger the tooth the more rigid is the gear. Even more important is that the rigidity is proportional to the *square* of the pitch circle radius, which should discourage the use of small diameter gears for cam systems, even if they are strong enough.

Similar results are obtained from the analysis of the rigidity of other types of gear pair, such as bevel gears and worm gears. The estimate of tooth stiffness S from design information may not be very accurate, so it is always best to obtain a measured rigidity value from the gear set manufacturer or from a bench test.

All types of gearing usually operate with a small amount of backlash which increases as the gears wear. This can cause a problem in cam transmissions when there is a reversal of torque in every motion cycle, for example in high speed or high inertia applications. Some gears are designed to have backlash adjustment that should be reset from time to time. and some have automatic self-adjustment. The aim in cam transmissions must always be to have minimum backlash commensurate with acceptable initial cost and acceptable power loss which may cause overheating in high speed drives.

As we have already seen, the overall elasticity of a power transmission is the sum of the elasticities of each section of it that is connected in series (Equation 11.2). In effect the elasticity of one section is 'transmitted' to the next. When there is gearing between the two sections, however, the transmitted elasticity is modified by the gear ratio. To study this effect assume a pair of gears with infinitely stiff teeth, the input gear having Z_i teeth and the output gear having Z_0 teeth. The gear ratio is Z_i/Z_0.

Let the rigidity of all sections before the input gear be R_i and the rigidity of all sections after the output gear be R_0. Also let the input and output torques of the gear pair be M_i and M_0 respectively.

The torsional deflection of the input shaft is M_i/R_i. This is transmitted as a deflection at output gear of $(M_i/R_i) . (Z_i/Z_0)$. The torsional deflection of the output sections of the transmission is M_0/R_0 which is of course added to the transmitted deflection.

The total deflection at the output end of the transmission is therefore:

$$\frac{M_i . Z_i}{R_i . Z_0} + \frac{M_0}{R_0}$$

and the overall *elasticity* as seen at the output is this deflection divided by the output torque:

$$\frac{1}{R} = \frac{M_i . Z_i}{R_i . Z_0 . M_0} + \frac{1}{R_0}$$

But, ignoring gear efficiency, $M_i/M_0 = Z_i/Z_0$, so the equation for overall elasticity (reciprocal rigidity) at output becomes:

$$\frac{1}{R} = \frac{1}{R_i . (Z_0/Z_i)^2} + \frac{1}{R_0} \qquad\qquad \text{Eq. (11.6)}$$

This is similar to Equation 11.2, except that the first term has been modified. The rigidity 'transmitted' by gearing is multiplied by the gear ratio *squared*.

In a similar way it can be shown that the backlash 'transmitted' by gearing is directly proportional to the gear ratio.

A reduction gear *increases* the rigidity and a step-up gear *reduces* it. This fact is important when long transmissions are unavoidable and reduction gearing is necessary. If possible the longest part of a geared transmission should be the high-speed shaft: not only does it carry less torque than the low-speed shaft, and can therefore be of smaller diameter based on strength, but its elasticity is considerably mollified by the gear ratio.

CHAINS AND BELTS

Equations 11.3, 11.4 and 11.5 can also be applied to chain drives, but in that case the stiffness S refers to the stiffness of the loaded length of chain between the chainwheels. For a given chain size the stiffness is inversely proportional to its length: very long chain drives should therefore be avoided as should small diameter sprockets. Belt drives behave in a similar way, but are generally less satisfactory than chain drives.

Flat belt and vee belt drives are seldom used in cam system transmissions (except at the very high speed end, e.g. the primary drive from the electric motor where their elasticity is not important). Timing belt drives which use toothed belts and toothed pulleys are, however, quite common because like chain drives they give an exact speed ratio for synchronising with other mechanisms in the machine. Timing belts are made of reinforced synthetic rubber and are rather elastic compared to metal chains of similar strength. This is partly because the rubber belt teeth tend to roll slightly in the pulley grooves under heavy loads. Nevertheless, timing belt drives are very successfully used in cam transmissions because they are silent and need no lubrication.

The backlash problem with chains and belts is similar to that with gears, but usually more severe. Slack chain drives are quite common in conventional steady torque transmissions, and not particularly detrimental to them. The use of chain tensioner devices, of which there are many types commercially available, is strongly recommended for all cam transmissions, and are essential for high speed or high inertia applications.

MOUNTINGS

One effect of using gears, chain drives, etc. in a transmission, which is often overlooked, is the flexibility of the supports. The tooth load produces an equal reaction force at the gear supports. When the gears are enclosed in a rigid casing the elastic deflections of the supports are usually small enough to be ignored, although the reaction torque on the structure supporting the casing may itself be important. A rigid gearbox is no advantage if it is not rigidly mounted.

Gears and chain wheels are sometimes unavoidably mounted on shafts rather far away from the shaft bearings so that the bending of the shaft due to the tooth load becomes a significant part of the overall rigidity of

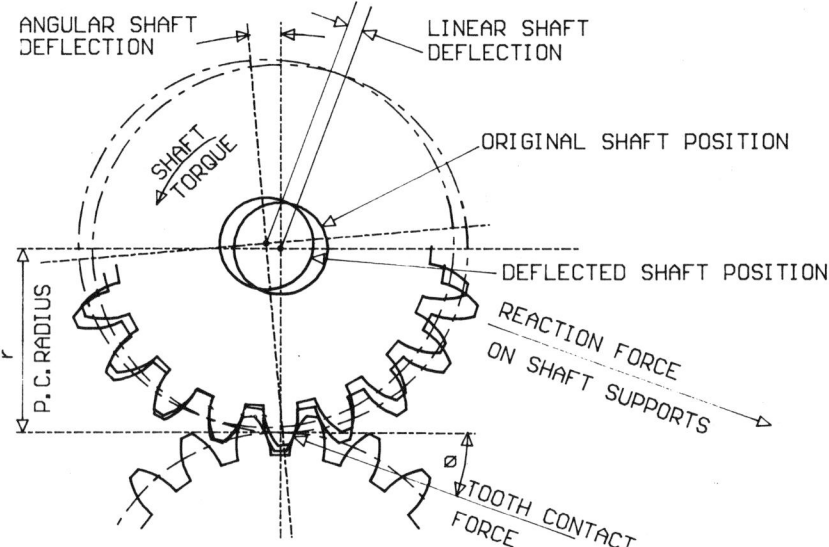

Fig. 11.3 Torsional elasticity due to shaft deflection

the transmission. Lateral deflection of the shaft has the same effect on angular displacement as tooth deflection and is, of course, in series with it. Figure 11.3 shows how the linear deflection of the shaft produces an angular deflection of the gear (or chain wheel) so that the effect is similar to torsional elasticity.

Here the shaft is displaced by the tooth contact force, $F . \sec \phi$, where F is the tangential force and ϕ is the involute gear pressure angle. This acts at a distance of $r . \cos \phi$ from the centre of the shaft, therefore the shaft torque is:

$$F . \sec \phi . r . \cos \phi = F . r \qquad \text{as expected.}$$

The angular deflection of the shaft, however, is $(\delta . \cos \phi)/r$, so that the effective torsional rigidity is:

$$R = \frac{F . r}{(\delta . \cos \phi)/r} = \frac{F . r^2}{\delta . \cos \phi} \qquad \text{Eq. (11.7)}$$

where δ is the lateral deflection of the shaft.

If the lateral stiffness of the shaft *in the plane of the gear* is S then

$$S = \text{force/deflection} = F \cdot \sec \phi / \delta$$

and the equivalent torsional rigidity is $R = S \cdot r^2$, which is the same as Equation 11.3. The relationship between lateral shaft stiffness and torsional rigidity is exactly the same as for tooth stiffness *and is independent of the gear tooth pressure angle.*

From beam bending theory we find that for a simply supported shaft of uniform cross-section the deflection at the sprocket *for a unit load* is:

$$\delta = \frac{l_1^2 \cdot l_2^2}{3 \cdot E \cdot J \cdot L} \quad \text{and its reciprocal is} \quad S = \frac{3 \cdot E \cdot J \cdot L}{l_1^2 \cdot l_2^2} \qquad \text{Eq. (11.8)}$$

where

$S = $ lateral stiffness of the shaft
l_1 *and* $l_2 = $ distances from the sprocket to end supports of the shaft
(bearings)
$E = $ Young's Modulus of elasticity of the shaft material.
$J = $ second moment of area of the shaft cross-section. For
bending of a circular shaft, $J = \pi \cdot (D^4 - d^4)/64$
$L = $ length of shaft between supports

If the shaft bearings are themselves mounted on a flexible structure the lateral deflection of the structure produced by the bearing reaction forces must also be taken into account in a similar way to the lateral deflection of the shaft. Structural flexibility is in series with all the other transmission elasticity.

Example 11.2

A camshaft is driven by a 2:1 reducing chain drive from a primary shaft which is driven by a motorised worm gear box, as shown in Fig. 11.4. Find the overall rigidity of the cam input transmission, from the output of the worm gear box. The shafts are made of medium carbon steel and the chainwheel and sprocket are solid discs of steel. The coupling is sufficiently rigid in torsion, and the shaft mountings are stiff enough to be ignored in the calculation. The transmission chain is 0.5 inch pitch with an assumed stiffness of 6×10^6 N/m for a 1 m length.

The rigidity of each section of the transmission will be considered separately and then finally combined.

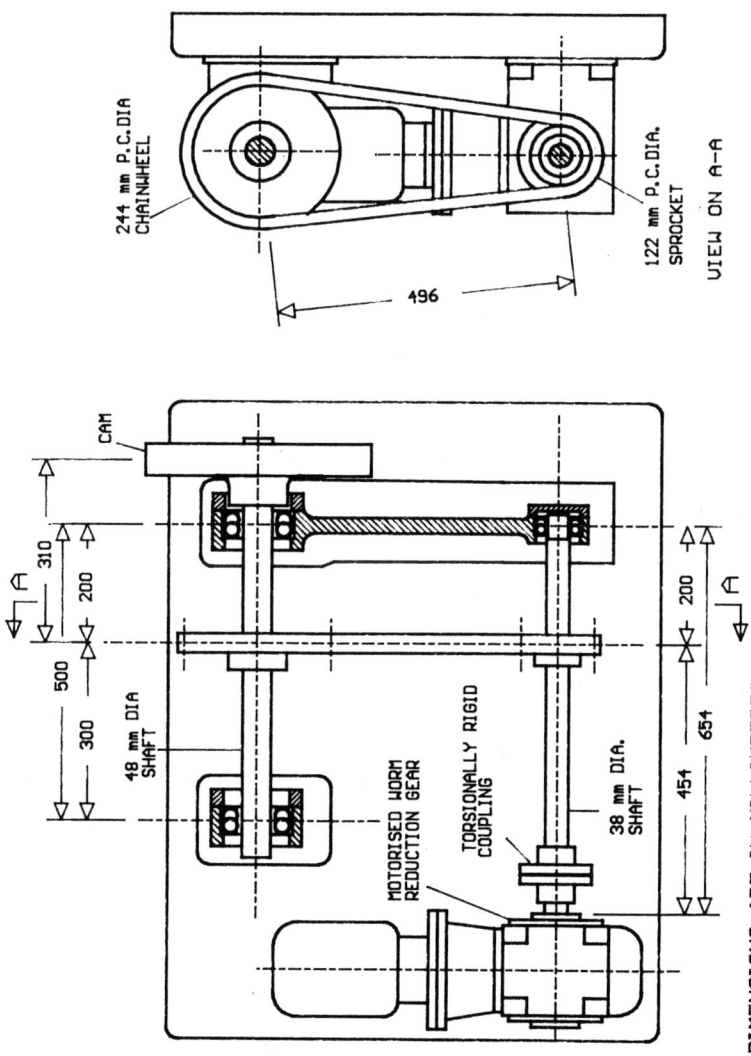

244 mm P.C.DIA CHAINWHEEL

122 mm P.C.DIA. SPROCKET

VIEW ON A-A

496

CAM

A

310

200

500

300

48 mm DIA SHAFT

MOTORISED WORM REDUCTION GEAR

TORSIONALLY RIGID COUPLING

38 mm DIA. SHAFT

200

654

454

A

Fig. 11.4   Example 11.2

ALL DIMENSIONS ARE IN MILLIMETRES

(1) Primary shaft torsion

This is similar to Example 11.1, but simpler because it is of uniform diameter. The section of the shaft transmitting torque is 38 mm diameter by 454 mm long, therefore its rigidity is

$$\frac{\pi \times 82.5 \times 10^9 \times 0.038^4}{32 \times 0.454} = 37{,}199 \text{ Nm/rad}$$

But this shaft is connected to the cam shaft by a 2:1 reduction drive, so its rigidity *referred to the camshaft* is

$$R_1 = 37{,}199 \times 2^2 = 148{,}796 \text{ Nm/rad}$$

(2) Primary shaft bending

For steel $E = 205 \times 10^9 \text{ N/m}^2$

$$J = \pi \cdot D^4/64 = \pi \times 0.038^4 \div 64 = 1.0235 \times 10^{-7} \text{ m}^4$$

$$S = \frac{3 \times 205 \times 10^9 \times 1.0235 \times 10^{-7} \times 0.654}{0.454^2 \times 0.200^2} = 4.993 \times 10^6 \text{ N/m}$$

The P.C. Radius of the sprocket is 61 mm $= 0.061$ m, therefore the equivalent torsional rigidity, from Equation 11.3, is

$$4.993 \times 10^6 \times 0.061^2 = 18{,}579 \text{ Nm/rad}$$

The rigidity *referred to the cam shaft* via the 2:1 reduction is

$$R_2 = 18{,}579 \times 2^2 = 74{,}316 \text{ Nm/rad}$$

(3) Drive chain

From the drawing we find that the length of chain that stretches under load is 496 mm and we are given that its stiffness is 6×10^6 N/m for a 1 m length, so $S = 6 \times 10^6 \div 0.496 = 12.096 \times 10^6$ N/m. The P.C. Radius of the chainwheel is 122 mm, therefore the rigidity of the chain referred to the cam shaft is

$$R_3 = 12.096 \times 10^6 \times 0.122^2 = 180{,}048 \text{ Nm/rad}$$

(4) Cam shaft bending

$$J = \pi \times 0.048^4 \div 64 = 2.606 \times 10^{-7} \text{ m}^4$$

$$S = \frac{3 \times 205 \times 10^9 \times 2.606 \times 10^{-7} \times 0.500}{0.300^2 \times 0.200^2} = 22.26 \times 10^6 \text{ N/m}$$

$$R_4 = 22.26 \times 10^6 \times 0.122^2 = 331{,}300 \text{ Nm/rad}$$

(5)   Cam shaft torsion (48 mm diameter × 310 mm long)

$$R_5 = \frac{\pi \times 82.5 \times 10^9 \times 0.048^4}{32 \times 0.310} = 138{,}694 \text{ Nm/rad}$$

These rigidities are combined using Equation 11.2 thus

$$\frac{1}{R} = \frac{1}{148{,}796} + \frac{1}{74{,}316} + \frac{1}{180{,}048} + \frac{1}{331{,}300} + \frac{1}{138{,}694}$$

$$= 3.596 \times 10^{-5}$$

The overall rigidity of the transmission referred to the cam is

$$R = 27{,}809 \text{ Nm/rad}$$

The most significant element of transmission elasticity is the bending of the primary shaft, which has the lowest rigidity R_2. This points to the possibility of a considerable improvement if the chain drive could be moved closer to the right-hand bearings or the sprockets could be increased in diameter.

LEVERS AND LINKS

These are similar to gears and chainwheels in that a linear deflection δ under force F at a distance r from the lever pivot (or fulcrum) can be expressed as a linear stiffness $S = F/\delta$ at that point. This is translated into a torsional rigidity at the pivot of $R = S \cdot r^2$ which is again the same general equation 11.5 given above. The designs of levers are many and varied, and the calculation of lever stiffness is therefore outside the scope of this book: it comes within the conventional theory of deflection of beams. In practice well designed levers are seldom a significant source of elasticity in transmissions, unless they are very long.

The elongation and compression of links (pull-rods and push-rods) are analogous to the elongation of a chain, described above, and the relationship between the stiffness of a link and the torsional rigidity at a lever pivot is exactly the same as between a chain and its chainwheel shaft. Equation 11.5 again applies.

OUTPUT TRANSMISSION

The estimation of rigidity of an output transmission is exactly the same as for an input transmission. Most inputs, of course, drive rotary cams and therefore rigidity is expressed as the overall *torsional* rigidity. With output transmissions, however, we are dealing with a payload that is driven by the cam follower and the motion can be either linear (reciprocating) or rotary (oscillating or indexing). If the follower has a translating – linear – motion then the overall rigidity of the output transmission is best expressed as a linear stiffness referred to the follower. If the follower motion is swinging or rotating then it is best expressed as a torsional rigidity referred to the follower axis.

Example 11.3

A cam-driven mechanism is operated by a lever and pull-rod transmission with a stroke-increasing ratio of 1.5:1 as shown in Fig. 11.5. A 60 mm long follower arm is keyed to one end of a 20 mm diameter steel pivot shaft on the other end of which is keyed a 90 mm long pull-rod lever. That lever pulls a payload by means of an 8 mm diameter × 120 mm long steel pull-rod. Find the overall transmission rigidity from the cam to the payload, assuming that the follower arm and pull-rod lever are stiff enough to be ignored in the calculation.

Working back from the payload to the cam,

(1)   Pull-rod

The stiffness of a bar in pure tension or compression is

$$S = E \cdot A/L \qquad\qquad\qquad \text{Eq. (11.9)}$$

where

E = Young's Modulus (say 205×10^9 N/m^2 for steel)

A = cross-section area

L = length under stress

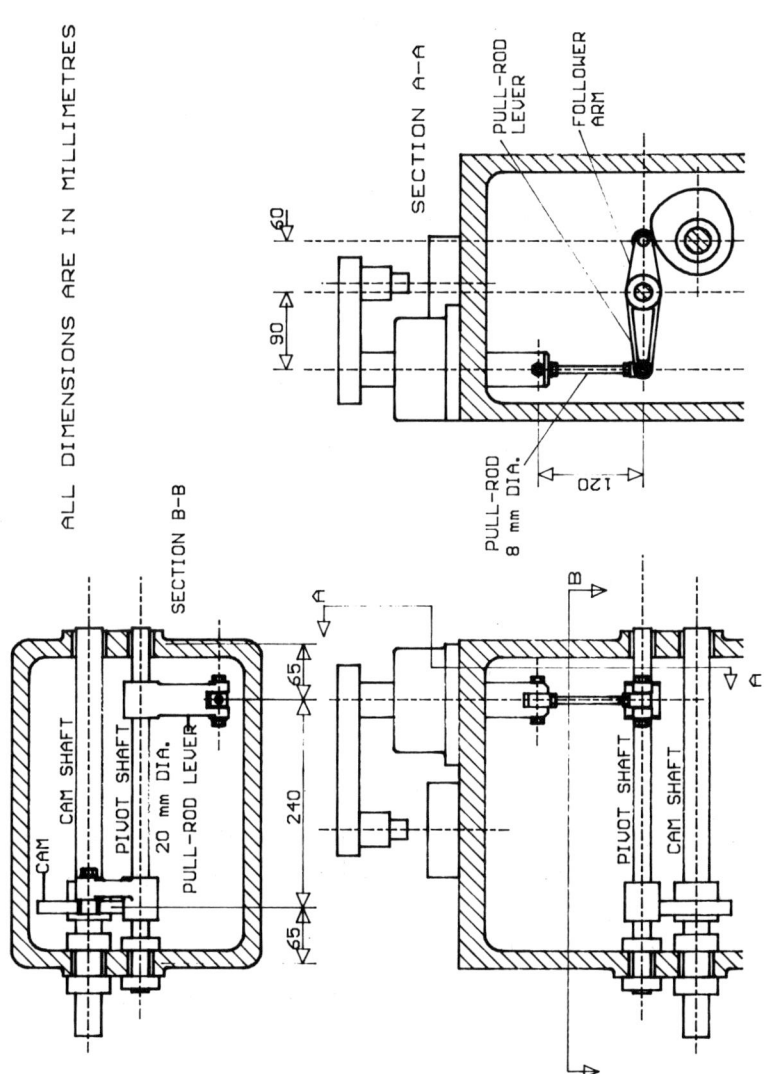

ALL DIMENSIONS ARE IN MILLIMETRES

SECTION A-A

PULL-ROD
LEVER

FOLLOWER
ARM

60

90

PULL-ROD
8 mm DIA.

120

SECTION B-B

CAM SHAFT

PIVOT SHAFT

20 mm DIA.

PULL-ROD LEVER

CAM

65

240

65

A

A

B

B

PIVOT SHAFT

CAM SHAFT

Fig. 11.5   Example 11.3

The cross-section area of the rod is

$$A = \pi \times 0.008^2 \div 4 = 5.026 \times 10^{-5} \, m^2$$

therefore

$$S = 205 \times 10^9 \times 5.026 \times 10^{-5} \div 0.120 = 85.87 \times 10^6 \, N/m$$

This acts at the end of a 90 mm long lever, so the torsional rigidity referred to the pivot shaft from Equation 11.5 is

$$R_1 = 85.87 \times 10^6 \times 0.090^2 = 695,500 \, Nm/rad$$

(2) Pivot shaft bending at pull-rod position

Using Equations 11.8 and 11.5:

$$J = \pi \times 0.020^4 \div 64 = 7.854 \times 10^{-9} \, m^4$$

$$S = \frac{3 \times 205 \times 10^9 \times 7.854 \times 10^{-9} \times 0.370}{0.305^2 \times 0.065^2} = 4.547 \times 10^6 \, N/m$$

Rigidity referred to the pivot shaft is

$$R_2 = 4.547 \times 10^6 \times 0.09^2 = 36,832 \, Nm/rad$$

(3) Pivot shaft torsion

The length of the shaft in torsion is 240 mm, so from Equation 11.1:

$$R_3 = \frac{\pi \times 82.5 \times 10^9 \times 0.020^4}{32 \times 0.240} = 5,400 \, Nm/rad$$

(4) Pivot shaft bending at cam follower position

The shaft is symmetrical along its length and its stiffness in bending S at this position is the same as at the pull-rod position. The rigidity referred to the pivot shaft is therefore:

$$R_4 = 4.547 \times 10^6 \times 0.060^2 = 16,369 \, Nm/rad$$

The overall torsional rigidity of the cam output transmission at the follower arm pivot is:

$$\frac{1}{R} = \frac{1}{695,500} + \frac{1}{36,832} + \frac{1}{5,400} + \frac{1}{16,369} = 2.7586 \times 10^{-4}$$

therefore

$R = 3,638$ Nm/rad

This could be expressed as a linear stiffness at the follower roller by transposing Equation 11.5

$S = R/r^2 = 3638 \div 0.060^2 = 1.01 \times 10^6$ N/m

It is clear from the above figures that torsion of the pivot shaft is by far the most elastic element, showing that the transmission rigidity can be considerably improved, if necessary, by increasing the shaft diameter. Because shaft rigidity and stiffness are proportional to the fourth power of diameter, an increase from 20 mm to only 24 mm would approximately double the overall rigidity.

COUPLINGS

There are many types and designs of shaft coupling available commercially, as well as special purpose designs. Some are 'rigid' in the sense that they are not intended to allow relative movement of any kind between the coupled shafts. Others are 'self-aligning' to allow limited movement between the shafts, usually to provide for either accidental or deliberate misalignment. Of the latter, some allow all degrees of freedom – these are the 'flexible' couplings – and some allow only one or two degrees of freedom. Various commercial types are illustrated in Fig. 11.6.

The flexible couplings shown are just a few of the many types available which transmit torque via a resilient medium (rubber, plastics or metal spring) and are *not* torsionally rigid: as such they are *not* recommended for cam systems. Of the torsionally rigid couplings shown the Cardan joint allows a large degree of angular misalignment, the Oldham coupling allows a large degree of parallel offset and some end float, and the gear and chain couplings allow a small degree of freedom in all directions except torsion. With the exception of the membrane type, the mechanical couplings are subject to gradual wear which results in rotary backlash, which may be a problem for cam drive transmission if the couplings are not large enough. A splined shaft (not shown) is a form of coupling which allows substantial longitudinal displacement of a shaft while retaining torsional rigidity. However, it must have clearance to allow for the relative movement, and because the pitch circle radius of splines is inevitably small the rotary backlash is potentially severe.

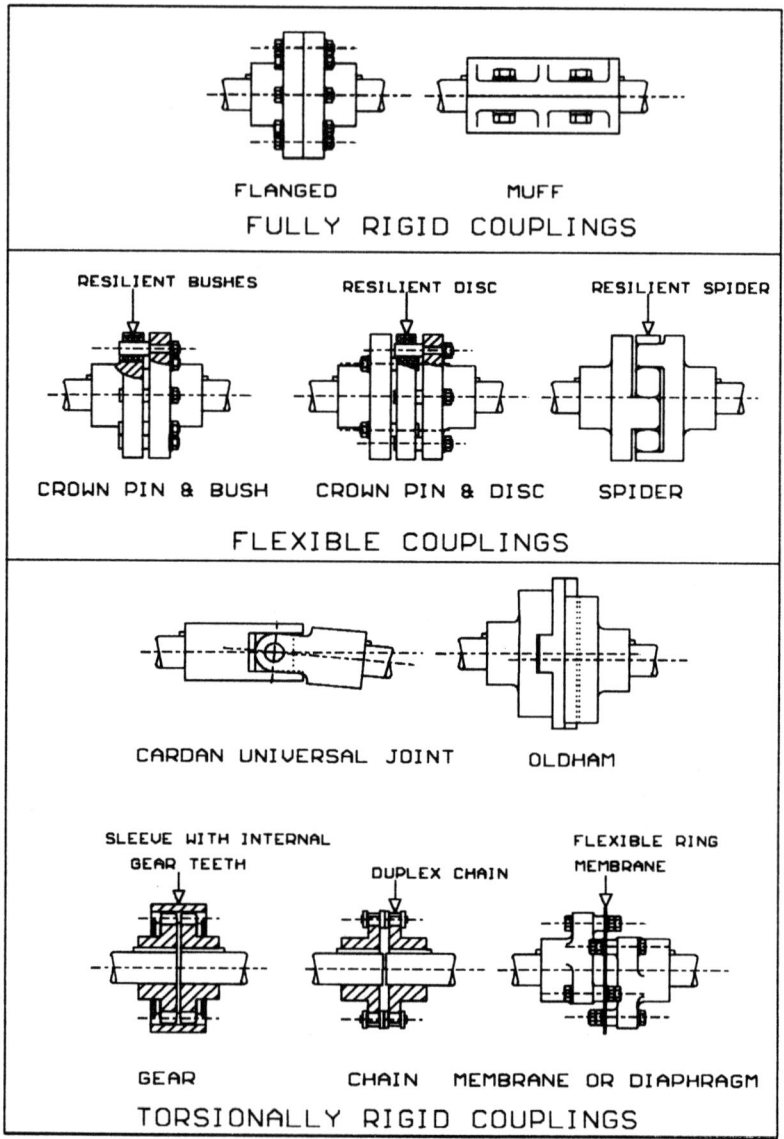

Fig. 11.6   Shaft couplings

The membrane or diaphragm coupling has a thin flexible membrane, sometimes laminated, attached alternately to the two hubs. This allows limited angular misalignment and end float, but virtually no torsional elasticity and no backlash. Lateral shaft offset, if required, can be achieved using two such couplings separated by a short length of shaft. This design of coupling is ideal for cam drives, for both input and output transmissions.

CHAPTER 12

Dynamic Response: Vibration and Backlash

When a payload is moved by a cam the action can be thought of in terms of command and response. The cam commands a particular motion and the payload responds. The response, however is neither immediate nor entirely true. Because the payload invariably has mass or inertia, a dynamic force or torque is required to move it and that force is transmitted through an elastic medium (the power transmission components discussed in the previous chapter) which must distort when subjected to the said force or torque. The resulting motion of the payload – its *dynamic response* – is a distortion of the command motion and takes the form of a vibration of more or less constant frequency, but of varying amplitude, superimposed on the command motion.

The mechanism can be represented by a simple model of a cam and a follower of negligible mass driving a payload of significant mass by means of a spring representing the power transmission, as illustrated in Fig. 12.1. The figure also shows an exaggerated diagram of the manner in which the motion of the payload (its dynamic response) departs from the motion of the cam (the command).

When the follower begins to move the payload cannot start moving until there is sufficient force to overcome its inertia, and that force must compress the spring by an amount proportional to it: the ratio of force to spring deflection is its stiffness, S. Consequently the payload displacement lags behind the cam follower displacement and continues to do so until the spring force imparts sufficient acceleration to the payload for its velocity to exceed the follower velocity. From then on the spring expands until it is at its original length and the payload displacement matches the follower displacement. But at that point the payload is moving faster than the follower and overshoots the follower displacement. As the spring expands more the payload velocity decreases relative to the follower velocity until the two displacements again match.

A similar cycle of events is repeated over and over again throughout the stroke, and the resulting payload motion is an oscillation of varying

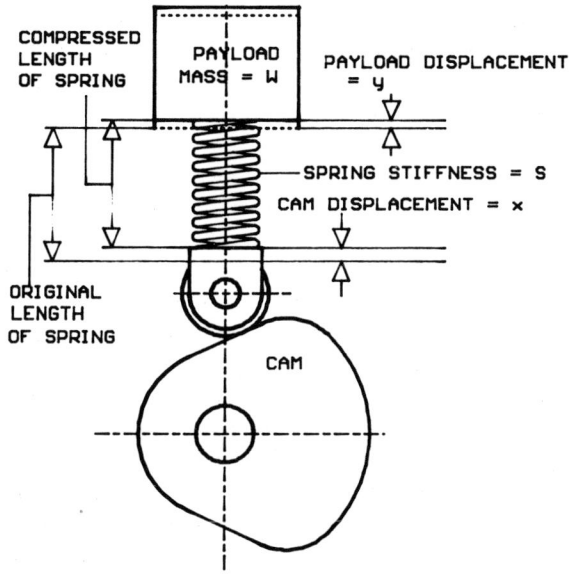

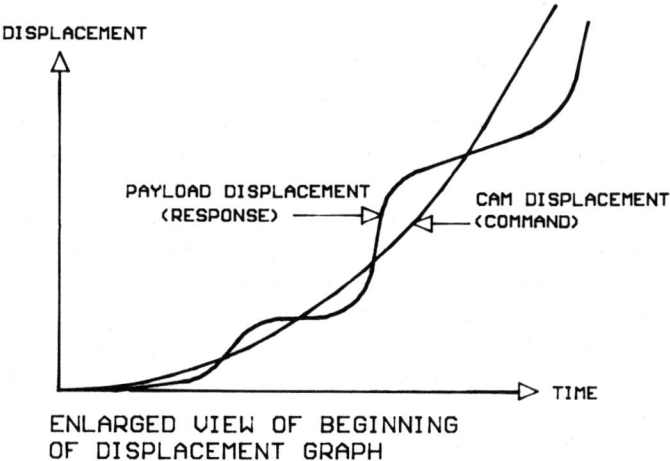

ENLARGED VIEW OF BEGINNING
OF DISPLACEMENT GRAPH

Fig. 12.1   Dynamic response model

amplitude superimposed on the basic cam motion. In practice the *displacement* amplitude of the oscillation is much less than illustrated in Fig. 12.1. The *acceleration* amplitude, however, is a large part of the total acceleration, and since inertia force is proportional to acceleration, the oscillation makes a significant contribution to the peak load on the system.

To avoid confusion with oscillating, or swinging, follower mechanisms this kind of oscillation is usually referred to in this book as a *vibration*, not to be confused with the vibrations that occur in machines, often of a random nature, from sources quite independent of the cam mechanism. The nature of the vibration, its amplitude in particular, is a function of the cam law, the motion period (the time occupied by the stroke), the mass of the payload and the elasticity of the mechanism. The frequency of the vibration is the *natural frequency* of the mass-spring system. All mass-spring systems, and therefore all cam operated mechanisms, have a natural frequency of vibration, which is the frequency with which the nominally static system will vibrate if strained from its equilibrium state and suddenly released. The natural frequency of a mechanism can be either measured or estimated with reasonable accuracy from the mass and stiffness involved (or in the case of rotary systems the moment of inertia and torsional rigidity).

For any given cam law the maximum dynamic response vibration amplitude is greatly dependent on the *period ratio* of the system. This is the ratio of the motion period (stroke time) to the vibration period. In other words it is the number of vibration cycles that occur in the motion period. It is designated as the parameter '*n*' and it is *one of the most important parameters* in estimating the peak load on a cam. Low values of *n* give high values of vibration amplitude and vice versa. Very low values of *n* can result in a disastrous mechanism performance.

When we talk of the speed of a cam mechanism as being 'high' or 'low' we usually relate this in our minds to the kind of task it is performing. For example, a cam-operated ten-tonnne press with a 400 mm stroke might be considered fast at 100 strokes per minute, but an automotive engine valve cam, with its light payload and short stroke, would be thought to be running very slowly at that speed. When selecting a cam law the more sensible guide to speed is the period ratio *n* of the application. High speed applications are those with a low period ratio, say less than 8, and slow speed applications (in which the choice of cam law is less critical) are those with a high period ratio, say more than 20. Most cam mechanisms in modern industrial machinery run with period ratios less than 20 and a few run with period ratios less than 5.

Figure 12.2 shows the dynamic responses of a mechanism with three popular cam laws and two different period ratios. The diagrams are all plotted to the same scale and show the output force plotted against time throughout one motion period and one dwell period, for a mechanism with a largely inertial payload. The diagrams in the top row represent a mechanism with a good, fairly rigid output transmission which has a period ratio of about 13.5. Those in the bottom row represent the same mechanism with the same payload, but with a bad, flexible output transmission which has a period ratio of about 3.25. The output transmission rigidity is the only difference between the two groups. In each case the nominal force (for a system with no vibration) is drawn with the dynamic response superimposed on it. The diagrams are computer-generated, but they agree well with measured responses from actual mechanisms.

A number of important points are illustrated by these diagrams.

(1) For all three cam laws a high period ratio produces a superimposed vibration of low amplitude. Conversely, a low period ratio can produce a high vibration amplitude (force or torque) which may be higher than the nominal maximum of the cam law, by a factor of more than 2.

(2) There is a residual vibration in the dwell period which may be damped down to zero before the next motion period if there is some friction load and the vibration amplitude is low, i.e. the transmission rigidity is good. The residual vibration of a bad transmission, however, may not be damped out quickly enough and could make the dynamic response of the subsequent motion worse still. The effects of damping are discussed in more detail later: the mechanism used for this illustration has a small friction load and a fair amount of viscous damping.

(3) If the transmissions are good the Mod Trap motion has a peak load less than either Mod Sine or Cycloidal, by virtue of its lower maximum acceleration. If transmissions are bad, however, Mod Sine is better than Mod Trap (note the *negative* peak of the Mod Trap) and Cycloidal is better than Mod Sine, especially as regards the residual vibration.

All three cam laws shown in Fig. 12.2 are considered to be 'good' laws for

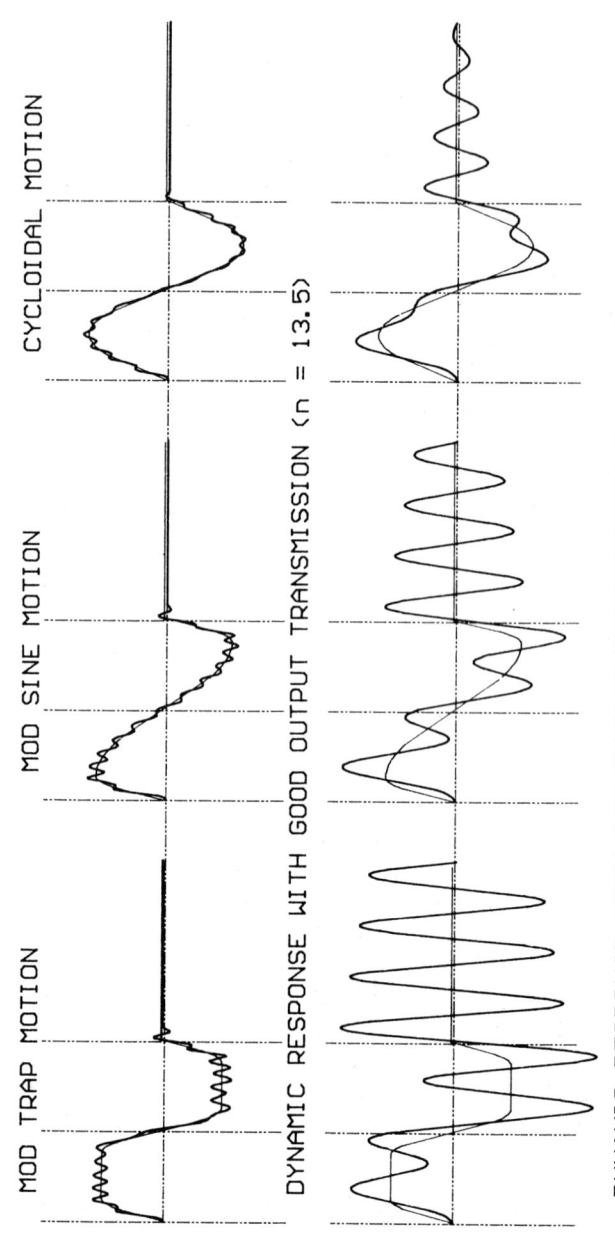

Fig. 12.2 Comparison of dynamic response of MT, MS, and CYC motions

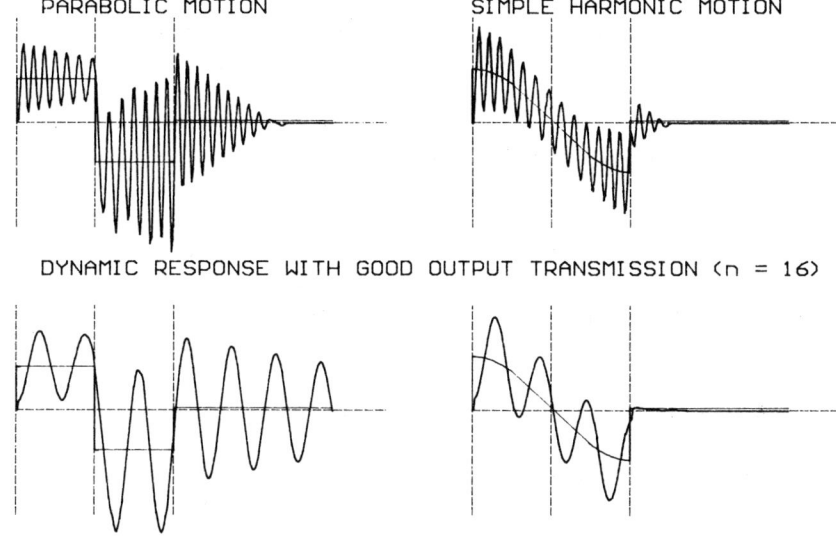

Fig. 12.3   Dynamic response of PAR and SH motions

high speed mechanisms. For comparison the dynamic response of mechanisms similar in all respects, except for the cam law, is illustrated in Fig. 12.3 with Parabolic and Simple Harmonic motions. These cam laws were popular originally because of their simple formulation, their low nominal maximum acceleration and the apparently smooth profiles that they produced. However, neither of them gives a satisfactory dynamic response, even at low speed with good transmissions (high value of period ratio).

The value of peak force or torque in both cases is much the same, at all period ratios, and is higher than the peak torque of the modern dynamically favourable cam laws.

TORSION FACTOR

The peak torque (positive or negative) of the superimposed vibration is, of course, a critical factor in estimating the load/life capacity of a cam mechanism. It is not possible to measure this at the mechanism design stage, although it can be estimated for a particular case using a fairly sophisticated computer program. More conveniently, it can be approximated by the use of a *torsion factor*, provided that the period ratio can be estimated for the application. The torsion factor, designated C_t, is the ratio of the actual peak force or torque (including the amplitude of the superimposed vibration) to the nominal peak force or torque of an undamped cam driven system with a wholly inertial load. The reason for it being known as a *torsion* factor is historical: it was first used in connection with cam driven *indexing* mechanisms in which the transmissions are rotary and subject to torsional vibration.

All cam driven mechanisms that can be fairly represented by the model of Fig. 12.1 and which have the same cam law have a torsion factor which is dependent *only* on the period ratio. This is true whether the mechanism is running fast or slow, is lightly or heavily loaded or is very rigid or very flexible: all those variable qualities of the mechanism affect the value of the period ratio, but once the period ratio and cam law are known a torsion factor can be established. Curves have been plotted of torsion factor C_t against period ratio n for the SCCA family of cam laws, and their peak values have been found empirically to be covered by the following equation:

$$C_t = \frac{p}{n^q} + r \qquad\qquad\qquad \text{Eq. (12.1)}$$

where p, q and r are constants peculiar to each cam law, as listed below:

Table 12.1   Torsion factor parameters for SCCA cam laws

Cam law	p	q	r
Parabolic	0	0	4
Simple Harmonic	0.2	0.65	2
Mod Trap	17.8	1.94	1.1
Mod Sine	3.87	1.13	1
MSC.20	4.5	0.96	1
MSC.33	4.91	0.94	1
MSC.50	7.75	1	1
MSC.66	9.66	0.9	1
Cycloidal	1.5	1	1
CYCC.50	8.57	1.07	1

Note: This table is valid for period ratios (n) of 3 and above.

It is apparent from the table that the Parabolic cam law has a very high torsion factor ($C_t = 4$) *irrespective of period ratio*, which makes it a non-preferred motion for any application, even slow speed ones with good transmissions. Simple Harmonic has a high torsion factor ($C_t > 2$) at all period ratios, making it less suitable than Mod Trap, Mod Sine or Cycloidal for many applications. It must be said, however, that the high torsion factor of SH motion is due to the sudden change of acceleration at the points of transition between dwell and motion. In applications where there are no dwells, therefore no sudden changes, SH motion is by far the best as regards vibration. In any case the low value of jerk at cross-over (transition between acceleration and deceleration) of SH motion minimises the deterimental effects of any backlash in the system.

NATURAL FREQUENCY AND PERIOD RATIO

For a simple spring-mass system, as shown in Fig. 12.1, the natural frequency is:

$$f = \frac{1}{2 \cdot \pi} \cdot \sqrt{\left(\frac{S}{W}\right)} \text{ for a linear system} \qquad \text{Eq. (12.2)}$$

or:

$$f = \frac{1}{2 \cdot \pi} \cdot \sqrt{\left(\frac{R}{I}\right)} \text{ for a rotary system} \qquad \text{Eq. (12.3)}$$

where

f = natural frequency [Hz]

S = linear stiffness [N/m]

W = mass [kg]

R = torsional rigidity [N.m/rad]

I = moment of inertia [kg.m^2]

and the period ratio is:

$$n = f.T \qquad \text{Eq. (12.4)}$$

where

T = motion period (time) [s]

When the input transmission is very rigid, or there is a high inertia (e.g. a flywheel) close to the cam, the S or R values in the above equations can be those of the output transmission alone. However, there are many applications in which the input transmission elasticity is significant and must be taken into account. The complexity of the various combinations of input and output rigidity, input and output inertia, etc. cannot be dealt with here, but in many cases the natural frequency can be estimated with sufficient accuracy for our purpose by combining the input and output transmission rigidities as if the cam mechanism were a gear pair.

The effective 'gear ratio' is:

$$\frac{\text{Output displacement}}{\text{Input displacement}} = C_v \cdot \frac{\text{Output stroke}}{\text{Input stroke}} \qquad \text{Eq. (12.5)}$$

This is strictly the *maximum* velocity ratio of the mechanism which applies only at one part of the motion: the 'gear ratio' of a cam mechanism varies throughout the motion from zero to maximum back to zero. There are so many adverse factors in a practical application that are not accounted for in this simplified treatment that the overstatement of elasticity implied by Equation 12.5 is not out of place in an estimate of natural frequency.

RESIDUAL VIBRATION

We have already seen that in most cases there is a residual vibration at the end of the motion, whose amplitude may be considerable. It is of course a continuation of the vibration that has built up during the motion period. However, it is possible for the residual vibration amplitude to be very small or zero, even though the in-motion amplitude be very large. This depends on exactly where in a vibration cycle the motion finishes. Slight variations in the value of n can make an enormous difference to the residual vibration amplitude.

The dynamic responses of two similar mechanisms are shown in Fig. 12.4. A high speed application (low period ratio) has been chosen for clarity of illustration. The only difference between the two mechanisms is the rigidity of the output transmission, so that there is a slightly different period ratio in each case. The diagram on the left has a value of $n = 5.25$ and the one on the right has a *worse* value of 4.5. Surprisingly, there is an extremely large residual vibration with the better period ratio, while the

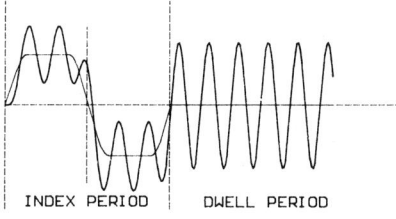 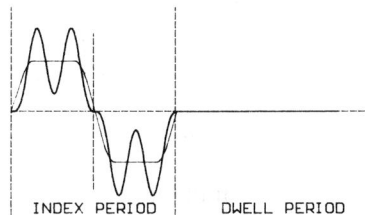

INDEX PERIOD | DWELL PERIOD   INDEX PERIOD | DWELL PERIOD

MAXIMUM RESIDUAL VIBRATION AMPLITUDE   MINIMUM RESIDUAL VIBRATION AMPLITUDE
PERIOD RATIO n = 5.25 APPROX.   PERIOD RATIO n = 4.5 APPROX.

Fig. 12.4 Residual vibration and period ratio

worse ratio produces practically no residual vibration. It is important to note that for these mechanism models no damping of any kind was used: the zero residual vibration is entirely due to the vibratory state of the mechanism at the beginning of the dwell period. The in-motion vibration is much the same for both mechanisms and is of similar amplitude to the worst residual vibration.

The value of n can seldom be predicted with great accuracy because of unforeseen errors in mass, rigidity or speed. In any case, there is a great variation of speed, and therefore of period ratio, during the run up to the full operating speed of the machine and during the shut down phase. Although residual vibration may be undesirable for many reasons, it is not in itself the criterion for estimating the load capacity of a cam system. The peak load on the system is always at the peak (positive or negative) of an in-motion vibration cycle.

DAMPING

Damping is the absorption of energy by the application of a force or torque which has the effect of reducing vibration, possibly eliminating it. There are three basic kinds of damping that apply to cam and follower systems: viscous damping (force proportional to velocity), hysteresis damping (in which more energy is used to stress a material than is released by unstressing it) and constant or Coulomb friction damping (in which a more or less constant force opposes the motion).

All of these are present to some degree in industrial machines, but the first two, viscous and hysteresis, are usually so low as to have little or no effect on vibrations. It is futile to increase viscous damping deliberately to dissipate vibration energy: to do any good near the beginning and end of the motion where the nominal velocity is very low, or in the dwell period, the damping factor would have to be so high that the energy wasted in the central high speed part of the motion would be unacceptable. Only that part of the viscous damping force that is related to the *change* of velocity contributes to the reduction of vibration, and the relative change of velocity due to vibration, even in the in-motion period, is quite small.

A constant friction damping force is often present in the form of friction in bearings, slideways, etc. This force, however, has no damping effect at all unless there is a change of its direction, that is a reversal of the direction of motion. Such reversals do not take place until the vibration velocity exceeds the nominal follower velocity, that is during and just before the dwell period. This form of damping is very effective in dissipating residual vibration, except in very high speed applications, but does not affect the majority of in-motion vibrations at all. If it is important in a particular application to eliminate residual vibration then the deliberate introduction of constant friction damping may be justified. However, it has the drawback that the follower can come to rest slightly out of position in the dwell period, being held by the friction force in a strained condition, either under- or over-shooting its target.

If peak load is used as a criterion for selecting a cam law – although it is by no means the only feasible criterion – then the product of torsion factor and coefficient of acceleration $(C_t \cdot C_a)$, which is a measure of the peak inertia load, indicates the best cam law for a particular range of period ratios. Using Equation 12.1 and the associated table the curves of $(C_t \cdot C_a)$ vs. n for Mod Trap, Mod Sine and Cycloidal laws are plotted in Fig. 12.5 to show which of the three is the best choice based on period ratio. This comparison is only valid where there is a very good input transmission and the peak load alone is used as a basis for cam law selection.

In these circumstances it can be seen that Mod Trap is the best choice for a period ratio above 6.4 and Cycloidal is best below 6.4. However, Mod Sine is by far the most useful of the three cam laws because, although it produces a slightly higher peak load than Mod Trap, it is very much more tolerant of an elastic input transmission. *The transmission systems of most industrial cam systems are such as to benefit from the use of the Mod Sine cam law in preference to Mod Trap.* Nevertheless, there is a positive advantage in using Cycloidal for systems with a low period ratio: It is recommended that Mod Sine be the first choice for period ratios above 6.4, and Cycloidal for ratios lower than that.

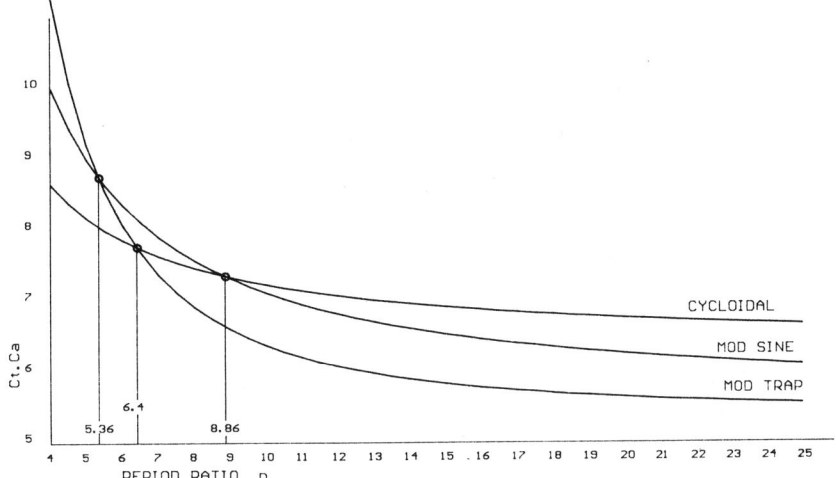

Fig. 12.5 Cam law selection criterion

BACKLASH*

Any clearance between the cam and follower or between any two force-transmitting components in the system gives rise to a shock load whenever there is a reversal of force or torque. The greater the clearance, the greater is the shock. The cumulative backlash is sum of all the clearance, play or slack in both the input and output transmissions, adjusted if necessary by gearing ratios, and in the cam track and cam follower. Typical examples are slack chain drives, gear tooth clearances, oversize enclosed cam tracks and worn follower roller bearings: there are many others.

Backlash can often be eliminated by applying sufficient external force with a spring or the payload weight, or even a friction force, to ensure that there is no force reversal at the operating speed. Spring-loaded open track cam systems are quite common, but to be fully effective in eliminating backlash the spring must be applied not just to the follower,

---

*Strictly speaking, 'backlash' is the shock effect of a reversal of foces in a system which has significant clearance between adjacent members that are transmitting the force. After many decades of usage it has come to mean the clearance itself, and that is how it is used throughout this book.

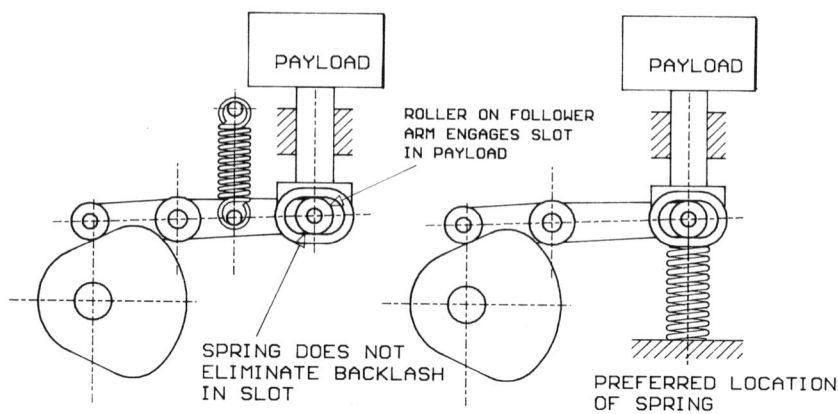

PAYLOAD

PAYLOAD

ROLLER ON FOLLOWER
ARM ENGAGES SLOT
IN PAYLOAD

SPRING DOES NOT
ELIMINATE BACKLASH
IN SLOT

PREFERRED LOCATION
OF SPRING

Fig. 12.6   Spring-loaded mechanism

but at a point in the output transmission that takes in *all* the significant clearances. This point is illustrated in Fig. 12.6.

When the direction of force reverses, typically at the cross-over in high inertia applications, there is a moment when the payload is in *free flight* after losing contact with positive force surfaces and before making contact with negative force surfaces. The magnitude of the force on making contact – the impact force – is related to the approach velocity of the contact surfaces – the impact velocity. This in turn is related to the values of cam acceleration and jerk at this point in the motion.

The process can be illustrated by the simplified and exaggerated diagram in Fig. 12.7. All the cumulative backlash in the system is represented by the separation of two cam profiles and the payload is represented as concentrated at a point running along the profiles. The lower profile can only exert a positive contact force on the payload and the upper profile only a negative contact force. It is assumed for the purpose of this description that dynamic response vibration and any other detrimental effects are not significant compared with the impact force.

The payload departs from the lower profile when its contact force becomes zero: that is when the inertia force due to profile deceleration is about to exceed any other retarding forces on the payload such as friction. From this point on, the payload follows a free flight path with a *natural deceleration* which is determined by the external forces (such as friction), until it makes contact with the upper profile. It strikes the profile with an

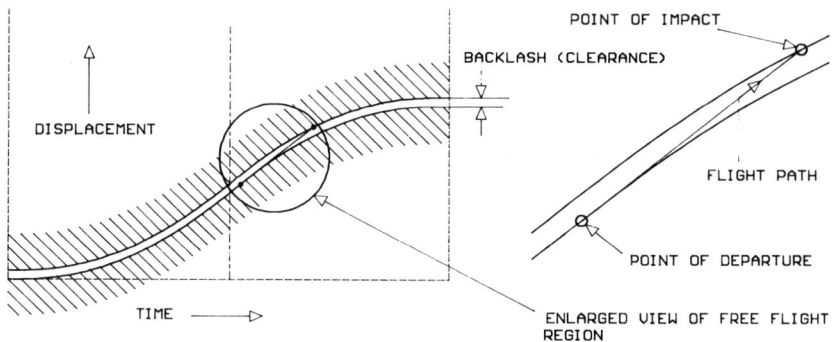

Fig. 12.7 Backlash effect

approach velocity – the impact velocity – which is the difference between its free-flight velocity and the profile velocity. The free-flight is now a little slower than at the point of departure, and the profile velocity is even slower because the impact point is so much farther along the profile. It is obvious from the diagram that both departure and impact points occur in the deceleration phase of the motion and are separated by a time and distance very much dependent on the amount of backlash. How far into the deceleration phase that these events occur depends on the magnitude of the external force in relation to the payload inertia force. In practice the possible range of impact positions is fairly wide and occurs just after the nominal cross-over point of the motion.

To minimise the backlash impact force, therefore, it is necessary to have a cam law with a low value of acceleration for a long period near the middle of the motion. In general, this condition is fulfilled by motions that have a low jerk at cross-over. The best 'low-impact' *standard* cam law is Simple Harmonic, but special low impact cam laws have been developed which improve on this. Here again the Mod Sine law offers a very good compromise: it has a lower jerk at cross-over than Mod Trap, or Cycloidal and a much better dynamic response vibration than Simple Harmonic.

It is difficult to quantify the impact effect of backlash in all circumstances, but an attempt is made here to indicate its importance in certain cases. To take account of various sizes, speeds and loads of all kinds of cam mechanisms it is convenient to normalise the variables on the same basis as before:

$$b_k = B_k/Y \qquad\qquad \text{Eq. (12.6)}$$

$$d_n = D_n . T^2/Y \qquad\qquad\qquad\qquad \text{Eq. (12.7)}$$

$$D_n = F_d/W \text{ for linear motion,} \qquad\qquad \text{Eq. (12.8)}$$

or

$$D_n = M_d/I \text{ for rotary motion.} \qquad\qquad \text{Eq. (12.9)}$$

$$v = V . T/Y \qquad\qquad\qquad\qquad\qquad \text{Eq. (12.10)}$$

where:

b_k = Normalised (dimensionless) backlash.

B_k = Real cumulative backlash referred, to the cam output.

d_n = Normalised (dimensionless) natural deceleration.

D_n = Real natural deceleration of the payload.

F_d = Decelerating force on the payload, referred to the cam output.

I = Inertia of the payload, referred to the cam output.

M_d = Decelerating torque on the payload, referred to the cam output.

T = The full motion period (time).

v = Normalised (dimensionless) impact velocity.

V = Actual impact velocity, linear or angular.

W = Mass of the payload, referred to the cam output.

Y = Output stroke of the motion, linear or angular.

The natural deceleration of the payload in free flight can vary between zero and the maximum deceleration of the cam law: a higher deceleration would maintain contact with the positive force cam surface. The higher the natural deceleration, the lower is the impact velocity. For any cam law a graph can be constructed to show how the normalised impact velocity varies with backlash and natural deceleration. Such graphs are shown in Fig. 12.8 for six standard cam laws from which the following observations can be made.

All the cam laws have similar impact velocities when the natural deceleration is zero and the backlash is large.

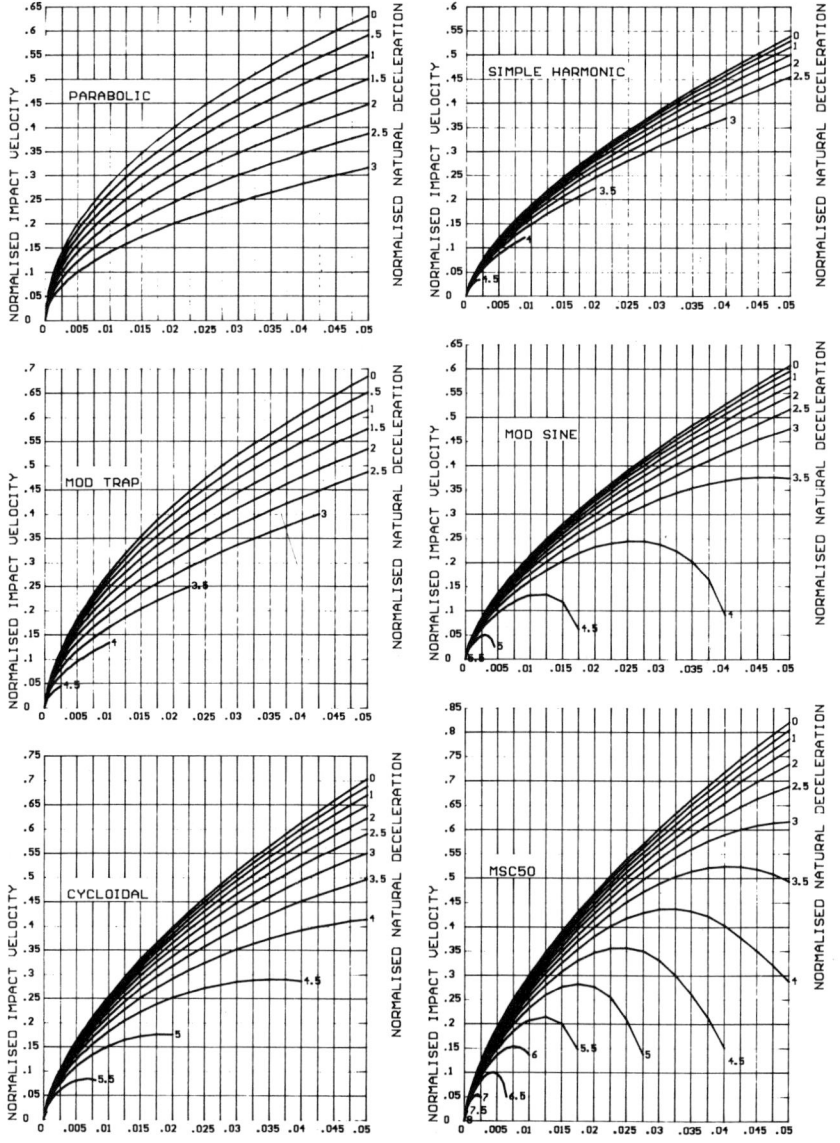

Fig. 12.8 Backlash of standard cam laws

There are considerable differences in impact velocities between cam laws when the natural deceleration is medium to high.

The cam laws with low jerk at cross-over (Simple Harmonic and Mod Sine) are better than the others at low values of backlash – the sort of backlash values likely to occur in practice.

For very high speed mechanisms the value of *normalised* natural deceleration must be low, because it is proportional to the square of the motion period, *T*, which can be very small. Also, the backlash is in practice kept to a minimum by precision manufacture. However, backlash cannot necessarily be totally eliminated, especially as the mechanism wears. For such applications, therefore, the impact velocity for high natural deceleration is irrelevant, and we need only look at the curves for normalised backlash up to say 0.01, and normalised natural deceleration up to a value of 1 or 2. Of the standard cam laws Mod Sine gives good results in these circumstances, but a specially designed 'Low-Impact' polynomial cam law is somewhat better and still has a good dynamic response. Fig. 12.9 compares the backlash of Mod Sine with that of the Low Impact Polynomial.

Values of normalised impact velocities, from which the above graphs were plotted are tabulated in Appendix E. These can be used to quantify an impact velocity for a given backlash, and from that can be derived an impact force.

A shock vibration is set up on impact, the amplitude of which depends on the impact velocity and the natural vibration frequency of the system.

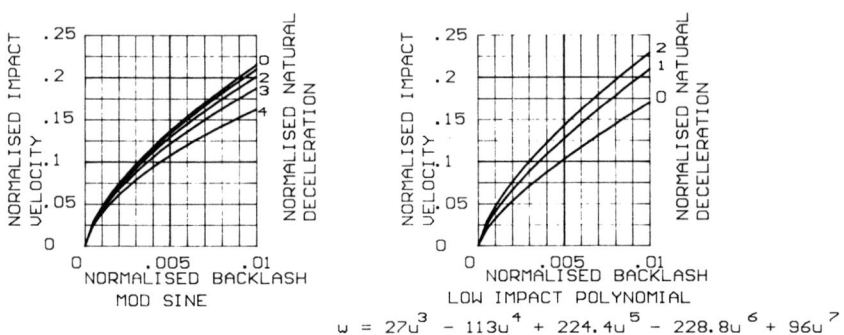

$$u = 27u^3 - 113u^4 + 224.4u^5 - 228.8u^6 + 96u^7$$

Fig. 12.9   Backlash of low impact polynomial and mod sine

We have already seen that natural frequency, f, can be calculated from the stiffness or rigidity of the system and the payload mass or inertia (Equations 12.2 and 12.3). The impact force or torque at the cam/follower contact point is given by:

$$F_s = \sqrt{\left[\left(\frac{V.S}{2.\pi.f}\right)^2 + F_d^2\right]} - F_d \text{ for linear motion} \qquad \text{Eq. (12.11)}$$

or

$$M_s = \sqrt{\left[\left(\frac{V.R}{2.\pi.f}\right)^2 + M_d^2\right]} - M_d \text{ for rotary motion} \qquad \text{Eq. (12.12)}$$

By substituting for f Equations 12.11 and 12.12 become respectively:

$$F_s = \sqrt{(W.S.V^2 + F_d^2)} - F_d \qquad \text{Eq. (12.13)}$$

and

$$M_s = \sqrt{(I.R.V^2 + M_d^2)} - M_d \qquad \text{Eq. (12.14)}$$

where:

f = natural frequency of the system

F_s = peak shock force *in the direction of follower motion* (note that the contact force on the cam is $F_s/\cos \phi$, where ϕ = pressure angle of the cam)

I = payload inertia

M_s = peak shock torque

R = angular rigidity of a rotary transmission

S = linear stiffness of a linear transmission

W = payload mass

The variables in the above equations must be in coherent units of measure (see Chapter 2).

In general, the best way to minimise the detrimental effects of dynamic response is to use a 'good' cam law and to ensure that both input and output transmissions are as rigid and free from backlash as possible.

The following example illustrates the calculation of dynamic response and backlash impact for a typical indexing mechanism application.

Example 12.1

A table with 20 work stations is indexed by an 8-stop globoidal cam indexing mechanism through a pair of gears with a 2.5:1 reduction ratio, as shown in Fig. 12.10. The stations weigh 1.5 kg each and are equally spaced on a 500 mm diameter pitch circle on a steel table 600 mm dia. × 16 mm thick. The table is attached to a main shaft which is 50 mm dia. × 320 mm long. The gears are approximately 128 dia. and 320 dia. × 40 mm wide, made of steel. The indexer turret is 163 mm dia. × 80 mm thick, with 8 rollers 2″ dia. × 1.25″ long pitched around a 195 mm dia. circle. The turret is mounted on a shaft 50 mm dia. × 208 mm long.

The length of the main shaft that is in torsion (reckoned from the under side of the table to the centre of the large gear wheel) is 275 mm, and the length of the turret shaft that is in torsion is 137 mm. The cam motion is

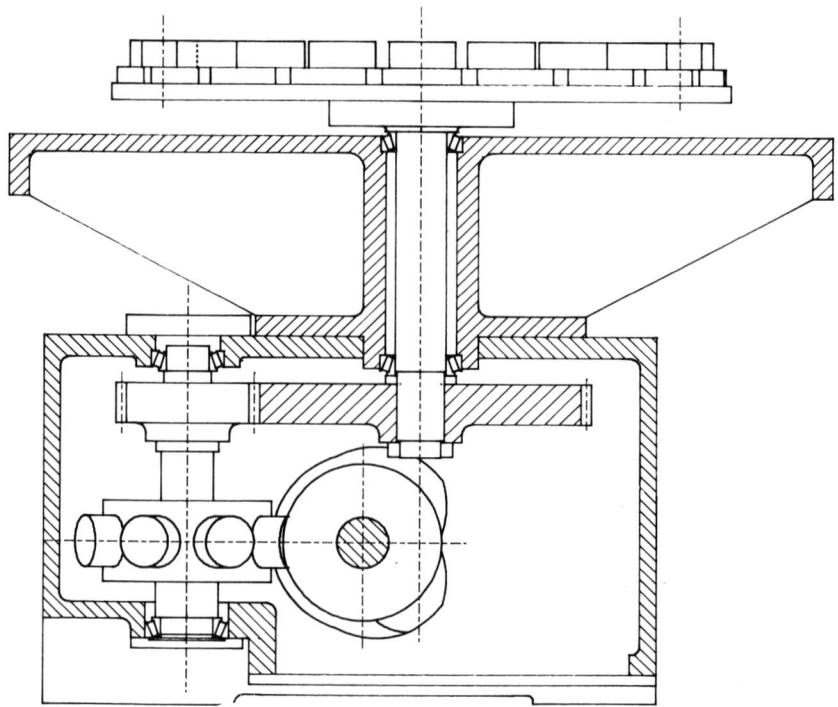

Fig. 12.10   Indexing mechanism – Example 12.1

Table 12.2 Example 12.1: Mass and inertia of components

Component	Mass (kg)	Inertia* (kg.m²)
Work stations, total for 20 off	30.00	1.899
Table, 600 mm dia. × 16 mm thick	35.28	1.588
Flange, 180 mm dia. × 25 mm thick	4.96	0.020
Main shaft, 50 mm dia. × 320 mm long	4.90	0.002
Large gear, 320 mm dia. × 40 mm thick	25.25	0.325
Slow speed assembly total	100.41	3.834
Small gear, 128 mm dia. × 40 mm thick	4.02	0.008
Turret shaft, 50 mm dia. × 208 mm long	3.19	0.001
Turret body, 163 mm dia. × 80 mm thick	13.02	0.043
Roller followers, total for 8 off	4.02	0.039
High speed assembly total	24.25	0.091
Combined total, referred to indexer turret	124.66	0.704

*'Inertia' in this context refers to the moment of inertia.

Mod Sine with an index period of 180° and runs at 150 cycles per minute. The cycle time is therefore 0.4 seconds with 0.2 seconds motion and 0.2 seconds dwell. The design of the input transmission (the cam drive) is good enough to assume that the cam runs at a truly constant speed. There is an estimated overall friction torque in the output transmission of 2 Nm referred to the indexer turret.

The maximum backlash permitted in the cam track when manufactured is 0.04 mm and it is considered that the gear teeth can wear to a backlash value of 0.1 mm before adjustment or replacement is necessary.

The mass and moment of inertia of each component of the indexing system are calculated as shown in the above table.

The final inertia figure in the above table takes into account the gear ratio. The inertia of the slow speed assembly is reduced by the square of the gear ratio when referred to the indexer turret shaft: $3.834/2.5^2 + 0.091 = 0.70444$.

To assess the dynamic response we must estimate the natural vibration frequency of the system, and from that the period ratio. The gears have a significant mass and inertia in an intermediate position in the transmission, which means that the system will not vibrate with a single, simple frequency, as required by the foregoing dynamic response theory. However, the high intertia of the table and work stations imposes a dominant frequency which gives a good approximation to the theoretical model: this is nearly always the case in practice.

Main (slow speed) shaft torsional rigidity, from Equation 11.1, is

$$R_1 = \frac{\pi \times 82.5 \times 10^9 \times 0.05^4}{32 \times 0.275} = 184{,}077 \text{ Nm/rad}$$

Second moment of area of shaft is

$$J = \pi \times 0.050^4 \div 64 = 3.068 \times 10^{-7} \text{ m}^4$$

Main shaft bending stiffness, from Equation 11.8, is

$$S = \frac{3 \times 205 \times 10^9 \times 3.068 \times 10^{-7} \times 0.272}{0.028^2 \times 0.244^2} = 1.1 \times 10^9 \text{ N/m}$$

and the equivalent torsional rigidity of this is

$$R_2 = 1.1 \times 10^9 \times 0.16^2 = 28{,}160{,}000 \text{ Nm/rad}$$

Indexer (high speed) shaft torsional rigidity is

$$R_3 = \frac{\pi \times 82.5 \times 10^9 \times 0.05^4}{32 \times 0.137} = 369{,}499 \text{ Nm/rad}$$

Turret shaft bending stiffness is

$$S = \frac{3 \times 205 \times 10^9 \times 3.068 \times 10^{-7} \times 0.24}{0.028^2 \times 0.212^2} = 1.3 \times 10^9 \text{ N/m}$$

and its equivalent torsional rigidity is

$$R_4 = 1.3 \times 10^9 \times 0.064^2 = 5{,}324{,}800 \text{ Nm/rad}$$

The combined rigidity, referred to the indexer turret, from Equation 11.6, is

$$R = \frac{1}{2.5^2 \div R_1 + 2.5^2 \div R_2 + 1 \div R_3 + 1 \div R_4} = 27{,}110 \text{ Nm/rad}$$

(Note that the bending stiffness of the shafts in this instance is so high that it could have been ignored.)

Approximate natural frequency , from Equation 12.3, is

$$f = \frac{1}{2 \times \pi} \cdot \sqrt{\left(\frac{27,110}{0.707}\right)} = 31.17 \text{ Hz}$$

The motion period is 0.2 seconds, so from Equation 12.4 the period ratio is

$$n = 31.17 \times 0.2 = 6.34$$

As a matter of interest, the natural frequency of the main shaft assembly alone is 36.46 Hz, which would give a period ratio of 7.29. Bearing in mind that the elasticity of bearing mountings has not been taken into account and that the inertia of the gears must reduce the primary natural frequency, it is reasonable (and safer) to use the lower value of 6.34.

Using the appropriate values of p, q and r for Mod Sine in Equation 12.1 we find that the torsion factor is

$$C_t = 3.87 \div 6.34^{1.13} + 1 = 1.48$$

The output stroke of the indexer is $Y = 2\pi \div 8 = 0.7854$ rad and the index period is $T = 0.2$ s. The coefficient of acceleration of Mod Sine is $C_a = 5.528$, therefore the nominal peak acceleration is

$$C_a \cdot Y/T^2 = 5.528 \times 0.7854 \div 0.2^2 = 108.54 \text{ rad/s}^2$$

From the table, the total intertia referred to the indexer is 0.704 kg . m^2, so the nominal peak inertia torque is 0.704 $\times$ 108.54 = 76.41 Nm, which is increased by the torsion factor to

$$M = 1.48 \times 0.704 \times 108.54 = 113.09 \text{ Nm}$$

and with the additional friction torque the peak torque for which the mechanism must be designed, ignoring the backlash impact effect, is 113.09 + 2 = 115.09 Nm.

Consider now the shock load that would be induced by the maximum permitted backlash. The total backlash expressed as an angle at the indexer turret is:

Gear teeth, 0.1 mm at 64 mm radius:

$$0.1 \div 64 = 0.00156$$

Indexer turret, 0.04 mm at 97.5 mm radius:

$$0.04 \div 97.5 = 0.00041$$

Total backlash = 0.00156 + 0.00041 = 0.00197 rad

The normalised backlash from Equation 12.6 is

b_k = 0.00197 ÷ 0.7854 = 0.00251

The natural deceleration of the system from Equation 12.9 is

D_n = 2 ÷ 0.704 = 2.84

and its normalised value from Equation 12.7 is

d_n = 2.84 × 0.2^2 ÷ 0.7854 = 0.148

which for our purpose can be taken as zero without incurring too much error.

From the table in Appendix E we find that for a normalised backlash of 0.0025 and zero natural deceleration the corresponding normalised impact velocity for Mod Sine is v = 0.0861 and therefore, from Equation 12.10, inverted, we find the *real* impact velocity is

$V = v . Y/T$ = 0.0861 × 0.7854 ÷ 0.2 = 0.338 rad/s

From Equation 12.14 the peak shock torque on the turret is therefore

$M_s = \sqrt{(0.704 \times 27{,}110 \times 0.338^2 + 2^2)} - 2 = 44.74$ Nm

At 40% of the peak torque without backlash this is a very significant load, and should not be ignored in the design of the mechanism. A safe way of taking it into account is simply to add it to the peak vibration torque (this assumes that the two peak torques occur at exactly the same point in the motion, which is quite possible):

Peak output torque, M = 115.09 + 44.74 = 159.83 Nm

Although it could be argued that this is too pessimistic, it does illustrate that to design the mechanism on the basis of the *nominal* dynamic torque of 78.41 Nm would be quite inadequate.

CHAPTER 13

Loading, Duty, Power

The loading imposed on a cam mechanism in service is known as the *duty* of the mechanism, as opposed to its *capacity*. The duty is determined by the payload, operating speed and dynamic response (see Chapter 12), whereas the capacity is the mechanism's ability to perform the duty for a specified lifetime. The effects on duty of the transmission design (elasticity, backlash) and the consequent dynamic response have been covered in previous chapters. Here we shall deal with the nominal loading, which can arise from a number of sources.

Nominal output loading, i.e. force in linear motion or torque in rotary motion, can be principally of the following kinds, either uniquely or in combination:

(a) constant, e.g. friction or gravity, or
(b) proportional to displacement, e.g. spring, or
(c) proportional to velocity, e.g. viscosity, or
(d) proportional to acceleration, e.g. inertia, or
(e) conforming to some special purpose function.

The majority of cam mechanism applications have a duty which is the sum of two or more of these kinds of loading. Indexing mechanisms, for example, are usually subjected to friction loading (a) and inertia loading (d). Spring-loaded reciprocating mechanisms usually have a duty composed of friction and spring pre-load (a), spring rate (b) and inertia loading (d). In very high speed applications viscous loading (c) may also be significant.

In all these, except (a), the load varies throughout the motion, and Fig. 13.1 shows diagrammatically these load patterns separately and in combination. The magnitude of each load component depends on the particular application, so that there is an infinite number of different possible load patterns. We cannot therefore find a simple general normalised expression for the peak loading on a cam, but we can tabulate normalised values for the most common combinations.

The diagrams show that there can be a period of negative loading when inertia loading (d) is involved. In such a case the follower would lose

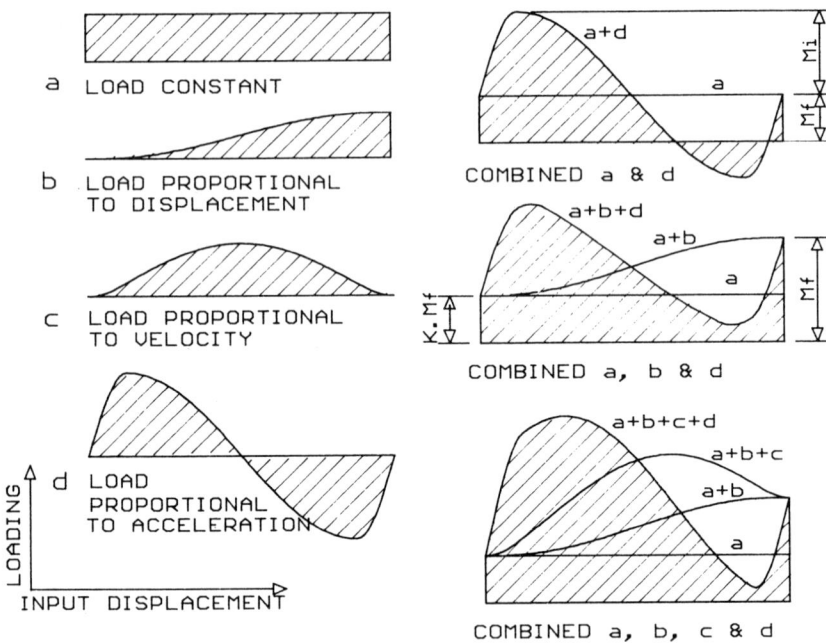

Fig. 13.1 **Loading patterns, separate and combined**

contact with an open track cam and control of the payload would be lost. Then either an enclosed track, a blade cam or a conjugate cam system must be used (see Chapter 1), or the constant loading or spring loading must be increased in order to retain control. The risk of loss of control is particularly high for an open track cam with falling motion, as illustrated in Fig. 13.2.

Virtually all indexing mechanisms and many reciprocating and oscillating mechanisms have a duty comprising a combination of constant load (a) and acceleration load (d), with insignificant amounts of the other kinds of load. This is the easiest combination to analyse and it will be used as a basis for normalising other combinations.

The constant load arises from friction forces, raising weights (gravitational force) and other work loads: these can be resolved into a single force or torque resisting output motion. The acceleration load arises from the inertia of the masses being moved and can be resolved into a single equivalent inertia connected to the output (cam follower).

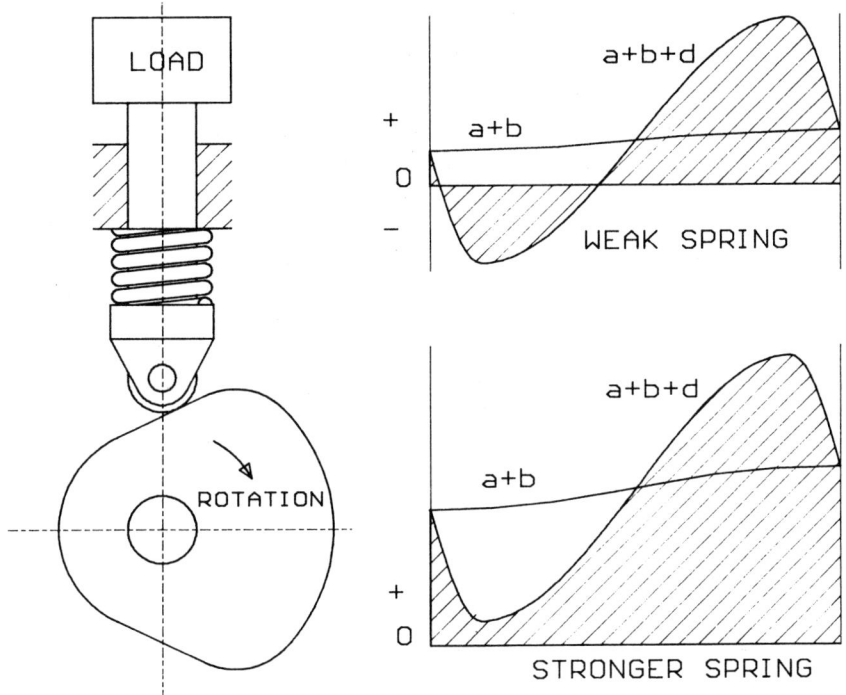

Fig. 13.2 Loading of open track, falling motion

Since the total load is the sum of inertia load and *constant* load, it follows that the peak load coincides with the maximum inertia load, as can be seen in Fig. 13.1. The maximum inertia load is:

$$F_i = W_e \cdot C_a \cdot Y/T^2 \text{ for linear motion.} \qquad \text{Eq. (13.1)}$$

and

$$M_i = I_e \cdot C_a \cdot B/T^2 \text{ for rotary motion.} \qquad \text{Eq. (13.2)}$$

where

B = Output stroke (rotation angle)   [rad]

C_a = Coefficient of acceleration of the cam law

F_i = Inertia force at the mechanism output.   [N]

I_e = Equivalent moment of inertia referred to the
　　mechanism output.   [kg . m^2]

M_i=Inertia torque at the mechanism output.   [N . m]

W_e=Equivalent mass referred to the mechanism output.   [kg]

These loadings are *nominal* and must be modified by the application of a torsion factor, C_t, defined in the previous chapter.

The constant load (such as friction) is F_f for linear motion and M_f for rotary motion, referred to the mechanism output in both cases, and this must be added to the inertia load. When there is also an output load component that is proportional to displacement, such as a spring force, it is convenient to include the *maximum value* of that component in the value of F_f or M_f, as if it were a constant load. In that case F_t or M_t would be overvalued, although it is a convenient *safe* value to use for preliminary design purposes.

For greater accuracy, to take all three loading patterns into account when appropriate, the total maximum load can be adjusted downwards by a 'load mix coefficient', C_m, thus:

$$F_t = C_m . (C_t . F_i + F_f)$$ Eq. (13.3)*

and

$$M_t = C_m . (C_t . M_i + M_f)$$ Eq. (13.4)*

where C_m depends on the relative values of constant load, spring load and inertia load. Values of this load mix coefficient are given for standard cam laws in Appendix C.

The spring rate effect is best defined by a 'pre-load coefficient' K, applicable to all kinds of cam systems, which is the ratio of the initial non-inertia loading (pre-load) to the final (maximum) non-inertia loading, thus:

$$K = \frac{F_p}{F_f} \quad \text{or} \quad \frac{M_p}{M_f}$$ Eq. (13.5)

where

F_p, M_p = minimum non-inertia load (pre-load)

---

*These equations exaggerate the peak load to the extent that the torsion factor is applied to the maximum positive inertia load, whereas the peak vibration quantified by the torsion factor usually occurs in the negative phase. The degree of exaggeration, however is not great.

F_f, M_f = maximum non-inertia load (full load)

When there is no spring load component $K = 1$ and $C_m = 1$.

The other cofficient needed to evaluate C_m is the ratio of peak inertia loading to maximum non-inertia loading, the inertia ratio Q:

$$Q = \frac{C_t \cdot F_i}{(C_t \cdot F_i + F_f)} \quad \text{or} \quad \frac{C_t \cdot M_i}{(C_t \cdot M_i + M_t)} \qquad \text{Eq. (13.6)}$$

The instantaneous output load for linear motion is

$$f_0 = \left\{ \frac{Q}{C_a} \cdot w'' + (1 - Q) \cdot [w + K \cdot (1 - w)] \right\} \cdot \{C_t \cdot F_i + F_f\}$$

Eq. (13.7)

It follows that the maximum value of this is given by Equation 13.3 where

$$C_m = \text{maximum value of } \frac{Q}{C_a} \cdot w'' + (1 - Q) \cdot [w + K \cdot (1 - w)]$$

Eq. (13.8)

The same value of C_m is valid for rotary motion. It can be evaluated for any cam law for a range of values of Q and K. The results for a number of standard laws are tabulated in Appendix C.

INPUT TORQUE

There is a direct relationship between the *output* load, defined by these equations, and the *input* torque of the cam for any given cam law. The input torque varies throughout the motion in a pattern that is a function of both the output load pattern and the cam law. Since the output load pattern varies with the mix of inertia load, spring load and constant load, so too does the input load pattern. For all cases the mix is defined by the factors Q and K, defined above.

For any cam law the instantaneous input torque on the cam can be expressed as a function of the instantaneous output load:

$$m_c = f_0 \cdot w' \cdot Y/A/E \text{ for linear motion} \qquad \text{Eq. (13.9)}$$

or

$$m_c = m_0 \cdot w' \cdot B/A/E \text{ for rotary motion} \qquad \text{Eq. (13.10)}$$

where

m_c = cam input torque   [N . m]
f_0 = output force   [N]
w' = normalised velocity factor of cam law
Y = linear output stroke [m]
A = angular input stroke   [rad]
m_o = output torque   [N . m]
B = angular output stroke   [rad]
E = efficiency of the cam and follower mechanism

The instantaneous output load, f_0 or m_0, is a function of the normalised cam law factors, the load mix and the payload (see Equation 13.7), so that the above equations can also be expressed as

$$m_c = w' \cdot \{w''Q/C_a + (1 - Q)[w + K \cdot (1 - w)]\}$$

$$\{C_t \cdot F_i + F_f\} \cdot Y/A/E$$

or

$$m_c = w' \cdot \{w''Q \ /C_u + (1 - Q)[w + K \cdot (1 - w)]\}$$

$$\{C_t \cdot M_i + M_f\} \cdot B/A/E$$

The maximum value of m_c occurs during the acceleration phase and can be simply expressed by introducing a normalised input torque coefficient, C_c thus:

$$M_c = C_c \cdot F_t \cdot Y/A/E \text{ for linear motion} \qquad \text{Eq. (13.11)}$$

or

$$M_c = C_c \cdot M_t \cdot B/A/E \text{ for rotary motion} \qquad \text{Eq. (13.12)}$$

where

M_c = Peak input torque, and

C_c = The maximum value of $w' \cdot \{w''Q \ ./C_a + (1 - Q)$

$$[w + K \cdot (1 - w)]\}/C_m \quad \text{Eq. (13.13)}$$

The value of C_c, the input torque coefficient, can be evaluated for any

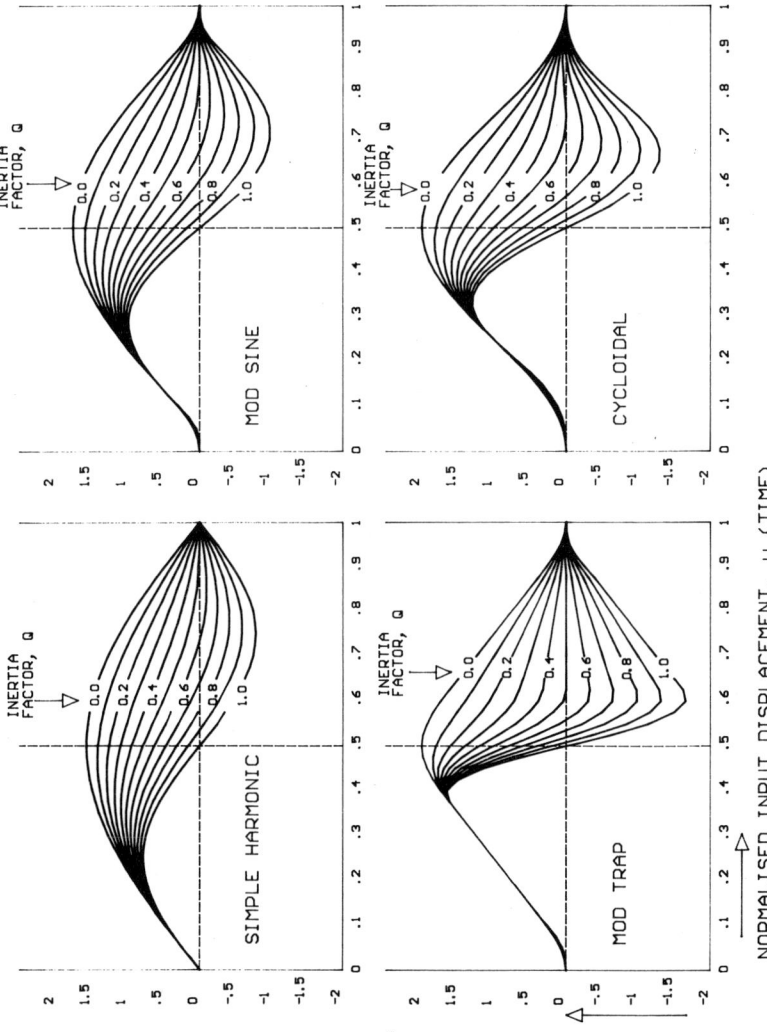

Fig. 13.3 Input torque patterns

cam law and a range of load mix factors. The results for several standard cam laws are tabulated in Appendix C.

The input torque patterns for some standard cam laws are shown in Fig. 13.3 for various values of Q when $K = 1$ (i.e. no spring loading).

From these patterns it can be seen that there is a wide disparity between the input torques of different cam laws, for all values of Q.

- Simple Harmonic's input torque varies smoothly and gradually from a low positive peak to a moderate negative: this is a good characteristic.
- Mod Trap's input torque varies quite suddenly from a high positive to a low negative: this is a bad characteristic which gives rise to severe overrun if the input transmission has low inertia and low rigidity.
- Cycloidal's input torque varies smoothly from a fairly high positive to a fairly low negative.
- Mod Sine's input torque, though not as good as Simple Harmonic's is better than Cycloidal's. This makes Mod Sine very tolerant of a bad input transmission which, combined with its good dynamic response properties, is the reason for its popularity.

The peak values of the curves are plotted in Fig. 13.4, giving graphs of C_c against Q for each cam law for the common mix of only inertia and

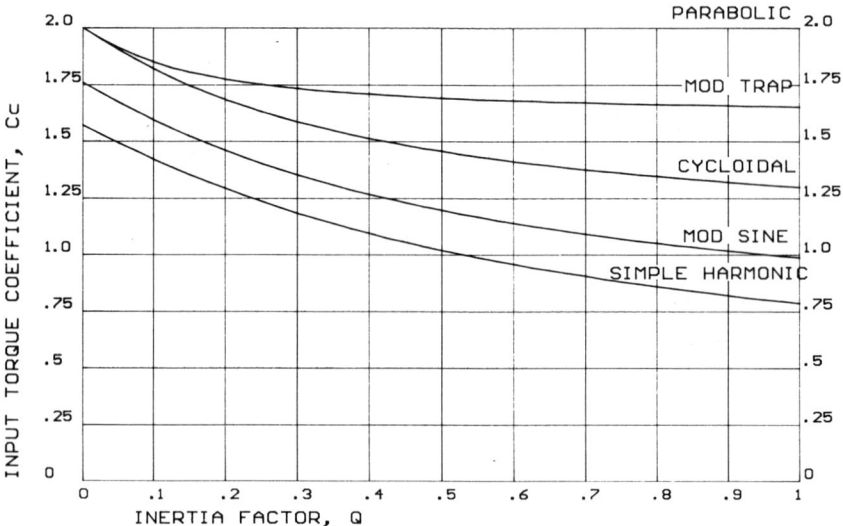

Fig. 13.4   Peak input torque coefficients

constant loading. The more complex loading pattern that includes a spring load component (where K is less than 1) gives rise to the graphs in Fig. 13.5, in which the values of C_m and C_c are plotted against both Q and K for four standard cam laws. From these plots we see that both the peak output load represented by C_m, and the peak input torque represented by C_c, are considerably reduced with low pre-load ratios.

OVERRUN

One important aspect of input torque revealed by these curves is the existence of a period of negative torque and the rapid drop from positive torque for some cam laws. This has a great influence on the design of the input transmission. If the transmission is excessively elastic and/or has a significant amount of backlash an overrun effect takes place. During the positive torque phase the elasticity of the transmission is wound up and its backlash is taken up. As the torque reduces, unwinding occurs until the torque is zero, then the system goes free while the clearance is taken up the other way, after which the transmission winds up in the opposite direction: the system overruns its drive.

The amount and suddenness of the overrun can be considerable, even in fairly well designed machines. During the overrun there can be a large increase in the cam speed associated with a significant cam acceleration. Referring back to Chapter 3, the importance of the transmitted acceleration Equation 3.8 is now apparent, repeated below.

real output acceleration
= (geometric output acceleration) × (input velocity)^2
+ (geometric output velocity) × (input acceleration)

The second term on the right-hand side is usually assumed to be zero because the input is of nominally constant velocity, zero acceleration. During overrun, however, there is high input acceleration; also the first term of the expression (usually assumed to be the *only* term) is increased in proportion to the square of the momentarily increased input velocity. These two effects can increase the output acceleration, and therefore the inertia loading, by as much as tenfold in bad cases. In practice, there have been instances where the damaging effect of overrun due to an unavoidably long slack chain drive has been eliminated by simply changing from Mod Trap motion (normally regarded as good) to Mod Sine.

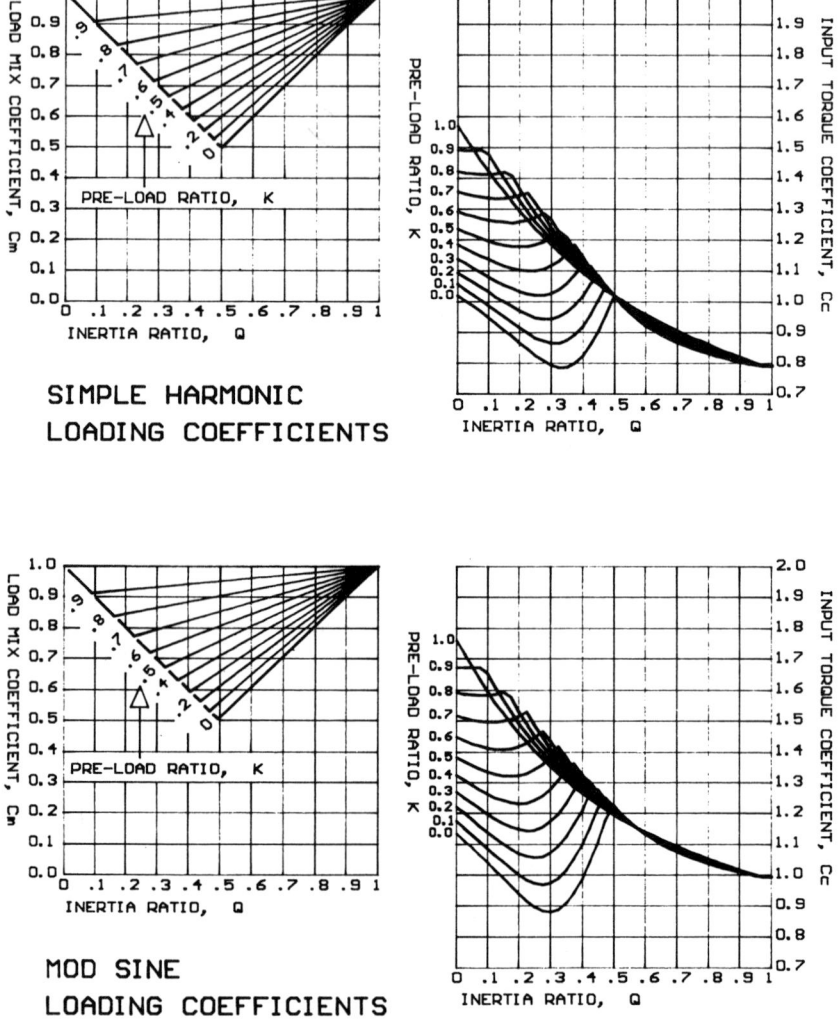

Fig. 13.5   Loading coefficients, Cm and Cc

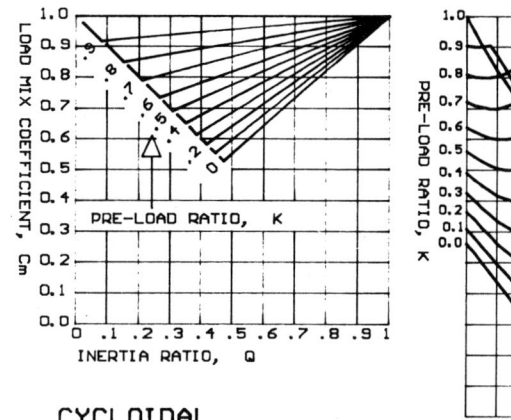

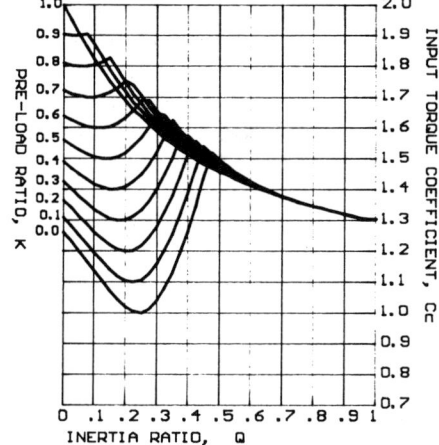

CYCLOIDAL
LOADING COEFFICIENTS

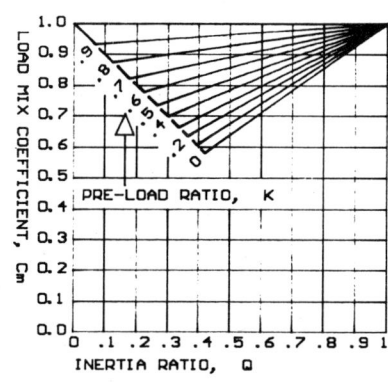

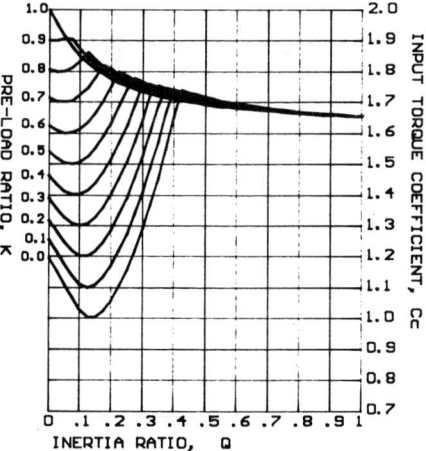

MOD TRAP
LOADING COEFFICIENTS

Good input transmission design can greatly mollify overrun. The first priority is to eliminate backlash as far as possible and then to increase rigidity. Even so, especially in high speed and/or high inertia applications, the elasticity of the input transmission can still be a problem. The introduction of high inertia (effectively a flywheel) closely connected to the cam is favourable, although not always practicable. A convenient and effective alternative to the direct flywheel is a high-ratio reduction gear close to the cam with a small flywheel on the high speed input, bearing in mind that the effective inertia of the flywheel is increased by the gear ratio squared when referred to the cam. A very useful form of gearing for this is worm gearing, probably in the form of a commercial motorised worm gear box (high quality, minimum backlash). The high speed motor armature is a very effective flywheel. Figure 13.6 shows two transmission designs for a bottle feeding mechanism that has a constant speed conveyor and an indexing table, both driven by the same motorised worm gear.

The original design has long shafts and a potentially slack chain drive between the worm gear and the indexing cam. The improved design keeps essentially the same mechanism layout, but has a better position for the motorised worm gear box, close to the cam to give a far better dynamic performance at high speed.

An important 'side effect' of using a worm gear is derived from its low reverse efficiency. The normal forward efficiency of a worm gear is fairly high, say 75% to 85%, depending on the gear ratio, but the reverse efficiency – when the worm wheel is driving the worm, i.e. when the input torque is negative – is very low, typically about 10% for gear ratios between 30:1 and 50:1 or about 55% for ratios between 15:1 and 25:1. This has a beneficial braking effect on the cam to reduce overrun during the negative torque phase.

It follows that, having chosen a good cam law, the greater the input inertia can be, the better it is. In the interests of economy, however, it is best not to have to introduce more inertia than is already available in the motor armature, the cam itself, transmission components, etc. Experience shows that there is a lower limit to the amount of inertia that will ensure an adequate performance. This is based on the more or less arbitrary rule that the cam speed should not increase by more than about 10% during the negative torque phase.

Ignoring any input braking effect and friction loading, there is a constant amount of kinetic energy in the system when it is running at full

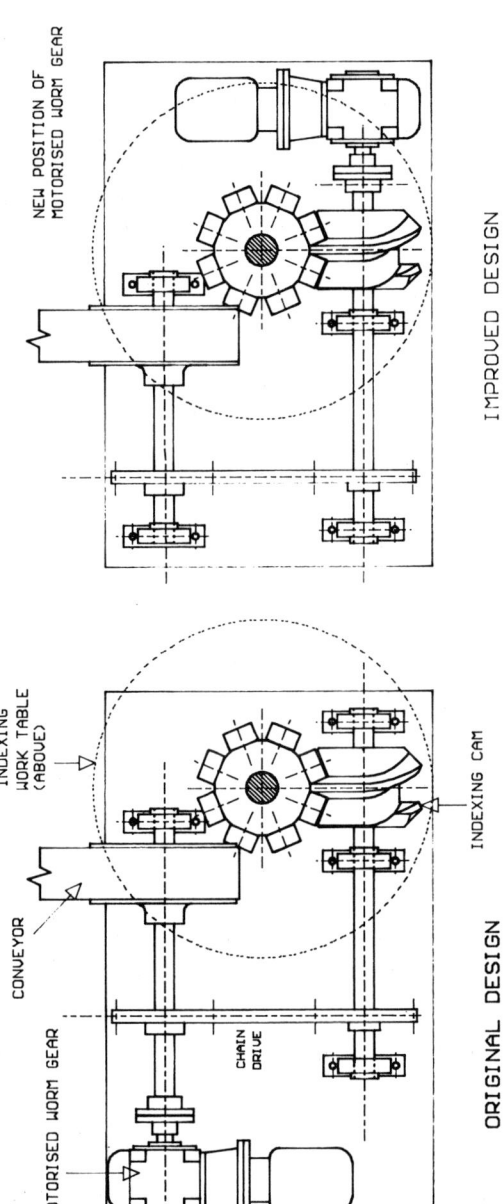

Fig. 13.6 Improved transmission design

speed. However, some of that energy is transferred from input to output when the output inertia is accelerated, and transferred back again when it is decelerated. The loss of energy from the input causes it to slow down and the regain of energy makes it speed up again. The output kinetic energy varies from zero when the load is at rest to a maximum when it is at its maximum speed, determined by the cam law. The input kinetic energy must fluctuate by the same amount, and from this we can find how much input inertia is required to limit the input speed fluctuation to 10%:

$$\text{Maximum output kinetic energy} = \frac{W_e}{2} \cdot \left(C_v \cdot \frac{Y}{T}\right)^2$$

$$\text{Variation in input kinetic energy} = \frac{I_c}{2} \cdot \left[\left(1 \cdot 1\frac{A}{T}\right)^2 - \left(\frac{A}{T}\right)^2\right]$$

Equating these two expressions gives:

$$I_c = \frac{W_e}{(1 \cdot 1^2 - 1)} \cdot \left(\frac{C_v \cdot Y}{A}\right)^2 \text{ for linear motion} \qquad \text{Eq. (13.14)}$$

and similarly

$$I_c = \frac{I_e}{(1 \cdot 1^2 - 1)} \cdot \left(\frac{C_v \cdot B}{A}\right)^2 \text{ for rotary motion} \qquad \text{Eq. (13.15)}$$

where

I_c = Minimum input inertia required (referred to the cam)

POWER

The power being transmitted at any moment at any point in the transmission is the instantaneous load multiplied by the instantaneous velocity at that point.

$$\text{Instantaneous power} = m_c \cdot v_i \qquad [\text{N} \cdot \text{m/sec}]$$

where

v_i = Angular velocity of the cam   [rad/sec]

For most applications the input velocity is virtually constant, and is usually expressed as the number of camshaft revolutions per minute.

The peak power is therefore $M_c . v_i$.

Input velocity $= A/T = 2 . \pi . N/60$ [rad/sec]

where

$N =$ camshaft rotation speed   [rev/min]

We can now give an expression for the peak input power of a cam system, in terms of the camshaft speed and the input torque, the latter being a function of the output loading, the input and output strokes and the input torque coefficient of the cam law, as already described.

$$P_{max} = M_c . A/T = 2 . \pi . M_c . N/60 \quad [\text{N . m/sec}] \qquad \text{Eq. (13.16)}$$

or

$$P_{max} = M_c . N/9549 \quad [\text{kW}] \qquad\qquad\qquad \text{Eq. (13.17)}^*$$

This is the *peak* power in the system used by the cam. To determine the size of an electric motor that is required to drive the mechanism, for example, we must add any power losses in the input transmission (friction in gearing, etc.), but we can take advantage of the fact that there is usually a significant amount of kinetic energy associated with the camshaft. The power fluctuates during the motion period, and is very low during a dwell period.

The reserve of energy, derived from the input inertia, supplies the extra power needed to cope with the peak requirement, provided that the input inertia is adequate, as described above. It is therefore necessary only to have a power source that can provide the *average* power of the motion cycle. The average power is never more than half of the peak power, and if the dwell period(s) are much longer than the motion period(s) the input energy reserve can be restored during one cycle with a power input of much less than half the maximum. Where the power source is an A.C. induction motor, quite apart from the benefit of a fairly high armature inertia (at high speed), there is the advantage that its peak torque is more than twice its normal torque rating, so that it is always safe to select such a motor with a nominal power rating of $P_{max}/2$ (plus of course any additional power required by other mechanisms in the machine).

---

*The SI standard unit of power is the Watt [W], which is exactly one Newton . metre per second [N . m/sec]. The common power unit in use is the kiloWatt [kW]: 1 kW = 1000 N . m/sec.

GEARING EFFECTS

Often there is some form of gearing in the transmissions, input or output, of a cam system. Chapter 11 deals with the effects of gearing on transmission elasticity and backlash. Loading is similarly affected by gearing.

The value of force or torque at one side of a gear pair is converted to a different value at the other side in direct proportion to the velocity ratio of the gears, and the value of inertia is converted in proportion to the square of the velocity ratio.

All the components of a system can therefore have equivalent values of load and inertia referred to the cam for input components or the follower for output components. The equivalent values are additive, so that a single value can be used to represent all components, and the system behaves the same as if there were no gearing in the transmissions, but the components had their equivalent values.

The loading and power equations above are valid for systems with gearing, provided that:

F_e=the sum of *all* equivalent forces referred to the follower output
W_e=the sum of *all* equivalent masses referred to the follower output
M_e=the sum of *all* equivalent torques referred to the follower output
I_e=the sum of *all* equivalent inertias referred to the follower output
I_c=the sum of *all* equivalent inertias referred to the cam input

The concept of equivalent force is based on the fact that at any instant the output power of a gear pair is equal to the input power times the efficiency. Power is force times velocity, so

$$F_2 . v_2 = F_1 . v_1 . E \qquad \text{or} \qquad F_1 = F_2 . v_2/v_1/E \qquad\qquad \text{Eq. (13.18)}$$

similarly

$$M_1 = M_2 . \omega_2/\omega_1/E \qquad\qquad\qquad\qquad\qquad \text{Eq. (13.19)}$$

The concept of equivalent mass or inertia is based on the fact that at any instant the kinetic energy of the equivalent mass must be the same as the kinetic energy of the actual mass, so

$$W_1 . v_1^2/2 = W_2 . v_2^2/2 \qquad \text{or} \qquad W_1 = W_2 . (v_2/v_1)^2 \qquad \text{Eq. (13.20)}$$

similarly

$$I_1 = I_2 . (\omega_2/\omega_1)^2 \qquad\qquad\qquad\qquad\qquad \text{Eq. (13.21)}$$

where v = linear velocity, ω = angular velocity and subscripts 1 and 2 refer to input and output sides of the gear pair respectively.

The gear ratio is v_2/v_1 or ω_2/ω_1

In the context of linear motion a lever is considered as a gear pair. Often output transmissions consist of a mixture of translating (linear motion) components and rotating components. The principles of equating power and kinetic energy still apply. For mixed systems:

$$F \cdot v = M \cdot \omega \cdot E \quad \text{or} \quad F \cdot v \cdot E = M \cdot \omega \qquad \text{Eq. (13.22)}$$

depending on whether linear motion is output and rotary motion is input or vice versa.

$$W \cdot v^2 = I \cdot \omega^2 \qquad \text{Eq. (13.23)}$$

(Note that in the case of a force F acting on a lever at radius r then $v = r \cdot \omega$ and these equations give $M = F \cdot r \cdot E$ or $M = F \cdot r/E$ for the pivot torque, as expected. Similarly, for a mass W concentrated at radius r we get $I = W \cdot r^2$ for its inertia about the pivot, as expected.)

The following example uses an imaginary mechanism to illustrate the use of most of the equations in this chapter.

Example 13.1

The machine shown in Fig. 13.7 has a single, open track disc cam and swinging roller follower with a Cycloidal angular motion, operating two mechanisms: a plunger with a linkage transmission and a workhead with a rack and gear segment transmission. The workhead and its rack and slide are lifted by the cam, but the plunger moves horizontally against a friction force. The roller follower is held in contact with the cam by a tension spring – the 'return spring' – which is strong enough to provide the operating force for both mechanisms.

(Note that for safety it is best to drive mechanisms positively out of the danger zone and spring load them into it. This is likely to avoid a crash if a mechanism becomes stiff.)

The cam is driven at 60 rev/min by a motorised worm reduction gear of 24:1 ratio: the motor is an A.C. squirrel cage running at 1440 rev/min. Find the peak loading in the rise period of the cam (the most heavily loaded period), which occupies 72° of camshaft rotation and therefore takes 0.2 sec. The linkage geometry is designed so as to incur negligible

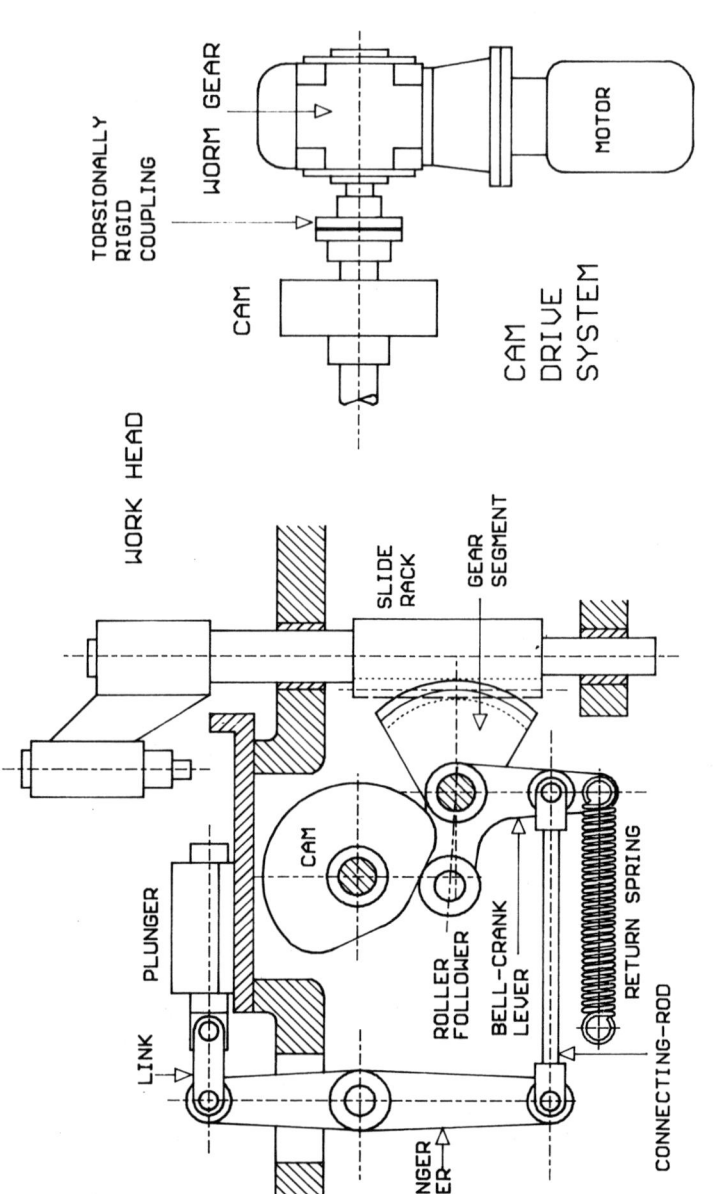

Fig. 13.7 Cam loading, Example 13.1

distortion of the Cycloidal motion law. The rigidity of the output transmission is estimated to be approximately 12,000 N . m/rad.

Find also the cam torque and the required size of an electric motor to provide sufficient power and input inertia.

Specification of output components

Bell-crank lever, roller follower and rocker shaft assembly	Pivot to conn-rod	100 mm
	Pivot to follower centre	100 mm
	Inertia about pivot	0.0231 kg . m^2
	Angular stroke	22.62° = 0.395 rad
	Efficiency	90%
Connecting-rod	Mass (including end joints)	0.75 kg
	Linear stroke	40 mm
Plunger lever	Pivot to conn-rod	200 mm
	Pivot to plunger	160 mm
	Inertia about pivot	0.0747 kg . m^2
	Angular stroke	11.31° = 0.197 rad
	Efficiency	95%
Plunger and link assembly	Mass	3.24 kg
	Slide friction force	3.15 N
	Linear stroke	32 mm
Gear segment	Pitch circle radius	110 mm
	Inertia about pivot	0.0855 kg . m^2
	Angular stroke	22.62° = 0.395 rad
	Efficiency	98%
Workhead, rack and slide assembly	Mass	18 kg
	Slide friction force	41 N
	Linear stroke	43.5 mm
Return spring	Spring-post to pivot	150 mm
	Force at extended working length	300 N
	Force at short working length (pre-load)	120 N
	Linear stroke	60 mm

Output loading

Convert all loadings to values referred to the bell-crank lever, to simulate a simple oscillating mechanism. The instantaneous velocity of a component is the cam law velocity factor at that point in the motion multiplied by the *stroke of the component* divided by the motion period (time). Since the period and velocity factor are the same for all components, the velocity of a component is proportional to its stroke, and *velocity ratios are the same as stroke ratios*.

Equivalent inertia calculation

		$[kg . m^2]$
Bell-crank	(no conversion needed)	0.0231
Connecting-rod	$0.75 \text{ kg} \times (0.04 \text{ m} \div 0.395 \text{ rad})^2$	0.0077
Plunger lever	$0.0747 \text{ kg . m}^2 \times (0.0197 \text{ rad} \div 0.0395 \text{ rad})^2$	0.0187
Plunger assembly	$3.24 \text{ kg} \times (0.032 \text{ m} \div 0.395 \text{ rad})^2$	0.0213
Gear segment	(no conversion needed)	0.0855
Workhead assembly	$18 \text{ kg} \times (0.0435 \text{ m} \div 0.395 \text{ rad})^2$	0.2183

$$I_e = 0.3746$$

Equivalent torque calculation

In the rise period C_a for Cycloidal motion is 6.263, the output stroke is $B = 0.395$ rad and the period is 0.2 sec., therefore the maximum acceleration of the bell-crank lever in the rise period is

$$C_a . B/T^2 = 6.283 \times 0.395 \div 0.2^2 = 62.046 \text{ [rad/s}^2]$$

The inertia torque, referred to the bell-crank, is

$$M_i = I_e . C_a . B/T^2 = 0.3746 \times 62.046 = 23.243 \text{ [N . m]}$$

The natural frequency of the system can be estimated from the inertia and rigidity to allow for dynamic response vibration (see Chapter 12):

$$f = \left(\frac{1}{2 . \pi}\right) . \sqrt{\left(\frac{12,000}{0.3746}\right)} = 28.5 \text{ [Hz]}$$

The period ratio is $n = f . T = 28.5 \times 0.2 = 5.7$ and the torsion factor C_t is found from Equation 11.1 using the Cycloidal constants:

$$C_t = 1.5/n + 1 = 1.5 \div 5.7 + 1 = 1.263$$

Non-inertia load

[N . m]

Plunger assembly friction 3.15 N × 0.032 m ÷ 0.395 rad ÷ 95%

0.268

Workhead slide friction 41 N × 0.0435 m ÷ 0.395 rad ÷ 98%

4.607

Spring force, max. 300 N × 0.06 m ÷ 0.395 rad

45.570

Maximum non-inertia load $M_f = 50.445$

Spring force, min. 120 N × 0.06 m ÷ 0.395 rad

18.228

Minimum non-inertia load (pre-load) $M_p = 23.103$

Pre-load ratio $K = M_p/M_f = 23.103 \div 50.445 = 0.458$

Inertia ratio $Q = \dfrac{C_t \cdot M_i}{(C_t \cdot M_i + M_f)} = \dfrac{1.263 \times 23.243}{(1.263 \times 23.243 + 50.445)} = 0.368$

By interpolation from the table in Appendix C:

$C_m = 0.705$ and $C_c = 1.571$

We can now use Equations 13.4 and 13.12 to find the peak loadings on the machine. The peak output loading (on the bell-crank) is

$M_t = C_m \cdot (C_t \cdot M_i + M_f) = 0.705 \times (1.263 \times 23.243 + 50.445)$

$= 56.260 \,[\text{N . m}]$

Input loading

The peak input torque (on the cam) allowing 90% cam and follower efficiency is

$M_c = C_c \cdot M_t \cdot B/A/E = 1.571 \times 56.260 \times 26.62° \div 72° \div 0.9$

$= 36.308 \,[\text{N . m}]$

The most economical drive unit is a motorised worm gear using a 4-pole squirrel-cage motor which runs at 1440 rev/min full-load speed (on a 50 Hz A.C. supply). This must have a reduction gear ratio of 1440/60 = 24:1. The 24:1 worm gear unit typically has a forward efficiency of about 75%

and a reverse efficiency of about 30%. The peak power is given by Equation 13.17:

$$P_{max} = M_c . N/9549 = 36.308 \times 60 \div 9549 \div 0.228 \, [kW]$$

A motor with sufficient full-load power to cover this peak must allow for the reduction gear efficiency of 75%:

$$P_{max} \div 0.75 = 0.228 \div 0.75 = 0.304 \, [kW]$$

This is covered by a standard 0.37 kW motor, but if we assume that there is sufficient input inertia to choose a motor for the average power only, and allowing again for the reduction gear efficiency of 75%, the minimum nominal motor power required is only

$$P_{max} \div 2 \div 0.75 = 0.228 \div 2 \div 0.75 = 0.152 \, [kW]$$

Choose a standard 0.25 kW motor and check the input inertia.

The estimated camshaft, cam, coupling and wormwheel inertia (all running at 60 rev/min) is 0.25 kg . m^2. The motor and wormshaft armature inertia is estimated at 0.0012 kg . m^2 (running at 1440 rev/min).

The equivalent input inertia referred to the cam is

$$I_c = 0.25 + 0.0012 \times (1440 \div 60)^2 = 0.941 \, [kg . m^2]$$

The minimum inertia required to ensure that speed fluctuation does not exceed 10% is given by Equation 13.15:

$$I_c = \frac{I_e}{(1 . 1^2 - 1)} \cdot \left(\frac{C_v . B}{A}\right)^2 = \frac{0.3746}{0.21} \times \left(\frac{2 \times 22.62°}{72°}\right)^2 = 0.704 \, [kg . m^2]$$

In this case the input inertia is adequate. It is interesting to note, however, that without the high speed motor inertia it is probable that the speed fluctuation would be unacceptable.

In this example the calculations have been carried out for rotary motion on the bell-crank lever. To design the cam and follower it is necessary to estimate the contact force. The peak force on the follower *in the direction of its motion* (i.e. the useful force) is the peak torque divided by the length of the bell-crank arm:

Useful follower force = 56.260 ÷ 0.1 = 562.6 [N]

This must be divided by the cosine of the cam pressure angle to get the *contact force*, which can then be checked against the load capacity of the cam. This process is described in Chapters 7, 8, 9 and 10.

CHAPTER 14

Stop/Start and Interrupted Drive

Although the effect of an accelerating cam system input has been explained in the context of overrun (Chapter 13), all the design criteria so far have assumed a nominally constant speed input, which is the normal running speed. That is the main basis for mechanism design, but it must not be overlooked that the machine has to run up to full speed, i.e. accelerated input, when started up and run down to zero, i.e. decelerated input, when it is shut down. These phases in the machine's operation can incur a critical loading on a cam system that is more severe than the normal full speed load. Furthermore, some cam mechanisms are deliberately started and stopped rapidly for functional reasons. A case in point is that of a cam driven indexing mechanism whose input is interrupted by an electro-magnetic clutch/brake unit, either to make the mechanism cycle on demand or to produce an extended dwell period.

A typical indexing system of this kind is shown in Fig. 14.1, which is used to explain the dynamics of an interrupted drive. The same principles apply to any means of accelerating or decelerating the input of any cam system, provided that a constant input torque is used for the purpose. Any friction clutch or friction brake can be assumed to apply a torque that is approximately constant. A common device used for interrupting a cam drive is a braked motor for which, although its start-up torque varies considerably as it speeds up, its peak start-up torque can be used as if it were constant to arrive at a safe design.

The torque and inertia diagram shown in Fig. 14.1 is a slightly simplified model of the system in which:

M_2 = clutch torque (positive value) or brake torque (negative value)
M_w = torque on the worm (or the input torque of any other kind of reduction gear used)*
M_b = torque on the wormwheel (or the output torque of any other kind of reduction gear)*
M_c = torque on the cam*
M_s = torque on the follower that starts or stops the payload*
M_f = constant component of output load, such as friction torque

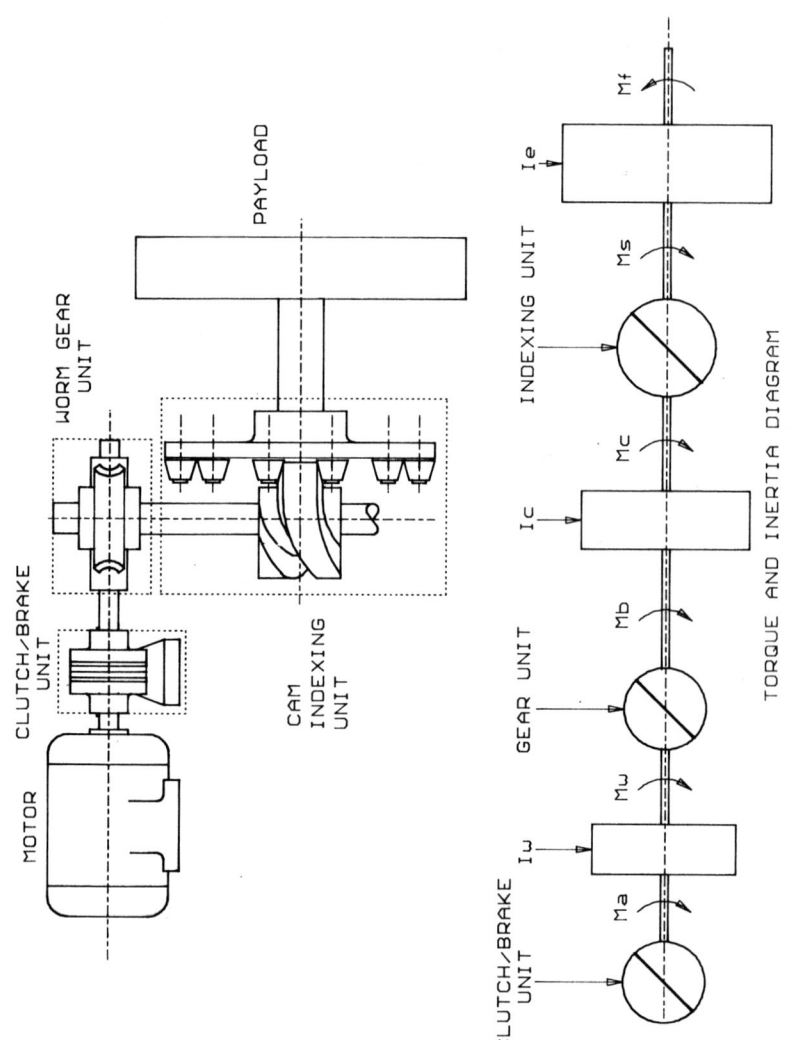

MOTOR

CLUTCH/BRAKE UNIT

WORM GEAR UNIT

PAYLOAD

CAM INDEXING UNIT

CLUTCH/BRAKE UNIT

GEAR UNIT

INDEXING UNIT

I_a M_a M_b I_c M_c M_s I_e M_f

TORQUE AND INERTIA DIAGRAM

Fig. 14.1   Typical interrupted indexing system

I_w = inertia of the wormshaft (or input shaft of any other kind of gearing) and all parts attached thereto, including the driven parts of the clutchbrake (which are usually significant)

I_c = inertia of the camshaft and all parts attached thereto, including the cam and the output parts of the reduction gear

I_e = equivalent inertia of all parts attached to the indexing mechanism output, including the payload and the follower

Other parameters of the system are:

A = input stroke of indexing mechanism
B = output stroke of indexing mechanism
v_a = instantaneous angular velocity of clutch/brake output
a_a = instantaneous angular acceleration of clutch/brake output
v_c = instantaneous angular velocity of indexer input (cam)
a_c = instantaneous angular acceleration of indexer input (cam)
v_t = instantaneous angular velocity of indexer output (turret)
a_t = instantaneous angular acceleration of indexer output (turret)
w' = normalised velocity factor of cam law
w'' = normalised acceleration factor of cam law
E_g = forward efficiency of the gear unit when accelerating or the reciprocal of its reverse efficiency when braking**
E_c = forward efficiency of the cam indexing unit when accelerating or the reciprocal of its reverse efficiency when braking**
G = velocity ratio of the gear unit: input velocity ÷ output velocity (this is the conventional way round for designating the ratio of a worm gear)

STOP/START DURING THE DWELL PERIOD

When a clutch/brake is used to interrupt an indexing drive it is usually arranged for it to operate during the dwell period. This is the safest

---

*These torques are taken as applying at the interface between the gear teeth or the interface between the cam track and follower, as appropriate, so that the inertias of the gears, cam and follower can be included with their adjoining parts.

**The reciprocal of the reverse efficiency is used when braking, to be mathematically consistent in all cases. For braking this value is therefore greater than 1, and for high-ratio worm gears it can also be negative. The reverse efficiency of a cam mechanism varies and is extremely low near the beginning and end of the stroke, but these are not critical positions for stopping the system, so a value of reverse efficiency at the mid-stroke position can be used.

strategy because no excessive loads are imposed on the system by increased acceleration or deceleration of the payload. In this case only the drive components up to and including the cam have to be accelerated or decelerated and the torques generated in the system are simply related to the clutch/brake torque M_a according to the equivalent inertias of those components thus:

$$M_w = M_a \cdot \left[\frac{I_c}{I_c + I_w \cdot G^2 \cdot E_g} \right]$$ Eq. (14.1)

$$M_b = M_w \cdot G \cdot E_g$$ Eq. (14.2)

And since the indexer is doing no work $M_c = 0$.

STOP/START DURING THE MOTION PERIOD

However, during the indexing period of the cam the payload is accelerated and decelerated according to the transmitted acceleration Equation 3.8 and the simple relationships to M_a no longer apply. It is possible, nevertheless, to formulate the instantaneous values of all the torques and accelerations in the system in terms of the known parameters (inertias etc.) and the cam law factors. Another complication is that start-up or braking may occur anywhere in the motion cycle, especially in the event of an emergency stop. For safe design it is necessary to know the worst values of the gear unit torques M_w and M_b and the indexing unit torques M_c and M_s both for start-up and braking. These torques vary as the cam position and with it the instantaneous cam law factors change, and also as the cam speed changes. It is necessary to choose the particular combination of cam law factors that results in the highest instantaneous torque at each part of the system.

The relationships between the input and output velocities and accelerations are:

$$v_a = v_c \cdot G$$ Eq. (14.3)

$$a_a = a_c \cdot G$$ Eq. (14.4)

$$v_t = \frac{B}{A} \cdot w' \cdot v_c$$ Eq. (14.5)

$$a_t = \frac{B}{A^2} \cdot w'' \cdot v_c^2 + \frac{B}{A} \cdot w' \cdot a_c$$ Eq. (14.6)

When the clutch/brake torque M_a is applied we therefore have the following torques in the system to accelerate or decelerate the inertias:

$$M_a - M_w = I_w . a_c . G \qquad \text{Eq. (14.7)}$$

$$M_b = M_w . G . E_g \qquad \text{Eq. (14.8)}$$

$$M_b - M_c = I_c . a_c \qquad \text{Eq. (14.9)}$$

$$M_s = M_c . (v_c / v_t) . E_c \qquad \text{Eq. (14.10)}$$

$$M_s - M_f = I_e . a_t \qquad \text{Eq. (14.11)}$$

Combining Equations 14.7, 14.8 and 14.9 we find the indexing mechanism *input* acceleration:

$$a_c = \frac{M_a . G . E_g - M_c}{I_w . G^2 . E_g + I_c} \qquad \text{Eq. (14.12)}$$

Eliminating M_s from Equations 14.10 and 14.11 we get:

$$M_c = (M_f + I_e . a_t) . (v_t / v_c) / E_c$$

Substituting for a_t and v_t / v_c from Equations 14.5 and 14.6 we get:

$$M_c = \{M_f + I_e . [(B/A^2) . w'' . v_c^2 + (B/A) . w' . a_c]\} . (B/A) . w' / E_c$$
$$\text{Eq. (14.13)}$$

Substituting this in Equation 14.12 and rearranging it we get an expression for a_c in terms of the clutch/brake torque, the cam law factors, the constant system parameters (inertias, etc.) and the *indexer input velocity* v_c:

$$a_c = \frac{M_a . G . E_g . E_c - [M_f + I_e . (B/A^2) . w'' . v_c^2](B/A) . w'}{I_w . G^2 . E_g . E_c + I_c . E_c + I_e . [(B/A) . w']^2} \qquad \text{Eq. (14.14)}$$

It is important to remember that the clutch/brake can be initiated at any moment in the indexing cycle, so that during the speed-up or braking process any speed v_c may coincide with any cam position and therefore with any combination of w' and w'', although w' and w'' are themselves interdependent.

Bearing in mind that w' is always positive and w'' may be either positive or negative, the denominator of Equation 14.14 is always positive, but the second term of the numerator may be of either sign. It may therefore take the same sign as the first term, with an appropriate sign of w'', to give either the largest positive numerator if M_a is positive (on start-up) or the largest negative one if M_a is negative (on braking). The worst cam position for the clutch/brake process to start must fulfil these conditions and it becomes apparent that the highest indexer input acceleration, and also therefore the highest output acceleration, occurs when the indexer input speed v_c is highest, i.e. full operating speed.

Equation 14.14 can be evaluated for every cam law position using the relevant values of w' and w'' and the full speed value of v_c to find the maximum acceleration and maximum deceleration, and from them the maximum torques. This is a tedious process, but it can be performed easily by a computer using a specially written program.

A simpler, but *approximate*, solution, however, can be found by using the maximum values of w' and w'' (i.e. C_v and C_a) in Equation 14.14 as if they occurred simultaneously, which in most cam laws they do not. The equation then becomes:

$$a_c = \frac{M_a.G.E_g.E_c - (M_f \pm C_t.M_i).(B/A).C_v}{I_w.G^2.E_g.E_c + I_c.E_c + C_t.I_e.[(B/A).C_v]^2}$$
Eq. (14.15)

where

M_f = non-inertia output torque on the indexing mechanism.

M_i = nominal peak output inertia torque on the indexing mechanism when running at full operating speed of $v_c = 2.\pi.N/60$ [rad/sec]. This is positive for start-up and negative for braking.

C_t = torsion factor.

The camshaft acceleration a_c, the worm gear output torque M_b and the indexer output torque M_s should all have a positive value on start-up and a negative one on braking, bearing in mind that M_a is positive for the former and negative for the latter. If M_a is not large enough to give a positive value on start-up it means that the camshaft may slow down at some point on the way up to full speed giving an unstable, erratic, or perhaps juddering start. In such a case it is recommended to increase the clutch torque to ensure a smooth start-up. Similarly a_c, M_b and M_s should all be negative during braking to avoid instability or jamming in that mode.

The other stop/start equations become:

$$a_t = (B/A^2).C_a.(2.\pi N/60)^2 + (B/A).C_v.a_c$$
Eq. (14.16)

$$M_s = C_t.I_e.a_t + M_f$$
Eq. (14.17)

$$M_c = M_s.(B/A).C_v/E_c$$
Eq. (14.18)

$$M_b = M_c + I_c.a_c$$
Eq. (14.19)

$$M_w = M_b/G/E_g \qquad \text{Eq. (14.20)}$$
$$M_a = M_w + I_w \cdot a_c \cdot G \qquad \text{Eq. (14.21)}$$

where N = full operating speed of the indexer [rev/min].

Example 14.1

A 12 stop indexing mechanism as shown in Fig. 14.1 drives a payload inertia of 2.45 kg.m² at 200 cycles per minute using a Mod Sine cam law with a 240° indexing period. A 1450 rev/min motor drives the system and a clutch/brake is used to interrupt the drive during the dwell period in order to operate the machine in a cycle-on-demand mode. The machine is liable to be stopped by the clutch/brake unit in an emergency and re-started anywhere in the indexing cycle. Find the normal operating loadings in the the system and the loadings in emergency stop/start. The transmissions are rigid enough to give a torsion factor of $C_t = 1.24$.

Properties of the drive components determined by the initial design are as follows:

Total output inertia including the indexer turret	$I_e = 2.93$ [kg.m²]
Non-inertia output torque	$M_f = 45$ [N.m]
Total camshaft inertia including cam and wormwheel	$I_c = 0.176$ [kg.m²]
Total wormshaft inertia including worm and output parts of clutch/brake	$I_w = 0.0008$ [kg.m²]
Worm gear ratio 1450 ÷ 200	$G = 7.25$
Worm gear forward efficiency	80%
Worm gear reverse efficiency	70%
Indexer forward efficiency	95%
Indexer reverse efficiency	85%
Indexer input stroke	$240° = 4.189$ [rad]
Indexer output stroke	$30° = 0.5236$ [rad]
Camshaft full speed $2 \times \pi \times 200 \div 60$	20.944 [rad/sec]
Index period (time) 60/200 sec × 240/360	$T = 0.2$ [sec]
Cam law coefficient of acceleration	$C_a = 5.528$
Cam law coefficient of velocity	$C_v = 1.759$
Clutch slipping torque*	$M_a = 10$ [N.m]
Brake slipping torque*	$M_a = -10$ [N.m]

---

*These values turn out to be unsatisfactory for stop/start during the indexing period and are subsequently revised to 15 N.m.

NOMINAL FULL SPEED LOADINGS (see Chapter 13)

$M_i = I_e . C_a . B/T^2 = 2.93 \times 5.528 \times 0.5236 \div 0.2^2$ $= 212.02 \, [\text{N.m}]$

$M_t = C_t . M_i + M_f = 1.24 \times 212.02 + 45 = 262.9 + 45 = 307.9 \, [\text{N.m}]$

The pre-load ratio is $K = 1$
and the inertia ratio is $Q = 262.9 \div 307.9$ $= 0.854$

From the Mod Sine table in Appendix C we find by interpolation $C_c = 1.033$, therefore

$M_c = C_c . M_t . B/A/E_c$
    $= 1.033 \times 307.9 \times 30 \div 240 \div 0.95$ $= 41.85 \, [\text{N.m}]$

$M_b = M_c$ (there is no acceleration of the inertia on this shaft)

$M_w = M_b/G/E_g = 41.85 \div 7.25 \div 0.8$ $= 7.216 \, [\text{N.m}]$

Torque transmitted through the clutch/brake $= M_w$ $= 7.216 \, [\text{N.m}]$
(there is no acceleration here either).

This is less than the slipping torque of the clutch/brake, proving that the size of clutch is adequate for normal operation. Overrun is not a problem in normal full speed operation because the inertia of the motor armature, which is attached to the wormshaft by the clutch, is sufficiently large.

STOP/START DURING THE DWELL PERIOD

For starting $E_g = 0.8$ and $M_a = 10 \, [\text{N.m}]$

$M_w = M_a . I_c/(I_c + I_w . G^2 . E_g)$
    $= 10 \times 0.176 \div (0.176 + 0.0008 \times 7.25^2 \times 0.8) = 8.395 \, [\text{N.m}]$

$M_b = M_w . G . E_g = 8.395 \times 7.25 \times 0.8$ $= 48.691 \, [\text{N.m}]$

For stopping $E_g = 1 \div 0.7 = 1.428$ and M_a $= -10 \, [\text{N.m}]$

$M_w = -10 \times 0.176 \div (0.176 + 0.0008 \times 7.25^2 \times 1.428)$
                                                        $= -7.456 \, [\text{N.m}]$

$M_b = M_w . G . E_g = -7.456 \times 7.25 \times 1.428$ $= -77.192 \, [\text{N.m}]$

The gear unit must have an output torque capacity greater than $77.192 \, \text{N.m}$ for interrupted drive operation with a 10 N.m clutch/brake, although the peak torque during indexing is only 41.85 N.m. Actually, since it is possible for the clutch to slip when the gear unit is stationary, for instance in the event of a blockage, the output torque on the gear unit could be the clutch torque times the gear ratio = $10 \times 7.25 = 72.5$ N.m. Later we find that it is advisable to increase the clutch torque to 15 N.m so the gear unit has to be strong enough for an output torque of $15 \times 7.25 = 108.75$ N.m rather than the normal operating torque of 77.192 N.m.

EMERGENCY STOP/START DURING THE MOTION PERIOD

This is evaluated separately for starting and stopping using Equations 14.14 to 14.21 inclusive, in sequence.

Start-up

(M_a and M_i both positive)

$$a_c = \frac{10 \times 7.25 \times 0.8 \times 0.95 - (45 + 1.24 \times 212.02) \times (30/240) \times 1.759}{0.0008 \times 7.25^2 \times 0.8 \times 0.95 + 0.176 \times 0.95 + 1.24 \times 2.93 \times [(30/240) \times 1.759]^2}$$

$$a_c = \frac{55.1 - 67.7}{0.032 + 0.1672 + 0.1757} = \frac{-12.6}{0.3749} \qquad = -33.61 \ [\text{rad/s}^2]$$

This is a negative value, which may result in an erratic start-up, so it is recommended to increase M_a to the next standard value of 15 N.m (the standard clutch/brake unit has a similar brake torque; -15 N.m).

Revised
$$a_c = (15 \times 7.25 \times 0.8 \times 0.95 - 67.7) \div 0.3409$$
$$= 14.95 \div 0.3749 \qquad = 39.88 \ [\text{rad/s}^2]$$
$$a_t = (0.5236/4.189^2) \times 5.528 \times 20.944^2 + (30/240) \times 1.759 \times 39.88$$
$$= 72.36 + 8.77 \qquad = 81.13 \ [\text{rad/s}^2]$$
$$M_s = 2.93 \times 1.24 \times 81.13 + 45 \qquad = 339.76 \ [\text{N.m}]$$
$$M_c = 339.76 \times (30/240) \times 1.759 \div 0.95 \qquad = 78.64 \ [\text{N.m}]$$
$$M_b = 78.64 + 0.176 \times 39.88 \qquad = 85.65 \ [\text{N.m}]$$
$$M_w = 85.65 \div 7.25 \div 0.8 \qquad = 14.77 \ [\text{N.m}]$$
$$M_a = 14.77 + 0.0008 \times 39.88 \times 7.25 \qquad = 14.999 \ [\text{N.m}]$$

Apart from rounding errors M_a, as derived from all the other calculated values, exactly matches the figure fed into the calculation, thus proving the arithmetic.

Stopping

$$a_c = \cfrac{\begin{array}{l} -15 \times 7.25 \times 1.428 \times 1.111 \\ \qquad - (45 - 1.24 \times 212.02) \times (30/240) \times 1.759 \end{array}}{\begin{array}{l} 0.0008 \times 7.25^2 \times 1.428 \times 1.111 + 0.176 \\ \qquad \times 1.111 + 1.24 \times 2.93 \times [(30/240) \times 1.759]^2 \end{array}}$$

$$a_c = \frac{-172.62 + 47.91}{0.0667 + 0.1955 + 0.1757} = \frac{-124.71}{0.4379} \qquad = -284.8 \,[\text{rad/s}^2]$$

$$\begin{aligned} a_t &= -(0.5236/4.189^2) \times 5.528 \times 20.944^2 - (30/240) \times 1.759 \times 284.8 \\ &= -72.36 - 62.62 \qquad\qquad\qquad\qquad\qquad = -134.98 \,[\text{rad/s}^2] \end{aligned}$$

$$M_s = -2.93 \times 1.24 \times 134.98 + 45 \qquad\qquad = -445.4 \,[\text{N.m}]$$

$$M_c = -445.4 \times (30/240) \times 1.759 \div 1.111 \qquad = -88.14 \,[\text{N.m}]$$

$$M_b = -88.14 + -0.176 \times 284.8 \qquad\qquad = -138.26 \,[\text{N.m}]$$

$$M_w = -138.26 \div 7.25 \div 1.428 \qquad\qquad\quad = -13.35 \,[\text{N.m}]$$

$$M_a = -13.35 + 0.0008 \times 284.8 \times 7.25 \qquad\quad = -15.006 \,[\text{N.m}]$$

Apart from rounding errors M_a, as derived from all the other calculated values, exactly matches the figure fed into the calculation, thus proving the arithmetic.

We now find that the system is subject to an emergency stop load of $M_s = 445.4$ N.m output torque on the indexing mechanism and $M_b = 138.26$ N.m output torque on the worm gear unit. These are the critical design values for the system, rather than the normal operating loads.

It could have been decided to ignore the fact that emergency start-up might be erratic, since it would seldom happen. This would permit the use of the originally chosen 10 N.m clutch/brake and a smaller gear unit because of the reduced emergency loadings. The loadings then become as follows, which are considerably more than the normal operating loads:

	start-up [N.m]	braking [N.m]	
M_a	10	−10	
M_w	10.19	−9.11	
M_b	59.13	−94.35*	(criterion for gear unit)
M_c	65.05	−67.36	
M_s	281.05	−340.42*	(criterion for indexer)

It has been mentioned that the reverse efficiency of worm gears can be very low and even be negative: this is particularly true of high ratios which may have small lead angles. A negative efficiency means that the wormwheel cannot drive the worm and the gear unit is locked unless the worm is moved by another means. Worm gears with negative reverse efficiency are sometimes used for hand powered hoists to make them self-locking, but the self-locking feature should never be relied on when personal safety is at risk because the slightest disturbance of the worm-shaft, e.g. vibration, can start it turning, and it may then run away when the dynamic friction at the gear teeth is low enough to make its reverse efficiency positive.

In cam drives a negative reverse efficiency need not cause problems if there is an adequate wormshaft inertia to ensure that the system comes smoothly to rest when stopped. The use of too high a braking torque, however will easily overload the system, and it may be possible to dispense with a brake altogether, letting the worm gear itself act as a brake. Very low or negative reverse efficiency in a worm gear cannot be accurately quantified and it is recommended that for *precise* stopping a gear unit with a more predictable reverse efficiency is chosen and the process controlled with a suitable brake. Other kinds of reduction gearing do not usually have negative reverse efficiency.

LINEAR MOTION MECHANISMS

Similar equations to those developed above for rotary motion systems can be developed in the same way for linear motion systems. The systems are identical up to and including the cam, but then the follower output has a force F_s to accelerate or decelerate a mass W_e with or against a non-inertia force F_f. The stop/start equations for a linear output motion are therefore:

$$a_c = \frac{M_a.G.E_g.E_c - (F_f \pm C_t.F_i).(Y/A).C_v}{I_w.G^2.E_g.E_c + I_c.E_c + C_t.W_e.[(Y/A).C_v]^2} \qquad \text{Eq. (14.22)}$$

$$a_t = (Y/A^2).C_a.(2.\pi.N/60)^2 + [Y/A].C_v.a_c \qquad \text{Eq. (14.23)}$$

$$F_s = C_t.W_e.a_t + F_f \qquad \text{Eq. (14.24)}$$

$$M_c = F_s.(Y/A).C_v/E_c \qquad \text{Eq. (14.25)}$$

$$M_b = M_c + I_c.a_c \qquad \text{Eq. (14.26)}$$

$$M_w = M_b/G/E_g \qquad\qquad\qquad\qquad \text{Eq. (14.27)}$$
$$M_a = M_w + I_w \cdot a_c \cdot G \qquad\qquad\qquad \text{Eq. (14.28)}$$

where

> a_t = *Linear* acceleration at the output of the follower instead of angular acceleration
>
> Y = *Linear* output stroke of cam follower
>
> F_f = Non-inertia output *force* on the follower
>
> F_i = Normal peak output inertia *force* on the follower at full operating speed
>
> F_s = Stopping or starting output *force*
>
> W_e = Equivalent *mass* of all parts attached to the indexing mechanism output, including the payload and the follower

All other symbols are the same as for rotary motion.

CHAPTER 15

Profile Machining Geometry

The profiles of industrial cams are formed on prepared cam blanks by an appropriate machine tool. The machine may use a metal removal technique such as milling, grinding, nibbling, EDM (electric discharge machining) or flame cutting. The choice of cutting technique depends on the type of cam, its material and hardness, whether roughing out or finishing, the accuracy required, etc. In some cases general purpose 'standard' machines can be used and sometimes it is better to use specialised cam profiling machines. In all cases the 'cutter', i.e. the metal removing tool, and the cam blank must be moved relative to each other in a prescribed pattern that is related to, but not necessarily the same as, the cam follower motion. This pattern can be produced by a 'master' cam or by Numerical Control or, more rarely, by a linkage or similar mechanism.

DISC CAMS

The simple relationship between the cam profile and the pitch curve of a disc cam is important for profile machining. The simplest way to machine a profile is to use a milling cutter or grinding wheel of exactly the same diameter as the roller follower and to simulate the follower motion by moving it along the follower pitch curve: this can be done whether the cam is stationary and the cutter moves or vice versa. The geometry of the cutter motion in this case is exactly the same as that of the follower pitch curve, fully described in Chapter 7.

However, the same profile can be machined with a different sized cutter moving on a different pitch – the *cutter* pitch curve. This enables a convenient standard size cutter to be used irrespective of the follower size. For example disc cams that use flat faced followers or small roller followers can be profile-ground using a large diameter high speed grinding wheel, provided that the radius of the wheel is smaller than the smallest *concave* radius of curvature of the profile.

Each of these curves – cam profile, follower pitch and cutter pitch – is parallel to, i.e. equidistant from, the others, so that a point on one curve

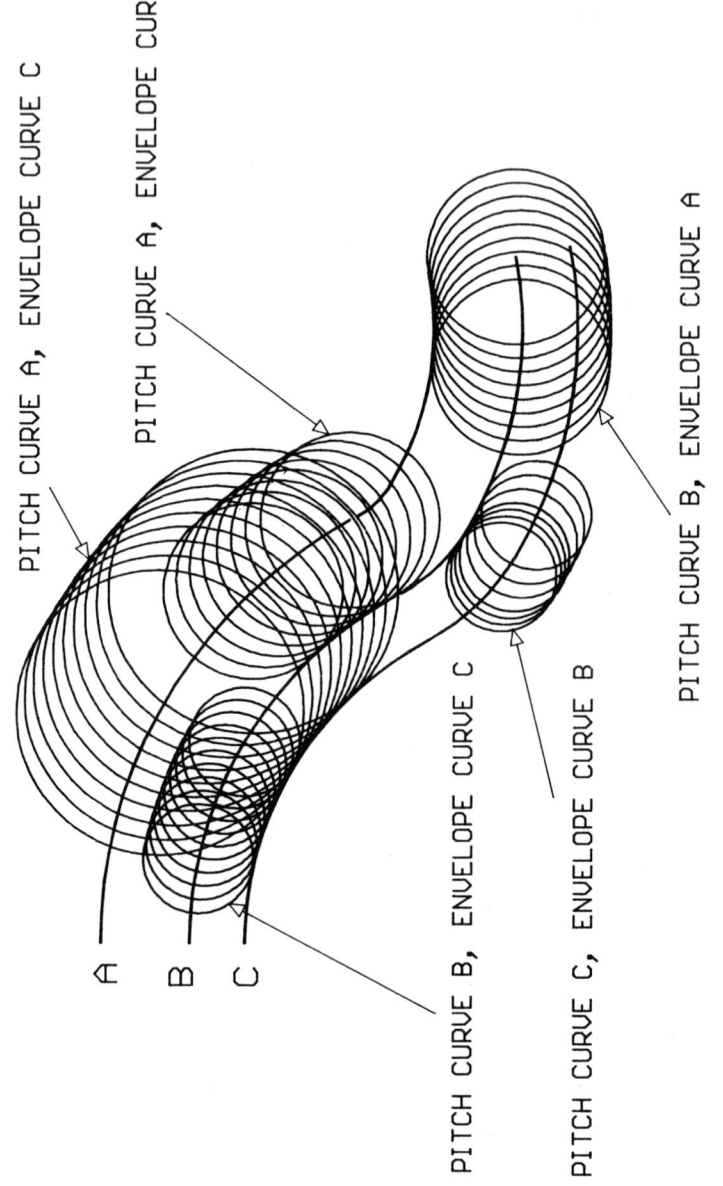

PITCH CURVE A, ENVELOPE CURVE C

PITCH CURVE A, ENVELOPE CURVE B

PITCH CURVE B, ENVELOPE CURVE A

PITCH CURVE B, ENVELOPE CURVE C

PITCH CURVE C, ENVELOPE CURVE B

A

B

C

Fig. 15.1   Relationship between plane parallel curves

is displaced from the relative point on another curve by a constant distance along the line of pressure (see Fig. 15.2). The distance between the follower pitch curve and the cutter pitch curve is the difference in the radii of the follower roller and the cutter. The distance between the cam profile curve and the cutter pitch curve is simply the cutter radius.

The three curves A, B and C in Fig. 15.1 are all parallel to each other and each curve can be considered as the envelope of a series of circles of appropriate radius pitched along one of the other curves, as illustrated. Thus, if one of the curves is the cam profile and another is the follower roller pitch curve, then the remaining curve could be a cutter pitch curve, provided that the cutter radius differs from the follower radius by an amount equal to the distance between the two pitch curves.

The cutter can be either larger or smaller than the roller, with the limitation, previously mentioned, that it must be smaller than the smallest *concave* radius of curvature of the cam profile to avoid under-cutting due to interference (see Chapter 7). The cutter pitch curve can be derived either from the cam profile itself or from the follower roller pitch curve. The latter is the case if a 'cutter diameter compensation' facility is used when cutting a cam on a Numerically Controlled machine. With this option the machine control automatically adjusts the cutter path to allow for a keyed-in change of the preset cutter diameter. This avoids re-programming a job when, for example, a reground cutter is used.

If the co-ordinates *and slopes* of a curve are known the co-ordinates of a parallel curve are found from the following equations (see Fig. 15.2).

$$x_2 = x_1 - j.\sin\mu \qquad\qquad \text{Eq. (15.1)}$$
$$y_2 = y_1 + j.\cos\mu \qquad\qquad \text{Eq. (15.2)}$$
$$r_2 = \surd[r_1^2 + j^2 + 2.r_1.j.\sin(\theta_1 - \mu)] \qquad\qquad \text{Eq. (15.3)}$$
$$\theta_2 = \theta_1 + \arcsin[j.\cos(\theta_1 - \mu)/r_2] \qquad\qquad \text{Eq. (15.4)}$$

where

$$\mu = \arctan(dy/dx) \qquad\qquad \text{Eq. (15.5)}$$

and

$$(\theta_1 - \mu) = \arctan\left(\frac{r_1}{dr_1/d\theta_1}\right) \qquad\qquad \text{Eq. (15.6)}$$

When these equations are used for cutter diameter compensation with numerical control profiling in a Cartesian co-ordinates mode (x, y), it is possible to use the small motion increments i_x and i_y which separate

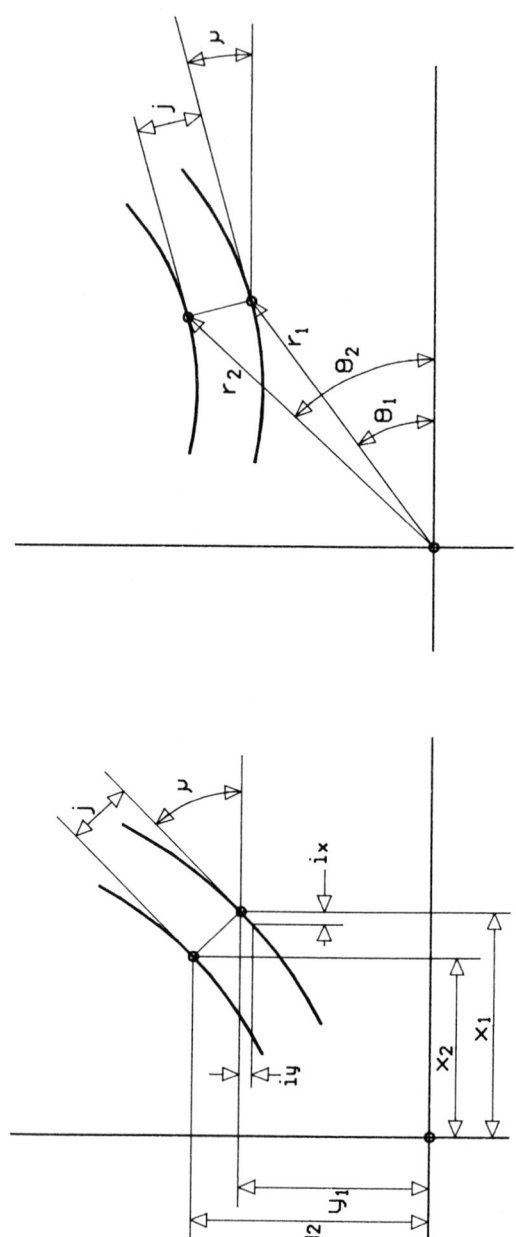

Fig. 15.2 Parallel curves: Cartesian and polar co-ordinates

consecutive command positions to evaluate the instantaneous slope of the profile during machining.

Thus, approximately:

$$\mu = \arctan (i_y/i_x) \hspace{3cm} \text{Eq. (15.7)}$$

This can be used in Equations 15.1 and 15.2 or x_2 and y_2 can be found from the following two equations:

$$x_2 = x_1 - j.i_y/\sqrt{(i_y^2 + i_x^2)} \hspace{2cm} \text{Eq. (15.8)}$$
$$y_2 = y_1 + j.i_x/\sqrt{(i_y^2 + i_x^2)} \hspace{2cm} \text{Eq. (15.9)}$$

Similarly, when the equations are used for cutter diameter compensation with numerical control profiling in a polar co-ordinates mode (r, θ), it is possible to use the small motion increments i_r and i_θ which separate consecutive command positions in that mode to evaluate the instantaneous slope of the profile during machining.

Thus, approximately:

$$(\theta_1 - \mu) = \arctan \left(\frac{r_1}{i_r/i\theta}\right) \hspace{2cm} \text{Eq. (15.10)}$$

This can be used in Equations 15.3 and 15.4 to find r_2 and θ_2.

The sign of j, the distance between the curves, can be positive or negative, depending whether curve 2 is as shown or on the opposite side of curve 1.

If the roller follower is tapered then the profile can still be machined with a cutter of different diameter, provided that it has the same taper angle as the roller.

Plane cams and disc cams are usually profile machined with rotary milling cutters or grinding wheels whose spindle axes are perpendicular to the plane of the cam blank. The cutter pitch curve can be generated in that plane in various modes of motion:

Cam blank motion	**Cutter spindle motion**
Stationary	Moves around curve in x- and y-directions simultaneously.
Moves in x-direction only	Moves in y-direction only
Moves in y-direction only	Moves in x-direction only
Moves around curve in x- and y-directions simultaneously	Stationary

Rotates only	Moves in x-direction only
Rotates only	Moves in x- and y-directions simultaneously (simulating arcuate path of swinging follower)
Rotates and moves in x-direction	Stationary
Rotates and moves in x- and y-directions simultaneously	Stationary

All these modes of motion can produce accurate profiles, depending on the quality of the machine tool and the cutting technology. However, the methods which use a rotating cam blank are preferred by some specialist cam manufacturers because of the inherent accuracy of the dwell zones. During dwell machining the cam blank axis and the cutter spindle are both stationary, held at a fixed distance apart, and the accuracy of the circular arc is mainly dependent on the quality of the machine tool bearings, which is usually of a very high standard.

CYLINDER CAMS

It is not possible to machine a cylinder cam with a cutter that is of a singificantly different size or shape from the roller follower. Only a very small difference is permissible for cutter diameter compensation (described above for disc cams). The motion of the cutter relative to the cam blank should be an exact simulation of the roller follower motion. This is dictated by the fact that both the pressure angle and radius of curvature vary across the face of the cam track wall, giving different, but confluent, roller pitch curves at different points on the wall. It is impossible for the various pitch curves of a cutter to be confluent unless the cutter is exactly the same as the roller. If the roller moves in a straight line then so must the cutter, and with the same offset, if any. If the roller moves on an arcuate path then so must the cutter, with the same arc radius and the same centre distance from the cam axis.

The modes of cutter motion for machining cylinder cams are therefore restricted in practice to the following (x-direction is parallel to cam axis):

Cam blank motion	**Cutter spindle motion**
Rotates only	Moves in x-direction only
Rotates only	Moves in x- and y-directions simultaneously (simulating arcuate path of swinging follower)
Rotates and moves in x-direction	Stationary
Rotates and moves in x- and y-directions simultaneously (simulating arcuate path of swinging follower)	Stationary

The geometry of the cutter motion is exactly the same as that of the follower pitch curve, fully described in Chapter 8. Slight deviations from this for the purpose of cutter diameter compensation can be calculated from Equations 15.1 and 15.2, where x_1 and x_2 are the arcuate displacements around the circumference of a development cylinder. An appropriate development cylinder diameter is midway between the outside diameter and the root diameter of the cam.

$$x_1 = r.\alpha_1 \qquad\qquad\qquad \text{Eq. (15.11)}$$

and

$$\alpha_2 = x_2/r \qquad\qquad\qquad \text{Eq. (15.12)}$$

where

α_1 and α_2 are angular displacements of the cam blank and $r =$ the development cylinder radius.

It should be noted that cylinder cams are often designed to use tapered follower rollers. This enables preload adjustment on a cam blade by simply moving the follower deeper into mesh with the cam. As mentioned, the profile must be machined with a similarly tapered cutter, and therefore cutter diameter compensation can be effected by moving the cutter deeper into mesh with the cam blank. Only very slight adjustments of this kind are permissible, otherwise the track geometry becomes too distorted and bad contact occurs between the roller and the cam.

If a blade cam is required to have preload on its blade and there are parts of its profile which are enclosed tracks, as in the case of an indexing cam for example, then special measures are needed to ensure track clearance.

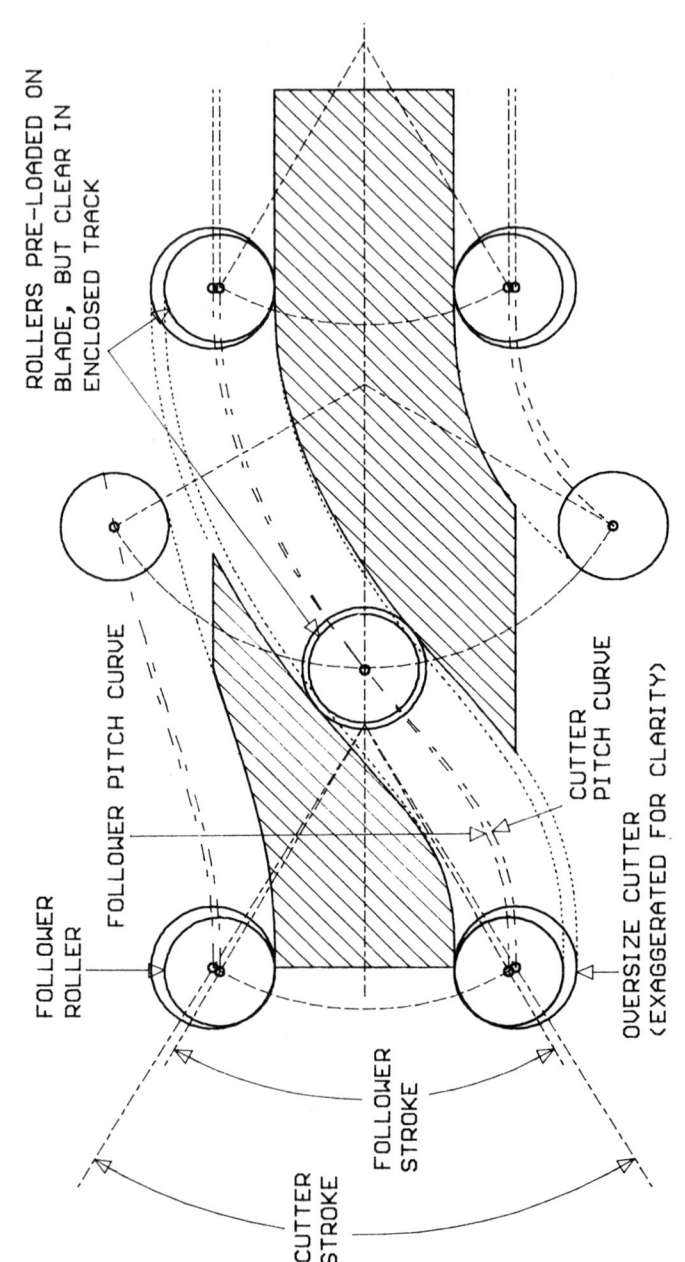

ROLLERS PRE-LOADED ON
BLADE, BUT CLEAR IN
ENCLOSED TRACK

FOLLOWER PITCH CURVE

FOLLOWER
ROLLER

CUTTER
PITCH CURVE

OVERSIZE CUTTER
(EXAGGERATED FOR CLARITY)

FOLLOWER
STROKE

CUTTER
STROKE

Fig. 15.3   Effect of oversize cutter with increased stroke

(1)   The cutter must be slightly bigger than the roller to provide the track clearance, or in the case of a tapered cutter it can be slightly deeper in mesh.

(2)   The cutter motion must be slightly different from the follower motion (see Fig. 15.3).

APPENDIX A

Cam Law Equations and Diagrams

All the cam laws that follow are *normalised* (see Chapters 3 and 4) whereby the input displacement, u, varies from 0 to 1, and the output displacement, w, a function of u, also varies from 0 to 1.

The derivatives of this function are the normalised *geometric* velocity, acceleration and jerk, designated w', w'' and w''' respectively. For convenience these values are usually termed the *cam law factors*: displacement factor, velocity factor, etc. The use of factors to calculate real displacements, velocities, etc. is explained by way of examples in Chapter 3. Generally the deceleration loop of a standard cam law is the reversed negative shape of the acceleration loop: the law is symmetrical, or 'reflective'. The *shape* of the acceleration loop of any cam law is its most characteristic feature: shape is independent of scale and time.

Several of the standard cam laws belong to the SCCA family: Sine–Constant–Cosine–Acceleration. They are defined by the general equations below, in which the period parameters a, b and c determine the shape of the acceleration and deceleration loops.

Note: These equations can also be used for *composite* SCCA cam laws, i.e. those with a period of constant velocity between the acceleration and deceleration periods: in such cases $(a + b + c)$ is less than 1. In no case may $(a + b + c)$ be greater than 1 and neither a, b nor c may be negative, but any of them may be zero.

The parameters for the standard laws that belong to this family are listed in Table 4.1, Chapter 4. The specific equations for some of the standard laws are also given in this Appendix. They are less complex than the general SCCA equations, but can, of course, be derived from them by substituting the appropriate values of a, b and c. Whenever a parameter is zero the relevant zone does not exist and the equations for the zone are ignored.

The following diagram shows the general form of SCCA cam laws.

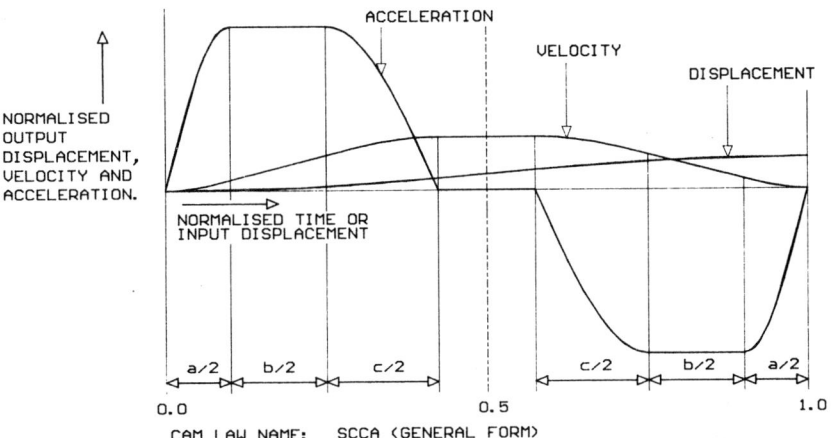

CAM LAW NAME: SCCA (GENERAL FORM)

SCCA CAM LAW EQUATIONS
(See diagram above)

$$C_a = \cfrac{1}{\left(\dfrac{a}{\pi} + \dfrac{b}{2} + \dfrac{c}{\pi}\right) + \dfrac{a^2}{4} + 2\dfrac{(c^2 - a^2)}{\pi^2} - \dfrac{(a + b)^2}{4} - \dfrac{c}{\pi}(a + b + c)}$$

$$C_V = C_a[(a + c)/\pi + b/2]$$

Zone 1: from $u = 0$ to $u = a/2$, but not when $a = 0$

$$w = C_a\left[\left(\frac{a}{\pi}\right)u - \left(\frac{a}{\pi}\right)^2 \sin\left(\frac{\pi u}{a}\right)\right]$$

$$w' = C_a\left[\left(\frac{a}{\pi}\right) - \left(\frac{a}{\pi}\right)\cos\left(\frac{\pi u}{a}\right)\right]$$

$$w'' = C_a \sin\left(\frac{\pi u}{a}\right)$$

$$w''' = C_a\left(\frac{\pi}{a}\right)\cos\left(\frac{\pi u}{a}\right)$$

Zone 2: from $u = a/2$ to $u = (a + b)/2$

$$w = C_a\left[\frac{u^2}{2} + \left(\frac{a}{\pi} - \frac{a}{2}\right)u + \frac{a^2}{8} - \left(\frac{a}{\pi}\right)^2\right]$$

$$w' = C_a\left[u + \left(\frac{a}{\pi} - \frac{a}{2}\right)\right]$$

$$w'' = C_a$$

$$w''' = 0$$

Zone 3: from $u = (a + b)/2$ to $u = (a + b + c)/2$

$$w = C_a\left\{\left(\frac{a}{\pi} + \frac{b}{2}\right)u + \frac{a^2}{8} + \frac{(c^2 - a^2)}{\pi^2} - \frac{(a + b)^2}{8} - \left(\frac{c}{\pi}\right)^2\cos\left[\frac{\pi}{c}\left(u - \frac{a}{2} - \frac{b}{2}\right)\right]\right\}$$

$$w' = C_a\left\{\left(\frac{a}{\pi} + \frac{b}{2}\right) + \left(\frac{c}{\pi}\right)\sin\left[\frac{\pi}{c}\left(u - \frac{a}{2} - \frac{b}{2}\right)\right]\right\}$$

$$w'' = C_a\cos\left[\frac{\pi}{c}\left(u - \frac{a}{2} - \frac{b}{2}\right)\right]$$

$$w''' = -C_a\frac{\pi}{c}\sin\left[\frac{\pi}{c}\left(u - \frac{a}{2} - \frac{b}{2}\right)\right]$$

Zone 4: from $u = (a + b + c)/2$ to $u = 1 - (a + b + c)/2$

$$w = C_a\left[\left(\frac{a}{\pi} + \frac{b}{2} + \frac{c}{\pi}\right)u + \frac{a^2}{8} + \frac{(c^2 - a^2)}{\pi^2} - \frac{(a + b)^2}{8} - \frac{c}{\pi}\frac{(a + b + c)}{2}\right]$$

$$w' = C_a\left(\frac{a}{\pi} + \frac{b}{2} + \frac{c}{\pi}\right)$$

$$w'' = 0 \qquad \text{and} \qquad w''' = 0$$

Zones 5, 6 and 7: from $u = 1 - (a + b + c)/2$ to $u = 1$

These zones in the last half of the motion are the 'mirror image' of those in the first half of the motion. The equations of zones 3, 2 and 1 can be used for zones 5, 6 and 7 respectively, but we must then replace u by $(1 - u)$, replace the resulting displacement w by $(1 - w)$ and change the sign of the acceleration w'' (make it negative).

PARABOLIC CAM LAW EQUATIONS
(Parabolic displacement)

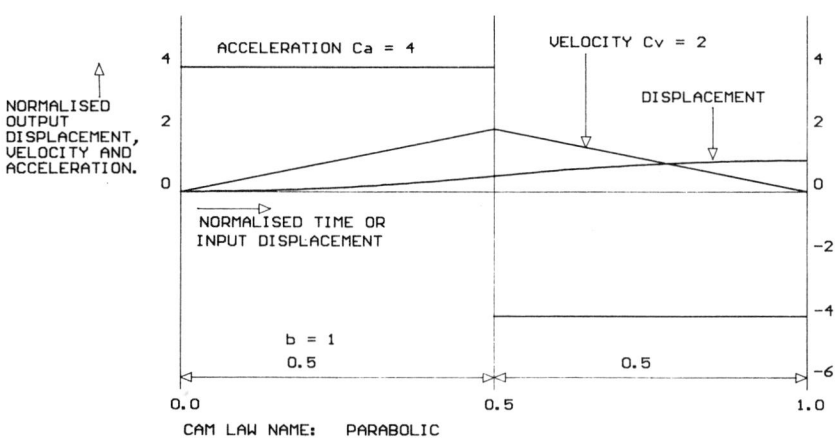

$C_a = 4$ $C_V = 2$

Zone 1: from $u = 0$ to $u = 0.5$

$w = 2u^2$

$w' = 4u$

$w'' = 4$ but discontinuous at beginning and end of zone

$w''' = 0$ but infinity (instantaneously) at beginning and end of zone

Zone 2: from $u = 0.5$ to $u = 1$

$w = 1 - 2(1 - u)^2$

$w' = 4(1 - u)$

$w'' = -4$ but discontinuous at beginning and end of zone

$w''' = 0$ but infinity (instantaneously) at beginning and end of zone

SIMPLE HARMONIC CAM LAW EQUATIONS

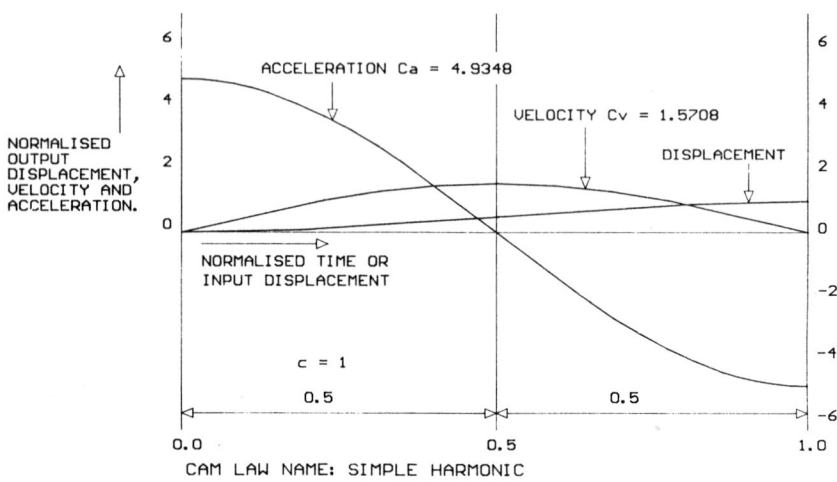

CAM LAW NAME: SIMPLE HARMONIC

$$C_a = \frac{\pi^2}{2} \qquad\qquad C_V = \frac{\pi}{2}$$

$$w = \frac{[1 - \cos(\pi u)]}{2}$$

$$w' = \frac{\pi}{2} \sin(\pi u)$$

$$w'' = \frac{\pi^2}{2} \cos(\pi u) \qquad \text{but discontinuous at beginning and end of zone}$$

$$w''' = -\frac{\pi^3}{2} \sin(\pi u) \qquad \text{but infinity (instantaneously) at beginning and end of zone}$$

MOD TRAP CAM LAW EQUATIONS
(Modified Trapezoid Acceleration)

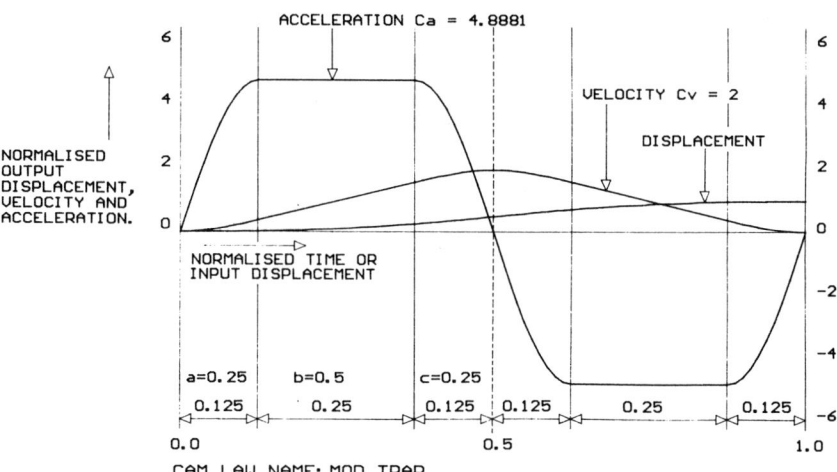

$$C_a = \frac{8\pi}{(\pi + 2)} \qquad\qquad C_V = 2$$

Zone 1: from $u = 0$ to $u = 0.125$

$$w = \frac{2u - (0.5/\pi)\sin(4\pi u)}{(\pi + 2)}$$

$$w' = \frac{2[1 - \cos(4\pi u)]}{(\pi + 2)}$$

$$w'' = \frac{8\pi \sin(4\pi u)}{(\pi + 2)}$$

$$w''' = \frac{32\pi^2 \cos(4\pi u)}{(\pi + 2)}$$

Zone 2: from $u = 0.125$ to $u = 0.375$

$$w = \frac{4\pi u^2 - (\pi - 2)u + \pi/16 - 0.5/\pi^2}{(\pi + 2)}$$

$$w' = \frac{8\pi u - \pi + 2}{(\pi + 2)}$$

$$w'' = \frac{8\pi}{(\pi + 2)}$$

$$w''' = 0$$

Zone 3: from $u = 0.375$ to $u = 0.625$

$$w = \frac{2(\pi + 1)u - 0.0625 - (0.5/\pi) \cos\left[4\pi(u - 0.375)\right]}{(\pi + 2)}$$

$$w' = \frac{2(\pi + 1) + 2 \sin\left[4\pi(u - 0.375)\right]}{(\pi + 2)}$$

$$w'' = \frac{8\pi \cos\left[4\pi(u - 0.375)\right]}{(\pi + 2)}$$

$$w''' = \frac{-32\pi^2 \sin\left[4\pi(u - 0.375)\right]}{(\pi + 2)}$$

Zone 4: from $u = 0.625$ to $u = 0.875$

Use the same equations as in zone 2, but replace u by $(1 - u)$, replace w by $(1 - w)$ and replace w'' by $-w''$.

Zone 5: from $u = 0.875$ to $u = 1$

Use the same equations as in zone 1, but replace u by $(1 - u)$, replace w by $(1 - w)$ and replace w'' by $-w''$.

MOD SINE CAM LAW EQUATIONS
(Modified Sine Acceleration)

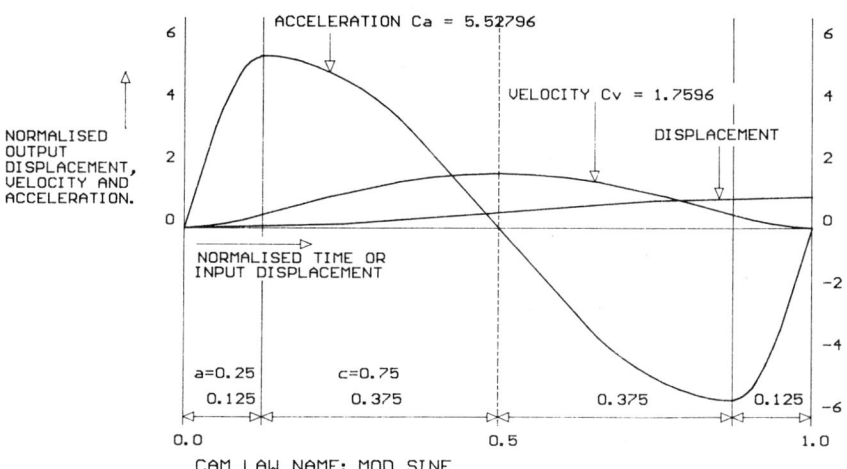

$$C_a = \frac{4\pi^2}{(\pi + 4)} \qquad\qquad C_V = \frac{4\pi}{(\pi + 4)}$$

Zone 1: from $u = 0$ to $u = 0.125$

$$w = \frac{\pi u - 0.25 \sin(4\pi u)}{(\pi + 4)}$$

$$w' = \frac{\pi[1 - \cos(4\pi u)]}{(\pi + 4)}$$

$$w'' = \frac{4\pi^2 \sin(4\pi u)}{(\pi + 4)}$$

$$w''' = \frac{16\pi^3 \cos(4\pi u)}{(\pi + 4)}$$

Zone 2: from $u = 0.125$ to $u = 0.875$

$$w = \frac{\pi u + 2 - 2.25 \cos\left[4\pi(u - 0.125)/3\right]}{(\pi + 4)}$$

$$w' = \frac{\pi + 3\pi \sin\left[4\pi(u - 0.125)/3\right]}{(\pi + 4)}$$

$$w'' = \frac{4\pi^2 \cos\left[4\pi(u - 0.125)/3\right]}{(\pi + 4)}$$

$$w''' = \frac{-16\pi^3/3 \sin\left[4\pi(u - 0.125)/3\right]}{(\pi + 4)}$$

Zone 3: from $u = 0.875$ to $u = 1$

$$w = \frac{\pi u + 4 - 0.25 \sin(4\pi u)}{(\pi + 4)}$$

$$w' = \frac{\pi[1 - \cos(4\pi u)]}{(\pi + 4)}$$

$$w'' = \frac{4\pi^2 \sin(4\pi u)}{(\pi + 4)}$$

$$w''' = \frac{16\pi^3 \cos(4\pi u)}{(\pi + 4)}$$

MSC.50 CAM LAW EQUATIONS

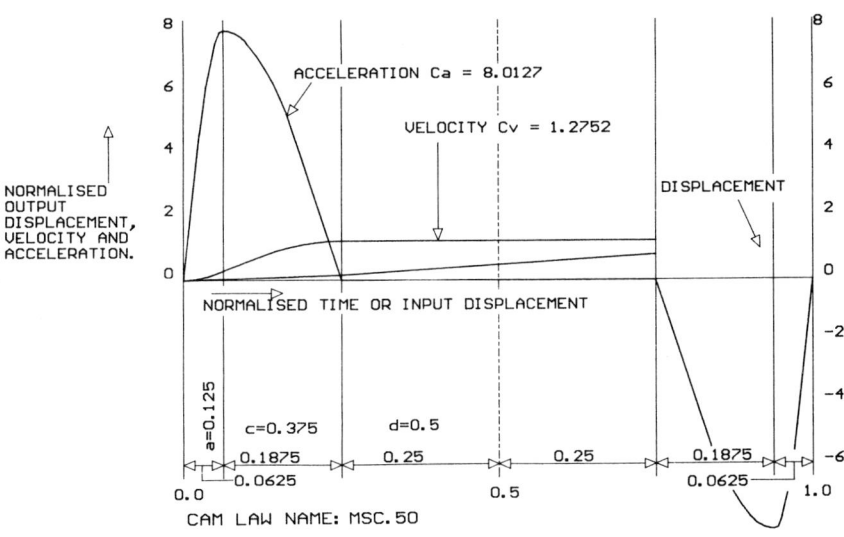

CAM LAW NAME: MSC.50

$$C_a = \frac{16\pi^2}{(5\pi + 4)} \qquad\qquad C_V = \frac{8\pi}{(5\pi + 4)}$$

Zone 1: from $u = 0$ to $u = 0.0625$

$$w = \frac{[2\pi u - 0.25 \sin (8\pi u)]}{(5\pi + 4)}$$

$$w' = \frac{2\pi[1 - \cos (8\pi u)]}{(5\pi + 4)}$$

$$w'' = \frac{16\pi^2 \sin (8\pi u)}{(5\pi + 4)}$$

$$w''' = \frac{128\pi^3 \cos (8\pi u)}{(5\pi + 4)}$$

Zone 2: from $u = 0.0625$ to $u = 0.25$

$$w = \frac{\{2\pi u + 2 - 2.25 \cos [8\pi(u - 0.0625)/3]\}}{(5\pi + 4)}$$

$$w' = \frac{\{2\pi + 6\pi \sin [8\pi(u - 0.0625)/3]\}}{(5\pi + 4)}$$

$$w'' = \frac{16\pi^2 \cos [8\pi(u - 0.0625)/3)}{(5\pi + 4)}$$

$$w''' = \frac{-128\pi^3/3 \sin [8\pi(u - 0.0625)/3]}{(5\pi + 4)}$$

Zone 3: from $u = 0.25$ to $u = 0.75$

$$w = \frac{(8\pi u + 2 - 1.5\pi)}{(5\pi + 4)}$$

$$w' = \frac{8\pi}{(5\pi + 4)}$$

$$w'' = 0$$

$$w''' = 0$$

Zone 4: from $u = 0.75$ to $u = 0.9375$

Use the same equations as in zone 2, but replace u by $(1 - u)$, replace w by $(1 - w)$ and replace w'' by $-w''$.

Zone 5: from $u = 0.9375$ to $u = 1$

Use the same equations as in zone 1, but replace u by $(1 - u)$, replace w by $(1 - w)$ and replace w'' by $-w''$.

CYCLOIDAL CAM LAW EQUATIONS

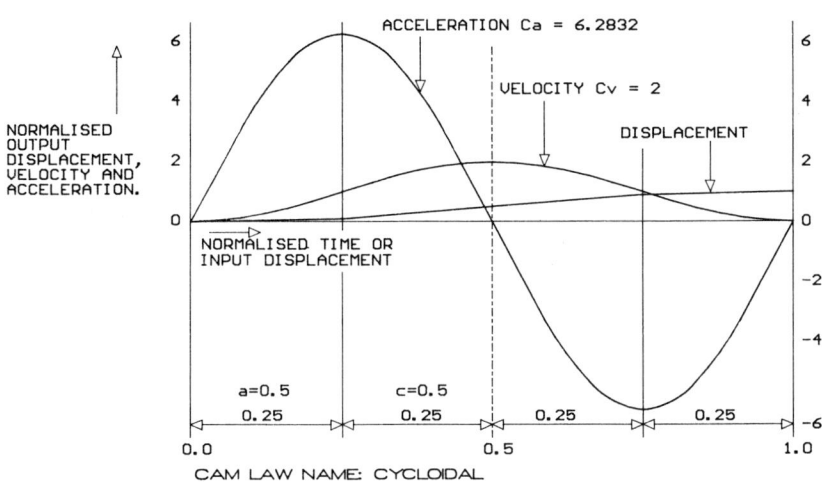

CAM LAW NAME: CYCLOIDAL

$$C_a = 2\pi \qquad\qquad C_V = 2$$
$$w = u - \frac{\sin(2\pi u)}{2\pi}$$
$$w' = 1 - \cos(2\pi u)$$
$$w'' = 2\pi \sin(2\pi u)$$
$$w''' = 4\pi^2 \cos(2\pi u)$$

CYCC.50 CAM LAW EQUATIONS

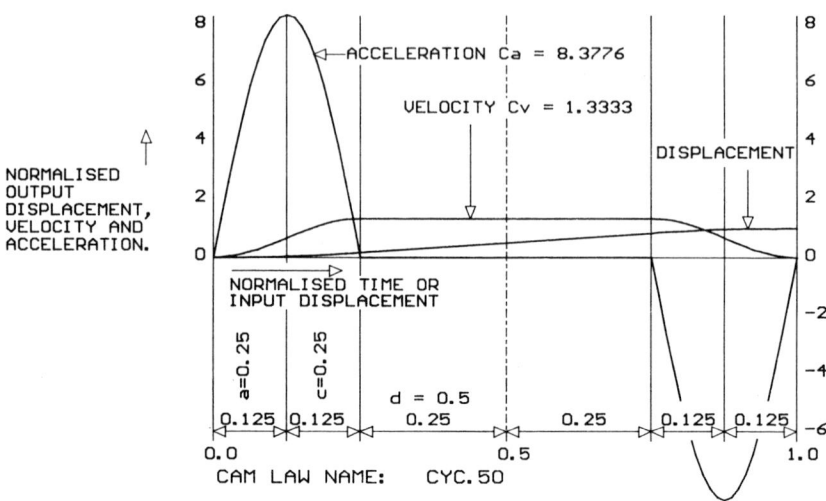

$$C_a = \frac{8\pi}{3} \qquad\qquad C_V = \frac{4}{3}$$

Zone 1: from $u = 0$ to $u = 0.25$

$$w = \frac{2}{3}u - \frac{1}{6\pi} \sin (4\pi u) \qquad\qquad w' = \frac{2}{3}[1 - \cos (4\pi u)]$$

$$w'' = \frac{8\pi}{3} \sin (4\pi u) \qquad\qquad w''' = \frac{32\pi^2}{3} \cos (4\pi u)$$

Zone 2: from $u = 0.25$ to $u = 0.75$

$$w = \frac{4u - 0.5}{3} \qquad\qquad w' = \frac{4}{3}$$

$$w'' = 0$$
$$w''' = 0$$

Zone 3: from $u = 0.75$ to $u = 1$

$$w = \frac{2u + 1 - 0.5/\pi \sin(4\pi u)}{3}$$

$$w' = \frac{2}{3}[1 - \cos(4\pi u)]$$

$$w'' = \frac{8\pi}{3} \sin(4\pi u)$$

$$w''' = \frac{32\pi^2}{3} \cos(4\pi u)$$

POLYNOMIAL CAM LAW EQUATIONS

The general forms of the polynomial equations are the infinite series:

$$w = A_0 + A_1 u + A_2 u^2 + A_3 u^3 + A_4 u^4 + A_5 u^5 + A_6 u^6 + \cdots$$
$$w' = A_1 + 2A_2 u + 3A_3 u^2 + 4A_4 u^3 + 5A_5 u^4 + 6A_6 u^5 + 7A_7 u^6 + \cdots$$
$$w'' = 2A_2 + 6A_3 u + 12A_4 u^2 + 20A_5 u^3 + 30A_6 u^4 + 42A_7 u^5 + \cdots$$
$$w''' = 6A_3 + 24A_4 u + 60A_5 u^2 + 120A_6 u^3 + 210A_7 u^4 + \cdots$$

Note: By putting $u = 0$ in the above equations, it can be seen that A_0, A_1 and $2A_2$ are the values of the normalised displacement, velocity and acceleration respectively at the beginning of the motion. For standard cam laws these are usually zero, therefore the terms containing the coefficients A_0, A_1 and A_2 disappear. Also the jerk at the beginning of the motion ($u = 0$) is $w''' = 6A_3$.

Any number of terms with various coefficients can be used to produce practical cam laws. The coefficients are chosen to meet particular conditions of an application for special motions, but for 'standard' motions are chosen to give particular *shapes* of the acceleration loop.

It is important to note that there are two distinct ways of using a polynomial for motion laws: *Reflective and Non-reflective*. The usefulness of a polynomial may be extended if the function applies only up to the mid-point of the motion, ignoring the equation after that point. The second half of the motion is the reversed reflection of the first half (as with most other standard cam laws). This can be termed a 'semi-polynomial' or *Reflective* polynomial as opposed to a full or *Non-reflective* polynomial. Some full polynomials are in fact symmetrical and could be treated as reflective if preferred, for instance the standard '3–4–5' cam law mentioned below. A reflective polynomial has to fulfil the conditions of zero displacement zero velocity and zero acceleration at the beginning; and

0.5 displacement and zero acceleration at the mid-point ($u = 0.5$, $w = 0.5$, $w'' = 0$).

Reflective polynomials offer much more variety of acceleration loop *shapes* than is possible with full ones, making them easier to design for specific dynamic characteristics.

STANDARD 3–4–5 POLYNOMIAL CAM LAW EQUATIONS
(Simulated Cycloidal)

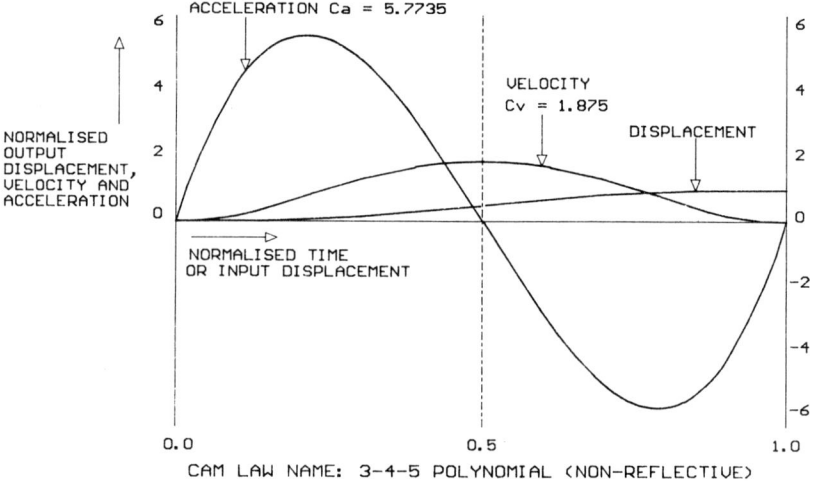

CAM LAW NAME: 3-4-5 POLYNOMIAL (NON-REFLECTIVE)

$$C_a = 5.7735 \qquad\qquad C_V = 1.875$$
$$w = 10 . u^3 - 15 . u^4 + 6 . u^5$$
$$w' = 30 . u^2 - 60 . u^3 + 30 . u^4$$
$$w'' = 60 . u - 180 . u^2 + 120 . u^3$$
$$w''' = 60 - 360 . u + 360 . u^2$$

This popular polynomial cam law may be regarded as a good substitute for the Cycloidal law and has virtually the same dynamic response properties.

3–4–5–6–R POLYNOMIAL (REFLECTIVE) CAM LAW EQUATIONS
(Simulated Mod Trap)

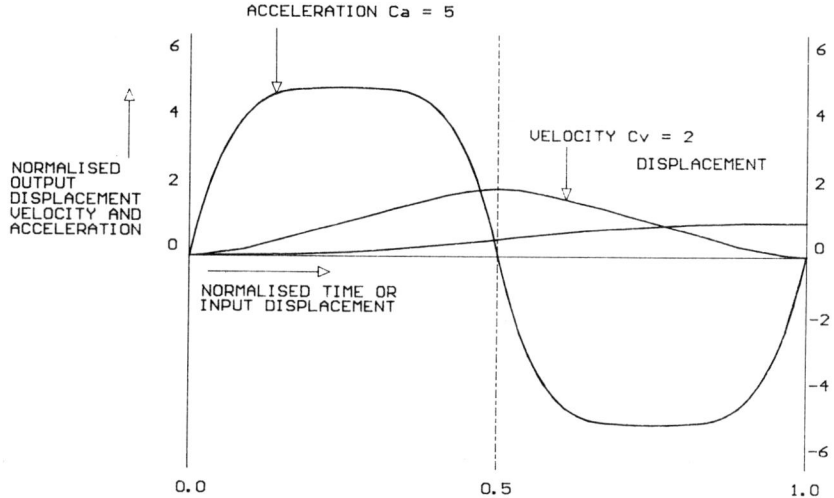

CAM LAW NAME: 3–4–5–6–R POLYNOMIAL (REFLECTIVE)

$$C_a = 5 \qquad C_V = 2$$
$$w = (40/3).u^3 - 40.u^4 + 64.u^5 - (128/3).u^6$$
$$w' = 40.u^2 - 160.u^3 + 320.u^4 - 256.u^5$$
$$w'' = 80.u - 480.u^2 + 1280.u^3 - 1280.u^4$$
$$w''' = 80 - 960.u + 3840.u^2 - 5120.u^3$$

This polynomial cam law may be regarded as a good substitute for Mod Trap law and has virtually the same dynamic response properties.

LOW IMPACT 3–4–5–6–7–R POLYNOMIAL CAM LAW EQUATIONS

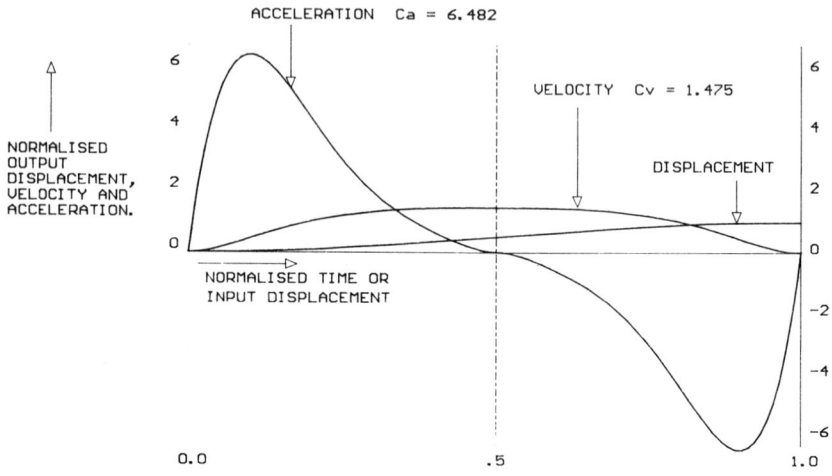

CAM LAW NAME: LOW IMPACT POLYNOMIAL (SYMMETRICAL REFLECTIVE POLYNOMIAL)

$$C_a = 6.48192 \qquad C_V = 1.475$$
$$w = 27.u^3 - 113.u^4 + 224.4.u^5 - 228.8.u^6 + 96.u^7$$
$$w' = 81.u^2 - 452.u^3 + 1122.u^4 - 1372.8.u^5 + 672.u^6$$
$$w'' = 162.u - 1356.u^2 + 4488.u^3 - 6864.u^4 + 4032.u^5$$
$$w''' = 162 - 2712.u + 13464.u^2 - 27456.u^3 + 20160.u^4$$

This polynomial cam law has been designed to have a lower impact velocity due to backlash than most standard laws. It is especially useful for those high speed mechanisms where a moderate amount of backlash cannot be avoided.

DOUBLE HARMONIC CAM LAW EQUATIONS

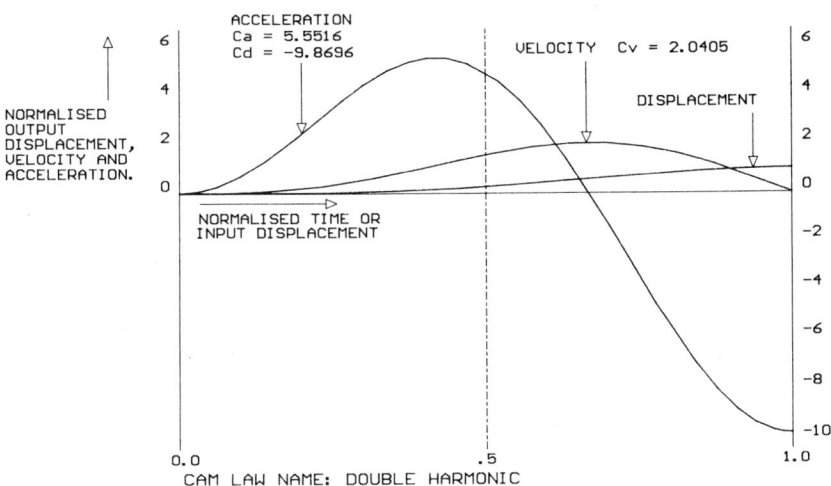

$$C_a = \frac{9}{16}\pi^2 \qquad\qquad C_d = -\pi^2 \qquad\qquad C_V = \frac{3\sqrt{3}}{8}\pi$$

$$w = \frac{1}{2}(1 - \cos \pi u) - \frac{1}{8}(1 - \cos 2\pi u)$$

$$w' = \frac{\pi}{2}\sin \pi u - \frac{\pi}{4}\sin 2\pi u$$

$$w'' = \frac{\pi^2}{2}\cos \pi u - \frac{\pi^2}{2}\cos 2\pi u$$

$$w''' = -\frac{\pi^3}{2}\sin \pi u + \pi^3 \sin 2\pi u$$

This cam law is suitable for a *symmetrical* D-R-R-D motion, but is not recommended for any other kind of motion because of the large step in the acceleration curve at the end of the motion.

APPENDIX B

Standard Cam Law Factor Tables

The following tables give the *normalised* displacement, velocity, acceleration, jerk, displacement ratio and blend factors for a number of standard cam laws. For a full explanation of these terms please read Chapters 3, 4 and 5.

Where SCCA parameters are given the cam law is a member of the general Sine–Constant–Cosine–Acceleration family (SCCA) and the parameters a, b, c and AF can be used in the SCCA equations, Appendix A, to calculate normalised displacement, w, velocity, w', acceleration, w'' and jerk, w''', for *any* point in the motion, e.g. a point not included in the table. The standard SCCA laws in the factor tables are symmetrical ($AF = 0.5$), but factors for asymmetrical motion can be calculated using an appropriate value of AF, or by cam law blending as described in Chapter 5.

For comparison purposes the factor tables for some external Geneva mechanisms have also been included, although they are not strictly *cam* laws as generally understood.

Geometric and (assuming constant input velocity*) *real* values of output displacement, velocity, etc. are found from the following equations:

	Geometric	**Real (time based)**
Displacement	$w \cdot Y$	$w \cdot Y$
Velocity	$w' \cdot Y/X$	$w' \cdot Y/T$
Acceleration	$w'' \cdot Y/X^2$	$w'' \cdot Y/T^2$
Jerk	$w''' \cdot Y/X^3$	$w''' \cdot Y/T^3$

where: w, w', w'' and w''' are the normalised displacement, velocity, acceleration and jerk respectively of the cam law.
T = time period of the full cam law motion.
X = input stroke (input displacement of the full cam law motion).

---

*If the input velocity is not constant the real output velocity, acceleration and jerk are significantly altered as described in Chapters 3 and 12.

Y = output stroke (output displacement of the full cam law motion).

LIST OF CAM LAW FACTOR TABLES

Table B.1   Cam law factors – Parabolic

PAR Factor Table		PARABOLIC					120 STEPS

SCCA PARAMETERS: a = 0   b = 1   c = 0   AF = .5   Ca = 4   Cd = 4   Cv = 2

STEP NO.	INPUT DISP. u	OUTPUT DISP. w	OUTPUT VELOC. w'	OUTPUT ACCEL. w''	O'PUT JERK w'''	DISP. RATIO w/u	BLEND FACTOR z = u.w'/w
0	0.00000	0.00000	0.00000	4.00000	INF.		
1	0.00833	0.00014	0.03333	4.00000	0.0	0.01667	2.0000
2	0.01667	0.00056	0.06667	4.00000	0.0	0.03333	2.0000
3	0.02500	0.00125	0.10000	4.00000	0.0	0.05000	2.0000
4	0.03333	0.00222	0.13333	4.00000	0.0	0.06667	2.0000
5	0.04167	0.00347	0.16667	4.00000	0.0	0.08333	2.0000
6	0.05000	0.00500	0.20000	4.00000	0.0	0.10000	2.00000
7	0.05833	0.00681	0.23333	4.00000	0.0	0.11667	2.00000
8	0.06667	0.00889	0.26667	4.00000	0.0	0.13333	2.00000
9	0.07500	0.01125	0.30000	4.00000	0.0	0.15000	2.00000
10	0.08333	0.01389	0.33333	4.00000	0.0	0.16667	2.00000
11	0.09167	0.01681	0.36667	4.00000	0.0	0.18333	2.00000
12	0.10000	0.02000	0.40000	4.00000	0.0	0.20000	2.00000
13	0.10833	0.02347	0.43333	4.00000	0.0	0.21667	2.00000
14	0.11667	0.02722	0.46667	4.00000	0.0	0.23333	2.00000
15	0.12500	0.03125	0.50000	4.00000	0.0	0.25000	2.00000
16	0.13333	0.03556	0.53333	4.00000	0.0	0.26667	2.00000
17	0.14167	0.04014	0.56667	4.00000	0.0	0.28333	2.00000
18	0.15000	0.04500	0.60000	4.00000	0.0	0.30000	2.00000
19	0.15833	0.05014	0.63333	4.00000	0.0	0.31667	2.00000
20	0.16667	0.05556	0.66667	4.00000	0.0	0.33333	2.00000
21	0.17500	0.06125	0.70000	4.00000	0.0	0.35000	2.00000
22	0.18333	0.06722	0.73333	4.00000	0.0	0.36667	2.00000
23	0.19167	0.07347	0.76667	4.00000	0.0	0.38333	2.00000
24	0.20000	0.08000	0.80000	4.00000	0.0	0.40000	2.00000
25	0.20833	0.08681	0.83333	4.00000	0.0	0.41667	2.00000
26	0.21667	0.09389	0.86667	4.00000	0.0	0.43333	2.00000
27	0.22500	0.10125	0.90000	4.00000	0.0	0.45000	2.00000
28	0.23333	0.10889	0.93333	4.00000	0.0	0.46667	2.00000
29	0.24167	0.11681	0.96667	4.00000	0.0	0.48333	2.00000
30	0.25000	0.12500	1.00000	4.00000	0.0	0.50000	2.00000
31	0.25833	0.13347	1.03333	4.00000	0.0	0.51667	2.00000
32	0.26667	0.14222	1.06667	4.00000	0.0	0.53333	2.00000
33	0.27500	0.15125	1.10000	4.00000	0.0	0.55000	2.00000
34	0.28333	0.16056	1.13333	4.00000	0.0	0.56667	2.00000
35	0.29167	0.17014	1.16667	4.00000	.0.0	0.58333	2.00000
36	0.30000	0.18000	1.20000	4.00000	0.0	0.60000	2.00000
37	0.30833	0.19014	1.23333	4.00000	0.0	0.61667	2.00000
38	0.31667	0.20056	1.26667	4.00000	0.0	0.63333	2.00000
39	0.32500	0.21125	1.30000	4.00000	0.0	0.65000	2.00000
40	0.33333	0.22222	1.33333	4.00000	0.0	0.66667	2.00000
41	0.34167	0.23347	1.36667	4.00000	0.0	0.68333	2.00000
42	0.35000	0.24500	1.40000	4.00000	0.0	0.70000	2.00000
43	0.35833	0.25681	1.43333	4.00000	0.0	0.71667	2.00000
44	0.36667	0.26889	1.46667	4.00000	0.0	0.73333	2.00000
45	0.37500	0.28125	1.50000	4.00000	0.0	0.75000	2.00000
46	0.38333	0.29389	1.53333	4.00000	0.0	0.76667	2.00000
47	0.39167	0.30681	1.56667	4.00000	0.0	0.78333	2.00000
48	0.40000	0.32000	1.60000	4.00000	0.0	0.80000	2.00000
49	0.40833	0.33347	1.63333	4.00000	0.0	0.81667	2.00000
50	0.41667	0.34722	1.66667	4.00000	0.0	0.83333	2.00000
51	0.42500	0.36125	1.70000	4.00000	0.0	0.85000	2.00000
52	0.43333	0.37556	1.73333	4.00000	0.0	0.86667	2.00000
53	0.44167	0.39014	1.76667	4.00000	0.0	0.88333	2.00000
54	0.45000	0.40500	1.80000	4.00000	0.0	0.90000	2.00000
55	0.45833	0.42014	1.83333	4.00000	0.0	0.91667	2.00000
56	0.46667	0.43556	1.86667	4.00000	0.0	0.93333	2.00000
57	0.47500	0.45125	1.90000	4.00000	0.0	0.95000	2.00000
58	0.48333	0.46722	1.93333	4.00000	0.0	0.96667	2.00000
59	0.49167	0.48347	1.96667	4.00000	0.0	0.98333	2.00000
60	0.50000	0.50000	2.00000	0.00000	INF.	1.00000	2.00000

STEP NO.	INPUT DISP. u	OUTPUT DISP. w	OUTPUT VELOC. w'	OUTPUT ACCEL. w''	O'PUT JERK w'''	DISP. RATIO w/u	BLEND FACTOR z = u.w'/w
61	0.50833	0.51653	1.96667	-4.00000	0.0	1.01612	1.93547
62	0.51667	0.53278	1.93333	-4.00000	0.0	1.03118	1.87487
63	0.52500	0.54875	1.90000	-4.00000	0.0	1.04524	1.81777
64	0.53333	0.56444	1.86667	-4.00000	0.0	1.05833	1.76378
65	0.54167	0.57986	1.83333	-4.00000	0.0	1.07051	1.71257
66	0.55000	0.59500	1.80000	-4.00000	0.0	1.08182	1.66387
67	0.55833	0.60986	1.76667	-4.00000	0.0	1.09229	1.61740
68	0.56667	0.62444	1.73333	-4.00000	0.0	1.10196	1.57295
69	0.57500	0.63875	1.70000	-4.00000	0.0	1.11087	1.53033
70	0.58333	0.65278	1.66667	-4.00000	0.0	1.11905	1.48936
71	0.59167	0.66653	1.63333	-4.00000	0.0	1.12653	1.44989
72	0.60000	0.68000	1.60000	-4.00000	0.0	1.13333	1.41176
73	0.60833	0.69319	1.56667	-4.00000	0.0	1.13950	1.37487
74	0.61667	0.70611	1.53333	-4.00000	0.0	1.14505	1.33910
75	0.62500	0.71875	1.50000	-4.00000	0.0	1.15000	1.30435
76	0.63333	0.73111	1.46667	-4.00000	0.0	1.15439	1.27052
77	0.64167	0.74319	1.43333	-4.00000	0.0	1.15823	1.23753
78	0.65000	0.75500	1.40000	-4.00000	0.0	1.16154	1.20530
79	0.65833	0.76653	1.36667	-4.00000	0.0	1.16435	1.17376
80	0.66667	0.77778	1.33333	-4.00000	0.0	1.16667	1.14286
81	0.67500	0.78875	1.30000	-4.00000	0.0	1.16852	1.11252
82	0.68333	0.79944	1.26667	-4.00000	0.0	1.16992	1.08270
83	0.69167	0.80986	1.23333	-4.00000	0.0	1.17088	1.05334
84	0.70000	0.82000	1.20000	-4.00000	0.0	1.17143	1.02439
85	0.70833	0.82986	1.16667	-4.00000	0.0	1.17157	0.99582
86	0.71667	0.83944	1.13333	-4.00000	0.0	1.17132	0.96757
87	0.72500	0.84875	1.10000	-4.00000	0.0	1.17069	0.93962
88	0.73333	0.85778	1.06667	-4.00000	0.0	1.16970	0.91192
89	0.74167	0.86653	1.03333	-4.00000	0.0	1.16835	0.88444
90	0.75000	0.87500	1.00000	-4.00000	0.0	1.16667	0.85714
91	0.75833	0.88319	0.96667	-4.00000	0.0	1.16465	0.83000
92	0.76667	0.89111	0.93333	-4.00000	0.0	1.16232	0.80299
93	0.77500	0.89875	0.90000	-4.00000	0.0	1.15968	0.77608
94	0.78333	0.90611	0.86667	-4.00000	0.0	1.15674	0.74923
95	0.79167	0.91319	0.83333	-4.00000	0.0	1.15351	0.72243
96	0.80000	0.92000	0.80000	-4.00000	0.0	1.15000	0.69565
97	0.80833	0.92653	0.76667	-4.00000	0.0	1.14622	0.66887
98	0.81667	0.93278	0.73333	-4.00000	0.0	1.14218	0.64205
99	0.82500	0.93875	0.70000	-4.00000	0.0	1.13788	0.61518
100	0.83333	0.94444	0.66667	-4.00000	0.0	1.13333	0.58824
101	0.84167	0.94986	0.63333	-4.00000	0.0	1.12855	0.56119
102	0.85000	0.95500	0.60000	-4.00000	0.0	1.12353	0.53403
103	0.85833	0.95986	0.56667	-4.00000	0.0	1.11828	0.50673
104	0.86667	0.96444	0.53333	-4.00000	0.0	1.11282	0.47926
105	0.87500	0.96875	0.50000	-4.00000	0.0	1.10714	0.45161
106	0.88333	0.97278	0.46667	-4.00000	0.0	1.10126	0.42376
107	0.89167	0.97653	0.43333	-4.00000	0.0	1.09517	0.39568
108	0.90000	0.98000	0.40000	-4.00000	0.0	1.08889	0.36735
109	0.90833	0.98319	0.36667	-4.00000	0.0	1.08242	0.33875
110	0.91667	0.98611	0.33333	-4.00000	0.0	1.07576	0.30986
111	0.92500	0.98875	0.30000	-4.00000	0.0	1.06892	0.28066
112	0.93333	0.99111	0.26667	-4.00000	0.0	1.06190	0.25112
113	0.94167	0.99319	0.23333	-4.00000	0.0	1.05472	0.22123
114	0.95000	0.99500	0.20000	-4.00000	0.0	1.04737	0.19095
115	0.95833	0.99653	0.16667	-4.00000	0.0	1.03986	0.16028
116	0.96667	0.99778	0.13333	-4.00000	0.0	1.03218	0.12918
117	0.97500	0.99875	0.10000	-4.00000	0.0	1.02436	0.09762
118	0.98333	0.99944	0.06667	-4.00000	0.0	1.01638	0.06559
119	0.99167	0.99986	0.03333	-4.00000	0.0	1.00826	0.03306
120	1.00000	1.00000	0.00000	-4.00000	INF.	1.00000	0.00000

Table B.2   Cam law factors – Simple harmonic

SH Factor table   SIMPLE HARMONIC   120 STEPS

SCCA PARAMETERS: a = 0   b = 0   c = 1   AF = .5   Ca = 4.934803   Cd = 4.934803   Cv = 1.570796

STEP NO.	INPUT DISP. u	OUTPUT DISP. w	OUTPUT VELOC. w'	OUTPUT ACCEL. w''	O'PUT JERK w'''	DISP. RATIO w/u	BLEND FACTOR z = u.w'/w	STEP NO.	INPUT DISP. u	OUTPUT DISP. w	OUTPUT VELOC. w'	OUTPUT ACCEL. w''	O'PUT JERK w'''	DISP. RATIO w/u	BLEND FACTOR z = u.w'/w
0	0.00000	0.00000	0.00000	4.93480	INF.			61	0.50833	0.51309	1.57026	-0.12918	-15.5	1.00935	1.55571
1	0.00833	0.00017	0.04112	4.93311	-0.4	0.02056	1.99992	62	0.51667	0.52617	1.56864	-0.25827	-15.5	1.01839	1.54032
2	0.01667	0.00069	0.08221	4.92804	-0.8	0.04111	1.99955	63	0.52500	0.53923	1.56595	-0.38718	-15.5	1.02710	1.52463
3	0.02500	0.00154	0.12324	4.91959	-1.2	0.06165	1.99895	64	0.53333	0.55226	1.56219	-0.51583	-15.4	1.03550	1.50864
4	0.03333	0.00274	0.16419	4.90777	-1.6	0.08217	1.99818	65	0.54167	0.56526	1.55736	-0.64412	-15.4	1.04356	1.49235
5	0.04167	0.00428	0.20503	4.89258	-2.0	0.10266	1.99715								
6	0.05000	0.00616	0.24573	4.87405	-2.4	0.12312	1.99589	66	0.55000	0.57822	1.55146	-0.77197	-15.3	1.05130	1.47575
7	0.05833	0.00837	0.28625	4.85217	-2.8	0.14353	1.99440	67	0.55833	0.59112	1.54449	-0.89930	-15.2	1.05872	1.45883
8	0.06667	0.01093	0.32659	4.82697	-3.2	0.16389	1.99268	68	0.56667	0.60396	1.53647	-1.02600	-15.2	1.06580	1.44161
9	0.07500	0.01382	0.36670	4.79845	-3.6	0.18420	1.99074	69	0.57500	0.61672	1.52740	-1.15201	-15.1	1.07256	1.42406
10	0.08333	0.01704	0.40655	4.76665	-4.0	0.20445	1.98856	70	0.58333	0.62941	1.51727	-1.27722	-15.0	1.07899	1.40620
11	0.09167	0.02059	0.44613	4.73159	-4.4	0.22462	1.98616	71	0.59167	0.64201	1.50611	-1.40156	-14.9	1.08508	1.38801
12	0.10000	0.02447	0.48540	4.69328	-4.8	0.24472	1.98352	72	0.60000	0.65451	1.49392	-1.52494	-14.7	1.09085	1.36950
13	0.10833	0.02868	0.52434	4.65175	-5.2	0.26473	1.98066	73	0.60833	0.66690	1.48070	-1.64727	-14.6	1.09628	1.35066
14	0.11667	0.03321	0.56292	4.60703	-5.6	0.28466	1.97756	74	0.61667	0.67918	1.46646	-1.76848	-14.5	1.10138	1.33148
15	0.12500	0.03806	0.60112	4.55916	-5.9	0.30448	1.97423	75	0.62500	0.69134	1.45123	-1.88847	-14.3	1.10615	1.31197
16	0.13333	0.04323	0.63890	4.50817	-6.3	0.32420	1.97067	76	0.63333	0.70337	1.43499	-2.00716	-14.2	1.11058	1.29211
17	0.14167	0.04871	0.67625	4.45408	-6.7	0.34382	1.96688	77	0.64167	0.71526	1.41778	-2.12449	-14.0	1.11468	1.27191
18	0.15000	0.05450	0.71313	4.39694	-7.0	0.36331	1.96285	78	0.65000	0.72700	1.39959	-2.24035	-13.8	1.11845	1.25136
19	0.15833	0.06059	0.74952	4.33679	-7.4	0.38268	1.95859	79	0.65833	0.73858	1.38044	-2.35468	-13.6	1.12189	1.23046
20	0.16667	0.06699	0.78540	4.27366	-7.8	0.40192	1.95410	80	0.66667	0.75000	1.36035	-2.46740	-13.4	1.12500	1.20920
21	0.17500	0.07368	0.82074	4.20761	-8.1	0.42103	1.94937	81	0.67500	0.76125	1.33932	-2.57843	-13.2	1.12778	1.18758
22	0.18333	0.08066	0.85552	4.13867	-8.4	0.43999	1.94440	82	0.68333	0.77232	1.31738	-2.68769	-13.0	1.13022	1.16559
23	0.19167	0.08794	0.88971	4.06690	-8.8	0.45880	1.93920	83	0.69167	0.78320	1.29453	-2.79510	-12.8	1.13234	1.14324
24	0.20000	0.09549	0.92329	3.99234	-9.1	0.47746	1.93377	84	0.70000	0.79389	1.27080	-2.90060	-12.5	1.13413	1.12050
25	0.20833	0.10332	0.95624	3.91504	-9.4	0.49595	1.92809	85	0.70833	0.80438	1.24620	-3.00412	-12.3	1.13560	1.09739
26	0.21667	0.11143	0.98853	3.83506	-9.8	0.51428	1.92218	86	0.71667	0.81466	1.22074	-3.10557	-12.0	1.13674	1.07390
27	0.22500	0.11980	1.02015	3.75245	-10.1	0.53243	1.91602	87	0.72500	0.82472	1.19444	-3.20490	-11.8	1.13755	1.05001
28	0.23333	0.12843	1.05107	3.66727	-10.4	0.55040	1.90963	88	0.73333	0.83457	1.16733	-3.30203	-11.5	1.13804	1.02573
29	0.24167	0.13731	1.08126	3.57958	-10.7	0.56819	1.90300	89	0.74167	0.84418	1.13942	-3.39689	-11.2	1.13822	1.00105
30	0.25000	0.14645	1.11072	3.48943	-11.0	0.58579	1.89612	90	0.75000	0.85355	1.11072	-3.48943	-11.0	1.13807	0.97597
31	0.25833	0.15582	1.13942	3.39689	-11.2	0.60318	1.88900	91	0.75833	0.86269	1.08126	-3.57958	-10.7	1.13761	0.95047
32	0.26667	0.16543	1.16733	3.30203	-11.5	0.62038	1.88164	92	0.76667	0.87157	1.05107	-3.66727	-10.4	1.13683	0.92456
33	0.27500	0.17528	1.19444	3.20490	-11.8	0.63737	1.87403	93	0.77500	0.88020	1.02015	-3.75245	-10.1	1.13575	0.89822
34	0.28333	0.18534	1.22074	3.10557	-12.0	0.65414	1.86617	94	0.78333	0.88857	0.98853	-3.83506	-9.8	1.13435	0.87146
35	0.29167	0.19562	1.24620	3.00412	-12.3	0.67069	1.85807	95	0.79167	0.89668	0.95624	-3.91504	-9.4	1.13264	0.84425
36	0.30000	0.20611	1.27080	2.90060	-12.5	0.68702	1.84972	96	0.80000	0.90451	0.92329	-3.99234	-9.1	1.13064	0.81661
37	0.30833	0.21680	1.29453	2.79510	-12.8	0.70313	1.84112	97	0.80833	0.91206	0.88971	-4.06690	-8.8	1.12833	0.78852
38	0.31667	0.22768	1.31738	2.68769	-13.0	0.71899	1.83226	98	0.81667	0.91934	0.85552	-4.13867	-8.4	1.12572	0.75998
39	0.32500	0.23875	1.33932	2.57843	-13.2	0.73462	1.82316	99	0.82500	0.92632	0.82074	-4.20761	-8.1	1.12281	0.73097
40	0.33333	0.25000	1.36035	2.46740	-13.4	0.75000	1.81380	100	0.83333	0.93301	0.78540	-4.27366	-7.8	1.11962	0.70149
41	0.34167	0.26142	1.38044	2.35468	-13.6	0.76513	1.80419	101	0.84167	0.93941	0.74952	-4.33679	-7.4	1.11613	0.67153
42	0.35000	0.27300	1.39959	2.24035	-13.8	0.78001	1.79431	102	0.85000	0.94550	0.71313	-4.39694	-7.0	1.11236	0.64110
43	0.35833	0.28474	1.41778	2.12449	-14.0	0.79464	1.78419	103	0.85833	0.95129	0.67625	-4.45408	-6.7	1.10830	0.61016
44	0.36667	0.29663	1.43499	2.00716	-14.2	0.80900	1.77380	104	0.86667	0.95677	0.63890	-4.50817	-6.3	1.10397	0.57873
45	0.37500	0.30866	1.45123	1.88847	-14.3	0.82309	1.76315	105	0.87500	0.96194	0.60112	-4.55916	-5.9	1.09936	0.54679
46	0.38333	0.32082	1.46646	1.76848	-14.5	0.83691	1.75223	106	0.88333	0.96679	0.56292	-4.60703	-5.6	1.09448	0.51433
47	0.39167	0.33310	1.48070	1.64727	-14.6	0.85046	1.74106	107	0.89167	0.97132	0.52434	-4.65175	-5.2	1.08933	0.48134
48	0.40000	0.34549	1.49392	1.52494	-14.7	0.86373	1.72961	108	0.90000	0.97553	0.48540	-4.69328	-4.8	1.08392	0.44782
49	0.40833	0.35799	1.50611	1.40156	-14.9	0.87672	1.71790	109	0.90833	0.97941	0.44613	-4.73159	-4.4	1.07825	0.41375
50	0.41667	0.37059	1.51727	1.27722	-15.0	0.88942	1.70592	110	0.91667	0.98296	0.40655	-4.76665	-4.0	1.07232	0.37913
51	0.42500	0.38328	1.52740	1.15201	-15.1	0.90183	1.69366	111	0.92500	0.98618	0.36670	-4.79845	-3.6	1.06615	0.34394
52	0.43333	0.39604	1.53647	1.02600	-15.2	0.91395	1.68114	112	0.93333	0.98907	0.32659	-4.82697	-3.2	1.05972	0.30818
53	0.44167	0.40888	1.54449	0.89930	-15.2	0.92577	1.66833	113	0.94167	0.99163	0.28626	-4.85217	-2.8	1.05306	0.27183
54	0.45000	0.42178	1.55146	0.77197	-15.3	0.93730	1.65525	114	0.95000	0.99384	0.24573	-4.87405	-2.4	1.04615	0.23489
55	0.45833	0.43474	1.55736	0.64412	-15.4	0.94852	1.64189	115	0.95833	0.99572	0.20503	-4.89258	-2.0	1.03901	0.19733
56	0.46667	0.44774	1.56219	0.51583	-15.4	0.95943	1.62824	116	0.96667	0.99726	0.16419	-4.90777	-1.6	1.03165	0.15916
57	0.47500	0.46077	1.56595	0.38718	-15.5	0.97004	1.61431	117	0.97500	0.99846	0.12324	-4.91959	-1.2	1.02406	0.12035
58	0.48333	0.47383	1.56864	0.25827	-15.5	0.98034	1.60010	118	0.98333	0.99931	0.08221	-4.92804	-0.8	1.01625	0.08089
59	0.49167	0.48691	1.57026	0.12918	-15.5	0.99033	1.58559	119	0.99167	0.99983	0.04112	-4.93311	-0.4	1.00823	0.04078
60	0.50000	0.50000	1.57080	-0.00000	-15.5	1.00000	1.57080	120	1.00000	1.00000	0.00000	-4.93480	INF.	1.00000	0.00000

Table B.3   Cam law factors – Modified trapezoid

| MT Factor table | | | | MODIFIED TRAPEZOID | | | | | | | | | | 120 STEPS | |

SCCA PARAMETERS: a = .25 b = .5 c = .25 AF = .5 Ca = 4.888124 Cd = 4.888124 Cv = 2

STEP NO.	INPUT DISP. u	OUTPUT DISP. w	OUTPUT VELOC. w'	OUTPUT ACCEL. w''	O'PUT JERK w'''	DISP. RATIO w/u	BLEND FACTOR z = u.w'/w	STEP NO.	INPUT DISP. u	OUTPUT DISP. w	OUTPUT VELOC. w'	OUTPUT ACCEL. w''	O'PUT JERK w'''	DISP. RATIO w/u	BLEND FACTOR z = u.w'/w
0	0.00000	0.00000	0.00000	0.00000	61.4			61	0.50833	0.51666	1.99787	-0.51095	-61.1	1.01638	1.96567
1	0.00833	0.00001	0.00213	0.51095	61.1	0.00071	2.99890	62	0.51667	0.53329	1.99150	-1.01630	-60.1	1.03217	1.92944
2	0.01667	0.00005	0.00850	1.01630	60.1	0.00284	2.99560	63	0.52500	0.54984	1.98096	-1.51051	-58.4	1.04732	1.89147
3	0.02500	0.00016	0.01904	1.51051	58.4	0.00637	2.99012	64	0.53333	0.56629	1.96637	-1.98818	-56.1	1.06180	1.85193
4	0.03333	0.00038	0.03363	1.98818	56.1	0.01128	2.98244	65	0.54167	0.58260	1.94789	-2.44407	-53.2	1.07557	1.81102
5	0.04167	0.00073	0.05211	2.44406	53.2	0.01753	2.97256								
6	0.05000	0.00125	0.07429	2.87317	49.7	0.02509	2.96048	66	0.55000	0.59875	1.92571	-2.87316	-49.7	1.08863	1.76893
7	0.05833	0.00198	0.09991	3.27079	45.6	0.03391	2.94620	67	0.55833	0.61469	1.90009	-3.27080	-45.6	1.10093	1.72589
8	0.06667	0.00293	0.12870	3.63258	41.1	0.04393	2.92970	68	0.56667	0.63040	1.87130	-3.63258	-41.1	1.11248	1.68210
9	0.07500	0.00413	0.16035	3.95458	36.1	0.05508	2.91099	69	0.57500	0.64587	1.83965	-3.95458	-36.1	1.12325	1.63780
10	0.08333	0.00561	0.19449	4.23324	30.7	0.06730	2.89006	70	0.58333	0.66106	1.80551	-4.23324	-30.7	1.13324	1.59322
11	0.09167	0.00738	0.23077	4.46552	25.0	0.08049	2.86691	71	0.59167	0.67595	1.76923	-4.46552	-25.0	1.14246	1.54862
12	0.10000	0.00946	0.26878	4.64888	19.0	0.09459	2.84152	72	0.60000	0.69054	1.73122	-4.64888	-19.0	1.15090	1.50423
13	0.10833	0.01186	0.30811	4.78131	12.8	0.10950	2.81391	73	0.60833	0.70480	1.69189	-4.78131	-12.8	1.15858	1.46031
14	0.11667	0.01460	0.34832	4.86135	6.4	0.12511	2.78405	74	0.61667	0.71874	1.65168	-4.86135	-6.4	1.16552	1.41712
15	0.12500	0.01767	0.38898	4.88812	0.0	0.14135	2.75194	75	0.62500	0.73233	1.61102	-4.88812	0.0	1.17173	1.37490
16	0.13333	0.02108	0.42972	4.88812	0.0	0.15810	2.71803	76	0.63333	0.74559	1.57028	-4.88812	0.0	1.17724	1.33386
17	0.14167	0.02483	0.47045	4.88812	0.0	0.17528	2.68408	77	0.64167	0.75850	1.52955	-4.88812	0.0	1.18208	1.29394
18	0.15000	0.02892	0.51119	4.88812	0.0	0.19281	2.65131	78	0.65000	0.77108	1.48881	-4.88812	0.0	1.18628	1.25503
19	0.15833	0.03335	0.55192	4.88812	0.0	0.21063	2.62029	79	0.65833	0.78332	1.44808	-4.88812	0.0	1.18985	1.21703
20	0.16667	0.03812	0.59266	4.88812	0.0	0.22872	2.59122	80	0.66667	0.79521	1.40734	-4.88812	0.0	1.19282	1.17984
21	0.17500	0.04323	0.63339	4.88812	0.0	0.24702	2.56415	81	0.67500	0.80677	1.36661	-4.88812	0.0	1.19522	1.14340
22	0.18333	0.04868	0.67413	4.88812	0.0	0.26551	2.53902	82	0.68333	0.81799	1.32587	-4.88812	0.0	1.19706	1.10761
23	0.19167	0.05446	0.71486	4.88812	0.0	0.28416	2.51572	83	0.69167	0.82887	1.28514	-4.88812	0.0	1.19837	1.07241
24	0.20000	0.06059	0.75559	4.88812	0.0	0.30295	2.49411	84	0.70000	0.83941	1.24441	-4.88812	0.0	1.19916	1.03773
25	0.20833	0.06706	0.79633	4.88812	0.0	0.32187	2.47405	85	0.70833	0.84961	1.20367	-4.88812	0.0	1.19945	1.00352
26	0.21667	0.07386	0.83706	4.88812	0.0	0.34090	2.45542	86	0.71667	0.85947	1.16294	-4.88812	0.0	1.19926	0.96971
27	0.22500	0.08101	0.87780	4.88812	0.0	0.36003	2.43809	87	0.72500	0.86899	1.12220	-4.88812	0.0	1.19861	0.93625
28	0.23333	0.08849	0.91853	4.88812	0.0	0.37925	2.42195	88	0.73333	0.87817	1.08147	-4.88812	0.0	1.19751	0.90310
29	0.24167	0.09632	0.95927	4.88812	0.0	0.39855	2.40688	89	0.74167	0.88702	1.04073	-4.88812	0.0	1.19598	0.87020
30	0.25000	0.10448	1.00000	4.88812	0.0	0.41792	2.39280	90	0.75000	0.89552	1.00000	-4.88812	0.0	1.19403	0.83750
31	0.25833	0.11298	1.04073	4.88812	0.0	0.43735	2.37961	91	0.75833	0.90368	0.95927	-4.88812	0.0	1.19167	0.80498
32	0.26667	0.12183	1.08147	4.88812	0.0	0.45685	2.36725	92	0.76667	0.91151	0.91853	-4.88812	0.0	1.18892	0.77257
33	0.27500	0.13101	1.12220	4.88812	0.0	0.47639	2.35563	93	0.77500	0.91899	0.87780	-4.88812	0.0	1.18580	0.74026
34	0.28333	0.14053	1.16294	4.88812	0.0	0.49599	2.34470	94	0.78333	0.92614	0.83706	-4.88812	0.0	1.18230	0.70799
35	0.29167	0.15039	1.20367	4.88812	0.0	0.51562	2.33440	95	0.79167	0.93294	0.79633	-4.88812	0.0	1.17845	0.67574
36	0.30000	0.16059	1.24441	4.88812	0.0	0.53530	2.32468	96	0.80000	0.93941	0.75559	-4.88812	0.0	1.17426	0.64346
37	0.30833	0.17113	1.28514	4.88812	0.0	0.55502	2.31550	97	0.80833	0.94554	0.71486	-4.88812	0.0	1.16974	0.61113
38	0.31667	0.18201	1.32587	4.88812	0.0	0.57477	2.30681	98	0.81667	0.95132	0.67413	-4.88812	0.0	1.16489	0.57870
39	0.32500	0.19323	1.36661	4.88812	0.0	0.59455	2.29857	99	0.82500	0.95677	0.63339	-4.88812	0.0	1.15972	0.54616
40	0.33333	0.20479	1.40734	4.88812	0.0	0.61436	2.29075	100	0.83333	0.96188	0.59266	-4.88812	0.0	1.15426	0.51349
41	0.34167	0.21668	1.44808	4.88812	0.0	0.63420	2.28333	101	0.84167	0.96665	0.55192	-4.88812	0.0	1.14849	0.48056
42	0.35000	0.22892	1.48881	4.88812	0.0	0.65406	2.27626	102	0.85000	0.97108	0.51119	-4.88812	0.0	1.14245	0.44745
43	0.35833	0.24150	1.52955	4.88812	0.0	0.67395	2.26954	103	0.85833	0.97517	0.47045	-4.88812	0.0	1.13612	0.41409
44	0.36667	0.25441	1.57028	4.88812	0.0	0.69385	2.26313	104	0.86667	0.97892	0.42972	-4.88812	0.0	1.12952	0.38044
45	0.37500	0.26767	1.61102	4.88812	0.0	0.71378	2.25701	105	0.87500	0.98233	0.38898	-4.88812	0.0	1.12266	0.34648
46	0.38333	0.28126	1.65168	4.86135	-6.4	0.73373	2.25107	106	0.88333	0.98540	0.34832	-4.86135	6.4	1.11555	0.31224
47	0.39167	0.29520	1.69189	4.78131	-12.8	0.75369	2.24481	107	0.89167	0.98814	0.30811	-4.78131	12.8	1.10819	0.27803
48	0.40000	0.30946	1.73122	4.64888	-19.0	0.77365	2.23773	108	0.90000	0.99054	0.26878	-4.64888	19.0	1.10060	0.24421
49	0.40833	0.32405	1.76923	4.46552	-25.0	0.79358	2.22943	109	0.90833	0.99262	0.23077	-4.46552	25.0	1.09279	0.21117
50	0.41667	0.33894	1.80551	4.23324	-30.7	0.81346	2.21954	110	0.91667	0.99439	0.19449	-4.23324	30.7	1.08479	0.17929
51	0.42500	0.35413	1.83965	3.95457	-36.1	0.83325	2.20781	111	0.92500	0.99587	0.16035	-3.95458	36.1	1.07661	0.14893
52	0.43333	0.36960	1.87130	3.63258	-41.1	0.85291	2.19401	112	0.93333	0.99707	0.12870	-3.63258	41.1	1.06829	0.12048
53	0.44167	0.38531	1.90009	3.27079	-45.6	0.87240	2.17799	113	0.94167	0.99802	0.09991	-3.27080	45.6	1.05985	0.09427
54	0.45000	0.40125	1.92571	2.87317	-49.7	0.89168	2.15965	114	0.95000	0.99875	0.07429	-2.87316	49.7	1.05131	0.07066
55	0.45833	0.41740	1.94789	2.44406	-53.2	0.91068	2.13892	115	0.95833	0.99927	0.05211	-2.44407	53.2	1.04272	0.04978
56	0.46667	0.43371	1.96637	1.98818	-56.1	0.92938	2.11579	116	0.96667	0.99962	0.03363	-1.98818	56.1	1.03409	0.03252
57	0.47500	0.45016	1.98096	1.51051	-58.4	0.94770	2.09028	117	0.97500	0.99984	0.01904	-1.51051	58.4	1.02548	0.01857
58	0.48333	0.46671	1.99150	1.01630	-60.1	0.96562	2.06242	118	0.98333	0.99995	0.00850	-1.01630	60.1	1.01690	0.00836
59	0.49167	0.48334	1.99787	0.51095	-61.1	0.98306	2.03229	119	0.99167	0.99999	0.00213	-0.51095	61.1	1.00840	0.00211
60	0.50000	0.50000	2.00000	-0.00000	-61.4	1.00000	2.00000	120	1.00000	1.00000	0.00000	0.00000	61.4	1.00000	0.00000

Table B.4　Cam law factors – Modified sine

MS Factor table　　　　　　　　　　　　　　　MODIFIED SINE　　　　　　　　120 STEPS

SCCA PARAMETERS: a = .25　b = 0　c = .75　AF = .5　Ca = 5.527957　Cd = 5.527957　Cv = 1.759603

STEP NO.	INPUT DISP. u	OUTPUT DISP. w	OUTPUT VELOC. w'	OUTPUT ACCEL. w''	O'PUT JERK w'''	DISP. RATIO w/u	BLEND FACTOR z = u.w'/w	STEP NO.	INPUT DISP. u	OUTPUT DISP. w	OUTPUT VELOC. w'	OUTPUT ACCEL. w''	O'PUT JERK w'''	DISP. RATIO w/u	BLEND FACTOR z = u.w'/w
0	0.00000	0.00000	0.00000	0.00000	69.5			61	0.50833	0.51466	1.75880	-0.19292	-23.1	1.01245	1.73717
1	0.00833	0.00001	0.00241	0.57783	69.1	0.00080	2.99890	62	0.51667	0.52931	1.75639	-0.38561	-23.1	1.02447	1.71444
2	0.01667	0.00005	0.00961	1.14933	67.9	0.00321	2.99560	63	0.52500	0.54393	1.75237	-0.57783	-23.0	1.03606	1.69139
3	0.02500	0.00018	0.02153	1.70823	66.1	0.00720	2.99012	64	0.53333	0.55851	1.74676	-0.76934	-22.9	1.04721	1.66802
4	0.03333	0.00043	0.03803	2.24842	63.5	0.01275	2.98244	65	0.54167	0.57304	1.73955	-0.95992	-22.8	1.05792	1.64432
5	0.04167	0.00083	0.05894	2.76398	60.2	0.01983	2.97256	66	0.55000	0.58750	1.73076	-1.14933	-22.6	1.06818	1.62029
6	0.05000	0.00142	0.08401	3.24925	56.2	0.02838	2.96048	67	0.55833	0.60188	1.72040	-1.33734	-22.5	1.07799	1.59593
7	0.05833	0.00224	0.11299	3.69893	51.6	0.03835	2.94620	68	0.56667	0.61617	1.70848	-1.52371	-22.3	1.08736	1.57123
8	0.06667	0.00331	0.14555	4.10807	46.5	0.04968	2.92970	69	0.57500	0.63035	1.69501	-1.70823	-22.0	1.09626	1.54618
9	0.07500	0.00467	0.18133	4.47221	40.8	0.06229	2.91099	70	0.58333	0.64441	1.68002	-1.89067	-21.8	1.10471	1.52078
10	0.08333	0.00634	0.21995	4.78735	34.7	0.07611	2.89006	71	0.59167	0.65835	1.66351	-2.07081	-21.5	1.11270	1.49502
11	0.09167	0.00834	0.26098	5.05004	28.3	0.09103	2.86691	72	0.60000	0.67213	1.64551	-2.24842	-21.2	1.12022	1.46891
12	0.10000	0.01070	0.30396	5.25740	21.5	0.10697	2.84153	73	0.60833	0.68577	1.62604	-2.42330	-20.8	1.12729	1.44244
13	0.10833	0.01341	0.34844	5.40716	14.4	0.12383	2.81391	74	0.61667	0.69923	1.60513	-2.59522	-20.4	1.13389	1.41560
14	0.11667	0.01651	0.39392	5.49767	7.3	0.14149	2.78405	75	0.62500	0.71252	1.58280	-2.76398	-20.1	1.14002	1.38839
15	0.12500	0.01998	0.43990	5.52796	0.0	0.15985	2.75194	76	0.63333	0.72561	1.55907	-2.92937	-19.6	1.14570	1.36081
16	0.13333	0.02384	0.48596	5.52459	-0.8	0.17879	2.71798	77	0.64167	0.73850	1.53398	-3.09119	-19.2	1.15090	1.33285
17	0.14167	0.02808	0.53196	5.51449	-1.6	0.19822	2.68374	78	0.65000	0.75117	1.50756	-3.24925	-18.7	1.15565	1.30452
18	0.15000	0.03270	0.57785	5.49767	-2.4	0.21803	2.65029	79	0.65833	0.76361	1.47984	-3.40335	-18.2	1.15993	1.27580
19	0.15833	0.03771	0.62357	5.47416	-3.2	0.23817	2.61812	80	0.66667	0.77583	1.45085	-3.55330	-17.7	1.16375	1.24671
20	0.16667	0.04310	0.66906	5.44397	-4.0	0.25858	2.58744	81	0.67500	0.78780	1.42063	-3.69892	-17.2	1.16711	1.21723
21	0.17500	0.04886	0.71428	5.40716	-4.8	0.27921	2.55826	82	0.68333	0.79950	1.38922	-3.84004	-16.7	1.17001	1.18736
22	0.18333	0.05500	0.75917	5.36375	-5.6	0.30000	2.53052	83	0.69167	0.81095	1.35664	-3.97648	-16.1	1.17245	1.15710
23	0.19167	0.06151	0.80366	5.31381	-6.4	0.32094	2.50411	84	0.70000	0.82211	1.32295	-4.10807	-15.5	1.17445	1.12645
24	0.20000	0.06839	0.84771	5.25740	-7.2	0.34197	2.47891	85	0.70833	0.83299	1.28819	-4.23466	-14.9	1.17599	1.09541
25	0.20833	0.07564	0.89127	5.19458	-7.9	0.36307	2.45479	86	0.71667	0.84358	1.25239	-4.35609	-14.3	1.17709	1.06397
26	0.21667	0.08325	0.93427	5.12543	-8.7	0.38422	2.43163	87	0.72500	0.85386	1.21560	-4.47221	-13.6	1.17774	1.03215
27	0.22500	0.09121	0.97667	5.05004	-9.4	0.40538	2.40930	88	0.73333	0.86384	1.17787	-4.58288	-12.9	1.17796	0.99992
28	0.23333	0.09952	1.01842	4.96849	-10.2	0.42653	2.38771	89	0.74167	0.87349	1.13924	-4.68797	-12.3	1.17774	0.96731
29	0.24167	0.10818	1.05946	4.88090	-10.9	0.44765	2.36674	90	0.75000	0.88282	1.09975	-4.78735	-11.6	1.17710	0.93429
30	0.25000	0.11718	1.09975	4.78735	-11.6	0.46871	2.34632	91	0.75833	0.89182	1.05946	-4.88090	-10.9	1.17602	0.90089
31	0.25833	0.12651	1.13924	4.68797	-12.3	0.48971	2.32635	92	0.76667	0.90048	1.01842	-4.96849	-10.2	1.17454	0.86708
32	0.26667	0.13616	1.17787	4.58288	-12.9	0.51061	2.30678	93	0.77500	0.90879	0.97667	-5.05004	-9.4	1.17263	0.83289
33	0.27500	0.14614	1.21560	4.47221	-13.6	0.53141	2.28752	94	0.78333	0.91675	0.93427	-5.12543	-8.7	1.17032	0.79830
34	0.28333	0.15642	1.25239	4.35609	-14.3	0.55207	2.26852	95	0.79167	0.92436	0.89127	-5.19458	-7.9	1.16761	0.76332
35	0.29167	0.16701	1.28819	4.23466	-14.9	0.57260	2.24973	96	0.80000	0.93161	0.84771	-5.25740	-7.2	1.16451	0.72796
36	0.30000	0.17789	1.32295	4.10807	-15.5	0.59296	2.23110	97	0.80833	0.93849	0.80366	-5.31381	-6.4	1.16101	0.69220
37	0.30833	0.18905	1.35664	3.97648	-16.1	0.61315	2.21259	98	0.81667	0.94500	0.75917	-5.36375	-5.6	1.15714	0.65607
38	0.31667	0.20050	1.38922	3.84004	-16.7	0.63314	2.19416	99	0.82500	0.95114	0.71428	-5.40716	-4.8	1.15290	0.61955
39	0.32500	0.21220	1.42063	3.69893	-17.2	0.65294	2.17576	100	0.83333	0.95690	0.66907	-5.44397	-4.0	1.14828	0.58267
40	0.33333	0.22417	1.45085	3.55330	-17.7	0.67251	2.15738	101	0.84167	0.96229	0.62357	-5.47416	-3.2	1.14331	0.54540
41	0.34167	0.23638	1.47984	3.40335	-18.2	0.69185	2.13897	102	0.85000	0.96730	0.57785	-5.49767	-2.4	1.13799	0.50778
42	0.35000	0.24883	1.50756	3.24925	-18.7	0.71094	2.12051	103	0.85833	0.97192	0.53196	-5.51449	-1.6	1.13233	0.46979
43	0.35833	0.26150	1.53398	3.09119	-19.2	0.72978	2.10199	104	0.86667	0.97616	0.48596	-5.52459	-0.8	1.12634	0.43145
44	0.36667	0.27439	1.55907	2.92937	-19.6	0.74834	2.08337	105	0.87500	0.98002	0.43990	-5.52796	0.0	1.12002	0.39276
45	0.37500	0.28748	1.58280	2.76398	-20.1	0.76663	2.06463	106	0.88333	0.98349	0.39392	-5.49767	7.3	1.11339	0.35380
46	0.38333	0.30077	1.60513	2.59522	-20.4	0.78461	2.04576	107	0.89167	0.98659	0.34844	-5.40716	14.4	1.10645	0.31492
47	0.39167	0.31423	1.62604	2.42330	-20.8	0.80230	2.02673	108	0.90000	0.98930	0.30396	-5.25740	21.5	1.09923	0.27653
48	0.40000	0.32787	1.64551	2.24842	-21.2	0.81966	2.00754	109	0.90833	0.99166	0.26098	-5.05004	28.3	1.09173	0.23905
49	0.40833	0.34165	1.66351	2.07081	-21.5	0.83670	1.98817	110	0.91667	0.99363	0.21995	-4.78735	34.7	1.08399	0.20271
50	0.41667	0.35559	1.68002	1.89067	-21.8	0.85341	1.96860	111	0.92500	0.99533	0.18133	-4.47221	40.8	1.07603	0.16852
51	0.42500	0.36965	1.69501	1.70823	-22.0	0.86976	1.94882	112	0.93333	0.99669	0.14555	-4.10807	46.5	1.06788	0.13630
52	0.43333	0.38383	1.70848	1.52371	-22.3	0.88577	1.92882	113	0.94167	0.99776	0.11299	-3.69893	51.6	1.05957	0.10664
53	0.44167	0.39812	1.72040	1.33733	-22.5	0.90140	1.90858	114	0.95000	0.99858	0.08401	-3.24925	56.2	1.05114	0.07993
54	0.45000	0.41250	1.73076	1.14933	-22.6	0.91667	1.88810	115	0.95833	0.99917	0.05894	-2.76398	60.2	1.04262	0.05653
55	0.45833	0.42696	1.73955	0.95992	-22.8	0.93155	1.86737	116	0.96667	0.99957	0.03803	-2.24842	63.5	1.03404	0.03678
56	0.46667	0.44149	1.74676	0.76934	-22.9	0.94605	1.84637	117	0.97500	0.99982	0.02153	-1.70823	66.1	1.02546	0.02100
57	0.47500	0.45607	1.75237	0.57783	-23.0	0.96015	1.82511	118	0.98333	0.99995	0.00961	-1.14933	67.9	1.01689	0.00945
58	0.48333	0.47069	1.75639	0.38561	-23.1	0.97384	1.80356	119	0.99167	0.99999	0.00241	-0.57783	69.1	1.00840	0.00239
59	0.49167	0.48534	1.75880	0.19292	-23.1	0.98713	1.78173	120	1.00000	1.00000	0.00000	0.00000	69.5	1.00000	0.00000
60	0.50000	0.50000	1.75960	-0.00000	-23.2	1.00000	1.75960								

Table B.5   Cam law factors – Modified sine with 20 per cent constant velocity

MSC.20 Factor table   MODIFIED SINE WITH 20% CONSTANT VELOCITY   120 STEPS

SCCA PARAMETERS: a = .2 b = 0 c = .6 AF = .5 Ca = 5.99863 Cd = 5.99863 Cv = 1.527539

STEP NO.	INPUT DISP. u	OUTPUT DISP. w	OUTPUT VELOC. w'	OUTPUT ACCEL. w''	O'PUT JERK w'''	DISP. RATIO w/u	BLEND FACTOR z = u.w'/w
0	0.00000	0.00000	0.00000	0.00000	94.2		
1	0.00833	0.00001	0.00327	0.78298	93.4	0.00109	2.99831
2	0.01667	0.00007	0.01301	1.55256	91.0	0.00435	2.99314
3	0.02500	0.00024	0.02907	2.29558	87.1	0.00974	2.98458
4	0.03333	0.00057	0.05116	2.99932	81.6	0.01721	2.97256
5	0.04167	0.00111	0.07892	3.65173	74.8	0.02669	2.95712
6	0.05000	0.00190	0.11185	4.24167	66.6	0.03807	2.93823
7	0.05833	0.00299	0.14941	4.75903	57.4	0.05124	2.91588
8	0.06667	0.00440	0.19094	5.19497	47.1	0.06607	2.89006
9	0.07500	0.00618	0.23574	5.54201	36.1	0.08241	2.86077
10	0.08333	0.00834	0.28305	5.79423	24.4	0.10009	2.82800
11	0.09167	0.01090	0.33204	5.94731	12.3	0.11894	2.79172
12	0.10000	0.01388	0.38188	5.99863	0.0	0.13877	2.75194
13	0.10833	0.01727	0.43186	5.99292	-1.4	0.15939	2.70939
14	0.11667	0.02107	0.48173	5.97580	-2.7	0.18064	2.66687
15	0.12500	0.02530	0.53142	5.94731	-4.1	0.20237	2.62603
16	0.13333	0.02993	0.58083	5.90750	-5.5	0.22448	2.58744
17	0.14167	0.03498	0.62985	5.85644	-6.8	0.24688	2.55119
18	0.15000	0.04043	0.67840	5.79423	-8.1	0.26951	2.51715
19	0.15833	0.04628	0.72639	5.72099	-9.4	0.29230	2.48510
20	0.16667	0.05253	0.77372	5.63687	-10.7	0.31519	2.45479
21	0.17500	0.05917	0.82031	5.54201	-12.0	0.33814	2.42597
22	0.18333	0.06620	0.86606	5.43660	-13.3	0.36110	2.39842
23	0.19167	0.07361	0.91089	5.32085	-14.5	0.38403	2.37193
24	0.20000	0.08138	0.95471	5.19497	-15.7	0.40692	2.34632
25	0.20833	0.08951	0.99744	5.05919	-16.9	0.42967	2.32143
26	0.21667	0.09800	1.03900	4.91379	-18.0	0.45231	2.29711
27	0.22500	0.10683	1.07931	4.75903	-19.1	0.47479	2.27325
28	0.23333	0.11599	1.11830	4.59522	-20.2	0.49708	2.24973
29	0.24167	0.12546	1.15588	4.42265	-21.2	0.51915	2.22647
30	0.25000	0.13525	1.19198	4.24167	-22.2	0.54098	2.20337
31	0.25833	0.14532	1.22655	4.05262	-23.2	0.56254	2.18036
32	0.26667	0.15568	1.25951	3.85584	-24.1	0.58381	2.15738
33	0.27500	0.16631	1.29079	3.65173	-24.9	0.60477	2.13436
34	0.28333	0.17719	1.32035	3.44067	-25.7	0.62538	2.11126
35	0.29167	0.18831	1.34812	3.22306	-26.5	0.64564	2.08803
36	0.30000	0.19966	1.37405	2.99931	-27.2	0.66552	2.06463
37	0.30833	0.21121	1.39809	2.76986	-27.9	0.68500	2.04102
38	0.31667	0.22295	1.42020	2.53513	-28.5	0.70406	2.01716
39	0.32500	0.23487	1.44033	2.29558	-29.0	0.72268	1.99303
40	0.33333	0.24695	1.45845	2.05165	-29.5	0.74086	1.96860
41	0.34167	0.25917	1.47451	1.80382	-30.0	0.75856	1.94384
42	0.35000	0.27152	1.48850	1.55256	-30.3	0.77578	1.91873
43	0.35833	0.28398	1.50038	1.29834	-30.7	0.79249	1.89324
44	0.36667	0.29652	1.51013	1.04165	-30.9	0.80870	1.86737
45	0.37500	0.30914	1.51774	0.78298	-31.1	0.82437	1.84108
46	0.38333	0.32181	1.52318	0.52282	-31.3	0.83951	1.81437
47	0.39167	0.33452	1.52645	0.26166	-31.4	0.85409	1.78722
48	0.40000	0.34725	1.52754	-0.00000	-31.4	0.86812	1.75960
49	0.40833	0.35998	1.52754	0.00000	0.0	0.88157	1.73274
50	0.41667	0.37271	1.52754	0.00000	0.0	0.89449	1.70772
51	0.42500	0.38543	1.52754	0.00000	0.0	0.90691	1.68434
52	0.43333	0.39816	1.52754	0.00000	0.0	0.91884	1.66246
53	0.44167	0.41089	1.52754	0.00000	0.0	0.93033	1.64194
54	0.45000	0.42362	1.52754	0.00000	0.0	0.94138	1.62265
55	0.45833	0.43635	1.52754	0.00000	0.0	0.95204	1.60449
56	0.46667	0.44908	1.52754	0.00000	0.0	0.96232	1.58734
57	0.47500	0.46181	1.52754	0.00000	0.0	0.97223	1.57116
58	0.48333	0.47454	1.52754	0.00000	0.0	0.98181	1.55584
59	0.49167	0.48727	1.52754	0.00000	0.0	0.99106	1.54132
60	0.50000	0.50000	1.52754	0.00000	0.0	1.00000	1.52754
61	0.50833	0.51273	1.52754	0.00000	0.0	1.00865	1.51444
62	0.51667	0.52546	1.52754	0.00000	0.0	1.01702	1.50198
63	0.52500	0.53819	1.52754	0.00000	0.0	1.02512	1.49011
64	0.53333	0.55092	1.52754	0.00000	0.0	1.03297	1.47878
65	0.54167	0.56365	1.52754	0.00000	0.0	1.04058	1.46797
66	0.55000	0.57638	1.52754	0.00000	0.0	1.04796	1.45763
67	0.55833	0.58911	1.52754	0.00000	0.0	1.05512	1.44774
68	0.56667	0.60184	1.52754	0.00000	0.0	1.06206	1.43827
69	0.57500	0.61457	1.52754	0.00000	0.0	1.06881	1.42920
70	0.58333	0.62729	1.52754	0.00000	0.0	1.07536	1.42049
71	0.59167	0.64002	1.52754	0.00000	0.0	1.08173	1.41212
72	0.60000	0.65275	1.52754	0.00000	0.0	1.08792	1.40409
73	0.60833	0.66548	1.52645	-0.26166	-31.4	1.09394	1.39537
74	0.61667	0.67819	1.52318	-0.52282	-31.3	1.09977	1.38500
75	0.62500	0.69086	1.51774	-0.78298	-31.1	1.10538	1.37305
76	0.63333	0.70348	1.51013	-1.04165	-30.9	1.11076	1.35956
77	0.64167	0.71602	1.50038	-1.29834	-30.7	1.11588	1.34457
78	0.65000	0.72848	1.48850	-1.55256	-30.3	1.12074	1.32815
79	0.65833	0.74083	1.47451	-1.80382	-30.0	1.12531	1.31032
80	0.66667	0.75305	1.45845	-2.05165	-29.5	1.12957	1.29115
81	0.67500	0.76513	1.44033	-2.29558	-29.0	1.13352	1.27067
82	0.68333	0.77705	1.42020	-2.53513	-28.5	1.13714	1.24892
83	0.69167	0.78879	1.39809	-2.76986	-27.9	1.14042	1.22594
84	0.70000	0.80034	1.37405	-2.99932	-27.2	1.14335	1.20178
85	0.70833	0.81169	1.34812	-3.22306	-26.5	1.14591	1.17646
86	0.71667	0.82281	1.32035	-3.44067	-25.7	1.14810	1.15003
87	0.72500	0.83369	1.29079	-3.65174	-24.9	1.14992	1.12251
88	0.73333	0.84432	1.25951	-3.85468	-24.1	1.15134	1.09395
89	0.74167	0.85468	1.22655	-4.05262	-23.2	1.15237	1.06437
90	0.75000	0.86475	1.19198	-4.24167	-22.2	1.15301	1.03381
91	0.75833	0.87454	1.15588	-4.42265	-21.2	1.15324	1.00229
92	0.76667	0.88401	1.11830	-4.59522	-20.2	1.15306	0.96985
93	0.77500	0.89317	1.07931	-4.75903	-19.1	1.15248	0.93651
94	0.78333	0.90200	1.03901	-4.91379	-18.0	1.15149	0.90231
95	0.79167	0.91049	0.99744	-5.05919	-16.9	1.15009	0.86728
96	0.80000	0.91862	0.95471	-5.19497	-15.7	1.14828	0.83143
97	0.80833	0.92639	0.91089	-5.32085	-14.5	1.14606	0.79480
98	0.81667	0.93380	0.86606	-5.43660	-13.3	1.14343	0.75742
99	0.82500	0.94083	0.82031	-5.54201	-12.0	1.14040	0.71932
100	0.83333	0.94747	0.77372	-5.63687	-10.7	1.13696	0.68052
101	0.84167	0.95372	0.72639	-5.72100	-9.4	1.13313	0.64105
102	0.85000	0.95957	0.67840	-5.79423	-8.1	1.12891	0.60094
103	0.85833	0.96502	0.62985	-5.85644	-6.8	1.12430	0.56021
104	0.86667	0.97007	0.58083	-5.90750	-5.5	1.11931	0.51891
105	0.87500	0.97470	0.53142	-5.94731	-4.1	1.11395	0.47706
106	0.88333	0.97893	0.48174	-5.97580	-2.7	1.10822	0.43469
107	0.89167	0.98273	0.43186	-5.99292	-1.4	1.10213	0.39184
108	0.90000	0.98612	0.38188	-5.99863	-0.0	1.09569	0.34853
109	0.90833	0.98910	0.33204	-5.94731	12.3	1.08891	0.30493
110	0.91667	0.99166	0.28305	-5.79423	24.4	1.08181	0.26164
111	0.92500	0.99382	0.23574	-5.54201	36.1	1.07440	0.21942
112	0.93333	0.99560	0.19094	-5.19496	47.1	1.06671	0.17900
113	0.94167	0.99701	0.14941	-4.75904	57.4	1.05877	0.14111
114	0.95000	0.99810	0.11185	-4.24167	66.6	1.05063	0.10646
115	0.95833	0.99889	0.07892	-3.65174	74.8	1.04232	0.07571
116	0.96667	0.99943	0.05116	-2.99931	81.6	1.03389	0.04949
117	0.97500	0.99976	0.02907	-2.29557	87.1	1.02539	0.02835
118	0.98333	0.99993	0.01301	-1.55256	91.0	1.01688	0.01280
119	0.99167	0.99999	0.00327	-0.78298	93.4	1.00839	0.00324
120	1.00000	1.00000	0.00000	0.00000	94.2	1.00000	0.00000

Table B.6   Cam law factors – Modified sine with $33\frac{1}{3}$ per cent constant velocity

MSC.33 Factor table   MODIFIED SINE WITH 33 $^1/_3$ % CONSTANT VELOCITY   120 STEPS

SCCA PARAMETERS: a = .1666667   b = 0   c = .5   AF = .5   Ca = 6.616604   Cd = 6.616604   Cv = 1.404087

STEP NO.	INPUT DISP. u	OUTPUT DISP. w	OUTPUT VELOC. w'	OUTPUT ACCEL. w''	O'PUT JERK w'''	DISP. RATIO w/u	BLEND FACTOR z = u.w'/w	STEP NO.	INPUT DISP. u	OUTPUT DISP. w	OUTPUT VELOC. w'	OUTPUT ACCEL. w''	O'PUT JERK w'''	DISP. RATIO w/u	BLEND FACTOR z = u.w'/w
0	0.00000	0.00000	0.00000	0.00000	124.7			61	0.50833	0.51170	1.40409	0.00000	0.0	1.00662	1.39485
1	0.00833	0.00001	0.00432	1.03506	123.2	0.00144	2.99762	62	0.51667	0.52340	1.40409	0.00000	0.0	1.01304	1.38602
2	0.01667	0.00010	0.01718	2.04464	118.6	0.00575	2.99014	63	0.52500	0.53510	1.40409	0.00000	0.0	1.01924	1.37758
3	0.02500	0.00032	0.03826	3.00387	111.1	0.01285	2.97778	64	0.53333	0.54680	1.40409	0.00000	0.0	1.02526	1.36950
4	0.03333	0.00075	0.06704	3.88914	100.9	0.02264	2.96049	65	0.54167	0.55850	1.40409	0.00000	0.0	1.03108	1.36176
5	0.04167	0.00146	0.10281	4.67864	88.2	0.03499	2.93823								
6	0.05000	0.00249	0.14470	5.35294	73.3	0.04971	2.91099	66	0.55000	0.57020	1.40409	0.00000	0.0	1.03674	1.35434
7	0.05833	0.00388	0.19166	5.89544	56.6	0.06658	2.87876	67	0.55833	0.58191	1.40409	0.00000	0.0	1.04222	1.34721
8	0.06667	0.00569	0.24255	6.29276	38.5	0.08536	2.84153	68	0.56667	0.59361	1.40409	0.00000	0.0	1.04754	1.34037
9	0.07500	0.00793	0.29611	6.53514	19.5	0.10578	2.79926	69	0.57500	0.60531	1.40409	0.00000	0.0	1.05271	1.33379
10	0.08333	0.01063	0.35102	6.61660	0.0	0.12755	2.75194	70	0.58333	0.61701	1.40409	0.00000	0.0	1.05773	1.32746
11	0.09167	0.01378	0.40613	6.60754	-2.2	0.15038	2.70081	71	0.59167	0.62871	1.40409	0.00000	0.0	1.06261	1.32136
12	0.10000	0.01740	0.46110	6.58036	-4.3	0.17398	2.65028	72	0.60000	0.64041	1.40409	0.00000	0.0	1.06735	1.31549
13	0.10833	0.02147	0.51576	6.53514	-6.5	0.19817	2.60259	73	0.60833	0.65211	1.40409	0.00000	0.0	1.07196	1.30983
14	0.11667	0.02599	0.56997	6.47202	-8.6	0.22279	2.55826	74	0.61667	0.66381	1.40409	0.00000	0.0	1.07645	1.30437
15	0.12500	0.03097	0.62357	6.39115	-10.8	0.24773	2.51715	75	0.62500	0.67551	1.40409	0.00000	0.0	1.08082	1.29910
16	0.13333	0.03638	0.67644	6.29276	-12.8	0.27288	2.47891	76	0.63333	0.68721	1.40409	0.00000	0.0	1.08507	1.29400
17	0.14167	0.04224	0.72841	6.17713	-14.9	0.29815	2.44310	77	0.64167	0.69891	1.40409	0.00000	0.0	1.08921	1.28908
18	0.15000	0.04852	0.77934	6.04457	-16.9	0.32347	2.40930	78	0.65000	0.71061	1.40409	0.00000	0.0	1.09325	1.28432
19	0.15833	0.05522	0.82910	5.89544	-18.9	0.34878	2.37715	79	0.65833	0.72231	1.40409	0.00000	0.0	1.09719	1.27972
20	0.16667	0.06234	0.87755	5.73015	-20.8	0.37401	2.34632	80	0.66667	0.73401	1.40409	-0.00000	-41.6	1.10102	1.27526
21	0.17500	0.06985	0.92456	5.54915	-22.6	0.39912	2.31652	81	0.67500	0.74571	1.40264	-0.34628	-41.5	1.10476	1.26964
22	0.18333	0.07774	0.97000	5.35294	-24.4	0.42404	2.28752	82	0.68333	0.75738	1.39832	-0.69163	-41.3	1.10837	1.26160
23	0.19167	0.08601	1.01374	5.14207	-26.2	0.44873	2.25911	83	0.69167	0.76901	1.39112	-1.03506	-41.1	1.11182	1.25121
24	0.20000	0.09463	1.05566	4.91710	-27.8	0.47316	2.23111	84	0.70000	0.78056	1.38107	-1.37567	-40.7	1.11509	1.23853
25	0.20833	0.10360	1.09565	4.67865	-29.4	0.49726	2.20337	85	0.70833	0.79202	1.36820	-1.71250	-40.2	1.11814	1.22364
26	0.21667	0.11289	1.13360	4.42737	-30.9	0.52101	2.17576	86	0.71667	0.80336	1.35255	-2.04464	-39.5	1.12096	1.20659
27	0.22500	0.12248	1.16941	4.16396	-32.3	0.54437	2.14818	87	0.72500	0.81456	1.33414	-2.37118	-38.8	1.12352	1.18746
28	0.23333	0.13237	1.20297	3.88914	-33.6	0.56730	2.12051	88	0.73333	0.82559	1.31304	-2.69122	-38.0	1.12580	1.16632
29	0.24167	0.14253	1.23420	3.60366	-34.9	0.58977	2.09269	89	0.74167	0.83643	1.28931	-3.00388	-37.0	1.12777	1.14324
30	0.25000	0.15293	1.26300	3.30830	-36.0	0.61173	2.06463	90	0.75000	0.84707	1.26300	-3.30830	-36.0	1.12942	1.11827
31	0.25833	0.16357	1.28931	3.00388	-37.0	0.63317	2.03627	91	0.75833	0.85747	1.23420	-3.60366	-34.9	1.13073	1.09150
32	0.26667	0.17441	1.31304	2.66922	-38.0	0.65406	2.00754	92	0.76667	0.86763	1.20297	-3.88914	-33.6	1.13169	1.06298
33	0.27500	0.18545	1.33414	2.37118	-38.8	0.67435	1.97841	93	0.77500	0.87752	1.16941	-4.16396	-32.3	1.13228	1.03279
34	0.28333	0.19664	1.35255	2.04464	-39.5	0.69403	1.94882	94	0.78333	0.88751	1.13360	-4.42737	-30.9	1.13249	1.00099
35	0.29167	0.20798	1.36820	1.71250	-40.2	0.71308	1.91873	95	0.79167	0.89640	1.09565	-4.67865	-29.4	1.13230	0.96763
36	0.30000	0.21944	1.38107	1.37567	-40.7	0.73146	1.88810	96	0.80000	0.90501	1.05566	-4.91709	-27.8	1.13171	0.93280
37	0.30833	0.23099	1.39112	1.03507	-41.1	0.74916	1.85691	97	0.80833	0.91399	1.01374	-5.14207	-26.2	1.13071	0.89655
38	0.31667	0.24262	1.39832	0.69162	-41.3	0.76616	1.82511	98	0.81667	0.92226	0.97000	-5.35294	-24.4	1.12930	0.85894
39	0.32500	0.25429	1.40264	0.34629	-41.5	0.78243	1.79268	99	0.82500	0.93015	0.92456	-5.54915	-22.6	1.12746	0.82004
40	0.33333	0.26599	1.40409	-0.00000	-41.6	0.79796	1.75960	100	0.83333	0.93766	0.87755	-5.73015	-20.8	1.12520	0.77991
41	0.34167	0.27769	1.40409	0.00000	0.0	0.81274	1.72760	101	0.84167	0.94478	0.82910	-5.89544	-18.9	1.12251	0.73862
42	0.35000	0.28939	1.40409	0.00000	0.0	0.82682	1.69818	102	0.85000	0.95148	0.77934	-6.04457	-16.9	1.11939	0.69622
43	0.35833	0.30109	1.40409	0.00000	0.0	0.84024	1.67105	103	0.85833	0.95776	0.72841	-6.17713	-14.9	1.11584	0.65279
44	0.36667	0.31279	1.40409	0.00000	0.0	0.85306	1.64594	104	0.86667	0.96362	0.67644	-6.29276	-12.8	1.11187	0.60838
45	0.37500	0.32449	1.40409	0.00000	0.0	0.86530	1.62265	105	0.87500	0.96903	0.62357	-6.39115	-10.8	1.10747	0.56306
46	0.38333	0.33619	1.40409	0.00000	0.0	0.87702	1.60098	106	0.88333	0.97401	0.56997	-6.47202	-8.6	1.10265	0.51691
47	0.39167	0.34789	1.40409	0.00000	0.0	0.88823	1.58077	107	0.89167	0.97853	0.51576	-6.53514	-6.5	1.09742	0.46997
48	0.40000	0.35959	1.40409	0.00000	0.0	0.89898	1.56187	108	0.90000	0.98260	0.46110	-6.58036	-4.3	1.09178	0.42234
49	0.40833	0.37129	1.40409	0.00000	0.0	0.90929	1.54416	109	0.90833	0.98622	0.40614	-6.60754	-2.2	1.08574	0.37406
50	0.41667	0.38299	1.40409	0.00000	0.0	0.91918	1.52754	110	0.91667	0.98937	0.35102	-6.61660	0.0	1.07931	0.32523
51	0.42500	0.39469	1.40409	0.00000	0.0	0.92869	1.51190	111	0.92500	0.99207	0.29611	-6.53514	19.5	1.07250	0.27609
52	0.43333	0.40639	1.40409	0.00000	0.0	0.93783	1.49716	112	0.93333	0.99431	0.24255	-6.29276	38.5	1.06533	0.22768
53	0.44167	0.41809	1.40409	0.00000	0.0	0.94663	1.48325	113	0.94167	0.99612	0.19166	-5.89544	56.6	1.05782	0.18118
54	0.45000	0.42980	1.40409	0.00000	0.0	0.95510	1.47009	114	0.95000	0.99751	0.14470	-5.35294	73.3	1.05002	0.13780
55	0.45833	0.44150	1.40409	0.00000	0.0	0.96326	1.45763	115	0.95833	0.99848	0.10281	-4.67865	88.2	1.04196	0.09867
56	0.46667	0.45320	1.40409	0.00000	0.0	0.97114	1.44582	116	0.96667	0.99925	0.06704	-3.88914	100.9	1.03370	0.06485
57	0.47500	0.46490	1.40409	0.00000	0.0	0.97873	1.43460	117	0.97500	0.99968	0.03826	-3.00387	111.1	1.02531	0.03731
58	0.48333	0.47660	1.40409	0.00000	0.0	0.98607	1.42393	118	0.98333	0.99990	0.01718	-2.04464	118.6	1.01685	0.01690
59	0.49167	0.48830	1.40409	0.00000	0.0	0.99315	1.41377	119	0.99167	0.99999	0.00432	-1.03506	123.2	1.00839	0.00429
60	0.50000	0.50000	1.40409	0.00000	0.0	1.00000	1.40409	120	1.00000	1.00000	0.00000	0.00000	124.7	1.00000	0.00000

Table B.7 Cam law factors – Modified sine with 50 per cent constant velocity

MSC.50 Factor table   MODIFIED SINE WITH 50% CONSTANT VELOCITY   120 STEPS

SCCA PARAMETERS: a = .125   b = 0   c = .375   AF = .5   Ca = 8.012684   Cd = 8.012684   Cv = 1.275258

STEP NO.	INPUT DISP. u	OUTPUT DISP. w	OUTPUT VELOC. w'	OUTPUT ACCEL. w''	O'PUT JERK w'''	DISP. RATIO w/u	BLEND FACTOR $z = u.w'/w$	STEP NO.	INPUT DISP. u	OUTPUT DISP. w	OUTPUT VELOC. w'	OUTPUT ACCEL. w''	O'PUT JERK w'''	DISP. RATIO w/u	BLEND FACTOR $z = u.w'/w$
0	0.00000	0.00000	0.00000	0.00000	201.4			61	0.50833	0.51063	1.27526	0.00000	0.0	1.00451	1.26953
1	0.00833	0.00002	0.00697	1.66593	197.0	0.00233	2.99560	62	0.51667	0.52125	1.27526	0.00000	0.0	1.00888	1.26403
2	0.01667	0.00015	0.02756	3.25905	184.0	0.00924	2.98244	63	0.52500	0.53188	1.27526	0.00000	0.0	1.01311	1.25876
3	0.02500	0.00051	0.06089	4.70974	162.9	0.02057	2.96048	64	0.53333	0.54251	1.27526	0.00000	0.0	1.01720	1.25369
4	0.03333	0.00120	0.10549	5.95459	134.8	0.03601	2.92970	65	0.54167	0.55314	1.27526	0.00000	0.0	1.02117	1.24882
5	0.04167	0.00230	0.15941	6.93919	100.7	0.05516	2.89006								
6	0.05000	0.00388	0.22030	7.62052	62.2	0.07753	2.84153	66	0.55000	0.56376	1.27526	0.00000	0.0	1.02502	1.24413
7	0.05833	0.00598	0.28549	7.96879	21.1	0.10254	2.78405	67	0.55833	0.57439	1.27526	0.00000	0.0	1.02876	1.23961
8	0.06667	0.00864	0.35219	8.00780	-2.3	0.12958	2.71798	68	0.56667	0.58502	1.27526	0.00000	0.0	1.03238	1.23526
9	0.07500	0.01185	0.41879	7.96879	-7.0	0.15802	2.65029	69	0.57500	0.59564	1.27526	0.00000	0.0	1.03590	1.23106
10	0.08333	0.01562	0.48490	7.89095	-11.7	0.18741	2.58744	70	0.58333	0.60627	1.27526	0.00000	0.0	1.03932	1.22701
11	0.09167	0.01993	0.55020	7.77467	-16.2	0.21743	2.53052	71	0.59167	0.61690	1.27526	0.00000	0.0	1.04265	1.22310
12	0.10000	0.02478	0.61437	7.62052	-20.7	0.24784	2.47891	72	0.60000	0.62753	1.27526	0.00000	0.0	1.04588	1.21932
13	0.10833	0.03017	0.67710	7.42923	-25.1	0.27846	2.43163	73	0.60833	0.63815	1.27526	0.00000	0.0	1.04902	1.21567
14	0.11667	0.03606	0.73809	7.20175	-29.4	0.30912	2.38771	74	0.61667	0.64878	1.27526	0.00000	0.0	1.05208	1.21214
15	0.12500	0.04246	0.79704	6.93919	-33.6	0.33970	2.34632	75	0.62500	0.65941	1.27526	0.00000	0.0	1.05505	1.20872
16	0.13333	0.04934	0.85365	6.64282	-37.5	0.37006	2.30678	76	0.63333	0.67003	1.27526	0.00000	0.0	1.05795	1.20541
17	0.14167	0.05668	0.90766	6.31408	-41.3	0.40011	2.26852	77	0.64167	0.68066	1.27526	0.00000	0.0	1.06077	1.20220
18	0.15000	0.06446	0.95880	5.95458	-44.9	0.42974	2.23110	78	0.65000	0.69129	1.27526	0.00000	0.0	1.06352	1.19909
19	0.15833	0.07265	1.00682	5.56608	-48.3	0.45887	2.19416	79	0.65833	0.70192	1.27526	0.00000	0.0	1.06620	1.19608
20	0.16667	0.08123	1.05149	5.15045	-51.4	0.48739	2.15738	80	0.66667	0.71254	1.27526	0.00000	0.0	1.06881	1.19315
21	0.17500	0.09017	1.09259	4.70974	-54.3	0.51525	2.12051	81	0.67500	0.72317	1.27526	0.00000	0.0	1.07136	1.19031
22	0.18333	0.09943	1.12992	4.24607	-56.9	0.54236	2.08337	82	0.68333	0.73380	1.27526	0.00000	0.0	1.07385	1.18756
23	0.19167	0.10899	1.16330	3.76173	-59.3	0.56864	2.04576	83	0.69167	0.74442	1.27526	0.00000	0.0	1.07628	1.18488
24	0.20000	0.11881	1.19257	3.25905	-61.3	0.59404	2.00754	84	0.70000	0.75505	1.27526	0.00000	0.0	1.07865	1.18228
25	0.20833	0.12885	1.21758	2.74050	-63.1	0.61850	1.96860	85	0.70833	0.76568	1.27526	0.00000	0.0	1.08096	1.17975
26	0.21667	0.13909	1.23821	2.20859	-64.5	0.64195	1.92882	86	0.71667	0.77631	1.27526	0.00000	0.0	1.08322	1.17729
27	0.22500	0.14948	1.25436	1.66593	-65.7	0.66435	1.88810	87	0.72500	0.78693	1.27526	0.00000	0.0	1.08543	1.17489
28	0.23333	0.15998	1.26595	1.11515	-66.5	0.68564	1.84637	88	0.73333	0.79756	1.27526	0.00000	0.0	1.08758	1.17256
29	0.24167	0.17056	1.27293	0.55894	-67.0	0.70579	1.80356	89	0.74167	0.80819	1.27526	0.00000	0.0	1.08969	1.17029
30	0.25000	0.18119	1.27526	-0.00000	-67.1	0.72474	1.75960	90	0.75000	0.81881	1.27526	0.00000	0.0	1.09175	1.16808
31	0.25833	0.19181	1.27526	0.00000	0.0	0.74250	1.71752	91	0.75833	0.82944	1.27293	-0.55894	-67.0	1.09376	1.16381
32	0.26667	0.20244	1.27526	0.00000	0.0	0.75915	1.67985	92	0.76667	0.84002	1.26595	-1.11515	-66.5	1.09567	1.15541
33	0.27500	0.21307	1.27526	0.00000	0.0	0.77479	1.64594	93	0.77500	0.85052	1.25436	-1.66593	-65.7	1.09745	1.14298
34	0.28333	0.22369	1.27526	0.00000	0.0	0.78951	1.61526	94	0.78333	0.86091	1.23821	-2.20859	-64.5	1.09903	1.12663
35	0.29167	0.23432	1.27526	0.00000	0.0	0.80339	1.58735	95	0.79167	0.87115	1.21758	-2.74050	-63.1	1.10039	1.10649
36	0.30000	0.24495	1.27526	0.00000	0.0	0.81649	1.56187	96	0.80000	0.88119	1.19257	-3.25905	-61.3	1.10149	1.08269
37	0.30833	0.25558	1.27526	0.00000	0.0	0.82889	1.53851	97	0.80833	0.89101	1.16330	-3.76173	-59.3	1.10228	1.05536
38	0.31667	0.26620	1.27526	0.00000	0.0	0.84064	1.51701	98	0.81667	0.90057	1.12993	-4.24607	-56.9	1.10274	1.02466
39	0.32500	0.27683	1.27526	0.00000	0.0	0.85178	1.49716	99	0.82500	0.90983	1.09259	-4.70974	-54.3	1.10283	0.99072
40	0.33333	0.28746	1.27526	0.00000	0.0	0.86237	1.47878	100	0.83333	0.91877	1.05149	-5.15045	-51.4	1.10252	0.95372
41	0.34167	0.29808	1.27526	0.00000	0.0	0.87244	1.46171	101	0.84167	0.92735	1.00682	-5.56608	-48.3	1.10180	0.91380
42	0.35000	0.30871	1.27526	0.00000	0.0	0.88203	1.44582	102	0.85000	0.93554	0.95880	-5.95459	-44.9	1.10063	0.87113
43	0.35833	0.31934	1.27526	0.00000	0.0	0.89118	1.43098	103	0.85833	0.94332	0.90766	-6.31408	-41.3	1.09901	0.82589
44	0.36667	0.32997	1.27526	0.00000	0.0	0.89991	1.41710	104	0.86667	0.95066	0.85365	-6.64282	-37.5	1.09691	0.77823
45	0.37500	0.34059	1.27526	0.00000	0.0	0.90825	1.40409	105	0.87500	0.95754	0.79704	-6.93919	-33.6	1.09433	0.72833
46	0.38333	0.35122	1.27526	0.00000	0.0	0.91623	1.39186	106	0.88333	0.96394	0.73809	-7.20175	-29.4	1.09125	0.67637
47	0.39167	0.36185	1.27526	0.00000	0.0	0.92386	1.38035	107	0.89167	0.96983	0.67710	-7.42923	-25.1	1.08766	0.62253
48	0.40000	0.37247	1.27526	0.00000	0.0	0.93119	1.36950	108	0.90000	0.97522	0.61437	-7.62051	-20.7	1.08357	0.56699
49	0.40833	0.38310	1.27526	0.00000	0.0	0.93821	1.35925	109	0.90833	0.98007	0.55020	-7.77467	-16.2	1.07898	0.50993
50	0.41667	0.39373	1.27526	0.00000	0.0	0.94495	1.34955	110	0.91667	0.98438	0.48490	-7.89095	-11.7	1.07387	0.45154
51	0.42500	0.40436	1.27526	0.00000	0.0	0.95143	1.34037	111	0.92500	0.98815	0.41879	-7.96879	-7.0	1.06827	0.39205
52	0.43333	0.41498	1.27526	0.00000	0.0	0.95765	1.33165	112	0.93333	0.99136	0.35219	-8.00780	-2.3	1.06217	0.33158
53	0.44167	0.42561	1.27526	0.00000	0.0	0.96365	1.32337	113	0.94167	0.99402	0.28549	-7.96879	21.0	1.05559	0.27041
54	0.45000	0.43624	1.27526	0.00000	0.0	0.96942	1.31549	114	0.95000	0.99612	0.22030	-7.62051	62.2	1.04855	0.21009
55	0.45833	0.44686	1.27526	0.00000	0.0	0.97498	1.30799	115	0.95833	0.99770	0.15941	-6.93920	100.7	1.04108	0.15312
56	0.46667	0.45749	1.27526	0.00000	0.0	0.98034	1.30083	116	0.96667	0.99880	0.10549	-5.95458	134.8	1.03324	0.10209
57	0.47500	0.46812	1.27526	0.00000	0.0	0.98551	1.29400	117	0.97500	0.99949	0.06089	-4.70973	162.9	1.02511	0.05940
58	0.48333	0.47875	1.27526	0.00000	0.0	0.99051	1.28748	118	0.98333	0.99985	0.02756	-3.25905	184.0	1.01679	0.02711
59	0.49167	0.48937	1.27526	0.00000	0.0	0.99533	1.28124	119	0.99167	0.99998	0.00697	-1.66593	197.0	1.00838	0.00691
60	0.50000	0.50000	1.27526	0.00000	0.0	1.00000	1.27526	120	1.00000	1.00000	0.00000	-0.00000	201.4	1.00000	0.00000

Table B.8 Cam law factors – Modified sine with $66\frac{2}{3}$ per cent constant velocity

MSC.66 Factor table   MODIFIED SINE WITH 66 $^2/_3$ % CONSTANT VELOCITY   120 STEPS

SCCA PARAMETERS: a = 8.333333E-02   b = 0   c = .25   AF = .5   Ca = 11.00893   Cd = 11.00893   Cv = 1.1680

STEP NO.	INPUT DISP. u	OUTPUT DISP. w	OUTPUT VELOC. w'	OUTPUT ACCEL. w''	O'PUT JERK w'''	DISP. RATIO w/u	BLEND FACTOR z = u.w'/w	STEP NO.	INPUT DISP. u	OUTPUT DISP. w	OUTPUT VELOC. w'	OUTPUT ACCEL. w''	O'PUT JERK w'''	DISP. RATIO w/u	BLEND FACTOR z = u.w'/w
0	0.00000	0.00000	0.00000	0.00000	415.0			61	0.50833	0.50973	1.16808	0.00000	0.0	1.00276	1.16487
1	0.00833	0.00004	0.01429	3.40195	394.7	0.00478	2.99012	62	0.51667	0.51947	1.16808	0.00000	0.0	1.00542	1.16178
2	0.01667	0.00031	0.05577	6.47089	335.8	0.01884	2.96048	63	0.52500	0.52920	1.16808	0.00000	0.0	1.00800	1.15881
3	0.02500	0.00103	0.12038	8.90641	243.9	0.04135	2.91099	64	0.53333	0.53894	1.16808	0.00000	0.0	1.01051	1.15594
4	0.03333	0.00237	0.20178	10.47011	128.3	0.07101	2.84153	65	0.54167	0.54867	1.16808	0.00000	0.0	1.01293	1.15317
5	0.04167	0.00442	0.29202	11.00893	-0.0	0.10611	2.75194								
6	0.05000	0.00724	0.38359	10.94862	-14.5	0.14474	2.65029	66	0.55000	0.55840	1.16808	0.00000	0.0	1.01528	1.15050
7	0.05833	0.01081	0.47416	10.76835	-28.8	0.18535	2.55826	67	0.55833	0.56814	1.16808	0.00000	0.0	1.01756	1.14792
8	0.06667	0.01513	0.56274	10.47011	-42.8	0.22701	2.47891	68	0.56667	0.57787	1.16808	0.00000	0.0	1.01977	1.14543
9	0.07500	0.02018	0.64835	10.05715	-56.3	0.26910	2.40930	69	0.57500	0.58761	1.16808	0.00000	0.0	1.02192	1.14302
10	0.08333	0.02593	0.73005	9.53401	-69.2	0.31115	2.34632	70	0.58333	0.59734	1.16808	0.00000	0.0	1.02401	1.14069
11	0.09167	0.03234	0.80696	8.90641	-81.3	0.35277	2.28752	71	0.59167	0.60707	1.16808	0.00000	0.0	1.02604	1.13844
12	0.10000	0.03936	0.87822	8.18123	-92.6	0.39363	2.23110	72	0.60000	0.61681	1.16808	0.00000	0.0	1.02801	1.13625
13	0.10833	0.04696	0.94306	7.36641	-102.8	0.43344	2.17576	73	0.60833	0.62654	1.16808	0.00000	0.0	1.02993	1.13414
14	0.11667	0.05506	1.00077	6.47088	-111.9	0.47195	2.12051	74	0.61667	0.63628	1.16808	0.00000	0.0	1.03180	1.13208
15	0.12500	0.06361	1.05071	5.50446	-119.8	0.50891	2.06463	75	0.62500	0.64601	1.16808	0.00000	0.0	1.03362	1.13009
16	0.13333	0.07255	1.09234	4.47773	-126.4	0.54412	2.00754	76	0.63333	0.65574	1.16808	0.00000	0.0	1.03539	1.12816
17	0.14167	0.08180	1.12521	3.40195	-131.6	0.57738	1.94882	77	0.64167	0.66548	1.16808	0.00000	0.0	1.03711	1.12629
18	0.15000	0.09128	1.14894	2.28888	-135.3	0.60852	1.88810	78	0.65000	0.67521	1.16808	0.00000	0.0	1.03879	1.12447
19	0.15833	0.10092	1.16328	1.15075	-137.6	0.63738	1.82511	79	0.65833	0.68495	1.16808	0.00000	0.0	1.04042	1.12270
20	0.16667	0.11064	1.16808	0.00000	0.0	0.66383	1.75960	80	0.66667	0.69468	1.16808	0.00000	0.0	1.04202	1.12098
21	0.17500	0.12037	1.16808	0.00000	0.0	0.68785	1.69818	81	0.67500	0.70441	1.16808	0.00000	0.0	1.04358	1.11931
22	0.18333	0.13011	1.16808	0.00000	0.0	0.70967	1.64594	82	0.68333	0.71415	1.16808	0.00000	0.0	1.04510	1.11768
23	0.19167	0.13984	1.16808	0.00000	0.0	0.72961	1.60098	83	0.69167	0.72388	1.16808	0.00000	0.0	1.04658	1.11610
24	0.20000	0.14957	1.16808	0.00000	0.0	0.74787	1.56187	84	0.70000	0.73362	1.16808	0.00000	0.0	1.04802	1.11456
25	0.20833	0.15931	1.16808	0.00000	0.0	0.76468	1.52754	85	0.70833	0.74335	1.16808	0.00000	0.0	1.04944	1.11306
26	0.21667	0.16904	1.16808	0.00000	0.0	0.78020	1.49716	86	0.71667	0.75308	1.16808	0.00000	0.0	1.05082	1.11160
27	0.22500	0.17878	1.16808	0.00000	0.0	0.79456	1.47009	87	0.72500	0.76282	1.16808	0.00000	0.0	1.05216	1.11017
28	0.23333	0.18851	1.16808	0.00000	0.0	0.80790	1.44582	88	0.73333	0.77255	1.16808	0.00000	0.0	1.05348	1.10878
29	0.24167	0.19825	1.16808	0.00000	0.0	0.82032	1.42393	89	0.74167	0.78228	1.16808	0.00000	0.0	1.05477	1.10743
30	0.25000	0.20798	1.16808	0.00000	0.0	0.83192	1.40409	90	0.75000	0.79202	1.16808	0.00000	0.0	1.05603	1.10611
31	0.25833	0.21771	1.16808	0.00000	0.0	0.84276	1.38602	91	0.75833	0.80175	1.16808	0.00000	0.0	1.05726	1.10482
32	0.26667	0.22745	1.16808	0.00000	0.0	0.85293	1.36950	92	0.76667	0.81149	1.16808	0.00000	0.0	1.05846	1.10356
33	0.27500	0.23718	1.16808	0.00000	0.0	0.86248	1.35434	93	0.77500	0.82122	1.16808	0.00000	0.0	1.05964	1.10234
34	0.28333	0.24692	1.16808	0.00000	0.0	0.87147	1.34037	94	0.78333	0.83096	1.16808	0.00000	0.0	1.06080	1.10114
35	0.29167	0.25665	1.16808	0.00000	0.0	0.87994	1.32746	95	0.79167	0.84069	1.16808	0.00000	0.0	1.06193	1.09997
36	0.30000	0.26638	1.16808	0.00000	0.0	0.88794	1.31549	96	0.80000	0.85042	1.16808	0.00000	0.0	1.06303	1.09882
37	0.30833	0.27612	1.16808	0.00000	0.0	0.89552	1.30437	97	0.80833	0.86016	1.16808	0.00000	0.0	1.06411	1.09770
38	0.31667	0.28585	1.16808	0.00000	0.0	0.90269	1.29400	98	0.81667	0.86989	1.16808	0.00000	0.0	1.06518	1.09661
39	0.32500	0.29559	1.16808	0.00000	0.0	0.90949	1.28432	99	0.82500	0.87963	1.16808	0.00000	0.0	1.06621	1.09554
40	0.33333	0.30532	1.16808	0.00000	0.0	0.91596	1.27526	100	0.83333	0.88936	1.16808	0.00000	0.0	1.06723	1.09450
41	0.34167	0.31505	1.16808	0.00000	0.0	0.92211	1.26675	101	0.84167	0.89908	1.16328	-1.15075	-137.6	1.06822	1.08900
42	0.35000	0.32479	1.16808	0.00000	0.0	0.92796	1.25876	102	0.85000	0.90872	1.14894	-2.28889	-135.3	1.06909	1.07469
43	0.35833	0.33452	1.16808	0.00000	0.0	0.93355	1.25123	103	0.85833	0.91820	1.12521	-3.40195	-131.6	1.06975	1.05184
44	0.36667	0.34426	1.16808	0.00000	0.0	0.93888	1.24413	104	0.86667	0.92745	1.09234	-4.47773	-126.4	1.07014	1.02075
45	0.37500	0.35399	1.16808	0.00000	0.0	0.94397	1.23741	105	0.87500	0.93639	1.05071	-5.50446	-119.8	1.07016	0.98183
46	0.38333	0.36372	1.16808	0.00000	0.0	0.94884	1.23106	106	0.88333	0.94494	1.00077	-6.47088	-111.9	1.06974	0.93552
47	0.39167	0.37346	1.16808	0.00000	0.0	0.95351	1.22504	107	0.89167	0.95304	0.94306	-7.36641	-102.8	1.06883	0.88233
48	0.40000	0.38319	1.16808	0.00000	0.0	0.95798	1.21932	108	0.90000	0.96064	0.87822	-8.18122	-92.6	1.06737	0.82279
49	0.40833	0.39293	1.16808	0.00000	0.0	0.96227	1.21389	109	0.90833	0.96766	0.80696	-8.90641	-81.3	1.06532	0.75748
50	0.41667	0.40266	1.16808	0.00000	0.0	0.96638	1.20872	110	0.91667	0.97407	0.73005	-9.53401	-69.2	1.06262	0.68703
51	0.42500	0.41239	1.16808	0.00000	0.0	0.97034	1.20379	111	0.92500	0.97982	0.64835	-10.05715	-56.3	1.05926	0.61208
52	0.43333	0.42213	1.16808	0.00000	0.0	0.97414	1.19909	112	0.93333	0.98487	0.56274	-10.47011	-42.8	1.05521	0.53329
53	0.44167	0.43186	1.16808	0.00000	0.0	0.97780	1.19460	113	0.94167	0.98919	0.47417	-10.76835	-28.8	1.05047	0.45139
54	0.45000	0.44160	1.16808	0.00000	0.0	0.98132	1.19031	114	0.95000	0.99276	0.38359	-10.94862	-14.5	1.04501	0.36707
55	0.45833	0.45133	1.16808	0.00000	0.0	0.98472	1.18621	115	0.95833	0.99558	0.29202	-11.00893	-0.0	1.03886	0.28110
56	0.46667	0.46106	1.16808	0.00000	0.0	0.98799	1.18228	116	0.96667	0.99763	0.20178	-10.47011	128.3	1.03203	0.19552
57	0.47500	0.47080	1.16808	0.00000	0.0	0.99115	1.17851	117	0.97500	0.99897	0.12038	-8.90640	243.9	1.02458	0.11749
58	0.48333	0.48053	1.16808	0.00000	0.0	0.99420	1.17489	118	0.98333	0.99969	0.05577	-6.47088	335.8	1.01663	0.05486
59	0.49167	0.49027	1.16808	0.00000	0.0	0.99715	1.17142	119	0.99167	0.99996	0.01429	-3.40194	394.7	1.00836	0.01417
60	0.50000	0.50000	1.16808	0.00000	0.0	1.00000	1.16808	120	1.00000	1.00000	0.00000	0.00000	415.0	1.00000	0.00000

Table B.9　Cam law factors – Cycloidal

CYC Factor table　　　　　　　　　　　　CYCLOIDAL　　　　　　　　　　120 STEPS

SCCA PARAMETERS: a = .5　b = 0　c = .5　AF = .5　Ca = 6.283185　Cd = 6.283185　Cv = 2

STEP NO.	INPUT DISP. u	OUTPUT DISP. w	OUTPUT VELOC. w'	OUTPUT ACCEL. w''	O'PUT JERK w'''	DISP. RATIO W/u	BLEND FACTOR z = u.w'/w
0	0.00000	0.00000	0.00000	0.00000	39.5		
1	0.00833	0.00000	0.00137	0.32884	39.4	0.00046	2.99974
2	0.01667	0.00003	0.00548	0.65677	39.3	0.00183	2.99890
3	0.02500	0.00010	0.01231	0.98291	39.0	0.00411	2.99751
4	0.03333	0.00024	0.02185	1.30635	38.6	0.00729	2.99560
5	0.04167	0.00047	0.03407	1.62621	38.1	0.01138	2.99312
6	0.05000	0.00082	0.04894	1.94161	37.5	0.01637	2.99012
7	0.05833	0.00130	0.06642	2.25169	36.9	0.02224	2.98656
8	0.06667	0.00193	0.08645	2.55560	36.1	0.02899	2.98244
9	0.07500	0.00275	0.10899	2.85251	35.2	0.03660	2.97778
10	0.08333	0.00376	0.13397	3.14159	34.2	0.04507	2.97256
11	0.09167	0.00498	0.16133	3.42207	33.1	0.05438	2.96680
12	0.10000	0.00645	0.19098	3.69316	31.9	0.06451	2.96048
13	0.10833	0.00817	0.22285	3.95414	30.7	0.07545	2.95362
14	0.11667	0.01017	0.25686	4.20427	29.3	0.08718	2.94620
15	0.12500	0.01246	0.29289	4.44288	27.9	0.09968	2.93822
16	0.13333	0.01506	0.33087	4.66932	26.4	0.11294	2.92970
17	0.14167	0.01798	0.37068	4.88295	24.8	0.12692	2.92062
18	0.15000	0.02124	0.41221	5.08320	23.2	0.14161	2.91099
19	0.15833	0.02485	0.45536	5.26952	21.5	0.15698	2.90081
20	0.16667	0.02883	0.50000	5.44140	19.7	0.17301	2.89006
21	0.17500	0.03319	0.54601	5.59836	17.9	0.18967	2.87876
22	0.18333	0.03794	0.59326	5.73998	16.1	0.20693	2.86691
23	0.19167	0.04308	0.64163	5.86586	14.1	0.22478	2.85450
24	0.20000	0.04863	0.69098	5.97566	12.2	0.24318	2.84153
25	0.20833	0.05460	0.74118	6.06909	10.2	0.26209	2.82800
26	0.21667	0.06099	0.79209	6.14588	8.2	0.28149	2.81391
27	0.22500	0.06780	0.84357	6.20583	6.2	0.30135	2.79926
28	0.23333	0.07505	0.89547	6.24877	4.1	0.32164	2.78405
29	0.24167	0.08273	0.94766	6.27457	2.1	0.34233	2.76827
30	0.25000	0.09085	1.00000	6.28319	0.0	0.36338	2.75194
31	0.25833	0.09940	1.05234	6.27457	-2.1	0.38476	2.73504
32	0.26667	0.10838	1.10453	6.24877	-4.1	0.40644	2.71758
33	0.27500	0.11780	1.15643	6.20583	-6.2	0.42838	2.69955
34	0.28333	0.12766	1.20791	6.14588	-8.2	0.45055	2.68096
35	0.29167	0.13793	1.25882	6.06909	-10.2	0.47292	2.66181
36	0.30000	0.14863	1.30902	5.97566	-12.2	0.49545	2.64208
37	0.30833	0.15975	1.35837	5.86586	-14.1	0.51811	2.62179
38	0.31667	0.17127	1.40674	5.73998	-16.1	0.54086	2.60094
39	0.32500	0.18319	1.45399	5.59836	-17.9	0.56367	2.57952
40	0.33333	0.19550	1.50000	5.44140	-19.7	0.58650	2.55753
41	0.34167	0.20819	1.54464	5.26952	-21.5	0.60929	2.53497
42	0.35000	0.22124	1.58779	5.08320	-23.2	0.63212	2.51185
43	0.35833	0.23465	1.62932	4.88295	-24.8	0.65483	2.48817
44	0.36667	0.24839	1.66913	4.66932	-26.4	0.67743	2.46391
45	0.37500	0.26246	1.70711	4.44288	-27.9	0.69989	2.43909
46	0.38333	0.27684	1.74314	4.20427	-29.3	0.72219	2.41371
47	0.39167	0.29151	1.77715	3.95414	-30.7	0.74427	2.38776
48	0.40000	0.30645	1.80902	3.69316	-31.9	0.76613	2.36125
49	0.40833	0.32165	1.83867	3.42207	-33.1	0.78772	2.33417
50	0.41667	0.33709	1.86603	3.14159	-34.2	0.80901	2.30654
51	0.42500	0.35275	1.89101	2.85251	-35.2	0.82999	2.27835
52	0.43333	0.36860	1.91355	2.55560	-36.1	0.85061	2.24961
53	0.44167	0.38463	1.93358	2.25169	-36.9	0.87086	2.22031
54	0.45000	0.40082	1.95106	1.94161	-37.5	0.89071	2.19046
55	0.45833	0.41714	1.96593	1.62621	-38.1	0.91013	2.16006
56	0.46667	0.43358	1.97815	1.30635	-38.6	0.92909	2.12912
57	0.47500	0.45010	1.98769	0.98291	-39.0	0.94758	2.09764
58	0.48333	0.46670	1.99452	0.65677	-39.3	0.96558	2.06562
59	0.49167	0.48334	1.99863	0.32884	-39.4	0.98306	2.03307
60	0.50000	0.50000	2.00000	-0.00000	-39.5	1.00000	2.00000

STEP NO.	INPUT DISP. u	OUTPUT DISP. w	OUTPUT VELOC. w'	OUTPUT ACCEL. w''	O'PUT JERK w'''	DISP. RATIO W/u	BLEND FACTOR z = u.w'/w
61	0.50833	0.51666	1.99863	-0.32884	-39.4	1.01639	1.96641
62	0.51667	0.53330	1.99452	-0.65677	-39.3	1.03220	1.93230
63	0.52500	0.54990	1.98769	-0.98291	-39.0	1.04742	1.89769
64	0.53333	0.56642	1.97815	-1.30635	-38.6	1.06204	1.86259
65	0.54167	0.58286	1.96593	-1.62621	-38.1	1.07605	1.82699
66	0.55000	0.59918	1.95106	-1.94161	-37.5	1.08942	1.79091
67	0.55833	0.61537	1.93358	-2.25169	-36.9	1.10215	1.75436
68	0.56667	0.63140	1.91355	-2.55560	-36.1	1.11424	1.71736
69	0.57500	0.64725	1.89101	-2.85251	-35.2	1.12566	1.67991
70	0.58333	0.66291	1.86603	-3.14159	-34.2	1.13642	1.64202
71	0.59167	0.67835	1.83867	-3.42207	-33.1	1.14650	1.60372
72	0.60000	0.69355	1.80902	-3.69316	-31.9	1.15591	1.56501
73	0.60833	0.70849	1.77715	-3.95414	-30.7	1.16465	1.52591
74	0.61667	0.72316	1.74314	-4.20427	-29.3	1.17270	1.48644
75	0.62500	0.73754	1.70711	-4.44288	-27.9	1.18006	1.44662
76	0.63333	0.75161	1.66913	-4.66932	-26.4	1.18675	1.40647
77	0.64167	0.76535	1.62932	-4.88295	-24.8	1.19276	1.36601
78	0.65000	0.77876	1.58779	-5.08320	-23.2	1.19809	1.32526
79	0.65833	0.79181	1.54464	-5.26952	-21.5	1.20275	1.28423
80	0.66667	0.80450	1.50000	-5.44140	-19.7	1.20675	1.24301
81	0.67500	0.81681	1.45399	-5.59836	-17.9	1.21009	1.20156
82	0.68333	0.82873	1.40674	-5.73998	-16.1	1.21277	1.15993
83	0.69167	0.84025	1.35837	-5.86586	-14.1	1.21482	1.11816
84	0.70000	0.85137	1.30902	-5.97566	-12.2	1.21624	1.07628
85	0.70833	0.86207	1.25882	-6.06909	-10.2	1.21703	1.03433
86	0.71667	0.87234	1.20791	-6.14588	-8.2	1.21722	0.99235
87	0.72500	0.88220	1.15643	-6.20583	-6.2	1.21682	0.95037
88	0.73333	0.89162	1.10453	-6.24877	-4.1	1.21584	0.90845
89	0.74167	0.90060	1.05234	-6.27457	-2.1	1.21430	0.86662
90	0.75000	0.90915	1.00000	-6.28319	0.0	1.21221	0.82494
91	0.75833	0.91727	0.94766	-6.27457	2.1	1.20959	0.78342
92	0.76667	0.92495	0.89547	-6.24877	4.1	1.20646	0.74223
93	0.77500	0.93220	0.84357	-6.20583	6.2	1.20283	0.70132
94	0.78333	0.93901	0.79209	-6.14588	8.2	1.19874	0.66077
95	0.79167	0.94540	0.74118	-6.06909	10.2	1.19419	0.62066
96	0.80000	0.95137	0.69098	-5.97567	12.2	1.18921	0.58105
97	0.80833	0.95692	0.64163	-5.86586	14.1	1.18382	0.54200
98	0.81667	0.96206	0.59326	-5.73998	16.1	1.17804	0.50360
99	0.82500	0.96681	0.54601	-5.59836	17.9	1.17189	0.46592
100	0.83333	0.97117	0.50000	-5.44140	19.7	1.16540	0.42904
101	0.84167	0.97515	0.45536	-5.26952	21.5	1.15859	0.39303
102	0.85000	0.97876	0.41221	-5.08320	23.2	1.15148	0.35799
103	0.85833	0.98202	0.37068	-4.88295	24.8	1.14410	0.32399
104	0.86667	0.98494	0.33087	-4.66932	26.4	1.13647	0.29114
105	0.87500	0.98754	0.29289	-4.44288	27.9	1.12862	0.25952
106	0.88333	0.98983	0.25686	-4.20427	29.3	1.12056	0.22922
107	0.89167	0.99183	0.22285	-3.95414	30.7	1.11233	0.20035
108	0.90000	0.99355	0.19098	-3.69316	31.9	1.10394	0.17300
109	0.90833	0.99502	0.16133	-3.42207	33.1	1.09543	0.14728
110	0.91667	0.99624	0.13397	-3.14159	34.2	1.08681	0.12327
111	0.92500	0.99725	0.10899	-2.85251	35.2	1.07811	0.10110
112	0.93333	0.99807	0.08645	-2.55560	36.1	1.06936	0.08085
113	0.94167	0.99870	0.06642	-2.25169	36.9	1.06057	0.06263
114	0.95000	0.99918	0.04894	-1.94161	37.5	1.05177	0.04653
115	0.95833	0.99953	0.03407	-1.62621	38.1	1.04298	0.03257
116	0.96667	0.99976	0.02185	-1.30635	38.6	1.03423	0.02113
117	0.97500	0.99990	0.01231	-0.98291	39.0	1.02554	0.01201
118	0.98333	0.99997	0.00548	-0.65677	39.3	1.01692	0.00539
119	0.99167	1.00000	0.00137	-0.32884	39.4	1.00840	0.00136
120	1.00000	1.00000	0.00000	0.00000	39.5	1.00000	0.00000

Table B.10 Cam law factors – Cycloidal with 50% constant velocity

CYCC.50 Factor table			CYCLOIDAL WITH 50% CONSTANT VELOCITY			120 STEPS

SCCA PARAMETERS: a = .25 b = 0 c = .25 AF = .5 Ca = 8.377581 Cd = 8.377581 Cv = 1.333333

STEP NO.	INPUT DISP. u	OUTPUT DISP. w	OUTPUT VELOC. w'	OUTPUT ACCEL. w''	O'PUT JERK w'''	DISP. RATIO w/u	BLEND FACTOR z = u.w'/w	STEP NO.	INPUT DISP. u	OUTPUT DISP. w	OUTPUT VELOC. w'	OUTPUT ACCEL. w''	O'PUT JERK w'''	DISP. RATIO w/u	BLEND FACTOR z = u.w'/w
0	0.00000	0.00000	0.00000	0.00000	105.3			61	0.50833	0.51111	1.33333	0.00000	0.0	1.00546	1.32609
1	0.00833	0.00001	0.00365	0.87570	104.7	0.00122	2.99890	62	0.51667	0.52222	1.33333	0.00000	0.0	1.01075	1.31915
2	0.01667	0.00008	0.01457	1.74180	103.0	0.00486	2.99560	63	0.52500	0.53333	1.33333	0.00000	0.0	1.01587	1.31250
3	0.02500	0.00027	0.03263	2.58881	100.1	0.01091	2.99012	64	0.53333	0.54444	1.33333	0.00000	0.0	1.02083	1.30612
4	0.03333	0.00064	0.05764	3.40747	96.2	0.01933	2.98244	65	0.54167	0.55556	1.33333	0.00000	0.0	1.02564	1.30000
5	0.04167	0.00125	0.08932	4.18879	91.2	0.03005	2.97256								
								66	0.55000	0.56667	1.33333	0.00000	0.0	1.03030	1.29412
6	0.05000	0.00215	0.12732	4.92422	85.2	0.04301	2.96048	67	0.55833	0.57778	1.33333	0.00000	0.0	1.03483	1.28846
7	0.05833	0.00339	0.17124	5.60570	78.2	0.05812	2.94620	68	0.56667	0.58889	1.33333	0.00000	0.0	1.03922	1.28302
8	0.06667	0.00502	0.22058	6.22576	70.4	0.07529	2.92970	69	0.57500	0.60000	1.33333	0.00000	0.0	1.04348	1.27778
9	0.07500	0.00708	0.27481	6.77761	61.9	0.09440	2.91099	70	0.58333	0.61111	1.33333	0.00000	0.0	1.04762	1.27273
10	0.08333	0.00961	0.33333	7.25520	52.6	0.11534	2.89006								
								71	0.59167	0.62222	1.33333	0.00000	0.0	1.05164	1.26786
11	0.09167	0.01265	0.39551	7.65330	42.8	0.13796	2.86691	72	0.60000	0.63333	1.33333	0.00000	0.0	1.05556	1.26316
12	0.10000	0.01621	0.46066	7.96755	32.5	0.16212	2.84153	73	0.60833	0.64444	1.33333	0.00000	0.0	1.05936	1.25862
13	0.10833	0.02033	0.52806	8.19451	21.9	0.18766	2.81391	74	0.61667	0.65556	1.33333	0.00000	0.0	1.06306	1.25424
14	0.11667	0.02502	0.59698	8.33169	11.0	0.21443	2.78405	75	0.62500	0.66667	1.33333	0.00000	0.0	1.06667	1.25000
15	0.12500	0.03028	0.66667	8.37758	0.0	0.24225	2.75194								
								76	0.63333	0.67778	1.33333	0.00000	0.0	1.07018	1.24590
16	0.13333	0.03613	0.73635	8.33169	-11.0	0.27096	2.71758	77	0.64167	0.68889	1.33333	0.00000	0.0	1.07359	1.24194
17	0.14167	0.04255	0.80527	8.19451	-21.9	0.30037	2.68096	78	0.65000	0.70000	1.33333	0.00000	0.0	1.07692	1.23810
18	0.15000	0.04954	0.87268	7.96755	-32.5	0.33030	2.64208	79	0.65833	0.71111	1.33333	0.00000	0.0	1.08017	1.23438
19	0.15833	0.05709	0.93782	7.65330	-42.8	0.36057	2.60094	80	0.66667	0.72222	1.33333	0.00000	0.0	1.08333	1.23077
20	0.16667	0.06517	1.00000	7.25520	-52.6	0.39100	2.55753								
								81	0.67500	0.73333	1.33333	0.00000	0.0	1.08642	1.22727
21	0.17500	0.07375	1.05852	6.77761	-61.9	0.42141	2.51185	82	0.68333	0.74444	1.33333	0.00000	0.0	1.08943	1.22388
22	0.18333	0.08280	1.11275	6.22576	-70.4	0.45162	2.46391	83	0.69167	0.75556	1.33333	0.00000	0.0	1.09237	1.22059
23	0.19167	0.09228	1.16210	5.60570	-78.2	0.48146	2.41371	84	0.70000	0.76667	1.33333	0.00000	0.0	1.09524	1.21739
24	0.20000	0.10215	1.20601	4.92422	-85.2	0.51075	2.36125	85	0.70833	0.77778	1.33333	0.00000	0.0	1.09804	1.21429
25	0.20833	0.11236	1.24402	4.18879	-91.2	0.53934	2.30654								
								86	0.71667	0.78889	1.33333	0.00000	0.0	1.10078	1.21127
26	0.21667	0.12287	1.27570	3.40747	-96.2	0.56708	2.24961	87	0.72500	0.80000	1.33333	0.00000	0.0	1.10345	1.20833
27	0.22500	0.13361	1.30070	2.58882	-100.1	0.59381	2.19046	88	0.73333	0.81111	1.33333	0.00000	0.0	1.10606	1.20548
28	0.23333	0.14453	1.31877	1.74180	-103.0	0.61940	2.12912	89	0.74167	0.82222	1.33333	0.00000	0.0	1.10861	1.20270
29	0.24167	0.15557	1.32968	0.87570	-104.7	0.64372	2.06562	90	0.75000	0.83333	1.33333	0.00000	0.0	1.11111	1.20000
30	0.25000	0.16667	1.33333	-0.00000	-105.3	0.66667	2.00000								
								91	0.75833	0.84443	1.32968	-0.87569	-104.7	1.11354	1.1941
31	0.25833	0.17778	1.33333	0.00000	0.0	0.68817	1.93750	92	0.76667	0.85547	1.31877	-1.74179	-103.0	1.11584	1.1818
32	0.26667	0.18889	1.33333	0.00000	0.0	0.70833	1.88235	93	0.77500	0.86639	1.30070	-2.58881	-100.1	1.11793	1.1635
33	0.27500	0.20000	1.33333	0.00000	0.0	0.72727	1.83333	94	0.78333	0.87713	1.27570	-3.40747	-96.2	1.11975	1.13927
34	0.28333	0.21111	1.33333	0.00000	0.0	0.74510	1.78947	95	0.79167	0.88764	1.24402	-4.18880	-91.2	1.12123	1.10951
35	0.29167	0.22222	1.33333	0.00000	0.0	0.76190	1.75000								
								96	0.80000	0.89785	1.20601	-4.92421	-85.2	1.12231	1.07458
36	0.30000	0.23333	1.33333	0.00000	0.0	0.77778	1.71429	97	0.80833	0.90772	1.16210	-5.60570	-78.2	1.12295	1.03486
37	0.30833	0.24444	1.33333	0.00000	0.0	0.79279	1.68182	98	0.81667	0.91720	1.11275	-6.22575	-70.4	1.12311	0.99078
38	0.31667	0.25556	1.33333	0.00000	0.0	0.80702	1.65217	99	0.82500	0.92625	1.05852	-6.77761	-61.9	1.12273	0.94281
39	0.32500	0.26667	1.33333	0.00000	0.0	0.82051	1.62500	100	0.83333	0.93483	1.00000	-7.25519	-52.6	1.12180	0.89143
40	0.33333	0.27778	1.33333	0.00000	0.0	0.83333	1.60000								
								101	0.84167	0.94291	0.93782	-7.65330	-42.8	1.12029	0.83713
41	0.34167	0.28889	1.33333	0.00000	0.0	0.84553	1.57692	102	0.85000	0.95046	0.87268	-7.96755	-32.5	1.11818	0.78044
42	0.35000	0.30000	1.33333	0.00000	0.0	0.85714	1.55556	103	0.85833	0.95745	0.80527	-8.19451	-21.9	1.11547	0.72191
43	0.35833	0.31111	1.33333	0.00000	0.0	0.86822	1.53571	104	0.86667	0.96387	0.73635	-8.33169	-11.0	1.11216	0.66209
44	0.36667	0.32222	1.33333	0.00000	0.0	0.87879	1.51724	105	0.87500	0.96972	0.66667	-8.37758	0.0	1.10825	0.60155
45	0.37500	0.33333	1.33333	0.00000	0.0	0.88889	1.50000								
								106	0.88333	0.97498	0.59698	-8.33169	11.0	1.10375	0.54086
46	0.38333	0.34444	1.33333	0.00000	0.0	0.89855	1.48387	107	0.89167	0.97967	0.52806	-8.19451	21.9	1.09870	0.48062
47	0.39167	0.35556	1.33333	0.00000	0.0	0.90780	1.46875	108	0.90000	0.98379	0.46066	-7.96755	32.5	1.09310	0.42142
48	0.40000	0.36667	1.33333	0.00000	0.0	0.91667	1.45455	109	0.90833	0.98735	0.39551	-7.65330	42.8	1.08700	0.36386
49	0.40833	0.37778	1.33333	0.00000	0.0	0.92517	1.44118	110	0.91667	0.99039	0.33333	-7.25519	52.6	1.08042	0.30852
50	0.41667	0.38889	1.33333	0.00000	0.0	0.93333	1.42857								
								111	0.92500	0.99292	0.27481	-6.77761	61.9	1.07343	0.25601
51	0.42500	0.40000	1.33333	0.00000	0.0	0.94118	1.41667	112	0.93333	0.99498	0.22058	-6.22575	70.4	1.06605	0.20691
52	0.43333	0.41111	1.33333	0.00000	0.0	0.94872	1.40541	113	0.94167	0.99661	0.17124	-5.60570	78.2	1.05835	0.16180
53	0.44167	0.42222	1.33333	0.00000	0.0	0.95597	1.39474	114	0.95000	0.99785	0.12732	-4.92421	85.2	1.05037	0.12122
54	0.45000	0.43333	1.33333	0.00000	0.0	0.96296	1.38462	115	0.95833	0.99870	0.08932	-4.18880	91.2	1.04217	0.08570
55	0.45833	0.44444	1.33333	0.00000	0.0	0.96970	1.37500								
								116	0.96667	0.99936	0.05764	-3.40747	96.2	1.03382	0.05575
56	0.46667	0.45556	1.33333	0.00000	0.0	0.97619	1.36585	117	0.97500	0.99973	0.03263	-2.58881	100.1	1.02536	0.03182
57	0.47500	0.46667	1.33333	0.00000	0.0	0.98246	1.35714	118	0.98333	0.99992	0.01457	-1.74180	103.0	1.01687	0.01433
58	0.48333	0.47778	1.33333	0.00000	0.0	0.98851	1.34884	119	0.99167	0.99999	0.00365	-0.87569	104.7	1.00839	0.00362
59	0.49167	0.48889	1.33333	0.00000	0.0	0.99435	1.34091	120	1.00000	1.00000	0.00000	0.00000	105.3	1.00000	0.00000
60	0.50000	0.50000	1.33333	0.00000	0.0	1.00000	1.33333								

Table B.11 Cam law factors – Polynomial 3–4–5

PNML 3-4-5 Factor table		POLYNOMIAL 3-4-5					120 STEPS
BASIC EQUATION $w = 10 u^3 - 15 u^4 + 6 u^5$		MOTION PARAMETERS:		Ca = 5.773503	Cd = 5.773503	Cv = 1.875	AF = .5

STEP NO.	INPUT DISP. u	OUTPUT DISP. w	OUTPUT VELOC. w'	OUTPUT ACCEL. w''	O'PUT JERK w'''	DISP. RATIO w/u	BLEND FACTOR z=u.w'/w
0	0.00000	0.00000	0.00000	0.00000	60.00		
1	0.00833	0.00001	0.00205	0.48757	57.03	0.00069	2.98743
2	0.01667	0.00005	0.00806	0.95056	54.10	0.00271	2.97471
3	0.02500	0.00015	0.01782	1.38938	51.22	0.00602	2.96183
4	0.03333	0.00035	0.03115	1.80444	48.40	0.01056	2.94881
5	0.04167	0.00068	0.04783	2.19618	45.63	0.01629	2.93563
6	0.05000	0.00116	0.06769	2.56500	42.90	0.02316	2.92229
7	0.05833	0.00182	0.09052	2.91132	40.22	0.03112	2.90879
8	0.06667	0.00267	0.11615	3.23556	37.60	0.04012	2.89513
9	0.07500	0.00376	0.14439	3.53813	35.03	0.05011	2.88130
10	0.08333	0.00509	0.17506	3.81944	32.50	0.06105	2.86730
11	0.09167	0.00668	0.20799	4.07993	30.02	0.07290	2.85313
12	0.10000	0.00856	0.24300	4.32000	27.60	0.08560	2.83878
13	0.10833	0.01074	0.27993	4.54007	25.23	0.09912	2.82426
14	0.11667	0.01323	0.31861	4.74056	22.90	0.11340	2.80956
15	0.12500	0.01605	0.35889	4.92188	20.63	0.12842	2.79468
16	0.13333	0.01922	0.40059	5.08444	18.40	0.14412	2.77960
17	0.14167	0.02273	0.44358	5.22868	16.23	0.16046	2.76434
18	0.15000	0.02661	0.48769	5.35500	14.10	0.17741	2.74889
19	0.15833	0.03086	0.53278	5.46382	12.03	0.19493	2.73324
20	0.16667	0.03549	0.57870	5.55556	10.00	0.21296	2.71739
21	0.17500	0.04051	0.62532	5.63063	8.03	0.23149	2.70134
22	0.18333	0.04592	0.67250	5.68944	6.10	0.25046	2.68508
23	0.19167	0.05172	0.72010	5.73243	4.23	0.26984	2.66861
24	0.20000	0.05792	0.76800	5.76000	2.40	0.28960	2.65193
25	0.20833	0.06452	0.81606	5.77257	0.63	0.30970	2.63504
26	0.21667	0.07152	0.86417	5.77056	-1.10	0.33010	2.61797
27	0.22500	0.07892	0.91220	5.75437	-2.77	0.35077	2.60058
28	0.23333	0.08672	0.96004	5.72444	-4.40	0.37167	2.58301
29	0.24167	0.09492	1.00757	5.68118	-5.97	0.39278	2.56521
30	0.25000	0.10352	1.05469	5.62500	-7.50	0.41406	2.54717
31	0.25833	0.11250	1.10128	5.55632	-8.97	0.43548	2.52889
32	0.26667	0.12187	1.14726	5.47555	-10.40	0.45701	2.51037
33	0.27500	0.13162	1.19251	5.38312	-11.78	0.47861	2.49161
34	0.28333	0.14174	1.23695	5.27944	-13.10	0.50026	2.47259
35	0.29167	0.15223	1.28047	5.16493	-14.37	0.52194	2.45331
36	0.30000	0.16308	1.32300	5.04000	-15.60	0.54360	2.43377
37	0.30833	0.17428	1.36444	4.90507	-16.77	0.56523	2.41397
38	0.31667	0.18582	1.40472	4.76056	-17.90	0.58679	2.39390
39	0.32500	0.19769	1.44376	4.60688	-18.97	0.60827	2.37356
40	0.33333	0.20988	1.48148	4.44444	-20.00	0.62964	2.35294
41	0.34167	0.22237	1.51781	4.27368	-20.98	0.65085	2.33204
42	0.35000	0.23517	1.55269	4.09500	-21.90	0.67191	2.31085
43	0.35833	0.24825	1.58604	3.90882	-22.77	0.69279	2.28937
44	0.36667	0.26160	1.61781	3.71555	-23.60	0.71345	2.26759
45	0.37500	0.27521	1.64795	3.51563	-24.38	0.73389	2.24551
46	0.38333	0.28906	1.67639	3.30945	-25.10	0.75407	2.22312
47	0.39167	0.30314	1.70309	3.09743	-25.78	0.77398	2.20043
48	0.40000	0.31744	1.72800	2.88000	-26.40	0.79360	2.17742
49	0.40833	0.33194	1.75108	2.65757	-26.98	0.81291	2.15409
50	0.41667	0.34662	1.77228	2.43056	-27.50	0.83189	2.13043
51	0.42500	0.36147	1.79157	2.19937	-27.98	0.85052	2.10645
52	0.43333	0.37647	1.80893	1.96444	-28.40	0.86879	2.08213
53	0.44167	0.39161	1.82431	1.72618	-28.78	0.88667	2.05748
54	0.45000	0.40687	1.83769	1.48500	-29.10	0.90416	2.03247
55	0.45833	0.42224	1.84905	1.24132	-29.37	0.92124	2.00713
56	0.46667	0.43768	1.85837	0.99555	-29.60	0.93790	1.98142
57	0.47500	0.45320	1.86564	0.74813	-29.78	0.95411	1.95537
58	0.48333	0.46877	1.87084	0.49945	-29.90	0.96988	1.92894
59	0.49167	0.48438	1.87396	0.24993	-29.97	0.98518	1.90216
60	0.50000	0.50000	1.87500	0.00000	-30.00	1.00000	1.87500
61	0.50833	0.51562	1.87396	-0.24993	-29.97	1.01434	1.84747
62	0.51667	0.53123	1.87084	-0.49944	-29.90	1.02818	1.81956
63	0.52500	0.54680	1.86564	-0.74812	-29.77	1.04152	1.79127
64	0.53333	0.56232	1.85837	-0.99556	-29.60	1.05434	1.76259
65	0.54167	0.57776	1.84905	-1.24132	-29.37	1.06664	1.73352
66	0.55000	0.59313	1.83769	-1.48500	-29.10	1.07841	1.70407
67	0.55833	0.60839	1.82431	-1.72618	-28.78	1.08965	1.67422
68	0.56667	0.62353	1.80893	-1.96444	-28.40	1.10034	1.64397
69	0.57500	0.63853	1.79157	-2.19937	-27.98	1.11049	1.61332
70	0.58333	0.65338	1.77228	-2.43056	-27.50	1.12008	1.58228
71	0.59167	0.66806	1.75108	-2.65757	-26.98	1.12912	1.55083
72	0.60000	0.68256	1.72800	-2.88000	-26.40	1.13760	1.51899
73	0.60833	0.69686	1.70309	-3.09743	-25.77	1.14552	1.48674
74	0.61667	0.71094	1.67639	-3.30945	-25.10	1.15288	1.45410
75	0.62500	0.72479	1.64795	-3.51563	-24.38	1.15967	1.42105
76	0.63333	0.73840	1.61781	-3.71555	-23.60	1.16590	1.38761
77	0.64167	0.75175	1.58604	-3.90882	-22.77	1.17156	1.35379
78	0.65000	0.76483	1.55269	-4.09500	-21.90	1.17666	1.31957
79	0.65833	0.77763	1.51781	-4.27368	-20.97	1.18120	1.28497
80	0.66667	0.79012	1.48148	-4.44444	-20.00	1.18519	1.25000
81	0.67500	0.80231	1.44376	-4.60688	-18.97	1.18861	1.21466
82	0.68333	0.81418	1.40472	-4.76056	-17.90	1.19149	1.17897
83	0.69167	0.82572	1.36444	-4.90507	-16.77	1.19381	1.14293
84	0.70000	0.83692	1.32300	-5.04000	-15.60	1.19560	1.10656
85	0.70833	0.84777	1.28047	-5.16493	-14.38	1.19685	1.06987
86	0.71667	0.85826	1.23695	-5.27944	-13.10	1.19757	1.03288
87	0.72500	0.86838	1.19251	-5.38312	-11.77	1.19777	0.99561
88	0.73333	0.87813	1.14726	-5.47556	-10.40	1.19745	0.95808
89	0.74167	0.88750	1.10128	-5.55632	-8.97	1.19663	0.92032
90	0.75000	0.89648	1.05469	-5.62500	-7.50	1.19531	0.88235
91	0.75833	0.90508	1.00757	-5.68118	-5.98	1.19351	0.84421
92	0.76667	0.91328	0.96004	-5.72444	-4.40	1.19123	0.80592
93	0.77500	0.92108	0.91220	-5.75438	-2.78	1.18849	0.76753
94	0.78333	0.92848	0.86417	-5.77056	-1.10	1.18529	0.72908
95	0.79167	0.93548	0.81606	-5.77257	0.63	1.18166	0.69061
96	0.80000	0.94208	0.76800	-5.76000	2.40	1.17760	0.65217
97	0.80833	0.94828	0.72010	-5.73243	4.23	1.17313	0.61383
98	0.81667	0.95408	0.67250	-5.68944	6.10	1.16826	0.57564
99	0.82500	0.95949	0.62532	-5.63063	8.02	1.16302	0.53767
100	0.83333	0.96451	0.57870	-5.55556	10.00	1.15741	0.50000
101	0.84167	0.96914	0.53278	-5.46382	12.02	1.15145	0.46270
102	0.85000	0.97339	0.48769	-5.35500	14.10	1.14516	0.42587
103	0.85833	0.97727	0.44358	-5.22868	16.23	1.13856	0.38959
104	0.86667	0.98078	0.40059	-5.08444	18.40	1.13167	0.35398
105	0.87500	0.98395	0.35889	-4.92188	20.63	1.12451	0.31915
106	0.88333	0.98677	0.31861	-4.74056	22.90	1.11710	0.28522
107	0.89167	0.98926	0.27993	-4.54007	25.22	1.10945	0.25231
108	0.90000	0.99144	0.24300	-4.32000	27.60	1.10160	0.22059
109	0.90833	0.99332	0.20799	-4.07993	30.03	1.09356	0.19019
110	0.91667	0.99491	0.17506	-3.81944	32.50	1.08536	0.16129
111	0.92500	0.99624	0.14439	-3.53813	35.03	1.07702	0.13406
112	0.93333	0.99733	0.11615	-3.23556	37.60	1.06856	0.10870
113	0.94167	0.99818	0.09052	-2.91132	40.22	1.06002	0.08540
114	0.95000	0.99884	0.06769	-2.56500	42.90	1.05141	0.06438
115	0.95833	0.99932	0.04783	-2.19618	45.62	1.04277	0.04567
116	0.96667	0.99965	0.03115	-1.80445	48.40	1.03412	0.03012
117	0.97500	0.99985	0.01782	-1.38937	51.23	1.02549	0.01738
118	0.98333	0.99995	0.00806	-0.95055	54.10	1.01690	0.00792
119	0.99167	0.99999	0.00205	-0.48757	57.03	1.00840	0.00203
120	1.00000	1.00000	0.00000	0.00000	60.00	1.00000	0.00000

Table B.12 Cam law factors – Reflective polynomial 3–4–5–6–R

PNML 3-4-5-6-R Factor table   REFLECTIVE POLYNOMIAL 3-4-5-6-R   120 STEPS

BASIC EQUATION $w = (40/3)\,u^3 - 40\,u^4 + 64\,u^5 - (128/3)\,u^6$ MOTION PARAMETERS: Ca = 5 Cd = 5 Cv = 2 AF = .5

STEP NO.	INPUT DISP. u	OUTPUT DISP. w	OUTPUT VELOC. w'	OUTPUT ACCEL. w''	O'PUT JERK w'''	DISP. RATIO w/u	BLEND FACTOR z=u.w'/w
0	0.00000	0.00000	0.00000	0.00000	80.00		
1	0.00833	0.00001	0.00269	0.63407	72.26	0.00090	2.97505
2	0.01667	0.00006	0.01039	1.20583	65.04	0.00352	2.95020
3	0.02500	0.00019	0.02262	1.71950	58.32	0.00773	2.92548
4	0.03333	0.00045	0.03890	2.17916	52.08	0.01341	2.90092
5	0.04167	0.00085	0.05880	2.58873	46.30	0.02044	2.87654
6	0.05000	0.00144	0.08192	2.95200	40.96	0.02872	2.85237
7	0.05833	0.00222	0.10788	3.27259	36.05	0.03814	2.82843
8	0.06667	0.00324	0.13635	3.55398	31.55	0.04862	2.80475
9	0.07500	0.00450	0.16702	3.79950	27.44	0.06005	2.78136
10	0.08333	0.00603	0.19959	4.01235	23.70	0.07236	2.75829
11	0.09167	0.00783	0.23381	4.19555	20.32	0.08547	2.73557
12	0.10000	0.00993	0.26944	4.35200	17.28	0.09931	2.71321
13	0.10833	0.01233	0.30627	4.48444	14.56	0.11380	2.69126
14	0.11667	0.01504	0.34412	4.59546	12.14	0.12890	2.66973
15	0.12500	0.01807	0.38281	4.68750	10.00	0.14453	2.64865
16	0.13333	0.02142	0.42220	4.76286	8.13	0.16065	2.62805
17	0.14167	0.02510	0.46215	4.82370	6.51	0.17721	2.60795
18	0.15000	0.02912	0.50256	4.87200	5.12	0.19416	2.58838
19	0.15833	0.03348	0.54332	4.90962	3.94	0.21146	2.56936
20	0.16667	0.03818	0.58436	4.93827	2.96	0.22908	2.55090
21	0.17500	0.04322	0.62561	4.95950	2.16	0.24698	2.53302
22	0.18333	0.04861	0.66700	4.97472	1.52	0.26513	2.51575
23	0.19167	0.05434	0.70851	4.98518	1.02	0.28351	2.49909
24	0.20000	0.06042	0.75008	4.99200	0.64	0.30208	2.48305
25	0.20833	0.06684	0.79170	4.99614	0.37	0.32083	2.46764
26	0.21667	0.07361	0.83334	4.99842	0.19	0.33974	2.45286
27	0.22500	0.08073	0.87500	4.99950	0.08	0.35880	2.43872
28	0.23333	0.08819	0.91667	4.99990	0.02	0.37798	2.42520
29	0.24167	0.09601	0.95833	4.99999	0.00	0.39727	2.41230
30	0.25000	0.10417	1.00000	5.00000	-0.00	0.41667	2.40000
31	0.25833	0.11267	1.04167	4.99999	-0.00	0.43616	2.38829
32	0.26667	0.12153	1.08333	4.99990	-0.02	0.45573	2.37714
33	0.27500	0.13073	1.12500	4.99950	-0.08	0.47538	2.36653
34	0.28333	0.14028	1.16666	4.99842	-0.19	0.49510	2.35642
35	0.29167	0.15017	1.20830	4.99614	-0.37	0.51488	2.34676
36	0.30000	0.16042	1.24992	4.99200	-0.64	0.53472	2.33752
37	0.30833	0.17101	1.29149	4.98518	-1.02	0.55461	2.32864
38	0.31667	0.18194	1.33300	4.97471	-1.52	0.57455	2.32007
39	0.32500	0.19322	1.37439	4.95950	-2.16	0.59453	2.31174
40	0.33333	0.20485	1.41564	4.93827	-2.96	0.61454	2.30357
41	0.34167	0.21681	1.45668	4.90962	-3.94	0.63458	2.29550
42	0.35000	0.22912	1.49744	4.87200	-5.12	0.65464	2.28743
43	0.35833	0.24177	1.53785	4.82370	-6.51	0.67471	2.27927
44	0.36667	0.25475	1.57780	4.76287	-8.13	0.69478	2.27093
45	0.37500	0.26807	1.61719	4.68750	-10.00	0.71484	2.26230
46	0.38333	0.28170	1.65588	4.59545	-12.14	0.73488	2.25326
47	0.39167	0.29566	1.69373	4.48444	-14.56	0.75488	2.24370
48	0.40000	0.30993	1.73056	4.35200	-17.28	0.77483	2.23348
49	0.40833	0.32450	1.76619	4.19555	-20.32	0.79470	2.22247
50	0.41667	0.33936	1.80041	4.01235	-23.70	0.81447	2.21053
51	0.42500	0.35450	1.83298	3.79950	-27.44	0.83413	2.19749
52	0.43333	0.36991	1.86365	3.55397	-31.55	0.85363	2.18319
53	0.44167	0.38556	1.89212	3.27258	-36.05	0.87296	2.16747
54	0.45000	0.40144	1.91808	2.95200	-40.96	0.89208	2.15012
55	0.45833	0.41752	1.94120	2.58873	-46.30	0.91095	2.13096
56	0.46667	0.43378	1.96110	2.17915	-52.08	0.92953	2.10977
57	0.47500	0.45019	1.97738	1.71950	-58.32	0.94778	2.08634
58	0.48333	0.46673	1.98960	1.20583	-65.04	0.96564	2.06040
59	0.49167	0.48334	1.99731	0.63406	-72.26	0.98307	2.03172
60	0.50000	0.50000	2.00000	-0.00000	-80.00	1.00000	2.00000
61	0.50833	0.51666	1.99731	-0.63406	-72.26	1.01638	1.96513
62	0.51667	0.53327	1.98961	-1.20582	-65.04	1.03214	1.92764
63	0.52500	0.54981	1.97738	-1.71950	-58.32	1.04725	1.88816
64	0.53333	0.56622	1.96110	-2.17916	-52.08	1.06166	1.84719
65	0.54167	0.58248	1.94120	-2.58873	-46.30	1.07535	1.80518
66	0.55000	0.59856	1.91808	-2.95200	-40.96	1.08830	1.76246
67	0.55833	0.61444	1.89212	-3.27258	-36.05	1.10049	1.71933
68	0.56667	0.63009	1.86365	-3.55397	-31.55	1.11193	1.67605
69	0.57500	0.64550	1.83298	-3.79950	-27.44	1.12260	1.63280
70	0.58333	0.66064	1.80041	-4.01235	-23.70	1.13252	1.58974
71	0.59167	0.67550	1.76619	-4.19555	-20.32	1.14169	1.54700
72	0.60000	0.69007	1.73056	-4.35200	-17.28	1.15012	1.50468
73	0.60833	0.70434	1.69373	-4.48444	-14.56	1.15782	1.46286
74	0.61667	0.71830	1.65588	-4.59545	-12.14	1.16480	1.42160
75	0.62500	0.73193	1.61719	-4.68750	-10.00	1.17109	1.38092
76	0.63333	0.74525	1.57780	-4.76287	-8.13	1.17671	1.34086
77	0.64167	0.75823	1.53785	-4.82370	-6.51	1.18166	1.30143
78	0.65000	0.77088	1.49744	-4.87200	-5.12	1.18596	1.26264
79	0.65833	0.78319	1.45668	-4.90962	-3.94	1.18965	1.22446
80	0.66667	0.79515	1.41564	-4.93827	-2.96	1.19273	1.18689
81	0.67500	0.80678	1.37439	-4.95950	-2.16	1.19523	1.14990
82	0.68333	0.81806	1.33300	-4.97471	-1.52	1.19716	1.11347
83	0.69167	0.82899	1.29149	-4.98518	-1.02	1.19855	1.07755
84	0.70000	0.83958	1.24992	-4.99200	-0.64	1.19941	1.04212
85	0.70833	0.84983	1.20830	-4.99614	-0.37	1.19976	1.00712
86	0.71667	0.85972	1.16666	-4.99842	-0.19	1.19961	0.97253
87	0.72500	0.86927	1.12500	-4.99950	-0.08	1.19899	0.93828
88	0.73333	0.87847	1.08333	-4.99990	-0.02	1.19792	0.90435
89	0.74167	0.88733	1.04167	-4.99999	-0.00	1.19640	0.87067
90	0.75000	0.89583	1.00000	-5.00000	0.00	1.19444	0.83721
91	0.75833	0.90399	0.95833	-4.99999	0.00	1.19208	0.80392
92	0.76667	0.91181	0.91667	-4.99990	0.02	1.18931	0.77075
93	0.77500	0.91927	0.87500	-4.99950	0.08	1.18616	0.73768
94	0.78333	0.92639	0.83334	-4.99842	0.19	1.18262	0.70466
95	0.79167	0.93316	0.79170	-4.99614	0.37	1.17873	0.67165
96	0.80000	0.93958	0.75008	-4.99200	0.64	1.17448	0.63865
97	0.80833	0.94566	0.70851	-4.98518	1.02	1.16989	0.60562
98	0.81667	0.95139	0.66700	-4.97472	1.52	1.16497	0.57255
99	0.82500	0.95678	0.62561	-4.95950	2.16	1.15973	0.53944
100	0.83333	0.96182	0.58436	-4.93827	2.96	1.15418	0.50630
101	0.84167	0.96652	0.54332	-4.90962	3.94	1.14834	0.47314
102	0.85000	0.97088	0.50256	-4.87200	5.12	1.14221	0.43999
103	0.85833	0.97490	0.46215	-4.82370	6.51	1.13580	0.40690
104	0.86667	0.97858	0.42220	-4.76286	8.13	1.12913	0.37392
105	0.87500	0.98193	0.38281	-4.68750	10.00	1.12221	0.34112
106	0.88333	0.98496	0.34412	-4.59546	12.14	1.11505	0.30861
107	0.89167	0.98767	0.30627	-4.48444	14.56	1.10767	0.27650
108	0.90000	0.99007	0.26944	-4.35200	17.28	1.10008	0.24493
109	0.90833	0.99217	0.23381	-4.19555	20.32	1.09229	0.21405
110	0.91667	0.99397	0.19959	-4.01234	23.70	1.08433	0.18407
111	0.92500	0.99550	0.16702	-3.79950	27.44	1.07621	0.15519
112	0.93333	0.99676	0.13635	-3.55398	31.55	1.06796	0.12768
113	0.94167	0.99778	0.10788	-3.27258	36.05	1.05958	0.10182
114	0.95000	0.99856	0.08192	-2.95200	40.96	1.05112	0.07794
115	0.95833	0.99915	0.05880	-2.58874	46.30	1.04259	0.05640
116	0.96667	0.99955	0.03890	-2.17915	52.08	1.03402	0.03762
117	0.97500	0.99981	0.02262	-1.71950	58.32	1.02544	0.02206
118	0.98333	0.99994	0.01039	-1.20583	65.04	1.01689	0.01022
119	0.99167	0.99999	0.00269	-0.63407	72.26	1.00840	0.00266
120	1.00000	1.00000	0.00000	0.00000	80.00	1.00000	0.00000

Table B.13   Cam law factors – Low impact reflective polynomial

PNML 3-4-5-6-7-R Factor table				LOW IMPACT REFLECTIVE POLYNOMIAL			120 STEPS
BASIC EQUATION w = $27u^3 - 113u^4 + 224.4u^5 - 228.8u^6 + 96u^7$				MOTION PARAMETERS: $C_a = C_d = 6.48192$		$Cv = 1.475$	$AF = .5$

STEP NO.	INPUT DISP. u	OUTPUT DISP. w	OUTPUT VELOC. w'	OUTPUT ACCEL. w''	O'PUT JERK w'''	DISP. RATIO w/u	BLEND FACTOR z=u.w'/w
0	0.00000	0.00000	0.00000	0.00000	162.00		
1	0.00833	0.00002	0.00537	1.25840	140.32	0.00181	2.96506
2	0.01667	0.00012	0.02049	2.34359	120.41	0.00699	2.93003
3	0.02500	0.00038	0.04399	3.26998	102.19	0.01519	2.89490
4	0.03333	0.00087	0.07459	4.05125	85.57	0.02608	2.85972
5	0.04167	0.00164	0.11114	4.70030	70.45	0.03935	2.82450
6	0.05000	0.00274	0.15259	5.22936	56.75	0.05471	2.78927
7	0.05833	0.00419	0.19800	5.64993	44.40	0.07189	2.75405
8	0.06667	0.00604	0.24649	5.97284	33.30	0.09066	2.71887
9	0.07500	0.00831	0.29730	6.20826	23.39	0.11078	2.68376
10	0.08333	0.01100	0.34974	6.36574	14.58	0.13204	2.64875
11	0.09167	0.01414	0.40320	6.45419	6.81	0.15426	2.61386
12	0.10000	0.01772	0.45714	6.48192	-0.00	0.17725	2.57912
13	0.10833	0.02176	0.51109	6.45667	-5.92	0.20085	2.54457
14	0.11667	0.02624	0.56463	6.38562	-11.00	0.22493	2.51024
15	0.12500	0.03117	0.61741	6.27539	-15.33	0.24934	2.47616
16	0.13333	0.03653	0.66913	6.13210	-18.95	0.27397	2.44236
17	0.14167	0.04232	0.71953	5.96136	-21.93	0.29870	2.40887
18	0.15000	0.04852	0.76842	5.76828	-24.32	0.32345	2.37573
19	0.15833	0.05512	0.81562	5.55753	-26.18	0.34812	2.34296
20	0.16667	0.06211	0.86101	5.33333	-27.56	0.37263	2.31060
21	0.17500	0.06946	0.90448	5.09947	-28.50	0.39693	2.27868
22	0.18333	0.07717	0.94598	4.85933	-29.07	0.42095	2.24723
23	0.19167	0.08522	0.98546	4.61591	-29.30	0.44465	2.21628
24	0.20000	0.09359	1.02291	4.37184	-29.23	0.46797	2.18586
25	0.20833	0.10227	1.05833	4.12941	-28.91	0.49088	2.15598
26	0.21667	0.11123	1.09175	3.89058	-28.37	0.51335	2.12669
27	0.22500	0.12046	1.12319	3.65700	-27.66	0.53537	2.09799
28	0.23333	0.12994	1.15271	3.43002	-26.79	0.55689	2.06991
29	0.24167	0.13966	1.18038	3.21074	-25.82	0.57792	2.04246
30	0.25000	0.14961	1.20625	3.00000	-24.75	0.59844	2.01567
31	0.25833	0.15976	1.23040	2.79841	-23.62	0.61844	1.98953
32	0.26667	0.17011	1.25292	2.60636	-22.46	0.63792	1.96407
33	0.27500	0.18064	1.27387	2.42407	-21.29	0.65688	1.93928
34	0.28333	0.19134	1.29335	2.25158	-20.12	0.67531	1.91518
35	0.29167	0.20219	1.31142	2.08876	-18.97	0.69323	1.89175
36	0.30000	0.21319	1.32818	1.93535	-17.86	0.71064	1.86900
37	0.30833	0.22433	1.34370	1.79101	-16.79	0.72754	1.84691
38	0.31667	0.23558	1.35806	1.65527	-15.79	0.74395	1.82547
39	0.32500	0.24696	1.37131	1.52760	-14.86	0.75987	1.80468
40	0.33333	0.25844	1.38354	1.40739	-14.00	0.77531	1.78450
41	0.34167	0.27001	1.39479	1.29405	-13.22	0.79028	1.76493
42	0.35000	0.28168	1.40512	1.18691	-12.51	0.80480	1.74593
43	0.35833	0.29343	1.41459	1.08534	-11.88	0.81887	1.72748
44	0.36667	0.30525	1.42323	0.98872	-11.32	0.83251	1.70956
45	0.37500	0.31715	1.43108	0.89647	-10.83	0.84573	1.69213
46	0.38333	0.32910	1.43818	0.80809	-10.39	0.85853	1.67516
47	0.39167	0.34112	1.44456	0.72313	-10.00	0.87093	1.65863
48	0.40000	0.35318	1.45024	0.64127	-9.65	0.88294	1.64250
49	0.40833	0.36528	1.45525	0.56227	-9.31	0.89457	1.62676
50	0.41667	0.37743	1.45962	0.48609	-8.97	0.90583	1.61136
51	0.42500	0.38961	1.46336	0.41279	-8.61	0.91673	1.59629
52	0.43333	0.40182	1.46651	0.34263	-8.21	0.92727	1.58153
53	0.44167	0.41405	1.46908	0.27610	-7.75	0.93747	1.56707
54	0.45000	0.42630	1.47112	0.21383	-7.18	0.94734	1.55290
55	0.45833	0.43857	1.47266	0.15675	-6.49	0.95687	1.53903
56	0.46667	0.45084	1.47375	0.10601	-5.65	0.96609	1.52547
57	0.47500	0.46313	1.47445	0.06307	-4.62	0.97501	1.51225
58	0.48333	0.47542	1.47483	0.02967	-3.36	0.98362	1.49938
59	0.49167	0.48771	1.47498	0.00783	-1.83	0.99195	1.48695
60	0.50000	0.50000	1.47500	-0.00001	0.00	1.00000	1.47500

STEP NO.	INPUT DISP. u	OUTPUT DISP. w	OUTPUT VELOC. w'	OUTPUT ACCEL. w''	O'PUT JERK w'''	DISP. RATIO w/u	BLEND FACTOR z=u.w'/w
61	0.50833	0.51229	1.47498	-0.00783	-1.83	1.00779	1.46358
62	0.51667	0.52458	1.47483	-0.02968	-3.36	1.01532	1.45257
63	0.52500	0.53687	1.47445	-0.06307	-4.62	1.02261	1.44184
64	0.53333	0.54916	1.47375	-0.10601	-5.65	1.02967	1.43129
65	0.54167	0.56143	1.47266	-0.15674	-6.50	1.03649	1.42081
66	0.55000	0.57370	1.47112	-0.21383	-7.18	1.04309	1.41035
67	0.55833	0.58595	1.46908	-0.27610	-7.75	1.04946	1.39984
68	0.56667	0.59818	1.46651	-0.34263	-8.21	1.05562	1.38924
69	0.57500	0.61039	1.46336	-0.41279	-8.61	1.06155	1.37852
70	0.58333	0.62257	1.45962	-0.48610	-8.97	1.06726	1.36763
71	0.59167	0.63472	1.45525	-0.56228	-9.31	1.07276	1.35655
72	0.60000	0.64682	1.45024	-0.64127	-9.65	1.07804	1.34526
73	0.60833	0.65888	1.44456	-0.72313	-10.00	1.08310	1.33373
74	0.61667	0.67090	1.43818	-0.80809	-10.39	1.08794	1.32193
75	0.62500	0.68285	1.43108	-0.89647	-10.83	1.09256	1.30984
76	0.63333	0.69475	1.42323	-0.98872	-11.32	1.09697	1.29742
77	0.64167	0.70657	1.41459	-1.08534	-11.88	1.10115	1.28465
78	0.65000	0.71832	1.40512	-1.18691	-12.51	1.10511	1.27148
79	0.65833	0.72999	1.39479	-1.29405	-13.22	1.10884	1.25788
80	0.66667	0.74156	1.38354	-1.40741	-14.00	1.11235	1.24380
81	0.67500	0.75304	1.37131	-1.52760	-14.86	1.11562	1.22920
82	0.68333	0.76442	1.35806	-1.65527	-15.79	1.11866	1.21401
83	0.69167	0.77567	1.34370	-1.79101	-16.79	1.12146	1.19818
84	0.70000	0.78681	1.32818	-1.93535	-17.86	1.12401	1.18165
85	0.70833	0.79781	1.31142	-2.08875	-18.97	1.12632	1.16435
86	0.71667	0.80866	1.29335	-2.25158	-20.12	1.12836	1.14621
87	0.72500	0.81936	1.27387	-2.42408	-21.29	1.13015	1.12717
88	0.73333	0.82989	1.25292	-2.60636	-22.46	1.13167	1.10714
89	0.74167	0.84024	1.23040	-2.79841	-23.62	1.13290	1.08606
90	0.75000	0.85039	1.20625	-3.00000	-24.75	1.13385	1.06385
91	0.75833	0.86034	1.18038	-3.21074	-25.82	1.13451	1.04043
92	0.76667	0.87006	1.15271	-3.43002	-26.79	1.13486	1.01573
93	0.77500	0.87954	1.12319	-3.65700	-27.66	1.13489	0.98969
94	0.78333	0.88877	1.09175	-3.89058	-28.37	1.13460	0.96223
95	0.79167	0.89773	1.05833	-4.12941	-28.91	1.13398	0.93329
96	0.80000	0.90641	1.02291	-4.37184	-29.23	1.13301	0.90283
97	0.80833	0.91478	0.98546	-4.61591	-29.30	1.13168	0.87080
98	0.81667	0.92283	0.94598	-4.85933	-29.07	1.12999	0.83716
99	0.82500	0.93054	0.90448	-5.09947	-28.50	1.12792	0.80190
100	0.83333	0.93789	0.86101	-5.33333	-27.56	1.12547	0.76502
101	0.84167	0.94488	0.81562	-5.55753	-26.18	1.12263	0.72653
102	0.85000	0.95148	0.76842	-5.76828	-24.32	1.11939	0.68646
103	0.85833	0.95768	0.71953	-5.96136	-21.93	1.11575	0.64489
104	0.86667	0.96347	0.66913	-6.13210	-18.95	1.11170	0.60190
105	0.87500	0.96883	0.61741	-6.27539	-15.33	1.10724	0.55761
106	0.88333	0.97376	0.56463	-6.38562	-11.00	1.10237	0.51220
107	0.89167	0.97824	0.51109	-6.45667	-5.92	1.09709	0.46586
108	0.90000	0.98228	0.45714	-6.48192	-0.00	1.09142	0.41885
109	0.90833	0.98586	0.40320	-6.45419	6.81	1.08535	0.37150
110	0.91667	0.98900	0.34974	-6.36574	14.58	1.07891	0.32416
111	0.92500	0.99169	0.29730	-6.20826	23.39	1.07210	0.27731
112	0.93333	0.99396	0.24649	-5.97284	33.30	1.06495	0.23145
113	0.94167	0.99581	0.19800	-5.64993	44.40	1.05749	0.18723
114	0.95000	0.99726	0.15259	-5.22936	56.75	1.04975	0.14536
115	0.95833	0.99836	0.11114	-4.70031	70.45	1.04177	0.10669
116	0.96667	0.99913	0.07459	-4.05125	85.57	1.03358	0.07217
117	0.97500	0.99962	0.04399	-3.26998	102.19	1.02525	0.04290
118	0.98333	0.99988	0.02049	-2.34359	120.41	1.01683	0.02015
119	0.99167	0.99998	0.00537	-1.25840	140.32	1.00839	0.00532
120	1.00000	1.00000	0.00000	0.00000	162.00	1.00000	0.00000

Table B.14   Cam law factors – Double harmonic

DH Factor table		DOUBLE HARMONIC					120 STEPS

MOTION PARAMETERS: Ca = 5.551653   Cd =-9.869605   Cv = 2.040524

STEP NO.	INPUT DISP. u	OUTPUT DISP. w	OUTPUT VELOC. w'	OUTPUT ACCEL. w''	O'PUT JERK w'''	DISP. RATIO w/u	BLEND FACTOR z=u.w'/w	STEP NO.	INPUT DISP. u	OUTPUT DISP. w	OUTPUT VELOC. w'	OUTPUT ACCEL. w''	O'PUT JERK w'''	DISP. RATIO w/u	BLEND FACTOR z=u.w'/w
0	0.00000	0.00000	0.00000	0.00000	0.00			61	0.50833	0.26326	1.61136	4.79886	-17.12	0.51789	3.11141
1	0.00833	0.00000	0.00001	0.00507	1.22	0.00000	3.99977	62	0.51667	0.27685	1.65074	4.64950	-18.72	0.53584	3.08064
2	0.01667	0.00000	0.00011	0.02027	2.43	0.00003	3.99909	63	0.52500	0.29077	1.68882	4.48687	-20.31	0.55384	3.04926
3	0.02500	0.00000	0.00038	0.04554	3.63	0.00010	3.99794	64	0.53333	0.30500	1.72549	4.31114	-21.86	0.57187	3.01728
4	0.03333	0.00001	0.00090	0.08080	4.83	0.00023	3.99634	65	0.54167	0.31952	1.76063	4.12253	-23.40	0.58989	2.98469
5	0.04167	0.00002	0.00175	0.12593	6.00	0.00044	3.99429								
6	0.05000	0.00004	0.00303	0.18077	7.16	0.00076	3.99177	66	0.55000	0.33434	1.79416	3.92130	-24.89	0.60788	2.95149
7	0.05833	0.00007	0.00479	0.24513	8.29	0.00120	3.98880	67	0.55833	0.34942	1.82595	3.70774	-26.36	0.62583	2.91767
8	0.06667	0.00012	0.00714	0.31880	9.39	0.00179	3.98537	68	0.56667	0.36476	1.85592	3.48216	-27.78	0.64370	2.88321
9	0.07500	0.00019	0.01013	0.40151	10.46	0.00254	3.98148	69	0.57500	0.38035	1.88396	3.24493	-29.15	0.66147	2.84813
10	0.08333	0.00029	0.01385	0.49299	11.49	0.00348	3.97713	70	0.58333	0.39616	1.90997	2.99644	-30.48	0.67913	2.81240
11	0.09167	0.00042	0.01837	0.59291	12.48	0.00462	3.97232	71	0.59167	0.41217	1.93387	2.73711	-31.75	0.69663	2.77603
12	0.10000	0.00060	0.02376	0.70094	13.43	0.00599	3.96705	72	0.60000	0.42838	1.95556	2.46740	-32.97	0.71397	2.73900
13	0.10833	0.00082	0.03008	0.81669	14.34	0.00759	3.96131	73	0.60833	0.44476	1.97496	2.18779	-34.13	0.73111	2.70131
14	0.11667	0.00110	0.03739	0.93976	15.19	0.00945	3.95512	74	0.61667	0.46129	1.99200	1.89880	-35.22	0.74804	2.66296
15	0.12500	0.00145	0.04576	1.06973	15.99	0.01159	3.94846	75	0.62500	0.47795	2.00659	1.60096	-36.25	0.76473	2.62393
16	0.13333	0.00187	0.05524	1.20614	16.74	0.01401	3.94134	76	0.63333	0.49473	2.01866	1.29486	-37.20	0.78115	2.58422
17	0.14167	0.00237	0.06588	1.34851	17.42	0.01675	3.93376	77	0.64167	0.51159	2.02815	0.98108	-38.09	0.79728	2.54382
18	0.15000	0.00297	0.07773	1.49634	18.05	0.01980	3.92570	78	0.65000	0.52852	2.03499	0.66025	-38.90	0.81311	2.50272
19	0.15833	0.00367	0.09083	1.64910	18.61	0.02319	3.91718	79	0.65833	0.54550	2.03913	0.33300	-39.63	0.82861	2.46092
20	0.16667	0.00449	0.10522	1.80626	19.10	0.02692	3.90819	80	0.66667	0.56250	2.04052	-0.00000	-40.28	0.84375	2.41840
21	0.17500	0.00543	0.12094	1.96726	19.53	0.03102	3.89874	81	0.67500	0.57950	2.03912	-0.33808	-40.85	0.85852	2.37516
22	0.18333	0.00651	0.13802	2.13151	19.88	0.03549	3.88881	82	0.68333	0.59648	2.03488	-0.68052	-41.33	0.87289	2.33119
23	0.19167	0.00773	0.15648	2.29842	20.17	0.04035	3.87841	83	0.69167	0.61341	2.02777	-1.02663	-41.72	0.88685	2.28647
24	0.20000	0.00912	0.17633	2.46740	20.38	0.04559	3.86753	84	0.70000	0.63027	2.01776	-1.37567	-42.03	0.90038	2.24101
25	0.20833	0.01068	0.19760	2.63782	20.51	0.05124	3.85618	85	0.70833	0.64703	2.00483	-1.72690	-42.25	0.91345	2.19479
26	0.21667	0.01242	0.22030	2.80906	20.57	0.05730	3.84435	86	0.71667	0.66367	1.98897	-2.07957	-42.38	0.92605	2.14780
27	0.22500	0.01435	0.24442	2.98048	20.56	0.06378	3.83205	87	0.72500	0.68017	1.97017	-2.43293	-42.41	0.93817	2.10003
28	0.23333	0.01649	0.26997	3.15145	20.46	0.07069	3.81926	88	0.73333	0.69650	1.94842	-2.78620	-42.36	0.94977	2.05147
29	0.24167	0.01885	0.29694	3.32131	20.29	0.07802	3.80599	89	0.74167	0.71264	1.92374	-3.13863	-42.21	0.96086	2.00211
30	0.25000	0.02145	0.32532	3.48943	20.04	0.08579	3.79224	90	0.75000	0.72855	1.89612	-3.48943	-41.97	0.97140	1.95194
31	0.25833	0.02428	0.35509	3.65516	19.72	0.09399	3.77800	91	0.75833	0.74423	1.86559	-3.83785	-41.64	0.98140	1.90094
32	0.26667	0.02737	0.38623	3.81786	19.32	0.10263	3.76327	92	0.76667	0.75964	1.83216	-4.18310	-41.21	0.99083	1.84912
33	0.27500	0.03072	0.41871	3.97687	18.84	0.11172	3.74805	93	0.77500	0.77476	1.79588	-4.52443	-40.69	0.99969	1.79644
34	0.28333	0.03435	0.45250	4.13158	18.28	0.12124	3.73234	94	0.78333	0.78956	1.75677	-4.86107	-40.09	1.00795	1.74291
35	0.29167	0.03827	0.48756	4.28134	17.65	0.13120	3.71614	95	0.79167	0.80403	1.71488	-5.19227	-39.39	1.01562	1.68851
36	0.30000	0.04248	0.52384	4.42554	16.95	0.14160	3.69943	96	0.80000	0.81814	1.67025	-5.51728	-38.60	1.02267	1.63322
37	0.30833	0.04700	0.56130	4.56358	16.17	0.15244	3.68223	97	0.80833	0.83186	1.62294	-5.83538	-37.73	1.02910	1.57704
38	0.31667	0.05184	0.59988	4.69485	15.32	0.16370	3.66453	98	0.81667	0.84518	1.57301	-6.14584	-36.77	1.03491	1.51995
39	0.32500	0.05700	0.63953	4.81878	14.41	0.17539	3.64632	99	0.82500	0.85807	1.52053	-6.44797	-35.73	1.04008	1.46193
40	0.33333	0.06250	0.68017	4.93480	13.43	0.18750	3.62760	100	0.83333	0.87051	1.46557	-6.74107	-34.60	1.04462	1.40298
41	0.34167	0.06834	0.72175	5.04237	12.38	0.20002	3.60837	101	0.84167	0.88249	1.40821	-7.02448	-33.40	1.04850	1.34307
42	0.35000	0.07453	0.76419	5.14096	11.27	0.21295	3.58863	102	0.85000	0.89398	1.34853	-7.29755	-32.12	1.05174	1.28219
43	0.35833	0.08108	0.80741	5.23006	10.10	0.22627	3.56837	103	0.85833	0.90496	1.28661	-7.55965	-30.77	1.05432	1.22033
44	0.36667	0.08799	0.85133	5.30919	8.88	0.23997	3.54759	104	0.86667	0.91541	1.22256	-7.81019	-29.35	1.05625	1.15746
45	0.37500	0.09527	0.89587	5.37790	7.60	0.25405	3.52629	105	0.87500	0.92533	1.15648	-8.04860	-27.86	1.05752	1.09358
46	0.38333	0.10292	0.94093	5.43575	6.27	0.26849	3.50447	106	0.88333	0.93468	1.08846	-8.27431	-26.30	1.05813	1.02866
47	0.39167	0.11095	0.98643	5.48233	4.90	0.28329	3.48211	107	0.89167	0.94346	1.01861	-8.48681	-24.69	1.05809	0.96269
48	0.40000	0.11936	1.03227	5.51728	3.48	0.29841	3.45922	108	0.90000	0.95166	0.94705	-8.68561	-23.02	1.05739	0.89564
49	0.40833	0.12816	1.07835	5.54023	2.02	0.31386	3.43580	109	0.90833	0.95924	0.87389	-8.87026	-21.29	1.05605	0.82751
50	0.41667	0.13734	1.12457	5.55089	0.53	0.32961	3.41184	110	0.91667	0.96622	0.79925	-9.04032	-19.52	1.05405	0.75826
51	0.42500	0.14690	1.17083	5.54895	-1.00	0.34565	3.38733	111	0.92500	0.97256	0.72326	-9.19540	-17.70	1.05142	0.68789
52	0.43333	0.15685	1.21702	5.53417	-2.55	0.36196	3.36227	112	0.93333	0.97827	0.64604	-9.33513	-15.83	1.04814	0.61636
53	0.44167	0.16718	1.26303	5.50633	-4.13	0.37853	3.33666	113	0.94167	0.98333	0.56772	-9.45920	-13.94	1.04424	0.54366
54	0.45000	0.17790	1.30876	5.46525	-5.73	0.39533	3.31050	114	0.95000	0.98773	0.48843	-9.56732	-12.01	1.03971	0.46977
55	0.45833	0.18900	1.35408	5.41077	-7.35	0.41236	3.28377	115	0.95833	0.99146	0.40831	-9.65924	-10.05	1.03457	0.39466
56	0.46667	0.20047	1.39890	5.34279	-8.97	0.42957	3.25649	116	0.96667	0.99453	0.32749	-9.73473	-8.07	1.02882	0.31831
57	0.47500	0.21231	1.44309	5.26123	-10.60	0.44697	3.22863	117	0.97500	0.99692	0.24611	-9.79364	-6.07	1.02248	0.24070
58	0.48333	0.22452	1.48655	5.16604	-12.24	0.46452	3.20020	118	0.98333	0.99863	0.16431	-9.83581	-4.05	1.01556	0.16179
59	0.49167	0.23708	1.52915	5.05722	-13.88	0.48220	3.17119	119	0.99167	0.99966	0.08222	-9.86115	-2.03	1.00806	0.08157
60	0.50000	0.25000	1.57080	4.93480	-15.50	0.50000	3.14159	120	1.00000	1.00000	-.00000	-9.86961	0.00	1.00000	-.00000

Table B.15 Cam law factors – 4 Station Geneva mechanism

GENEVA 4 Factor table   4 STATION GENEVA MECHANISM   120 STEPS
PARAMETERS: Ca = 8.493265   Cd = 8.493265   Cv = 2.414213   Cam angle(Input stroke) = 90 degrees, Output stroke = 90 degrees

STEP NO.	INPUT DISP. u	OUTPUT DISP. w	OUTPUT VELOC. w'	OUTPUT ACCEL. w''	O'PUT JERK w'''	DISP. RATIO w/u	BLEND FACTOR z=u.w'/w	STEP NO.	INPUT DISP. u	OUTPUT DISP. w	OUTPUT VELOC. w'	OUTPUT ACCEL. w''	O'PUT JERK w'''	DISP. RATIO w/u	BLEND FACTOR z=u.w'/w
0	0.00000	0.00000	-.00000	1.57080	7.40										
1	0.00833	0.00006	0.01335	1.63399	7.77	0.00663	2.01241	61	0.50833	0.52011	2.41010	-0.98500	-117.5	1.02316	2.35555
2	0.01667	0.00022	0.02724	1.70030	8.15	0.01344	2.02665	62	0.51667	0.54015	2.39784	-1.95327	-114.5	1.04544	2.29361
3	0.02500	0.00051	0.04170	1.76992	8.56	0.02044	2.04046	63	0.52500	0.56005	2.37764	-2.88879	-109.7	1.06676	2.22884
4	0.03333	0.00092	0.05675	1.84300	8.99	0.02762	2.05466	64	0.53333	0.57975	2.34983	-3.77688	-103.2	1.08703	2.16169
5	0.04167	0.00146	0.07243	1.91976	9.44	0.03500	2.06911	65	0.54167	0.59919	2.31486	-4.60476	-95.29	1.10620	2.09262
6	0.05000	0.00213	0.08876	2.00037	9.91	0.04259	2.08391	66	0.55000	0.61831	2.27328	-5.36198	-86.28	1.12421	2.02212
7	0.05833	0.00294	0.10578	2.08506	10.42	0.05039	2.09900	67	0.55833	0.63706	2.22571	-6.04065	-76.49	1.14101	1.95065
8	0.06667	0.00389	0.12352	2.17404	10.94	0.05842	2.11441	68	0.56667	0.65539	2.17283	-6.63550	-66.22	1.15658	1.87867
9	0.07500	0.00500	0.14202	2.26753	11.50	0.06667	2.13017	69	0.57500	0.67327	2.11536	-7.14388	-55.78	1.17090	1.80662
10	0.08333	0.00626	0.16133	2.36577	12.08	0.07517	2.14626	70	0.58333	0.69064	2.05401	-7.56551	-45.45	1.18395	1.73487
11	0.09167	0.00769	0.18147	2.46900	12.70	0.08391	2.16270	71	0.59167	0.70749	1.98950	-7.90224	-35.44	1.19576	1.66380
12	0.10000	0.00929	0.20249	2.57747	13.34	0.09291	2.17949	72	0.60000	0.72379	1.92253	-8.15768	-25.96	1.20632	1.59372
13	0.10833	0.01107	0.22444	2.69144	14.02	0.10217	2.19663	73	0.60833	0.73953	1.85375	-8.33684	-17.15	1.21566	1.52489
14	0.11667	0.01303	0.24736	2.81116	14.72	0.11172	2.21412	74	0.61667	0.75468	1.78378	-8.44572	-9.11	1.22381	1.45756
15	0.12500	0.01519	0.27131	2.93691	15.46	0.12156	2.23197	75	0.62500	0.76926	1.71317	-8.49102	-1.90	1.23081	1.39191
16	0.13333	0.01756	0.29633	3.06894	16.23	0.13169	2.25017	76	0.63333	0.78324	1.64242	-8.47976	4.46	1.23669	1.32808
17	0.14167	0.02014	0.32248	3.20752	17.03	0.14214	2.26873	77	0.64167	0.79663	1.57198	-8.41904	9.97	1.24150	1.26619
18	0.15000	0.02294	0.34981	3.35289	17.86	0.15291	2.28764	78	0.65000	0.80944	1.50222	-8.31581	14.67	1.24529	1.20632
19	0.15833	0.02597	0.37838	3.50531	18.72	0.16402	2.30688	79	0.65833	0.82167	1.43348	-8.17674	18.59	1.24811	1.14852
20	0.16667	0.02925	0.40825	3.66500	19.61	0.17548	2.32647	80	0.66667	0.83333	1.36603	-8.00804	21.79	1.25000	1.09282
21	0.17500	0.03278	0.43948	3.83216	20.51	0.18730	2.34638	81	0.67500	0.84444	1.30008	-7.81543	24.34	1.25102	1.03921
22	0.18333	0.03658	0.47214	4.00695	21.44	0.19950	2.36659	82	0.68333	0.85501	1.23582	-7.60407	26.30	1.25123	0.98769
23	0.19167	0.04065	0.50629	4.18951	22.39	0.21209	2.38710	83	0.69167	0.86504	1.17338	-7.37856	27.74	1.25066	0.93821
24	0.20000	0.04502	0.54199	4.37991	23.32	0.22509	2.40788	84	0.70000	0.87457	1.11287	-7.14294	28.74	1.24938	0.89074
25	0.20833	0.04969	0.57931	4.57815	24.26	0.23851	2.42890	85	0.70833	0.88360	1.05435	-6.90071	29.34	1.24743	0.84522
26	0.21667	0.05468	0.61831	4.78414	25.18	0.25236	2.45013	86	0.71667	0.89215	0.99787	-6.65486	29.61	1.24485	0.80160
27	0.22500	0.06000	0.65906	4.99769	26.07	0.26666	2.47154	87	0.72500	0.90023	0.94344	-6.40791	29.61	1.24170	0.75980
28	0.23333	0.06567	0.70162	5.21850	26.87	0.28143	2.49308	88	0.73333	0.90788	0.89107	-6.16199	29.38	1.23801	0.71976
29	0.24167	0.07170	0.74606	5.44610	27.69	0.29668	2.51469	89	0.74167	0.91509	0.84073	-5.91883	28.95	1.23383	0.68140
30	0.25000	0.07811	0.79241	5.67983	28.38	0.31243	2.53631	90	0.75000	0.92189	0.79241	-5.67983	28.38	1.22919	0.64466
31	0.25833	0.08491	0.84073	5.91883	28.95	0.32868	2.55789	91	0.75833	0.92830	0.74606	-5.44610	27.69	1.22414	0.60946
32	0.26667	0.09212	0.89107	6.16200	29.38	0.34547	2.57932	92	0.76667	0.93433	0.70162	-5.21850	26.92	1.21870	0.57572
33	0.27500	0.09977	0.94344	6.40791	29.61	0.36279	2.60053	93	0.77500	0.94000	0.65906	-4.99769	26.07	1.21291	0.54338
34	0.28333	0.10785	0.99787	6.65486	29.61	0.38066	2.62141	94	0.78333	0.94532	0.61831	-4.78413	25.18	1.20679	0.51236
35	0.29167	0.11640	1.05435	6.90071	29.34	0.39910	2.64184	95	0.79167	0.95031	0.57814	-4.57814	24.26	1.20039	0.48260
36	0.30000	0.12543	1.11287	7.14294	28.74	0.41811	2.66169	96	0.80000	0.95498	0.54199	-4.37991	23.32	1.19373	0.45403
37	0.30833	0.13496	1.17338	7.37856	27.74	0.43770	2.68081	97	0.80833	0.95935	0.50629	-4.18951	22.38	1.18682	0.42659
38	0.31667	0.14499	1.23582	7.60407	26.30	0.45788	2.69903	98	0.81667	0.96342	0.47214	-4.00695	21.44	1.17970	0.40022
39	0.32500	0.15556	1.30008	7.81543	24.34	0.47864	2.71618	99	0.82500	0.96722	0.43948	-3.83216	20.51	1.17239	0.37486
40	0.33333	0.16667	1.36603	8.00804	21.79	0.50000	2.73205	100	0.83333	0.97403	0.40825	-3.66500	19.61	1.16490	0.35046
41	0.34167	0.17833	1.43348	8.17674	18.59	0.52194	2.74643	101	0.84167	0.97403	0.37838	-3.50531	18.72	1.15726	0.32696
42	0.35000	0.19056	1.50222	8.31581	14.67	0.54446	2.75910	102	0.85000	0.97706	0.34981	-3.35289	17.86	1.14949	0.30432
43	0.35833	0.20337	1.57198	8.41903	9.97	0.56754	2.76979	103	0.85833	0.97986	0.32248	-3.20752	17.03	1.14159	0.28248
44	0.36667	0.21676	1.64242	8.47976	4.46	0.59117	2.77824	104	0.86667	0.98244	0.29633	-3.06894	16.23	1.13359	0.26141
45	0.37500	0.23074	1.71317	8.49102	-1.90	0.61532	2.78420	105	0.87500	0.98481	0.27131	-2.93691	15.46	1.12549	0.24106
46	0.38333	0.24532	1.78378	8.44572	-9.11	0.63995	2.78736	106	0.88333	0.98697	0.24736	-2.81116	14.72	1.11732	0.22139
47	0.39167	0.26047	1.85376	8.33684	-17.15	0.66504	2.78745	107	0.89167	0.98893	0.22444	-2.69144	14.02	1.10908	0.20237
48	0.40000	0.27621	1.92253	8.15768	-25.96	0.69052	2.78418	108	0.90000	0.99071	0.20249	-2.57747	13.34	1.10079	0.18395
49	0.40833	0.29251	1.98950	7.90224	-35.44	0.71635	2.77728	109	0.90833	0.99231	0.18147	-2.46900	12.70	1.09245	0.16611
50	0.41667	0.30936	2.05401	7.56551	-45.45	0.74246	2.76648	110	0.91667	0.99374	0.16132	-2.36577	12.08	1.08408	0.14881
51	0.42500	0.32673	2.11536	7.14388	-55.78	0.76879	2.75155	111	0.92500	0.99500	0.14202	-2.26753	11.50	1.07568	0.13203
52	0.43333	0.34461	2.17283	6.63550	-66.22	0.79524	2.73229	112	0.93333	0.99611	0.12352	-2.17404	10.94	1.06726	0.11574
53	0.44167	0.36294	2.22571	6.04065	-76.49	0.82174	2.70853	113	0.94167	0.99706	0.10578	-2.08506	10.42	1.05883	0.09990
54	0.45000	0.38169	2.27328	5.36198	-86.28	0.84819	2.68015	114	0.95000	0.99787	0.08876	-2.00037	9.91	1.05039	0.08450
55	0.45833	0.40081	2.31486	4.60476	-95.29	0.87449	2.64710	115	0.95833	0.99854	0.07243	-1.91976	9.44	1.04196	0.06951
56	0.46667	0.42025	2.34983	3.77687	-103.2	0.90053	2.60938	116	0.96667	0.99908	0.05675	-1.84300	8.99	1.03353	0.05491
57	0.47500	0.43995	2.37764	2.88879	-109.7	0.92621	2.56706	117	0.97500	0.99949	0.04170	-1.76992	8.56	1.02512	0.04068
58	0.48333	0.45985	2.39784	1.95328	-114.5	0.95142	2.52027	118	0.98333	0.99978	0.02724	-1.70030	8.15	1.01672	0.02679
59	0.49167	0.47989	2.41010	0.98500	-117.5	0.97605	2.46923	119	0.99167	0.99994	0.01335	-1.63399	7.77	1.00835	0.01324
60	0.50000	0.50000	2.41421	0.00000	-118.5	1.00000	2.41421	120	1.00000	1.00000	-.00000	-1.57080	7.40	1.00000	-.00000

Table B.16   Cam law factors – 6 Station Geneva mechanism

```
GENEVA 6 Factor table                6 STATION GENEVA MECHANISM                        120 STEPS
MOTION PARAMETERS: Ca = 5.653347  Cd = 5.653347  Cv = 2   Cam angle(Input stroke) = 120 degrees,  Output stroke = 60 degrees
```

STEP NO.	INPUT DISP. u	OUTPUT DISP. w	OUTPUT VELOC. w'	OUTPUT ACCEL. w''	O'PUT JERK w'''	DISP. RATIO w/u	BLEND FACTOR z=u.w'/w	STEP NO.	INPUT DISP. u	OUTPUT DISP. w	OUTPUT VELOC. w'	OUTPUT ACCEL. w''	O'PUT JERK w'''	DISP. RATIO w/u	BLEND FACTOR z=u.w'/w
0	0.00000	0.00000	-.00000	2.41840	8.77			61	0.50833	0.51666	1.99817	-0.43809	-52.44	1.01638	1.96596
1	0.00833	0.00008	0.02046	2.49262	9.04	0.01018	2.00917	62	0.51667	0.53329	1.99271	-0.87286	-51.84	1.03218	1.93058
2	0.01667	0.00034	0.04155	2.56908	9.31	0.02057	2.02035	63	0.52500	0.54986	1.98364	-1.30104	-50.86	1.04736	1.89395
3	0.02500	0.00078	0.06329	2.64782	9.59	0.03117	2.03058	64	0.53333	0.56634	1.97105	-1.71950	-49.51	1.06189	1.85616
4	0.03333	0.00140	0.08569	2.72886	9.86	0.04198	2.04104	65	0.54167	0.58270	1.95502	-2.12527	-47.82	1.07576	1.81734
5	0.04167	0.00221	0.10877	2.81220	10.14	0.05302	2.05154								
6	0.05000	0.00321	0.13256	2.89788	10.42	0.06428	2.06213	66	0.55000	0.59892	1.93567	-2.51563	-45.82	1.08894	1.77757
7	0.05833	0.00442	0.15708	2.98587	10.70	0.07578	2.07278	67	0.55833	0.61496	1.91314	-2.88812	-43.54	1.10142	1.73699
8	0.06667	0.00583	0.18233	3.07617	10.97	0.08751	2.08349	68	0.56667	0.63080	1.88759	-3.24059	-41.02	1.11317	1.69569
9	0.07500	0.00746	0.20835	3.16875	11.25	0.09949	2.09427	69	0.57500	0.64641	1.85919	-3.57122	-38.30	1.12419	1.65381
10	0.08333	0.00931	0.23515	3.26357	11.51	0.11171	2.10508	70	0.58333	0.66177	1.82814	-3.87854	-35.43	1.13447	1.61144
11	0.09167	0.01138	0.26275	3.36057	11.77	0.12418	2.11593	71	0.59167	0.67687	1.79462	-4.16142	-32.44	1.14401	1.56871
12	0.10000	0.01369	0.29117	3.45968	12.02	0.13690	2.12680	72	0.60000	0.69168	1.75885	-4.41909	-29.39	1.15280	1.52572
13	0.10833	0.01624	0.32042	3.56080	12.25	0.14989	2.13769	73	0.60833	0.70618	1.72104	-4.65110	-26.29	1.16084	1.48258
14	0.11667	0.01903	0.35052	3.66381	12.47	0.16314	2.14858	74	0.61667	0.72036	1.68140	-4.85732	-23.20	1.16815	1.43938
15	0.12500	0.02208	0.38149	3.76857	12.67	0.17666	2.15944	75	0.62500	0.73420	1.64016	-5.03794	-20.15	1.17472	1.39621
16	0.13333	0.02539	0.41333	3.87488	12.84	0.19045	2.17027	76	0.63333	0.74769	1.59751	-5.19337	-17.17	1.18056	1.35317
17	0.14167	0.02897	0.44607	3.98254	12.99	0.20452	2.18105	77	0.64167	0.76082	1.55367	-5.32429	-14.27	1.18569	1.31034
18	0.15000	0.03283	0.47971	4.09132	13.11	0.21887	2.19176	78	0.65000	0.77358	1.50884	-5.43155	-11.49	1.19013	1.26780
19	0.15833	0.03697	0.51426	4.20091	13.19	0.23350	2.20236	79	0.65833	0.78597	1.46320	-5.51620	-8.85	1.19387	1.22560
20	0.16667	0.04140	0.54973	4.31098	13.22	0.24843	2.21285	80	0.66667	0.79797	1.41696	-5.57940	-6.35	1.19695	1.18381
21	0.17500	0.04614	0.58611	4.42117	13.21	0.26364	2.22319	81	0.67500	0.80958	1.37027	-5.62241	-4.00	1.19938	1.14248
22	0.18333	0.05118	0.62341	4.53104	13.15	0.27914	2.23335	82	0.68333	0.82080	1.32330	-5.64657	-1.82	1.20118	1.10167
23	0.19167	0.05653	0.66163	4.64011	13.02	0.29493	2.24331	83	0.69167	0.83163	1.27621	-5.65327	0.19	1.20236	1.06142
24	0.20000	0.06221	0.70075	4.74783	12.82	0.31103	2.25302	84	0.70000	0.84207	1.22913	-5.64391	2.03	1.20296	1.02175
25	0.20833	0.06821	0.74075	4.85359	12.55	0.32744	2.26246	85	0.70833	0.85212	1.18219	-5.61990	3.71	1.20299	0.98270
26	0.21667	0.07455	0.78163	4.95673	12.19	0.34409	2.27158	86	0.71667	0.86178	1.13550	-5.58262	5.22	1.20248	0.94430
27	0.22500	0.08124	0.82336	5.05649	11.74	0.36107	2.28034	87	0.72500	0.87105	1.08917	-5.53342	6.57	1.20144	0.90655
28	0.23333	0.08828	0.86589	5.15207	11.18	0.37834	2.28870	88	0.73333	0.87993	1.04330	-5.47360	7.76	1.19991	0.86949
29	0.24167	0.09567	0.90921	5.24257	10.52	0.39589	2.29661	89	0.74167	0.88844	0.99797	-5.40441	8.82	1.19789	0.83311
30	0.25000	0.10343	0.95325	5.32703	9.73	0.41373	2.30402	90	0.75000	0.89657	0.95325	-5.32703	9.73	1.19542	0.79742
31	0.25833	0.11156	0.99797	5.40441	8.82	0.43186	2.31088	91	0.75833	0.90433	0.90921	-5.24257	10.52	1.19252	0.76243
32	0.26667	0.12007	1.04331	5.47360	7.76	0.45026	2.31714	92	0.76667	0.91172	0.86589	-5.15207	11.18	1.18920	0.72813
33	0.27500	0.12895	1.08917	5.53342	6.57	0.46892	2.32272	93	0.77500	0.91876	0.82336	-5.05649	11.74	1.18550	0.69452
34	0.28333	0.13822	1.13550	5.58262	5.22	0.48784	2.32759	94	0.78333	0.92545	0.78163	-4.95673	12.19	1.18142	0.66160
35	0.29167	0.14788	1.18218	5.61990	3.71	0.50701	2.33166	95	0.79167	0.93179	0.74075	-4.85359	12.55	1.17700	0.62936
36	0.30000	0.15793	1.22913	5.64391	2.03	0.52642	2.33488	96	0.80000	0.93779	0.70075	-4.74783	12.82	1.17224	0.59778
37	0.30833	0.16837	1.27621	5.65327	0.19	0.54605	2.33717	97	0.80833	0.94347	0.66163	-4.64011	13.02	1.16718	0.56686
38	0.31667	0.17920	1.32330	5.64657	-1.82	0.56588	2.33847	98	0.81667	0.94882	0.62341	-4.53104	13.15	1.16183	0.53658
39	0.32500	0.19042	1.37027	5.62241	-4.00	0.58591	2.33872	99	0.82500	0.95386	0.58611	-4.42117	13.21	1.15620	0.50693
40	0.33333	0.20203	1.41696	5.57940	-6.35	0.60610	2.33783	100	0.83333	0.95860	0.54973	-4.31098	13.22	1.15031	0.47789
41	0.34167	0.21403	1.46320	5.51620	-8.85	0.62644	2.33574	101	0.84167	0.96303	0.51426	-4.20091	13.19	1.14419	0.44946
42	0.35000	0.22642	1.50884	5.43155	-11.49	0.64691	2.33237	102	0.85000	0.96717	0.47971	-4.09132	13.11	1.13785	0.42160
43	0.35833	0.23918	1.55367	5.32429	-14.27	0.66748	2.32767	103	0.85833	0.97103	0.44607	-3.98254	12.99	1.13129	0.39430
44	0.36667	0.25231	1.59751	5.19337	-17.17	0.68812	2.32156	104	0.86667	0.97461	0.41333	-3.87488	12.84	1.12455	0.36756
45	0.37500	0.26580	1.64016	5.03794	-20.15	0.70880	2.31398	105	0.87500	0.97792	0.38149	-3.76857	12.67	1.11762	0.34134
46	0.38333	0.27964	1.68140	4.85732	-23.20	0.72950	2.30487	106	0.88333	0.98097	0.35052	-3.66382	12.47	1.11053	0.31563
47	0.39167	0.29382	1.72104	4.65109	-26.29	0.75018	2.29417	107	0.89167	0.98376	0.32042	-3.56080	12.25	1.10328	0.29042
48	0.40000	0.30832	1.75885	4.41909	-29.39	0.77080	2.28184	108	0.90000	0.98631	0.29117	-3.45968	12.02	1.09590	0.26569
49	0.40833	0.32313	1.79462	4.16142	-32.44	0.79134	2.26784	109	0.90833	0.98863	0.26275	-3.36057	11.77	1.08839	0.24141
50	0.41667	0.33823	1.82814	3.87854	-35.43	0.81174	2.25212	110	0.91667	0.99069	0.23515	-3.26357	11.51	1.08075	0.21758
51	0.42500	0.35359	1.85919	3.57122	-38.30	0.83198	2.23467	111	0.92500	0.99254	0.20835	-3.16875	11.25	1.07301	0.19418
52	0.43333	0.36920	1.88759	3.24059	-41.02	0.85201	2.21546	112	0.93333	0.99413	0.18233	-3.07617	10.97	1.06518	0.17118
53	0.44167	0.38504	1.91314	2.88812	-43.54	0.87180	2.19449	113	0.94167	0.99558	0.15708	-2.98587	10.70	1.05725	0.14857
54	0.45000	0.40108	1.93567	2.51563	-45.82	0.89129	2.17176	114	0.95000	0.99679	0.13256	-2.89788	10.42	1.04925	0.12634
55	0.45833	0.41730	1.95502	2.12527	-47.82	0.91046	2.14728	115	0.95833	0.99779	0.10877	-2.81220	10.14	1.04117	0.10447
56	0.46667	0.43366	1.97105	1.71950	-49.51	0.92926	2.12109	116	0.96667	0.99860	0.08569	-2.72886	9.86	1.03304	0.08295
57	0.47500	0.45014	1.98364	1.30104	-50.86	0.94766	2.09321	117	0.97500	0.99922	0.06329	-2.64782	9.59	1.02484	0.06175
58	0.48333	0.46671	1.99271	0.87286	-51.84	0.96560	2.06370	118	0.98333	0.99966	0.04155	-2.56908	9.31	1.01660	0.04087
59	0.49167	0.48334	1.99817	0.43809	-52.44	0.98306	2.03260	119	0.99167	0.99992	0.02046	-2.49262	9.04	1.00832	0.02029
60	0.50000	0.50000	2.00000	0.00000	-52.64	1.00000	2.00000	120	1.00000	1.00000	-.00000	-2.41840	8.77	1.00000	-.00000

Table B.17   Cam law factors – 8 Station Geneva mechanism

GENEVA 8 Factor table   8 STATION GENEVA MECHANISM   120 STEPS
PARAMETERS: Ca = 4.946291 Cd = 4.946291 Cv = 1.859743 Cam angle(Input stroke) = 135 degrees, Output stroke = 45 degrees

STEP NO.	INPUT DISP. u	OUTPUT DISP. w	OUTPUT VELOC. w'	OUTPUT ACCEL. w''	O'PUT JERK w'''	DISP. RATIO w/u	BLEND FACTOR z=u.w'/w	STEP NO.	INPUT DISP. u	OUTPUT DISP. w	OUTPUT VELOC. w'	OUTPUT ACCEL. w''	O'PUT JERK w'''	DISP. RATIO w/u	BLEND FACTOR z=u.w'/w
0	0.00000	0.00000	-.00000	2.92790	8.57			61	0.50833	0.51549	1.85844	-0.31192	-37.37	1.01409	1.83263
1	0.00833	0.00010	0.02470	2.99993	8.71	0.01230	2.00819	62	0.51667	0.53097	1.85455	-0.62226	-37.09	1.02768	1.80460
2	0.01667	0.00041	0.05000	3.07311	8.85	0.02480	2.01636	63	0.52500	0.54640	1.84808	-0.92950	-36.62	1.04075	1.77571
3	0.02500	0.00094	0.07592	3.14737	8.97	0.03750	2.02441	64	0.53333	0.56176	1.83907	-1.23212	-35.98	1.05330	1.74600
4	0.03333	0.00168	0.10246	3.22266	9.09	0.05041	2.03250	65	0.54167	0.57704	1.82756	-1.52867	-35.17	1.06531	1.71553
5	0.04167	0.00265	0.12963	3.29890	9.20	0.06353	2.04060								
6	0.05000	0.00384	0.15744	3.37599	9.30	0.07685	2.04862	66	0.55000	0.59221	1.81361	-1.81777	-34.19	1.07675	1.68434
7	0.05833	0.00527	0.18590	3.45383	9.38	0.09039	2.05662	67	0.55833	0.60726	1.79729	-2.09813	-33.07	1.08763	1.65248
8	0.06667	0.00694	0.21501	3.53231	9.45	0.10414	2.06456	68	0.56667	0.62216	1.77867	-2.36858	-31.81	1.09793	1.62002
9	0.07500	0.00886	0.24478	3.61129	9.50	0.11811	2.07245	69	0.57500	0.63690	1.75784	-2.62802	-30.43	1.10765	1.58700
10	0.08333	0.01102	0.27520	3.69062	9.54	0.13229	2.08025	70	0.58333	0.65145	1.73490	-2.87553	-28.95	1.11678	1.55349
11	0.09167	0.01345	0.30629	3.77015	9.55	0.14669	2.08796	71	0.59167	0.66581	1.70995	-3.11026	-27.37	1.12531	1.51954
12	0.10000	0.01613	0.33804	3.84968	9.54	0.16131	2.09558	72	0.60000	0.67995	1.68310	-3.33155	-25.72	1.13325	1.48521
13	0.10833	0.01908	0.37045	3.92901	9.50	0.17615	2.10307	73	0.60833	0.69386	1.65447	-3.53882	-24.01	1.14058	1.45054
14	0.11667	0.02231	0.40352	4.00791	9.43	0.19120	2.11043	74	0.61667	0.70752	1.62416	-3.73165	-22.26	1.14733	1.41561
15	0.12500	0.02581	0.43724	4.08614	9.34	0.20648	2.11764	75	0.62500	0.72092	1.59231	-3.90976	-20.48	1.15347	1.38045
16	0.13333	0.02960	0.47162	4.16341	9.21	0.22197	2.12469	76	0.63333	0.73405	1.55904	-4.07298	-18.69	1.15903	1.34513
17	0.14167	0.03367	0.50663	4.23945	9.04	0.23768	2.13155	77	0.64167	0.74690	1.52447	-4.22126	-16.90	1.16400	1.30968
18	0.15000	0.03804	0.54227	4.31392	8.83	0.25361	2.13820	78	0.65000	0.75946	1.48873	-4.35467	-15.12	1.16840	1.27417
19	0.15833	0.04271	0.57852	4.38648	8.58	0.26975	2.14463	79	0.65833	0.77171	1.45194	-4.47337	-13.37	1.17222	1.23862
20	0.16667	0.04769	0.61537	4.45674	8.28	0.28611	2.15081	80	0.66667	0.78365	1.41421	-4.57764	-11.66	1.17548	1.20309
21	0.17500	0.05297	0.65280	4.52431	7.93	0.30268	2.15672	81	0.67500	0.79528	1.37568	-4.66780	-9.99	1.17819	1.16762
22	0.18333	0.05857	0.69077	4.58875	7.53	0.31946	2.16233	82	0.68333	0.80658	1.33645	-4.74430	-8.38	1.18036	1.13224
23	0.19167	0.06448	0.72926	4.64960	7.07	0.33643	2.16763	83	0.69167	0.81755	1.29665	-4.80760	-6.83	1.18200	1.09699
24	0.20000	0.07072	0.76825	4.70636	6.55	0.35361	2.17257	84	0.70000	0.82819	1.25636	-4.85824	-5.34	1.18313	1.06190
25	0.20833	0.07729	0.80769	4.75853	5.96	0.37099	2.17715	85	0.70833	0.83849	1.21571	-4.89680	-3.93	1.18375	1.02700
26	0.21667	0.08419	0.84755	4.80555	5.31	0.38855	2.18132	86	0.71667	0.84845	1.17478	-4.92389	-2.59	1.18388	0.99231
27	0.22500	0.09142	0.88777	4.84686	4.59	0.40629	2.18505	87	0.72500	0.85807	1.13367	-4.94014	-1.33	1.18354	0.95787
28	0.23333	0.09898	0.92831	4.88186	3.80	0.42421	2.18833	88	0.73333	0.86734	1.09247	-4.94622	-0.14	1.18274	0.92368
29	0.24167	0.10689	0.96911	4.90993	2.93	0.44230	2.19110	89	0.74167	0.87628	1.05126	-4.94277	0.96	1.18150	0.88977
30	0.25000	0.11513	1.01012	4.93045	1.98	0.46054	2.19334	90	0.75000	0.88487	1.01012	-4.93045	1.98	1.17982	0.85616
31	0.25833	0.12372	1.05126	4.94277	0.96	0.47893	2.19502	91	0.75833	0.89311	0.96911	-4.90993	2.93	1.17773	0.82286
32	0.26667	0.13266	1.09247	4.94622	-0.14	0.49746	2.19611	92	0.76667	0.90102	0.92831	-4.88186	3.80	1.17524	0.78989
33	0.27500	0.14193	1.13367	4.94014	-1.33	0.51612	2.19655	93	0.77500	0.90858	0.88777	-4.84686	4.59	1.17237	0.75724
34	0.28333	0.15155	1.17478	4.92389	-2.59	0.53488	2.19633	94	0.78333	0.91581	0.84754	-4.80555	5.31	1.16913	0.72494
35	0.29167	0.16151	1.21571	4.89680	-3.93	0.55375	2.19541	95	0.79167	0.92271	0.80769	-4.75853	5.96	1.16553	0.69298
36	0.30000	0.17181	1.25636	4.85824	-5.34	0.57270	2.19374	96	0.80000	0.92928	0.76825	-4.70636	6.55	1.16160	0.66137
37	0.30833	0.18245	1.29665	4.80760	-6.83	0.59173	2.19129	97	0.80833	0.93552	0.72926	-4.64960	7.07	1.15734	0.63012
38	0.31667	0.19342	1.33645	4.74430	-8.38	0.61080	2.18803	98	0.81667	0.94143	0.69077	-4.58875	7.53	1.15278	0.59922
39	0.32500	0.20472	1.37568	4.66780	-9.99	0.62991	2.18392	99	0.82500	0.94703	0.65280	-4.52431	7.93	1.14792	0.56868
40	0.33333	0.21635	1.41421	4.57764	-11.66	0.64904	2.17893	100	0.83333	0.95231	0.61537	-4.45674	8.28	1.14278	0.53849
41	0.34167	0.22829	1.45194	4.47337	-13.37	0.66817	2.17302	101	0.84167	0.95729	0.57852	-4.38648	8.58	1.13737	0.50865
42	0.35000	0.24054	1.48873	4.35467	-15.12	0.68727	2.16616	102	0.85000	0.96196	0.54227	-4.31392	8.83	1.13172	0.47916
43	0.35833	0.25310	1.52447	4.22126	-16.90	0.70632	2.15832	103	0.85833	0.96633	0.50663	-4.23945	9.04	1.12582	0.45001
44	0.36667	0.26595	1.55904	4.07298	-18.69	0.72531	2.14948	104	0.86667	0.97040	0.47162	-4.16341	9.21	1.11970	0.42120
45	0.37500	0.27908	1.59231	3.90976	-20.48	0.74421	2.13960	105	0.87500	0.97419	0.43724	-4.08614	9.34	1.11336	0.39272
46	0.38333	0.29248	1.62416	3.73165	-22.26	0.76300	2.12866	106	0.88333	0.97769	0.40352	-4.00791	9.43	1.10682	0.36457
47	0.39167	0.30614	1.65447	3.53882	-24.01	0.78165	2.11665	107	0.89167	0.98092	0.37045	-3.92901	9.50	1.10009	0.33674
48	0.40000	0.32005	1.68310	3.33155	-25.72	0.80013	2.10354	108	0.90000	0.98387	0.33804	-3.84968	9.54	1.09319	0.30922
49	0.40833	0.33419	1.70995	3.11027	-27.37	0.81843	2.08932	109	0.90833	0.98655	0.30629	-3.77015	9.55	1.08611	0.28200
50	0.41667	0.34855	1.73490	2.87553	-28.95	0.83651	2.07398	110	0.91667	0.98898	0.27520	-3.69062	9.54	1.07888	0.25508
51	0.42500	0.36310	1.75784	2.62802	-30.43	0.85435	2.05751	111	0.92500	0.99114	0.24478	-3.61129	9.50	1.07150	0.22844
52	0.43333	0.37784	1.77867	2.36858	-31.81	0.87193	2.03994	112	0.93333	0.99306	0.21501	-3.53231	9.45	1.06399	0.20208
53	0.44167	0.39274	1.79729	2.09813	-33.07	0.88922	2.02120	113	0.94167	0.99473	0.18590	-3.45383	9.38	1.05635	0.17599
54	0.45000	0.40779	1.81361	1.81777	-34.19	0.90619	2.00136	114	0.95000	0.99616	0.15744	-3.37599	9.30	1.04859	0.15015
55	0.45833	0.42296	1.82756	1.52867	-35.17	0.92282	1.98041	115	0.95833	0.99735	0.12963	-3.29890	9.20	1.04072	0.12456
56	0.46667	0.43824	1.83907	1.23212	-35.98	0.93908	1.95837	116	0.96667	0.99832	0.10246	-3.22266	9.09	1.03274	0.09921
57	0.47500	0.45360	1.84808	0.92950	-36.62	0.95494	1.93525	117	0.97500	0.99906	0.07592	-3.14737	8.97	1.02468	0.07407
58	0.48333	0.46903	1.85455	0.62227	-37.09	0.97041	1.91109	118	0.98333	0.99959	0.05000	-3.07311	8.85	1.01653	0.04919
59	0.49167	0.48451	1.85844	0.31192	-37.37	0.98544	1.88591	119	0.99167	0.99990	0.02470	-2.99993	8.71	1.00830	0.02450
60	0.50000	0.50000	1.85974	0.00000	-37.46	1.00000	1.85974	120	1.00000	1.00000	-.00000	-2.92790	8.57	1.00000	-.00000

APPENDIX C

Loading Coefficients

The loading coefficients in the following tables are used to calculate the peak output load and the peak input torque on a cam system that has an output loading pattern comprising any combination of:

- constant load, e.g. friction force or gravity force
- load proportional to displacement, e.g. a spring force
- load proportional to acceleration, e.g. inertia force

LOAD MIX COEFFICIENT, C_m

After calculating the *nominal* maximum output load by simply adding the peak inertia load to the maximum non-inertia load, the *actual* maximum in-motion load is found by multiplying that sum by C_m.

INPUT TORQUE COEFFICIENT, C_c

The input torque on the cam is found by multiplying the output load as found above by the output/input stroke ratio, and multiplying the product by C_c.

Both coefficients, C_m and C_c, are dependent on the pre-load ratio K and the inertia ratio Q which define the load pattern and must be estimated first.

Refer to Chapter 13 for full details.

Table C.1

PARABOLIC MOTION LOADING COEFFICIENTS										
PRE-LOAD RATIO K	INERTIA RATIO, Q									
	0.00	0.10	0.20	0.30	0.40	0.50	0.60	0.70	0.80	0.90	1.00
	LOAD MIX COEFFICIENT Cm									
1.0	1.000	1.000	1.000	1.000	1.000	1.000	1.000	1.000	1.000	1.000	1.000
0.9	1.000	0.955	0.960	0.965	0.970	0.975	0.980	0.985	0.990	0.995	1.000
0.8	1.000	0.900	0.920	0.930	0.940	0.950	0.960	0.970	0.980	0.990	1.000
0.7	1.000	0.900	0.880	0.895	0.910	0.925	0.940	0.955	0.970	0.985	1.000
0.6	1.000	0.900	0.800	0.860	0.880	0.900	0.920	0.940	0.960	0.980	1.000
0.5	1.000	0.900	0.800	0.825	0.850	0.875	0.900	0.925	0.950	0.975	1.000
0.4	1.000	0.900	0.800	0.790	0.820	0.850	0.880	0.910	0.940	0.970	1.000
0.3	1.000	0.900	0.800	0.700	0.790	0.825	0.860	0.895	0.930	0.965	1.000
0.2	1.000	0.900	0.800	0.700	0.760	0.800	0.840	0.880	0.920	0.960	1.000
0.1	1.000	0.900	0.800	0.700	0.730	0.775	0.820	0.865	0.910	0.955	1.000
0.0	1.000	0.900	0.800	0.700	0.600	0.750	0.800	0.850	0.900	0.950	1.000

PRE-LOAD RATIO K	INERTIA RATIO, Q									
	0.00	0.10	0.20	0.30	0.40	0.50	0.60	0.70	0.80	0.90	1.00
	INPUT TORQUE COEFFICIENT Cc									
1.0	2.000	2.000	2.000	2.000	2.000	2.000	2.000	2.000	2.000	2.000	2.000
0.9	1.900	2.000	2.000	2.000	2.000	2.000	2.000	2.000	2.000	2.000	2.000
0.8	1.800	2.022	2.000	2.000	2.000	2.000	2.000	2.000	2.000	2.000	2.000
0.7	1.700	1.922	2.000	2.000	2.000	2.000	2.000	2.000	2.000	2.000	2.000
0.6	1.600	1.822	2.100	2.000	2.000	2.000	2.000	2.000	2.000	2.000	2.000
0.5	1.500	1.722	2.000	2.000	2.000	2.000	2.000	2.000	2.000	2.000	2.000
0.4	1.400	1.622	1.900	2.000	2.000	2.000	2.000	2.000	2.000	2.000	2.000
0.3	1.301	1.522	1.800	2.157	2.000	2.000	2.000	2.000	2.000	2.000	2.000
0.2	1.217	1.422	1.700	2.057	2.000	2.000	2.000	2.000	2.000	2.000	2.000
0.1	1.148	1.322	1.600	1.957	2.000	2.000	2.000	2.000	2.000	2.000	2.000
0.0	1.089	1.222	1.500	1.857	2.333	2.000	2.000	2.000	2.000	2.000	2.000

Table C.2

SIMPLE HARMONIC LOADING COEFFICIENTS										
PRE-LOAD RATIO K	INERTIA RATIO, Q									
	0.00	0.10	0.20	0.30	0.40	0.50	0.60	0.70	0.80	0.90	1.00
	LOAD MIX COEFFICIENT Cm									
1.0	1.000	1.000	1.000	1.000	1.000	1.000	1.000	1.000	1.000	1.000	1.000
0.9	1.000	0.910	0.920	0.930	0.940	0.950	0.960	0.970	0.980	0.990	1.000
0.8	1.000	0.900	0.840	0.860	0.880	0.900	0.920	0.940	0.960	0.980	1.000
0.7	1.000	0.900	0.800	0.790	0.820	0.850	0.880	0.910	0.940	0.970	1.000
0.6	1.000	0.900	0.800	0.720	0.760	0.800	0.840	0.880	0.920	0.960	1.000
0.5	1.000	0.900	0.800	0.700	0.700	0.750	0.800	0.850	0.900	0.950	1.000
0.4	1.000	0.900	0.800	0.700	0.640	0.700	0.760	0.820	0.880	0.940	1.000
0.3	1.000	0.900	0.800	0.700	0.600	0.650	0.720	0.790	0.860	0.930	1.000
0.2	1.000	0.900	0.800	0.700	0.600	0.600	0.680	0.760	0.840	0.920	1.000
0.1	1.000	0.900	0.800	0.700	0.600	0.550	0.640	0.730	0.820	0.910	1.000
0.0	1.000	0.900	0.800	0.700	0.600	0.500	0.600	0.700	0.800	0.900	1.000

PRE-LOAD RATIO K	INERTIA RATIO, Q									
	0.00	0.10	0.20	0.30	0.40	0.50	0.60	0.70	0.80	0.90	1.00
	INPUT TORQUE COEFFICIENT Cc									
1.0	1.571	1.422	1.293	1.185	1.095	1.020	0.958	0.905	0.859	0.820	0.785
0.9	1.494	1.479	1.325	1.200	1.100	1.020	0.955	0.902	0.857	0.819	0.785
0.8	1.422	1.414	1.364	1.218	1.106	1.020	0.953	0.899	0.854	0.817	0.785
0.7	1.355	1.337	1.344	1.240	1.113	1.020	0.950	0.895	0.852	0.816	0.785
0.6	1.293	1.264	1.259	1.267	1.121	1.020	0.947	0.892	0.849	0.814	0.785
0.5	1.237	1.198	1.178	1.209	1.131	1.020	0.944	0.888	0.846	0.812	0.785
0.4	1.185	1.137	1.102	1.117	1.143	1.020	0.940	0.884	0.843	0.811	0.785
0.3	1.138	1.081	1.033	1.028	1.120	1.020	0.936	0.880	0.840	0.809	0.785
0.2	1.095	1.032	0.970	0.944	1.020	1.020	0.932	0.875	0.836	0.807	0.785
0.1	1.056	0.987	0.914	0.865	0.922	1.020	0.927	0.871	0.833	0.806	0.785
0.0	1.020	0.947	0.865	0.793	0.824	1.020	0.922	0.866	0.829	0.804	0.785

Table C.3

MOD TRAP LOADING COEFFICIENTS										
PRE-LOAD RATIO K	INERTIA RATIO, Q									
	0.00	0.10	0.20	0.30	0.40	0.50	0.60	0.70	0.80	0.90	1.00
	LOAD MIX COEFFICIENT Cm										
1.0	1.000	1.000	1.000	1.000	1.000	1.000	1.000	1.000	1.000	1.000	1.000
0.9	1.000	0.935	0.942	0.949	0.956	0.963	0.971	0.978	0.985	0.993	1.000
0.8	1.000	0.900	0.884	0.898	0.912	0.927	0.941	0.956	0.971	0.985	1.000
0.7	1.000	0.900	0.827	0.847	0.869	0.891	0.912	0.934	0.956	0.978	1.000
0.6	1.000	0.900	0.800	0.797	0.825	0.854	0.883	0.912	0.941	0.971	1.000
0.5	1.000	0.900	0.800	0.747	0.782	0.818	0.854	0.890	0.927	0.963	1.000
0.4	1.000	0.900	0.800	0.700	0.739	0.782	0.825	0.869	0.912	0.956	1.000
0.3	1.000	0.900	0.800	0.700	0.696	0.746	0.796	0.847	0.898	0.949	1.000
0.2	1.000	0.900	0.800	0.700	0.653	0.710	0.767	0.825	0.883	0.941	1.000
0.1	1.000	0.900	0.800	0.700	0.611	0.674	0.738	0.803	0.869	0.934	1.000
0.0	1.000	0.900	0.800	0.700	0.600	0.638	0.709	0.781	0.854	0.927	1.000
PRE-LOAD RATIO K	INERTIA RATIO, Q									
	0.00	0.10	0.20	0.30	0.40	0.50	0.60	0.70	0.80	0.90	1.00
	INPUT TORQUE COEFFICIENT Cc										
1.0	2.000	1.851	1.776	1.735	1.710	1.693	1.681	1.672	1.665	1.660	1.655
0.9	1.901	1.872	1.786	1.740	1.713	1.695	1.682	1.673	1.666	1.660	1.655
0.8	1.806	1.833	1.796	1.746	1.717	1.697	1.684	1.674	1.666	1.660	1.655
0.7	1.714	1.722	1.807	1.752	1.720	1.700	1.685	1.675	1.667	1.660	1.655
0.6	1.625	1.614	1.751	1.758	1.724	1.702	1.687	1.676	1.667	1.660	1.655
0.5	1.541	1.506	1.634	1.765	1.728	1.704	1.688	1.676	1.668	1.661	1.655
0.4	1.462	1.402	1.518	1.765	1.731	1.706	1.690	1.677	1.668	1.661	1.655
0.3	1.388	1.300	1.402	1.646	1.735	1.709	1.691	1.678	1.669	1.661	1.655
0.2	1.319	1.202	1.286	1.527	1.739	1.711	1.693	1.679	1.669	1.661	1.655
0.1	1.255	1.110	1.171	1.408	1.743	1.714	1.694	1.680	1.670	1.661	1.655
0.0	1.196	1.027	1.057	1.290	1.655	1.716	1.695	1.681	1.670	1.662	1.655

Table C.4

MOD SINE LOADING COEFFICIENTS										
PRE-LOAD RATIO K	INERTIA RATIO, Q									
	0.00	0.10	0.20	0.30	0.40	0.50	0.60	0.70	0.80	0.90	1.00
	LOAD MIX COEFFICIENT Cm										
1.0	1.000	1.000	1.000	1.000	1.000	1.000	1.000	1.000	1.000	1.000	1.000
0.9	1.000	0.912	0.922	0.931	0.941	0.951	0.961	0.971	0.980	0.990	1.000
0.8	1.000	0.900	0.844	0.863	0.883	0.902	0.922	0.941	0.961	0.980	1.000
0.7	1.000	0.900	0.800	0.795	0.824	0.853	0.883	0.912	0.941	0.971	1.000
0.6	1.000	0.900	0.800	0.728	0.766	0.805	0.843	0.883	0.922	0.961	1.000
0.5	1.000	0.900	0.800	0.700	0.708	0.756	0.804	0.853	0.902	0.951	1.000
0.4	1.000	0.900	0.800	0.700	0.650	0.707	0.765	0.824	0.883	0.941	1.000
0.3	1.000	0.900	0.800	0.700	0.600	0.659	0.726	0.795	0.863	0.931	1.000
0.2	1.000	0.900	0.800	0.700	0.600	0.610	0.688	0.765	0.843	0.922	1.000
0.1	1.000	0.900	0.800	0.700	0.600	0.562	0.649	0.736	0.824	0.912	1.000
0.0	1.000	0.900	0.800	0.700	0.600	0.514	0.610	0.707	0.804	0.902	1.000
PRE-LOAD RATIO K	INERTIA RATIO, Q									
	0.00	0.10	0.20	0.30	0.40	0.50	0.60	0.70	0.80	0.90	1.00
	INPUT TORQUE COEFFICIENT Cc										
1.0	1.760	1.596	1.461	1.353	1.266	1.197	1.140	1.092	1.052	1.017	0.987
0.9	1.674	1.655	1.494	1.369	1.273	1.199	1.140	1.091	1.051	1.017	0.987
0.8	1.593	1.585	1.533	1.388	1.281	1.201	1.139	1.090	1.049	1.016	0.987
0.7	1.517	1.496	1.516	1.410	1.290	1.204	1.139	1.088	1.048	1.015	0.987
0.6	1.447	1.412	1.417	1.437	1.300	1.206	1.138	1.087	1.046	1.014	0.987
0.5	1.382	1.335	1.322	1.385	1.311	1.209	1.138	1.085	1.045	1.013	0.987
0.4	1.323	1.263	1.232	1.279	1.325	1.212	1.137	1.083	1.043	1.012	0.987
0.3	1.268	1.198	1.147	1.174	1.322	1.215	1.136	1.082	1.042	1.011	0.987
0.2	1.219	1.140	1.070	1.072	1.210	1.219	1.135	1.080	1.040	1.010	0.987
0.1	1.174	1.087	1.001	0.973	1.098	1.223	1.134	1.078	1.038	1.009	0.987
0.0	1.133	1.040	0.939	0.880	0.986	1.227	1.133	1.075	1.036	1.008	0.987

Table C.5

MSC.20 LOADING COEFFICIENTS										
PRE-LOAD RATIO K	INERTIA RATIO, Q									
	0.00	0.10	0.20	0.30	0.40	0.50	0.60	0.70	0.80	0.90	1.00

LOAD MIX COEFFICIENT Cm

K	0.00	0.10	0.20	0.30	0.40	0.50	0.60	0.70	0.80	0.90	1.00
1.0	1.000	1.000	1.000	1.000	1.000	1.000	1.000	1.000	1.000	1.000	1.000
0.9	1.000	0.912	0.921	0.931	0.941	0.951	0.961	0.970	0.980	0.990	1.000
0.8	1.000	0.900	0.843	0.862	0.882	0.901	0.921	0.941	0.961	0.980	1.000
0.7	1.000	0.900	0.800	0.793	0.823	0.852	0.882	0.911	0.941	0.970	1.000
0.6	1.000	0.900	0.800	0.725	0.764	0.803	0.842	0.882	0.921	0.961	1.000
0.5	1.000	0.900	0.800	0.700	0.705	0.754	0.803	0.852	0.901	0.951	1.000
0.4	1.000	0.900	0.800	0.700	0.646	0.705	0.764	0.823	0.882	0.941	1.000
0.3	1.000	0.900	0.800	0.700	0.600	0.656	0.724	0.793	0.862	0.931	1.000
0.2	1.000	0.900	0.800	0.700	0.600	0.607	0.685	0.764	0.842	0.921	1.000
0.1	1.000	0.900	0.800	0.700	0.600	0.558	0.646	0.734	0.823	0.911	1.000
0.0	1.000	0.900	0.800	0.700	0.600	0.509	0.606	0.705	0.803	0.901	1.000

PRE-LOAD RATIO K	INERTIA RATIO, Q										
	0.00	0.10	0.20	0.30	0.40	0.50	0.60	0.70	0.80	0.90	1.00

INPUT TORQUE COEFFICIENT Cc

K	0.00	0.10	0.20	0.30	0.40	0.50	0.60	0.70	0.80	0.90	1.00
1.0	1.528	1.386	1.268	1.174	1.099	1.039	0.990	0.948	0.913	0.883	0.857
0.9	1.475	1.456	1.280	1.178	1.098	1.036	0.986	0.945	0.911	0.882	0.857
0.8	1.425	1.421	1.350	1.183	1.097	1.033	0.982	0.942	0.908	0.881	0.857
0.7	1.377	1.368	1.368	1.207	1.096	1.029	0.978	0.938	0.906	0.879	0.857
0.6	1.331	1.315	1.315	1.270	1.095	1.025	0.974	0.934	0.903	0.878	0.857
0.5	1.287	1.264	1.262	1.262	1.093	1.021	0.969	0.931	0.901	0.877	0.857
0.4	1.246	1.214	1.209	1.209	1.123	1.016	0.964	0.927	0.898	0.875	0.857
0.3	1.207	1.168	1.156	1.156	1.156	1.010	0.959	0.922	0.895	0.874	0.857
0.2	1.170	1.123	1.103	1.103	1.103	1.004	0.953	0.918	0.892	0.873	0.857
0.1	1.136	1.082	1.050	1.050	1.050	0.998	0.947	0.913	0.889	0.871	0.857
0.0	1.104	1.043	1.000	0.997	0.997	0.980	0.940	0.908	0.886	0.870	0.857

Table C.6

MSC.33 LOADING COEFFICIENTS										
PRE-LOAD RATIO K	INERTIA RATIO, Q									
	0.00	0.10	0.20	0.30	0.40	0.50	0.60	0.70	0.80	0.90	1.00

LOAD MIX COEFFICIENT Cm

K	0.00	0.10	0.20	0.30	0.40	0.50	0.60	0.70	0.80	0.90	1.00
1.0	1.000	1.000	1.000	1.000	1.000	1.000	1.000	1.000	1.000	1.000	1.000
0.9	1.000	0.911	0.921	0.931	0.941	0.951	0.960	0.970	0.980	0.990	1.000
0.8	1.000	0.900	0.842	0.862	0.881	0.901	0.921	0.941	0.960	0.980	1.000
0.7	1.000	0.900	0.800	0.792	0.822	0.852	0.881	0.911	0.941	0.970	1.000
0.6	1.000	0.900	0.800	0.723	0.763	0.802	0.842	0.881	0.921	0.960	1.000
0.5	1.000	0.900	0.800	0.700	0.704	0.753	0.802	0.852	0.901	0.951	1.000
0.4	1.000	0.900	0.800	0.700	0.644	0.704	0.763	0.822	0.881	0.941	1.000
0.3	1.000	0.900	0.800	0.700	0.600	0.654	0.723	0.792	0.862	0.931	1.000
0.2	1.000	0.900	0.800	0.700	0.600	0.605	0.684	0.763	0.842	0.921	1.000
0.1	1.000	0.900	0.800	0.700	0.600	0.556	0.644	0.733	0.822	0.911	1.000
0.0	1.000	0.900	0.800	0.700	0.600	0.506	0.605	0.703	0.802	0.901	1.000

PRE-LOAD RATIO K	INERTIA RATIO, Q										
	0.00	0.10	0.20	0.30	0.40	0.50	0.60	0.70	0.80	0.90	1.00

INPUT TORQUE COEFFICIENT Cc

K	0.00	0.10	0.20	0.30	0.40	0.50	0.60	0.70	0.80	0.90	1.00
1.0	1.404	1.274	1.166	1.079	1.011	0.955	0.910	0.872	0.840	0.812	0.788
0.9	1.367	1.350	1.187	1.078	1.006	0.950	0.905	0.868	0.837	0.810	0.788
0.8	1.331	1.329	1.263	1.080	1.001	0.944	0.899	0.863	0.834	0.809	0.788
0.7	1.297	1.292	1.292	1.141	0.996	0.938	0.894	0.859	0.831	0.807	0.788
0.6	1.263	1.255	1.254	1.214	0.987	0.932	0.888	0.855	0.828	0.806	0.788
0.5	1.230	1.217	1.217	1.217	1.038	0.925	0.882	0.850	0.825	0.804	0.788
0.4	1.199	1.181	1.180	1.179	1.098	0.917	0.876	0.845	0.822	0.803	0.788
0.3	1.169	1.145	1.142	1.142	1.142	0.908	0.869	0.840	0.818	0.801	0.788
0.2	1.141	1.111	1.105	1.105	1.104	0.899	0.861	0.835	0.815	0.800	0.788
0.1	1.113	1.078	1.068	1.067	1.067	0.889	0.853	0.829	0.812	0.798	0.788
0.0	1.087	1.046	1.031	1.030	1.029	1.017	0.845	0.823	0.808	0.797	0.788

Table C.7

MSC.50 LOADING COEFFICIENTS										
PRE-LOAD RATIO K	INERTIA RATIO, Q									
	0.00	0.10	0.20	0.30	0.40	0.50	0.60	0.70	0.80	0.90	1.00
	LOAD MIX COEFFICIENT Cm										
1.0	1.000	1.000	1.000	1.000	1.000	1.000	1.000	1.000	1.000	1.000	1.000
0.9	1.000	0.911	0.921	0.931	0.940	0.950	0.960	0.970	0.980	0.990	1.000
0.8	1.000	0.900	0.841	0.861	0.881	0.901	0.921	0.940	0.960	0.980	1.000
0.7	1.000	0.900	0.800	0.792	0.821	0.851	0.881	0.911	0.940	0.970	1.000
0.6	1.000	0.900	0.800	0.722	0.762	0.802	0.841	0.881	0.921	0.960	1.000
0.5	1.000	0.900	0.800	0.700	0.702	0.752	0.801	0.851	0.901	0.950	1.000
0.4	1.000	0.900	0.800	0.700	0.643	0.702	0.762	0.821	0.881	0.940	1.000
0.3	1.000	0.900	0.800	0.700	0.600	0.653	0.722	0.792	0.861	0.931	1.000
0.2	1.000	0.900	0.800	0.700	0.600	0.603	0.682	0.762	0.841	0.921	1.000
0.1	1.000	0.900	0.800	0.700	0.600	0.554	0.643	0.732	0.821	0.911	1.000
0.0	1.000	0.900	0.800	0.700	0.600	0.500	0.603	0.702	0.801	0.901	1.000
PRE-LOAD RATIO K	INERTIA RATIO, Q									
	0.00	0.10	0.20	0.30	0.40	0.50	0.60	0.70	0.80	0.90	1.00
	INPUT TORQUE COEFFICIENT Cc										
1.0	1.275	1.157	1.059	0.980	0.918	0.867	0.826	0.792	0.762	0.737	0.716
0.9	1.252	1.237	1.088	0.975	0.911	0.860	0.820	0.787	0.759	0.736	0.716
0.8	1.230	1.229	1.169	0.999	0.903	0.853	0.814	0.782	0.756	0.734	0.716
0.7	1.208	1.206	1.206	1.066	0.894	0.845	0.807	0.777	0.753	0.733	0.716
0.6	1.186	1.183	1.183	1.146	0.932	0.837	0.800	0.772	0.750	0.731	0.716
0.5	1.165	1.160	1.160	1.160	0.991	0.827	0.793	0.767	0.746	0.729	0.716
0.4	1.144	1.137	1.137	1.137	1.061	0.809	0.785	0.761	0.743	0.728	0.716
0.3	1.124	1.114	1.114	1.114	1.114	0.853	0.777	0.756	0.739	0.726	0.716
0.2	1.105	1.091	1.090	1.090	1.090	0.904	0.769	0.750	0.735	0.724	0.716
0.1	1.086	1.068	1.067	1.067	1.067	0.964	0.664	0.744	0.732	0.723	0.716
0.0	1.067	1.046	1.044	1.044	1.044	1.044	0.693	0.737	0.728	0.721	0.716

Table C.8

MSC.66 LOADING COEFFICIENTS										
PRE-LOAD RATIO K	INERTIA RATIO, Q									
	0.00	0.10	0.20	0.30	0.40	0.50	0.60	0.70	0.80	0.90	1.00
	LOAD MIX COEFFICIENT Cm										
1.0	1.000	1.000	1.000	1.000	1.000	1.000	1.000	1.000	1.000	1.000	1.000
0.9	1.000	0.900	0.920	0.930	0.940	0.950	0.960	0.970	0.980	0.990	1.000
0.8	1.000	0.900	0.800	0.861	0.881	0.900	0.920	0.940	0.960	0.980	1.000
0.7	1.000	0.900	0.800	0.791	0.821	0.851	0.881	0.910	0.940	0.970	1.000
0.6	1.000	0.900	0.800	0.700	0.761	0.801	0.841	0.881	0.920	0.960	1.000
0.5	1.000	0.900	0.800	0.700	0.701	0.751	0.801	0.851	0.900	0.950	1.000
0.4	1.000	0.900	0.800	0.700	0.600	0.701	0.761	0.821	0.881	0.940	1.000
0.3	1.000	0.900	0.800	0.700	0.600	0.652	0.721	0.791	0.861	0.930	1.000
0.2	1.000	0.900	0.800	0.700	0.600	0.500	0.681	0.761	0.841	0.920	1.000
0.1	1.000	0.900	0.800	0.700	0.600	0.500	0.642	0.731	0.821	0.910	1.000
0.0	1.000	0.900	0.800	0.700	0.600	0.500	0.602	0.701	0.801	0.900	1.000
PRE-LOAD RATIO K	INERTIA RATIO, Q									
	0.00	0.10	0.20	0.30	0.40	0.50	0.60	0.70	0.80	0.90	1.00
	INPUT TORQUE COEFFICIENT Cc										
1.0	1.168	1.051	0.934	0.818	0.823	0.790	0.757	0.725	0.698	0.675	0.655
0.9	1.152	1.152	1.001	0.867	0.817	0.783	0.750	0.720	0.695	0.674	0.655
0.8	1.136	1.136	1.136	0.924	0.811	0.776	0.743	0.715	0.692	0.672	0.655
0.7	1.120	1.120	1.120	0.991	0.819	0.768	0.736	0.710	0.688	0.670	0.655
0.6	1.104	1.104	1.104	1.104	0.870	0.759	0.728	0.704	0.685	0.669	0.655
0.5	1.088	1.088	1.088	1.088	0.930	0.749	0.720	0.699	0.681	0.667	0.655
0.4	1.071	1.071	1.071	1.071	1.071	0.764	0.712	0.693	0.678	0.665	0.655
0.3	1.055	1.055	1.055	1.055	1.055	0.810	0.703	0.687	0.674	0.664	0.655
0.2	1.039	1.039	1.039	1.039	1.039	1.039	0.694	0.680	0.670	0.662	0.655
0.1	1.023	1.023	1.023	1.023	1.023	1.023	0.684	0.674	0.666	0.660	0.655
0.0	1.007	1.007	1.007	1.007	1.007	1.007	0.674	0.667	0.662	0.658	0.655

Table C.9

CYCLOIDAL LOADING COEFFICIENTS											
PRE-LOAD RATIO K	INERTIA RATIO, Q										
	0.00	0.10	0.20	0.30	0.40	0.50	0.60	0.70	0.80	0.90	1.00
	LOAD MIX COEFFICIENT Cm										
1.0	1.000	1.000	1.000	1.000	1.000	1.000	1.000	1.000	1.000	1.000	1.000
0.9	1.000	0.919	0.928	0.937	0.946	0.955	0.964	0.973	0.982	0.991	1.000
0.8	1.000	0.900	0.856	0.874	0.891	0.909	0.927	0.946	0.964	0.982	1.000
0.7	1.000	0.900	0.800	0.811	0.837	0.864	0.891	0.918	0.946	0.973	1.000
0.6	1.000	0.900	0.800	0.749	0.784	0.819	0.855	0.891	0.927	0.964	1.000
0.5	1.000	0.900	0.800	0.700	0.730	0.774	0.819	0.864	0.909	0.955	1.000
0.4	1.000	0.900	0.800	0.700	0.678	0.730	0.783	0.837	0.891	0.946	1.000
0.3	1.000	0.900	0.800	0.700	0.625	0.685	0.747	0.810	0.873	0.936	1.000
0.2	1.000	0.900	0.800	0.700	0.600	0.641	0.711	0.783	0.855	0.927	1.000
0.1	1.000	0.900	0.800	0.700	0.600	0.597	0.676	0.756	0.837	0.918	1.000
0.0	1.000	0.900	0.800	0.700	0.600	0.553	0.640	0.729	0.819	0.909	1.000
PRE-LOAD RATIO K	INERTIA RATIO, Q										
	0.00	0.10	0.20	0.30	0.40	0.50	0.60	0.70	0.80	0.90	1.00
	INPUT TORQUE COEFFICIENT Cc										
1.0	2.000	1.821	1.687	1.588	1.514	1.457	1.412	1.376	1.346	1.321	1.299
0.9	1.902	1.873	1.714	1.602	1.521	1.460	1.414	1.376	1.346	1.321	1.299
0.8	1.809	1.805	1.746	1.618	1.529	1.464	1.415	1.377	1.346	1.320	1.299
0.7	1.721	1.701	1.751	1.635	1.537	1.468	1.416	1.377	1.346	1.320	1.299
0.6	1.638	1.601	1.635	1.654	1.546	1.471	1.418	1.377	1.345	1.320	1.299
0.5	1.562	1.506	1.521	1.648	1.555	1.475	1.419	1.377	1.345	1.320	1.299
0.4	1.491	1.418	1.410	1.525	1.565	1.479	1.420	1.378	1.345	1.320	1.299
0.3	1.426	1.337	1.302	1.402	1.576	1.483	1.422	1.378	1.345	1.319	1.299
0.2	1.366	1.263	1.200	1.281	1.516	1.488	1.423	1.378	1.345	1.319	1.299
0.1	1.311	1.196	1.105	1.160	1.390	1.492	1.424	1.378	1.344	1.319	1.299
0.0	1.261	1.137	1.019	1.041	1.264	1.497	1.426	1.378	1.344	1.319	1.299

Table C.10

CYCC.50 LOADING COEFFICIENTS											
PRE-LOAD RATIO K	INERTIA RATIO, Q										
	0.00	0.10	0.20	0.30	0.40	0.50	0.60	0.70	0.80	0.90	1.00
	LOAD MIX COEFFICIENT Cm										
1.0	1.000	1.000	1.000	1.000	1.000	1.000	1.000	1.000	1.000	1.000	1.000
0.9	1.000	0.913	0.922	0.932	0.942	0.952	0.961	0.971	0.981	0.990	1.000
0.8	1.000	0.900	0.845	0.864	0.884	0.903	0.922	0.942	0.961	0.981	1.000
0.7	1.000	0.900	0.800	0.797	0.826	0.855	0.884	0.913	0.942	0.971	1.000
0.6	1.000	0.900	0.800	0.729	0.767	0.806	0.845	0.884	0.922	0.961	1.000
0.5	1.000	0.900	0.800	0.700	0.709	0.758	0.806	0.855	0.903	0.952	1.000
0.4	1.000	0.900	0.800	0.700	0.651	0.709	0.767	0.826	0.884	0.942	1.000
0.3	1.000	0.900	0.800	0.700	0.600	0.661	0.729	0.796	0.864	0.932	1.000
0.2	1.000	0.900	0.800	0.700	0.600	0.613	0.690	0.767	0.845	0.922	1.000
0.1	1.000	0.900	0.800	0.700	0.600	0.564	0.651	0.738	0.826	0.913	1.000
0.0	1.000	0.900	0.800	0.700	0.600	0.500	0.613	0.709	0.806	0.903	1.000
PRE-LOAD RATIO K	INERTIA RATIO, Q										
	0.00	0.10	0.20	0.30	0.40	0.50	0.60	0.70	0.80	0.90	1.00
	TORQUE COEFFICIENT Cc										
1.0	1.333	1.214	1.124	1.058	1.009	0.971	0.942	0.917	0.897	0.880	0.866
0.9	1.311	1.293	1.137	1.053	1.004	0.966	0.937	0.914	0.895	0.879	0.866
0.8	1.289	1.289	1.220	1.048	0.998	0.961	0.933	0.911	0.893	0.879	0.866
0.7	1.268	1.267	1.267	1.113	0.991	0.956	0.929	0.908	0.891	0.878	0.866
0.6	1.247	1.244	1.244	1.195	0.985	0.950	0.924	0.905	0.889	0.877	0.866
0.5	1.226	1.222	1.222	1.222	1.034	0.944	0.920	0.901	0.887	0.875	0.866
0.4	1.206	1.200	1.200	1.200	1.105	0.937	0.915	0.898	0.885	0.874	0.866
0.3	1.186	1.178	1.178	1.178	1.178	0.930	0.910	0.894	0.883	0.873	0.866
0.2	1.166	1.156	1.156	1.156	1.156	0.943	0.904	0.891	0.880	0.872	0.866
0.1	1.147	1.133	1.133	1.133	1.133	1.004	0.899	0.887	0.878	0.871	0.866
0.0	1.128	1.111	1.111	1.111	1.111	1.111	0.893	0.883	0.876	0.870	0.866

Table C.11

	POLYNOMIAL 3-4-5 LOADING COEFFICIENTS										
PRE-LOAD RATIO K	INERTIA RATIO, Q										
	0.00	0.10	0.20	0.30	0.40	0.50	0.60	0.70	0.80	0.90	1.00
	LOAD MIX COEFFICIENT Cm										
1.0	1.000	1.000	1.000	1.000	1.000	1.000	1.000	1.000	1.000	1.000	1.000
0.9	1.000	0.917	0.926	0.935	0.944	0.953	0.963	0.972	0.981	0.991	1.000
0.8	1.000	0.900	0.852	0.870	0.888	0.907	0.925	0.944	0.963	0.981	1.000
0.7	1.000	0.900	0.800	0.806	0.833	0.861	0.888	0.916	0.944	0.972	1.000
0.6	1.000	0.900	0.800	0.742	0.778	0.814	0.851	0.888	0.925	0.963	1.000
0.5	1.000	0.900	0.800	0.700	0.723	0.768	0.814	0.860	0.907	0.953	1.000
0.4	1.000	0.900	0.800	0.700	0.668	0.722	0.777	0.833	0.888	0.944	1.000
0.3	1.000	0.900	0.800	0.700	0.613	0.676	0.740	0.805	0.870	0.935	1.000
0.2	1.000	0.900	0.800	0.700	0.600	0.630	0.703	0.777	0.851	0.925	1.000
0.1	1.000	0.900	0.800	0.700	0.600	0.585	0.666	0.749	0.832	0.916	1.000
0.0	1.000	0.900	0.800	0.700	0.600	0.539	0.630	0.721	0.814	0.907	1.000
PRE-LOAD RATIO K	INERTIA RATIO, Q										
	0.00	0.10	0.20	0.30	0.40	0.50	0.60	0.70	0.80	0.90	1.00
	INPUT TORQUE COEFFICIENT Cc										
1.0	1.875	1.705	1.570	1.468	1.391	1.330	1.282	1.243	1.210	1.183	1.159
0.9	1.783	1.757	1.599	1.483	1.397	1.333	1.282	1.242	1.210	1.183	1.159
0.8	1.697	1.690	1.632	1.499	1.404	1.335	1.283	1.242	1.209	1.182	1.159
0.7	1.615	1.594	1.629	1.517	1.412	1.338	1.283	1.242	1.208	1.182	1.159
0.6	1.539	1.502	1.521	1.537	1.420	1.341	1.284	1.241	1.208	1.181	1.159
0.5	1.468	1.416	1.416	1.513	1.429	1.343	1.284	1.240	1.207	1.181	1.159
0.4	1.404	1.337	1.315	1.398	1.439	1.346	1.284	1.240	1.206	1.180	1.159
0.3	1.344	1.264	1.219	1.283	1.450	1.350	1.285	1.239	1.206	1.180	1.159
0.2	1.290	1.198	1.129	1.170	1.364	1.353	1.285	1.238	1.205	1.179	1.159
0.1	1.240	1.139	1.047	1.059	1.245	1.357	1.285	1.238	1.204	1.179	1.159
0.0	1.195	1.085	0.974	0.951	1.126	1.360	1.285	1.237	1.203	1.178	1.159

Table C.12

	POLYNOMIAL 3-4-5-6-R LOADING COEFFICIENTS										
PRE-LOAD RATIO K	INERTIA RATIO, Q										
	0.00	0.10	0.20	0.30	0.40	0.50	0.60	0.70	0.80	0.90	1.00
	LOAD MIX COEFFICIENT Cm										
1.0	1.000	1.000	1.000	1.000	1.000	1.000	1.000	1.000	1.000	1.000	1.000
0.9	1.000	0.928	0.934	0.941	0.949	0.957	0.966	0.974	0.983	0.991	1.000
0.8	1.000	0.900	0.872	0.885	0.900	0.916	0.932	0.949	0.965	0.983	1.000
0.7	1.000	0.900	0.812	0.830	0.852	0.875	0.899	0.923	0.948	0.974	1.000
0.6	1.000	0.900	0.800	0.777	0.805	0.835	0.866	0.898	0.932	0.965	1.000
0.5	1.000	0.900	0.800	0.725	0.759	0.795	0.834	0.874	0.915	0.957	1.000
0.4	1.000	0.900	0.800	0.700	0.713	0.756	0.802	0.849	0.898	0.948	1.000
0.3	1.000	0.900	0.800	0.700	0.668	0.717	0.770	0.825	0.882	0.940	1.000
0.2	1.000	0.900	0.800	0.700	0.624	0.679	0.739	0.801	0.865	0.931	1.000
0.1	1.000	0.900	0.800	0.700	0.600	0.641	0.707	0.777	0.849	0.923	1.000
0.0	1.000	0.900	0.800	0.700	0.600	0.604	0.676	0.753	0.832	0.915	1.000
PRE-LOAD RATIO K	INERTIA RATIO, Q										
	0.00	0.10	0.20	0.30	0.40	0.50	0.60	0.70	0.80	0.90	1.00
	INPUT TORQUE COEFFICIENT Cc										
1.0	2.000	1.850	1.762	1.702	1.659	1.624	1.597	1.574	1.554	1.537	1.522
0.9	1.901	1.886	1.788	1.721	1.672	1.635	1.604	1.579	1.557	1.539	1.522
0.8	1.805	1.836	1.810	1.737	1.685	1.644	1.611	1.584	1.560	1.540	1.522
0.7	1.711	1.727	1.830	1.753	1.696	1.652	1.617	1.588	1.563	1.542	1.522
0.6	1.622	1.619	1.742	1.767	1.707	1.660	1.623	1.592	1.566	1.543	1.522
0.5	1.536	1.512	1.628	1.781	1.717	1.668	1.629	1.596	1.568	1.544	1.522
0.4	1.456	1.407	1.514	1.727	1.727	1.676	1.634	1.600	1.571	1.545	1.522
0.3	1.382	1.303	1.401	1.610	1.737	1.683	1.640	1.604	1.573	1.547	1.522
0.2	1.313	1.201	1.288	1.493	1.746	1.690	1.645	1.608	1.576	1.548	1.522
0.1	1.251	1.101	1.176	1.377	1.697	1.697	1.650	1.611	1.578	1.549	1.522
0.0	1.194	1.029	1.064	1.260	1.578	1.704	1.655	1.615	1.580	1.550	1.522

Table C.13

LOW IMPACT POLYNOMIAL LOADING COEFFICIENTS										

PRE-LOAD RATIO K	INERTIA RATIO, Q									
	0.00	0.10	0.20	0.30	0.40	0.50	0.60	0.70	0.80	0.90	1.00
	LOAD MIX COEFFICIENT Cm									
1.0	1.000	1.000	1.000	1.000	1.000	1.000	1.000	1.000	1.000	1.000	1.000
0.9	1.000	0.912	0.921	0.931	0.941	0.951	0.961	0.971	0.980	0.990	1.000
0.8	1.000	0.900	0.843	0.863	0.882	0.902	0.921	0.941	0.961	0.980	1.000
0.7	1.000	0.900	0.800	0.794	0.823	0.853	0.882	0.912	0.941	0.971	1.000
0.6	1.000	0.900	0.800	0.725	0.764	0.804	0.843	0.882	0.921	0.961	1.000
0.5	1.000	0.900	0.800	0.700	0.706	0.755	0.804	0.853	0.902	0.951	1.000
0.4	1.000	0.900	0.800	0.700	0.647	0.705	0.764	0.823	0.882	0.941	1.000
0.3	1.000	0.900	0.800	0.700	0.600	0.656	0.725	0.794	0.862	0.931	1.000
0.2	1.000	0.900	0.800	0.700	0.600	0.607	0.686	0.764	0.843	0.921	1.000
0.1	1.000	0.900	0.800	0.700	0.600	0.558	0.647	0.735	0.823	0.912	1.000
0.0	1.000	0.900	0.800	0.700	0.600	0.509	0.607	0.705	0.804	0.902	1.000

PRE-LOAD RATIO K	INERTIA RATIO, Q									
	0.00	0.10	0.20	0.30	0.40	0.50	0.60	0.70	0.80	0.90	1.00
	INPUT TORQUE COEFFICIENT Cc									
1.0	1.475	1.329	1.190	1.061	0.956	0.883	0.830	0.790	0.759	0.733	0.712
0.9	1.408	1.387	1.218	1.066	0.949	0.875	0.823	0.785	0.755	0.731	0.712
0.8	1.349	1.340	1.267	1.081	0.941	0.866	0.817	0.780	0.752	0.730	0.712
0.7	1.297	1.282	1.271	1.115	0.932	0.857	0.809	0.775	0.749	0.728	0.712
0.6	1.250	1.230	1.212	1.158	0.936	0.847	0.802	0.770	0.746	0.727	0.712
0.5	1.207	1.182	1.158	1.139	0.958	0.837	0.794	0.764	0.742	0.725	0.712
0.4	1.169	1.139	1.109	1.082	0.986	0.826	0.786	0.759	0.739	0.724	0.712
0.3	1.134	1.100	1.065	1.031	1.002	0.814	0.777	0.753	0.735	0.722	0.712
0.2	1.103	1.065	1.024	0.983	0.946	0.801	0.768	0.747	0.732	0.721	0.712
0.1	1.073	1.032	0.987	0.940	0.894	0.787	0.759	0.741	0.728	0.719	0.712
0.0	1.046	1.001	0.952	0.899	0.845	0.785	0.749	0.735	0.725	0.717	0.712

APPENDIX D

Moment of Inertia

'Inertia' in common language means a tendency to remain stationary. In physics it means the reluctance of a body to change its state of rest or its *state of motion*. It is measured by mass for linear motion and by moment of inertia for rotary motion. Generally in this book the term 'mass' is used for inertia in a linear motion context and the term 'inertia' is used as a short form of 'moment of inertia' in a rotary motion context.

The moment of inertia is defined as

$$I = W . k^2 \qquad\qquad \text{Eq. (D.1)}$$

where:

W is the mass of the body (for all practical purposes numerically equivalent to its weight if coherent units are used, see Chapter 2).

k is the radius of gyration, which is a radius at which all the mass may be considered to be concentrated to give the same relationship between torque and angular acceleration.

Whenever rotary motion is involved in a cam-driven mechanism it is necessary to estimate the inertia of rotating or swinging parts in order to calculate the load on the mechanism. This can be a complicated process, but it is made easier by breaking down each part into components of simple shape: cylinders, discs, rectangular blocks, etc. The inertias of the components (all referred to the same axis of rotation) can be added together to find the inertia of the whole.

When the part contains a void that conforms to a simple shape it is sometimes convenient to calculate the inertia of the solid part, ignoring the void, calculate the inertia of the void as if it were solid and then subtract one from the other. Usually inertia calculations are sufficiently accurate if small details such as fillets, bolt holes, etc. are ignored. Similarly, if a dimension is very small, such as the thickness of a very thin plate, it may be eliminated from the inertia equation without much loss of accuracy.

A body has minimum (basic) inertia when it rotates about an axis that passes through its centre of gravity (its central axis). When a body rotates about some other axis parallel to the central axis, offset by a distance C, its inertia is

$$I = W . k^2 + W . C^2 \qquad \text{or} \qquad W . (k^2 + C^2) \qquad \text{Eq. (D.2)}$$

The following equations allow for an offset C, but the basic inertia can be found by putting $C = 0$. The dimensions in the equations are defined by reference to the accompanying numbered diagrams.

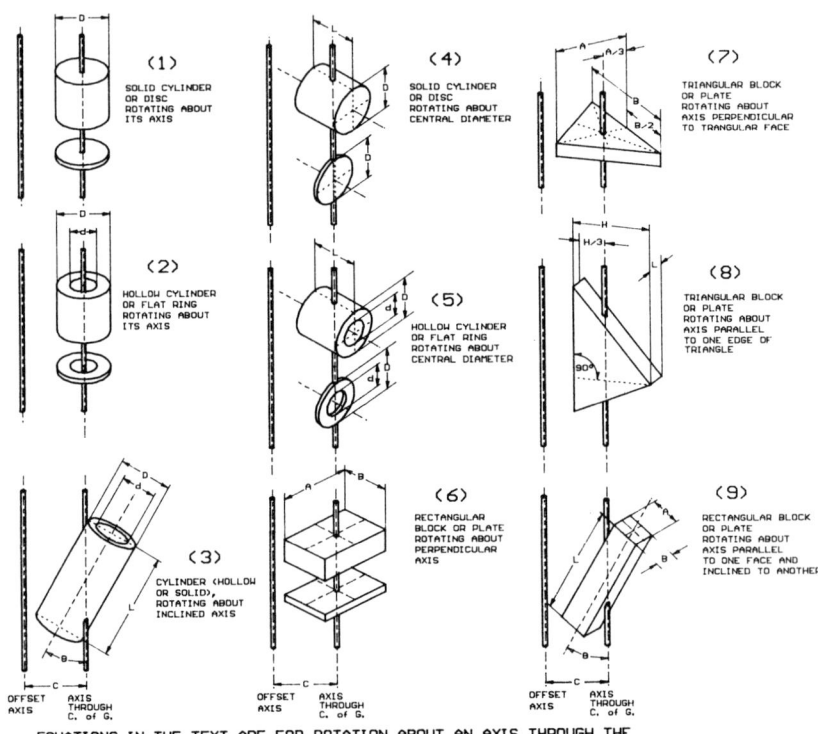

EQUATIONS IN THE TEXT ARE FOR ROTATION ABOUT AN AXIS THROUGH THE CENTRE OF GRAVITY OR ROTATION ABOUT AN OFFSET PARALLEL AXIS

Fig. D.1   Inertia of simple shapes

(1) Solid cylinder or disc rotating about its axis

$$I = W . \left(\frac{D^2}{8} + C^2 \right) \qquad \text{Eq. (D.3)}$$

(2) Hollow cylinder or disc rotating about its axis

$$I = W.\left[\frac{(D^2 + d^2)}{8} + C^2\right]$$ Eq. (D.4)

Note that by putting $d = 0$ this becomes the same as Equation (D.3).

(3) Hollow cylinder rotating about an inclined axis

$$I = W.\left[\frac{(D^2 + d^2)}{16}.(1 + \cos^2 \Theta) + \frac{L^2}{12}.\sin^2 \Theta + C^2\right]$$ Eq. D.5

Note that by putting $\Theta = 0$ this becomes the same as Equation (D.4) and by putting $\Theta = 90°$ it becomes the same as Equation (D.7).

(4) Solid cylinder rotating about a diameter

$$I = W.\left[\frac{D^2}{16} + \frac{L^2}{12} + C^2\right]$$ Eq. (D.6)

(5) Hollow cylinder rotating about a diameter

$$I = W.\left[\frac{(D^2 + d^2)}{16} + \frac{L^2}{12} + C^2\right]$$ Eq. (D.7)

(6) Rectangular block or plate

$$I = W.\left[(\frac{A^2 + B^2}{12} + C^2\right]$$ Eq. (D.8)

(7) Triangular block or plate rotating about an axis perpendicular to the trianglular face

$$I = W.\left[\frac{A^2}{18} + \frac{B^2}{24} + C^2\right]$$ Eq. (D.9)

A is the length of a *median* of the triangle, i.e. the line from an apex to the middle of the opposite side. The centre of gravity of the triangle lies on the median 2/3 of its length from the apex. This is the point where all three medians intersect.

(8) Triangular block or plate rotating about an axis parallel to one side of the triangle

$$I = W.\left[\frac{H^2}{18} + \frac{B^2}{12} + C^2\right]$$ Eq. (D.10)

H is the perpendicular height of the apex of the triangle from its base. The base is the side to which the rotation axis is parallel.

(9) Rectangular block or plate rotating about an inclined axis

$$I = W.\left[\frac{(A^2 + B^2.\cos^2\Theta + L^2.\sin^2\Theta)}{12} + C^2\right]$$ Eq. (D.11)

Note that by putting $\Theta = 0$ this becomes the same as Equation (D.8).

GEARING EFFECTS

It is necessary to calculate the 'equivalent inertia' I_e of a system when various parts are connected by gearing and run at different speeds. Usually it is required to know the equivalent inertia at the immediate output of the cam follower. Each part of the system can have its inertia converted to an equivalent value at another part of the system by multiplying it by the gear ratio squared.

The gear ratio can be expressed as a speed ratio, ω_1/ω_2 giving the equation

$$I_2 = I_1.(\omega_1/\omega_2)^2$$ Eq. (D.12)

where subscripts 1 and 2 refer to the respective parts of the system. All such converted values referred to one part of the system can be added together to give a total equivalent inertia at that point.

The above applies to conventional transmission gearing, but there is the special case of *planetary* gearing that occurs in some machines. A typical case is an indexing table with a number of workheads spaced on a pitch circle which are rotated about their own axes by a constant speed sun gear as shown in the drawing below. It is necessary to calculate the equivalent inertia of the planetary workheads, referred to the centre of the table, to estimate the load on the indexing mechanism.

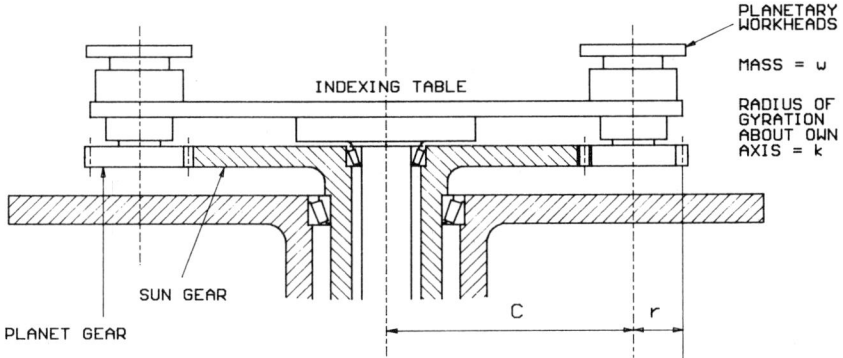

Fig. D.2   Planetary gearing

When the table is accelerated by the indexing cam, not only do the workheads accelerate around their pitch circle, but they also accelerate (or decelerate) about their own axes by virtue of the planetary gear ratio. This increases the equivalent inertia to

$$I_e = W . \left[\left(\frac{C.k}{r} \right)^2 + C^2 \right]$$ Eq. (D.13)

In effect the basic inertia of the workhead *about its own axis* is increased by a factor of $(C/r)^2$ where C is the workhead pitch circle radius and r is the planet gear radius.

APPENDIX E

Backlash Impact Velocities

For a full explanation of backlash impact velocities see Chapter 12.

To take account of various sizes, speeds and loads of all kinds of cam mechanisms it is convenient to use normalised cam law factors and to use the following normalised variables.

$$b_k = B_k/Y \qquad \text{Eq. (E.1)}$$
$$d_n = D_n \cdot T^2/Y \qquad \text{Eq. (E.2)}$$
$$v = V \cdot T/Y \qquad \text{Eq. (E.3)}$$

To calculate the impact velocity due to backlash first calculate the normalised backlash b_k and normalised natural deceleration d_n. Look up the value of the normalised impact velocity v that corresponds to these in the one of the following tables for the appropriate cam law, and calculate the real impact velocity thus:

$$V = v \cdot Y/T \qquad \text{Eq. (E.4)}$$
$$F_s = \sqrt{(W \cdot S \cdot V^2 + F_d^2)} - F_d \quad \text{for linear motion} \qquad \text{Eq. (E.5)}$$
or
$$M_s = \sqrt{(I \cdot R \cdot V^2 + M_d^2)} - M_d \quad \text{for rotary motion} \qquad \text{Eq. (E.6)}$$

where:

b_k = Normalised (dimensionless) backlash.
B_k = Real cumulative backlash referred to the cam output.
d_n = Normalised (dimensionless) natural deceleration.
D_n = Real natural deceleration of the payload = F_d/W for linear motion or M_d/I for rotary motion.
F_d = Decelerating force on the payload, referred to the cam output.
F_s = Peak shock force *in the direction of follower motion* (note that the contact force on the cam is $F_s/\cos\phi$, where ϕ = pressure angle of the cam).
I = Inertia of the payload, referred to the cam output.
M_d = Decelerating torque on the payload, referred to the cam output.
M_s = Peak shock torque.
R = Angular rigidity of a rotary transmission.

S = Linear stiffness of a linear transmission.
T = The full motion period (time).
v = Normalised (dimensionless) impact velocity.
V = Actual impact velocity, linear or angular.
W = Mass of the payload, referred to the cam output.
Y = Output stroke of the motion, linear or angular.

Table E.1

BACKLASH IMPACT VELOCITIES				CAM LAW: PARABOLIC		
SCCA parameters: a = 0 , b = 1 , c = 0 , Cv = 2 , Ca = 4						

Normalised Backlash	Normalised Natural Deceleration						
	0	.5	1	1.5	2	2.5	3
	Normalised Impact Velocity						
0.0000	0.0000	0.0000	0.0000	0.0000	0.0000	0.0000	0.0000
0.0005	0.0633	0.0592	0.0548	0.0500	0.0447	0.0387	0.0316
0.0010	0.0894	0.0837	0.0775	0.0707	0.0632	0.0548	0.0447
0.0015	0.1095	0.1025	0.0949	0.0866	0.0775	0.0671	0.0548
0.0020	0.1265	0.1183	0.1095	0.1000	0.0894	0.0775	0.0632
0.0025	0.1414	0.1323	0.1225	0.1118	0.1000	0.0866	0.0707
0.0030	0.1549	0.1449	0.1342	0.1225	0.1095	0.0949	0.0775
0.0035	0.1673	0.1565	0.1449	0.1323	0.1183	0.1025	0.0837
0.0040	0.1789	0.1673	0.1549	0.1414	0.1265	0.1095	0.0894
0.0045	0.1897	0.1775	0.1643	0.1500	0.1342	0.1162	0.0949
0.0050	0.2000	0.1871	0.1732	0.1581	0.1414	0.1225	0.1000
0.0055	0.2098	0.1962	0.1817	0.1658	0.1483	0.1285	0.1049
0.0060	0.2191	0.2049	0.1897	0.1732	0.1549	0.1342	0.1095
0.0065	0.2280	0.2133	0.1975	0.1803	0.1613	0.1396	0.1140
0.0070	0.2366	0.2214	0.2049	0.1871	0.1673	0.1449	0.1183
0.0075	0.2449	0.2291	0.2121	0.1937	0.1732	0.1500	0.1225
0.0080	0.2530	0.2366	0.2191	0.2000	0.1789	0.1549	0.1265
0.0085	0.2608	0.2439	0.2258	0.2062	0.1844	0.1597	0.1304
0.0090	0.2683	0.2510	0.2324	0.2121	0.1897	0.1643	0.1342
0.0095	0.2757	0.2579	0.2387	0.2179	0.1949	0.1688	0.1378
0.0100	0.2828	0.2646	0.2449	0.2236	0.2000	0.1732	0.1414
0.0125	0.3162	0.2958	0.2739	0.2500	0.2236	0.1937	0.1581
0.0150	0.3464	0.3240	0.3000	0.2739	0.2449	0.2121	0.1732
0.0175	0.3742	0.3500	0.3240	0.2958	0.2646	0.2291	0.1871
0.0200	0.4000	0.3742	0.3464	0.3162	0.2828	0.2449	0.2000
0.0225	0.4243	0.3969	0.3674	0.3354	0.3000	0.2598	0.2121
0.0250	0.4472	0.4183	0.3873	0.3536	0.3162	0.2739	0.2236
0.0275	0.4690	0.4387	0.4062	0.3708	0.3317	0.2872	0.2345
0.0300	0.4899	0.4583	0.4243	0.3873	0.3464	0.3000	0.2450
0.0325	0.5099	0.4770	0.4416	0.4031	0.3606	0.3122	0.2550
0.0350	0.5292	0.4950	0.4583	0.4183	0.3742	0.3240	0.2646
0.0375	0.5477	0.5123	0.4743	0.4330	0.3873	0.3354	0.2739
0.0400	0.5657	0.5292	0.4899	0.4472	0.4000	0.3464	0.2828
0.0425	0.5831	0.5454	0.5050	0.4610	0.4123	0.3571	0.2915
0.0450	0.6000	0.5612	0.5196	0.4743	0.4243	0.3674	0.3000
0.0475	0.6164	0.5766	0.5339	0.4873	0.4359	0.3775	0.3082
0.0500	0.6325	0.5916	0.5477	0.5000	0.4472	0.3873	0.3162

Table E.2

BACKLASH IMPACT VELOCITIES				CAM LAW: SIMPLE HARMONIC					

SCCA parameters: a = 0 , b = 0 , c = 1 , Cv = 1.570796 , Ca = 4.934803

Normalised Backlash	0	.5	1	Normalised Natural Deceleration 1.5	2	2.5	3	3.5	4	4.5
Backlash				Normalised Impact Velocity						
0.0000	0.0000	0.0000	0.0000	0.0000	0.0000	0.0000	0.0000	0.0000	0.0000	0.0000
0.0005	0.0259	0.0258	0.0256	0.0252	0.0248	0.0242	0.0233	0.0222	0.0205	0.0172
0.0010	0.0411	0.0408	0.0404	0.0399	0.0391	0.0381	0.0367	0.0348	0.0319	0.0263
0.0015	0.0538	0.0534	0.0529	0.0521	0.0511	0.0497	0.0479	0.0453	0.0413	0.0334
0.0020	0.0651	0.0646	0.0640	0.0630	0.0617	0.0600	0.0577	0.0544	0.0494	0.0340
0.0025	0.0755	0.0749	0.0741	0.0730	0.0714	0.0694	0.0666	0.0628	0.0568	
0.0030	0.0852	0.0845	0.0836	0.0822	0.0804	0.0781	0.0749	0.0705	0.0636	
0.0035	0.0943	0.0936	0.0925	0.0910	0.0890	0.0863	0.0827	0.0777	0.0699	
0.0040	0.1031	0.1022	0.1010	0.0993	0.0971	0.0941	0.0901	0.0845	0.0758	
0.0045	0.1114	0.1105	0.1091	0.1073	0.1048	0.1015	0.0972	0.0910	0.0813	
0.0050	0.1194	0.1184	0.1169	0.1149	0.1122	0.1087	0.1039	0.0972	0.0866	
0.0055	0.1272	0.1261	0.1245	0.1223	0.1194	0.1155	0.1104	0.1032	0.0917	
0.0060	0.1347	0.1335	0.1318	0.1294	0.1263	0.1222	0.1167	0.1089	0.0965	
0.0065	0.1421	0.1408	0.1389	0.1363	0.1330	0.1286	0.1228	0.1144	0.1012	
0.0070	0.1492	0.1478	0.1458	0.1431	0.1395	0.1349	0.1286	0.1198	0.1057	
0.0075	0.1561	0.1546	0.1525	0.1497	0.1459	0.1410	0.1344	0.1250	0.1100	
0.0080	0.1629	0.1613	0.1591	0.1561	0.1521	0.1469	0.1399	0.1300	0.1141	
0.0085	0.1696	0.1679	0.1655	0.1623	0.1582	0.1527	0.1454	0.1349	0.1181	
0.0090	0.1761	0.1743	0.1718	0.1685	0.1641	0.1584	0.1507	0.1397	0.1220	
0.0095	0.1825	0.1806	0.1780	0.1745	0.1699	0.1639	0.1559	0.1444	0.1120	
0.0100	0.1887	0.1868	0.1840	0.1804	0.1756	0.1693	0.1609	0.1489	0.0924	
0.0125	0.2186	0.2161	0.2128	0.2083	0.2026	0.1950	0.1849	0.1703		
0.0150	0.2464	0.2434	0.2395	0.2343	0.2275	0.2187	0.2068	0.1897		
0.0175	0.2725	0.2692	0.2646	0.2586	0.2509	0.2408	0.2272	0.2076		
0.0200	0.2974	0.2935	0.2884	0.2817	0.2730	0.2617	0.2464	0.2241		
0.0225	0.3212	0.3168	0.3111	0.3036	0.2940	0.2814	0.2644	0.2260		
0.0250	0.3440	0.3392	0.3329	0.3247	0.3141	0.3003	0.2816	0.1833		
0.0275	0.3660	0.3607	0.3538	0.3449	0.3333	0.3183	0.2979	0.1268		
0.0300	0.3873	0.3815	0.3740	0.3643	0.3519	0.3356	0.3135			
0.0325	0.4079	0.4017	0.3936	0.3832	0.3698	0.3523	0.3284			
0.0350	0.4280	0.4213	0.4126	0.4014	0.3871	0.3684	0.3428			
0.0375	0.4475	0.4403	0.4310	0.4191	0.4039	0.3839	0.3566			
0.0400	0.4665	0.4589	0.4490	0.4364	0.4202	0.3990	0.3699			
0.0425	0.4851	0.4770	0.4665	0.4531	0.4360	0.4136	0.3526			
0.0450	0.5033	0.4947	0.4836	0.4695	0.4514	0.4278	0.3306			
0.0475	0.5211	0.5120	0.5003	0.4855	0.4665	0.4416	0.3071			
0.0500	0.5385	0.5290	0.5167	0.5011	0.4812	0.4550	0.2816			

Table E.3

BACKLASH IMPACT VELOCITIES					CAM LAW: MOD TRAP				

SCCA parameters: a = .25 , b = .5 , c = .25 ,   Cv = 2 , Ca = 4.888124

Normalised Backlash	Normalised Natural Deceleration									
	0	.5	1	1.5	2	2.5	3	3.5	4	4.5
	Normalised Impact Velocity									
0.0000	0.0000	0.0000	0.0000	0.0000	0.0000	0.0000	0.0000	0.0000	0.0000	0.0000
0.0005	0.0406	0.0402	0.0396	0.0389	0.0379	0.0365	0.0347	0.0322	0.0280	0.0195
0.0010	0.0640	0.0633	0.0622	0.0608	0.0590	0.0566	0.0534	0.0487	0.0409	0.0277
0.0015	0.0835	0.0823	0.0808	0.0787	0.0762	0.0728	0.0681	0.0613	0.0506	0.0340
0.0020	0.1006	0.0991	0.0971	0.0944	0.0911	0.0867	0.0806	0.0717	0.0587	0.0393
0.0025	0.1162	0.1143	0.1118	0.1086	0.1045	0.0991	0.0916	0.0808	0.0658	0.0440
0.0030	0.1307	0.1284	0.1254	0.1216	0.1167	0.1103	0.1014	0.0890	0.0723	
0.0035	0.1443	0.1416	0.1381	0.1337	0.1280	0.1206	0.1103	0.0965	0.0782	
0.0040	0.1571	0.1540	0.1501	0.1450	0.1386	0.1301	0.1185	0.1034	0.0837	
0.0045	0.1693	0.1658	0.1614	0.1557	0.1485	0.1390	0.1263	0.1099	0.0888	
0.0050	0.1809	0.1771	0.1721	0.1659	0.1579	0.1473	0.1335	0.1161	0.0937	
0.0055	0.1921	0.1878	0.1824	0.1756	0.1668	0.1552	0.1404	0.1219	0.0983	
0.0060	0.2029	0.1982	0.1923	0.1848	0.1752	0.1627	0.1470	0.1275	0.1027	
0.0065	0.2133	0.2082	0.2018	0.1937	0.1833	0.1699	0.1533	0.1328	0.1070	
0.0070	0.2234	0.2179	0.2109	0.2022	0.1910	0.1768	0.1593	0.1379	0.1110	
0.0075	0.2331	0.2272	0.2198	0.2104	0.1984	0.1834	0.1651	0.1429	0.1150	
0.0080	0.2426	0.2362	0.2283	0.2183	0.2056	0.1898	0.1708	0.1477	0.1188	
0.0085	0.2518	0.2450	0.2366	0.2259	0.2125	0.1960	0.1762	0.1523	0.1224	
0.0090	0.2608	0.2536	0.2446	0.2333	0.2192	0.2020	0.1815	0.1568	0.1260	
0.0095	0.2695	0.2619	0.2524	0.2405	0.2257	0.2078	0.1866	0.1611	0.1295	
0.0100	0.2780	0.2700	0.2600	0.2474	0.2320	0.2135	0.1916	0.1654	0.1329	
0.0125	0.3179	0.3077	0.2950	0.2796	0.2612	0.2398	0.2148	0.1852		
0.0150	0.3539	0.3415	0.3263	0.3084	0.2876	0.2636	0.2358	0.2031		
0.0175	0.3869	0.3722	0.3548	0.3347	0.3117	0.2853	0.2550	0.2195		
0.0200	0.4173	0.4006	0.3813	0.3591	0.3340	0.3055	0.2729	0.2348		
0.0225	0.4456	0.4271	0.4060	0.3820	0.3550	0.3245	0.2897	0.2491		
0.0250	0.4722	0.4521	0.4292	0.4036	0.3748	0.3424	0.3055			
0.0275	0.4974	0.4757	0.4513	0.4240	0.3936	0.3594	0.3206			
0.0300	0.5214	0.4982	0.4724	0.4436	0.4115	0.3756	0.3350			
0.0325	0.5443	0.5198	0.4925	0.4623	0.4287	0.3912	0.3488			
0.0350	0.5663	0.5405	0.5119	0.4802	0.4452	0.4062	0.3621			
0.0375	0.5875	0.5604	0.5305	0.4976	0.4611	0.4206	0.3749			
0.0400	0.6080	0.5797	0.5485	0.5143	0.4766	0.4346	0.3873			
0.0425	0.6277	0.5983	0.5660	0.5305	0.4915	0.4481	0.3993			
0.0450	0.6469	0.6164	0.5829	0.5462	0.5059	0.4612				
0.0475	0.6655	0.6339	0.5993	0.5615	0.5200	0.4740				
0.0500	0.6837	0.6510	0.6153	0.5764	0.5337	0.4864				

Table E.4

BACKLASH IMPACT VELOCITIES				CAM LAW: MOD SINE						

SCCA parameters: a = .25 , b = 0 , c = .75 , Cv = 1.759603 , Ca = 5.527957

Normalised Backlash	Normalised Natural Deceleration									
	0	.5	1	1.5	2	2.5	3	3.5	4	4.5	5

Normalised Backlash	Normalised Impact Velocity										
0.0000	0.0000	0.0000	0.0000	0.0000	0.0000	0.0000	0.0000	0.0000	0.0000	0.0000	0.0000
0.0005	0.0296	0.0294	0.0292	0.0289	0.0285	0.0279	0.0272	0.0262	0.0250	0.0231	0.0196
0.0010	0.0469	0.0466	0.0463	0.0457	0.0450	0.0440	0.0428	0.0412	0.0390	0.0358	0.0298
0.0015	0.0614	0.0610	0.0604	0.0597	0.0587	0.0574	0.0557	0.0535	0.0506	0.0461	0.0376
0.0020	0.0743	0.0738	0.0731	0.0721	0.0709	0.0692	0.0672	0.0644	0.0607	0.0550	0.0437
0.0025	0.0861	0.0855	0.0846	0.0835	0.0820	0.0800	0.0776	0.0743	0.0698	0.0630	0.0477
0.0030	0.0972	0.0964	0.0954	0.0941	0.0923	0.0901	0.0872	0.0835	0.0783	0.0703	0.0492
0.0035	0.1076	0.1068	0.1056	0.1041	0.1021	0.0995	0.0963	0.0920	0.0861	0.0770	0.0476
0.0040	0.1175	0.1166	0.1153	0.1135	0.1113	0.1085	0.1049	0.1001	0.0935	0.0833	0.0414
0.0045	0.1270	0.1260	0.1245	0.1226	0.1202	0.1171	0.1131	0.1078	0.1006	0.0892	0.0246
0.0050	0.1362	0.1350	0.1334	0.1313	0.1286	0.1253	0.1209	0.1152	0.1072	0.0948	
0.0055	0.1450	0.1437	0.1420	0.1397	0.1368	0.1332	0.1285	0.1223	0.1136	0.1001	
0.0060	0.1535	0.1522	0.1503	0.1478	0.1447	0.1408	0.1357	0.1291	0.1198	0.1051	
0.0065	0.1618	0.1604	0.1583	0.1557	0.1524	0.1482	0.1428	0.1356	0.1257	0.1098	
0.0070	0.1699	0.1683	0.1662	0.1634	0.1598	0.1554	0.1496	0.1420	0.1314	0.1141	
0.0075	0.1778	0.1761	0.1738	0.1708	0.1671	0.1623	0.1562	0.1482	0.1369	0.1181	
0.0080	0.1855	0.1837	0.1813	0.1781	0.1742	0.1691	0.1627	0.1542	0.1422	0.1217	
0.0085	0.1931	0.1911	0.1885	0.1852	0.1811	0.1758	0.1690	0.1600	0.1474	0.1248	
0.0090	0.2004	0.1984	0.1957	0.1922	0.1878	0.1822	0.1751	0.1657	0.1524	0.1276	
0.0095	0.2077	0.2055	0.2027	0.1990	0.1944	0.1886	0.1811	0.1712	0.1573	0.1299	
0.0100	0.2148	0.2125	0.2095	0.2057	0.2009	0.1948	0.1870	0.1767	0.1620	0.1317	
0.0125	0.2485	0.2457	0.2420	0.2374	0.2315	0.2241	0.2147	0.2020	0.1840	0.1333	
0.0150	0.2799	0.2766	0.2722	0.2667	0.2598	0.2512	0.2400	0.2251	0.2034	0.1180	
0.0175	0.3095	0.3055	0.3005	0.2942	0.2863	0.2763	0.2635	0.2464	0.2195	0.0614	
0.0200	0.3375	0.3330	0.3273	0.3201	0.3112	0.3000	0.2855	0.2661	0.2319		
0.0225	0.3642	0.3592	0.3528	0.3448	0.3349	0.3224	0.3062	0.2845	0.2402		
0.0250	0.3899	0.3843	0.3772	0.3684	0.3575	0.3437	0.3259	0.3016	0.2443		
0.0275	0.4146	0.4084	0.4007	0.3911	0.3791	0.3640	0.3445	0.3173	0.2435		
0.0300	0.4385	0.4317	0.4233	0.4128	0.3999	0.3835	0.3623	0.3314	0.2371		
0.0325	0.4616	0.4543	0.4451	0.4339	0.4199	0.4022	0.3793	0.3436	0.2239		
0.0350	0.4840	0.4761	0.4663	0.4542	0.4392	0.4202	0.3956	0.3540	0.2013		
0.0375	0.5058	0.4973	0.4869	0.4739	0.4579	0.4376	0.4113	0.3625	0.1637		
0.0400	0.5270	0.5180	0.5068	0.4931	0.4760	0.4544	0.4261	0.3690	0.0900		
0.0425	0.5477	0.5381	0.5263	0.5117	0.4936	0.4707	0.4401	0.3734			
0.0450	0.5679	0.5578	0.5453	0.5298	0.5107	0.4865	0.4531	0.3756			
0.0475	0.5877	0.5770	0.5638	0.5475	0.5273	0.5018	0.4652	0.3755			
0.0500	0.6071	0.5958	0.5819	0.5647	0.5435	0.5166	0.4762	0.3729			

Cams for Industry

Table E.5

BACKLASH IMPACT VELOCITIES					CAM LAW: MSC.20							

SCCA parameters: a = .2 , b = 0 , c = .6 , Cv = 1.527539 , Ca = 5.998631

Normalised Backlash	Normalised Natural Deceleration											
	0	.5	1	1.5	2	2.5	3	3.5	4	4.5	5	5.5
	Normalised Impact Velocity											
0.0000	0.0000	0.0000	0.0000	0.0000	0.0000	0.0000	0.0000	0.0000	0.0000	0.0000	0.0000	0.0000
0.0005	0.0327	0.0326	0.0323	0.0321	0.0316	0.0311	0.0304	0.0296	0.0284	0.0269	0.0246	0.0204
0.0010	0.0519	0.0516	0.0512	0.0506	0.0499	0.0490	0.0478	0.0464	0.0444	0.0418	0.0379	0.0301
0.0015	0.0679	0.0674	0.0669	0.0661	0.0651	0.0638	0.0622	0.0602	0.0576	0.0540	0.0485	0.0346
0.0020	0.0821	0.0815	0.0808	0.0798	0.0786	0.0770	0.0750	0.0724	0.0691	0.0646	0.0576	0.0314
0.0025	0.0951	0.0945	0.0936	0.0924	0.0908	0.0889	0.0866	0.0835	0.0796	0.0741	0.0656	
0.0030	0.1073	0.1065	0.1055	0.1041	0.1023	0.1001	0.0973	0.0938	0.0892	0.0828	0.0728	
0.0035	0.1188	0.1179	0.1167	0.1151	0.1130	0.1105	0.1074	0.1034	0.0982	0.0909	0.0793	
0.0040	0.1297	0.1287	0.1273	0.1255	0.1233	0.1205	0.1170	0.1125	0.1067	0.0985	0.0850	
0.0045	0.1402	0.1390	0.1375	0.1355	0.1330	0.1299	0.1261	0.1211	0.1147	0.1056	0.0899	
0.0050	0.1503	0.1490	0.1473	0.1451	0.1424	0.1390	0.1348	0.1294	0.1223	0.1123	0.0937	
0.0055	0.1600	0.1586	0.1567	0.1543	0.1514	0.1477	0.1431	0.1373	0.1296	0.1188	0.0965	
0.0060	0.1694	0.1678	0.1658	0.1633	0.1601	0.1561	0.1512	0.1449	0.1366	0.1249	0.0982	
0.0065	0.1785	0.1768	0.1747	0.1719	0.1685	0.1643	0.1590	0.1523	0.1434	0.1307	0.0986	
0.0070	0.1874	0.1856	0.1833	0.1803	0.1767	0.1722	0.1666	0.1594	0.1499	0.1363	0.0977	
0.0075	0.1961	0.1941	0.1917	0.1885	0.1847	0.1799	0.1739	0.1663	0.1562	0.1417	0.0952	
0.0080	0.2045	0.2025	0.1998	0.1965	0.1924	0.1874	0.1811	0.1730	0.1623	0.1468	0.0908	
0.0085	0.2128	0.2106	0.2078	0.2043	0.2000	0.1947	0.1880	0.1795	0.1682	0.1517	0.0838	
0.0090	0.2209	0.2186	0.2156	0.2120	0.2074	0.2018	0.1948	0.1859	0.1739	0.1562	0.0732	
0.0095	0.2288	0.2264	0.2233	0.2194	0.2147	0.2088	0.2014	0.1921	0.1795	0.1604	0.0560	
0.0100	0.2366	0.2341	0.2308	0.2268	0.2218	0.2156	0.2079	0.1981	0.1850	0.1643		
0.0125	0.2736	0.2704	0.2664	0.2614	0.2553	0.2478	0.2384	0.2264	0.2102	0.1782		
0.0150	0.3080	0.3041	0.2993	0.2934	0.2862	0.2774	0.2663	0.2520	0.2325	0.1820		
0.0175	0.3402	0.3357	0.3302	0.3234	0.3151	0.3049	0.2921	0.2756	0.2511	0.1735		
0.0200	0.3708	0.3656	0.3593	0.3516	0.3422	0.3306	0.3161	0.2974	0.2657	0.1462		
0.0225	0.3999	0.3941	0.3870	0.3784	0.3679	0.3550	0.3387	0.3175	0.2758	0.0687		
0.0250	0.4278	0.4213	0.4135	0.4039	0.3923	0.3780	0.3601	0.3358	0.2811			
0.0275	0.4546	0.4475	0.4389	0.4284	0.4157	0.4000	0.3803	0.3519	0.2812			
0.0300	0.4804	0.4727	0.4633	0.4519	0.4381	0.4210	0.3995	0.3656	0.2754			
0.0325	0.5054	0.4970	0.4869	0.4746	0.4596	0.4411	0.4177	0.3769	0.2625			
0.0350	0.5297	0.5206	0.5097	0.4964	0.4803	0.4604	0.4347	0.3857	0.2403			
0.0375	0.5532	0.5435	0.5317	0.5176	0.5003	0.4790	0.4504	0.3918	0.2045			
0.0400	0.5761	0.5657	0.5532	0.5380	0.5197	0.4970	0.4647	0.3952	0.1482			
0.0425	0.5984	0.5873	0.5740	0.5579	0.5384	0.5142	0.4775	0.3956	0.0445			
0.0450	0.6201	0.6084	0.5943	0.5773	0.5566	0.5308	0.4888	0.3929				
0.0475	0.6414	0.6290	0.6140	0.5961	0.5742	0.5465	0.4986	0.3868				
0.0500	0.6621	0.6491	0.6333	0.6144	0.5913	0.5613	0.5068	0.3768				

Table E.6

BACKLASH IMPACT VELOCITIES				CAM LAW: MSC.33									

SCCA parameters: a = .1666667 , b = 0 , c = .5 , Cv = 1.404087 , Ca = 6.616604

Normalised Backlash	0	.5	1	1.5	2	2.5	3	3.5	4	4.5	5	5.5	6
			Normalised Natural Deceleration										
			Normalised Impact Velocity										
0.0000	0.0000	0.0000	0.0000	0.0000	0.0000	0.0000	0.0000	0.0000	0.0000	0.0000	0.0000	0.0000	0.0000
0.0005	0.0359	0.0358	0.0356	0.0352	0.0348	0.0343	0.0337	0.0330	0.0320	0.0308	0.0292	0.0268	0.0227
0.0010	0.0569	0.0566	0.0562	0.0556	0.0549	0.0541	0.0531	0.0518	0.0501	0.0481	0.0453	0.0412	0.0334
0.0015	0.0744	0.0740	0.0734	0.0726	0.0717	0.0705	0.0690	0.0673	0.0650	0.0621	0.0583	0.0525	0.0378
0.0020	0.0900	0.0894	0.0887	0.0877	0.0865	0.0850	0.0832	0.0809	0.0781	0.0745	0.0695	0.0621	0.0326
0.0025	0.1043	0.1036	0.1027	0.1015	0.1000	0.0982	0.0960	0.0933	0.0899	0.0856	0.0796	0.0705	
0.0030	0.1176	0.1168	0.1157	0.1143	0.1126	0.1105	0.1079	0.1048	0.1009	0.0958	0.0888	0.0778	
0.0035	0.1302	0.1292	0.1279	0.1263	0.1244	0.1220	0.1191	0.1155	0.1111	0.1053	0.0972	0.0839	
0.0040	0.1421	0.1410	0.1396	0.1378	0.1356	0.1329	0.1297	0.1257	0.1207	0.1142	0.1051	0.0884	
0.0045	0.1536	0.1523	0.1507	0.1487	0.1463	0.1434	0.1398	0.1353	0.1298	0.1226	0.1125	0.0914	
0.0050	0.1646	0.1632	0.1614	0.1592	0.1566	0.1533	0.1494	0.1446	0.1385	0.1306	0.1195	0.0926	
0.0055	0.1752	0.1736	0.1717	0.1693	0.1664	0.1629	0.1587	0.1534	0.1468	0.1382	0.1261	0.0917	
0.0060	0.1854	0.1838	0.1817	0.1791	0.1760	0.1722	0.1676	0.1619	0.1548	0.1455	0.1322	0.0885	
0.0065	0.1954	0.1936	0.1913	0.1886	0.1852	0.1812	0.1762	0.1702	0.1625	0.1525	0.1379	0.0822	
0.0070	0.2051	0.2032	0.2007	0.1978	0.1942	0.1899	0.1846	0.1781	0.1699	0.1592	0.1430	0.0713	
0.0075	0.2145	0.2125	0.2099	0.2068	0.2029	0.1983	0.1927	0.1858	0.1771	0.1657	0.1476	0.0515	
0.0080	0.2238	0.2215	0.2188	0.2155	0.2114	0.2066	0.2006	0.1933	0.1841	0.1719	0.1515		
0.0085	0.2328	0.2304	0.2275	0.2240	0.2197	0.2146	0.2083	0.2006	0.1909	0.1780	0.1549		
0.0090	0.2416	0.2391	0.2360	0.2323	0.2278	0.2224	0.2158	0.2077	0.1974	0.1838	0.1577		
0.0095	0.2502	0.2476	0.2444	0.2405	0.2358	0.2301	0.2232	0.2146	0.2038	0.1895	0.1597		
0.0100	0.2587	0.2559	0.2526	0.2485	0.2435	0.2375	0.2303	0.2214	0.2100	0.1949	0.1611		
0.0125	0.2989	0.2955	0.2913	0.2862	0.2802	0.2728	0.2640	0.2530	0.2390	0.2183	0.1564		
0.0150	0.3362	0.3321	0.3271	0.3211	0.3139	0.3052	0.2947	0.2817	0.2649	0.2345	0.1219		
0.0175	0.3712	0.3663	0.3605	0.3536	0.3452	0.3352	0.3231	0.3079	0.2876	0.2427			
0.0200	0.4042	0.3987	0.3921	0.3841	0.3747	0.3633	0.3495	0.3323	0.3063	0.2419			
0.0225	0.4357	0.4295	0.4220	0.4131	0.4025	0.3898	0.3743	0.3547	0.3207	0.2299			
0.0250	0.4658	0.4589	0.4506	0.4407	0.4290	0.4149	0.3977	0.3749	0.3306	0.2027			
0.0275	0.4947	0.4871	0.4779	0.4671	0.4542	0.4387	0.4198	0.3927	0.3356	0.1475			
0.0300	0.5226	0.5142	0.5042	0.4924	0.4784	0.4615	0.4406	0.4077	0.3353				
0.0325	0.5494	0.5403	0.5295	0.5168	0.5015	0.4832	0.4600	0.4199	0.3292				
0.0350	0.5755	0.5656	0.5540	0.5402	0.5238	0.5041	0.4777	0.4292	0.3162				
0.0375	0.6007	0.5901	0.5776	0.5629	0.5453	0.5240	0.4936	0.4355	0.2947				
0.0400	0.6252	0.6139	0.6006	0.5848	0.5660	0.5430	0.5078	0.4386	0.2619				
0.0425	0.6491	0.6370	0.6228	0.6060	0.5860	0.5609	0.5200	0.4383	0.2205				
0.0450	0.6723	0.6595	0.6444	0.6266	0.6054	0.5776	0.5303	0.4345	0.1692				
0.0475	0.6950	0.6814	0.6654	0.6466	0.6241	0.5931	0.5385	0.4268	0.0928				
0.0500	0.7171	0.7028	0.6859	0.6660	0.6421	0.6073	0.5448	0.4147					

Table E.7

BACKLASH IMPACT VELOCITIES					CAM LAW: MSC.50								

SCCA parameters: a = .125 , b = 0 , c = .375 , Cv = 1.275258 , Ca = 8.012684

Normalised Backlash	Normalised Natural Deceleration												
	0	.5	1	1.5	2	2.5	3	3.5	4	4.5	5	5.5	6
	Normalised Impact Velocity												
0.0000	0.0000	0.0000	0.0000	0.0000	0.0000	0.0000	0.0000	0.0000	0.0000	0.0000	0.0000	0.0000	0.0000
0.0005	0.0421	0.0419	0.0417	0.0414	0.0411	0.0407	0.0402	0.0396	0.0388	0.0380	0.0369	0.0357	0.0340
0.0010	0.0666	0.0663	0.0659	0.0654	0.0648	0.0641	0.0632	0.0621	0.0609	0.0594	0.0577	0.0554	0.0526
0.0015	0.0871	0.0867	0.0861	0.0854	0.0845	0.0835	0.0822	0.0808	0.0791	0.0770	0.0746	0.0715	0.0675
0.0020	0.1053	0.1047	0.1040	0.1030	0.1019	0.1006	0.0991	0.0972	0.0951	0.0925	0.0893	0.0854	0.0803
0.0025	0.1220	0.1212	0.1203	0.1192	0.1178	0.1162	0.1144	0.1122	0.1095	0.1064	0.1026	0.0979	0.0916
0.0030	0.1375	0.1366	0.1355	0.1342	0.1326	0.1307	0.1286	0.1260	0.1229	0.1193	0.1148	0.1093	0.1019
0.0035	0.1522	0.1511	0.1498	0.1483	0.1465	0.1444	0.1419	0.1389	0.1354	0.1313	0.1262	0.1198	0.1113
0.0040	0.1661	0.1649	0.1634	0.1617	0.1596	0.1573	0.1544	0.1512	0.1472	0.1426	0.1369	0.1296	0.1200
0.0045	0.1794	0.1780	0.1764	0.1745	0.1722	0.1695	0.1664	0.1628	0.1585	0.1533	0.1469	0.1389	0.1279
0.0050	0.1921	0.1906	0.1889	0.1867	0.1842	0.1813	0.1779	0.1739	0.1691	0.1635	0.1565	0.1476	0.1349
0.0055	0.2044	0.2028	0.2009	0.1985	0.1958	0.1926	0.1889	0.1846	0.1794	0.1732	0.1656	0.1559	0.1408
0.0060	0.2164	0.2146	0.2124	0.2099	0.2070	0.2035	0.1995	0.1948	0.1892	0.1825	0.1743	0.1637	0.1457
0.0065	0.2279	0.2260	0.2237	0.2210	0.2178	0.2141	0.2098	0.2047	0.1987	0.1915	0.1826	0.1712	0.1494
0.0070	0.2392	0.2371	0.2346	0.2317	0.2283	0.2243	0.2197	0.2143	0.2079	0.2002	0.1906	0.1781	0.1519
0.0075	0.2501	0.2479	0.2452	0.2421	0.2385	0.2343	0.2294	0.2236	0.2168	0.2085	0.1983	0.1846	0.1532
0.0080	0.2608	0.2584	0.2556	0.2523	0.2484	0.2440	0.2387	0.2326	0.2254	0.2166	0.2058	0.1905	0.1531
0.0085	0.2712	0.2687	0.2657	0.2622	0.2581	0.2534	0.2479	0.2414	0.2337	0.2245	0.2130	0.1957	0.1515
0.0090	0.2814	0.2787	0.2755	0.2719	0.2676	0.2626	0.2568	0.2500	0.2419	0.2321	0.2199	0.2004	0.1482
0.0095	0.2914	0.2885	0.2852	0.2813	0.2768	0.2716	0.2655	0.2583	0.2498	0.2395	0.2266	0.2045	0.1429
0.0100	0.3012	0.2982	0.2947	0.2906	0.2859	0.2804	0.2740	0.2664	0.2575	0.2467	0.2330	0.2079	0.1352
0.0125	0.3475	0.3438	0.3394	0.3344	0.3285	0.3217	0.3138	0.3045	0.2935	0.2800	0.2595	0.2144	
0.0150	0.3904	0.3859	0.3806	0.3746	0.3676	0.3595	0.3501	0.3390	0.3258	0.3084	0.2765	0.1996	
0.0175	0.4304	0.4252	0.4191	0.4120	0.4039	0.3945	0.3836	0.3707	0.3550	0.3309	0.2827	0.1481	
0.0200	0.4682	0.4622	0.4552	0.4472	0.4379	0.4272	0.4147	0.3999	0.3805	0.3467	0.2766		
0.0225	0.5041	0.4973	0.4894	0.4803	0.4699	0.4579	0.4438	0.4269	0.4016	0.3556	0.2548		
0.0250	0.5383	0.5307	0.5219	0.5119	0.5003	0.4869	0.4712	0.4510	0.4181	0.3567	0.2092		
0.0275	0.5711	0.5627	0.5530	0.5419	0.5291	0.5143	0.4968	0.4719	0.4298	0.3493	0.1370		
0.0300	0.6026	0.5933	0.5827	0.5706	0.5566	0.5405	0.5203	0.4895	0.4363	0.3317			
0.0325	0.6329	0.6228	0.6113	0.5981	0.5830	0.5652	0.5414	0.5036	0.4375	0.3010			
0.0350	0.6622	0.6513	0.6388	0.6246	0.6082	0.5885	0.5601	0.5142	0.4328	0.2610			
0.0375	0.6905	0.6788	0.6654	0.6501	0.6324	0.6099	0.5763	0.5210	0.4216	0.2136			
0.0400	0.7179	0.7054	0.6910	0.6746	0.6554	0.6295	0.5897	0.5239	0.4029	0.1521			
0.0425	0.7446	0.7312	0.7158	0.6983	0.6772	0.6472	0.6005	0.5227	0.3775	0.0251			
0.0450	0.7704	0.7562	0.7399	0.7212	0.6975	0.6629	0.6084	0.5172	0.3500				
0.0475	0.7956	0.7805	0.7632	0.7431	0.7163	0.6765	0.6135	0.5070	0.3201				
0.0500	0.8201	0.8041	0.7858	0.7639	0.7336	0.6880	0.6157	0.4916	0.2872				

Table E.8

BACKLASH IMPACT VELOCITIES					CAM LAW: MSC.66							

SCCA parameters: a = 8.333334E-02 , b = 0 , c = .25 , Cv = 1.168083 , Ca = 11.00893

Normalised Backlash	Normalised Natural Deceleration												
	0	.5	1	1.5	2	2.5	3	3.5	4	4.5	5	5.5	6
	Normalised Impact Velocity												
0.0000	0.0000	0.0000	0.0000	0.0000	0.0000	0.0000	0.0000	0.0000	0.0000	0.0000	0.0000	0.0000
0.0005	0.0535	0.0533	0.0531	0.0529	0.0526	0.0522	0.0519	0.0514	0.0509	0.0504	0.0497	0.0490	0.0482
0.0010	0.0845	0.0842	0.0839	0.0834	0.0829	0.0823	0.0816	0.0808	0.0800	0.0790	0.0779	0.0766	0.0752
0.0015	0.1105	0.1100	0.1094	0.1088	0.1080	0.1072	0.1062	0.1051	0.1039	0.1025	0.1010	0.0992	0.0972
0.0020	0.1334	0.1328	0.1321	0.1312	0.1302	0.1291	0.1279	0.1265	0.1250	0.1232	0.1212	0.1190	0.1164
0.0025	0.1544	0.1536	0.1527	0.1517	0.1505	0.1492	0.1477	0.1460	0.1441	0.1419	0.1395	0.1368	0.1337
0.0030	0.1740	0.1730	0.1719	0.1707	0.1693	0.1677	0.1659	0.1640	0.1617	0.1592	0.1564	0.1532	0.1496
0.0035	0.1924	0.1913	0.1900	0.1886	0.1870	0.1851	0.1831	0.1808	0.1783	0.1754	0.1722	0.1685	0.1643
0.0040	0.2098	0.2086	0.2071	0.2055	0.2037	0.2016	0.1993	0.1967	0.1938	0.1906	0.1870	0.1829	0.1781
0.0045	0.2265	0.2251	0.2235	0.2216	0.2196	0.2173	0.2147	0.2118	0.2086	0.2051	0.2010	0.1964	0.1911
0.0050	0.2425	0.2409	0.2391	0.2371	0.2348	0.2323	0.2295	0.2263	0.2228	0.2188	0.2144	0.2093	0.2035
0.0055	0.2579	0.2561	0.2541	0.2519	0.2495	0.2467	0.2436	0.2402	0.2363	0.2320	0.2271	0.2216	0.2152
0.0060	0.2727	0.2708	0.2687	0.2663	0.2636	0.2606	0.2572	0.2535	0.2493	0.2446	0.2394	0.2334	0.2265
0.0065	0.2872	0.2851	0.2828	0.2801	0.2772	0.2740	0.2704	0.2664	0.2619	0.2568	0.2511	0.2447	0.2372
0.0070	0.3011	0.2989	0.2964	0.2936	0.2905	0.2870	0.2831	0.2788	0.2740	0.2686	0.2625	0.2555	0.2475
0.0075	0.3147	0.3124	0.3097	0.3067	0.3033	0.2996	0.2955	0.2909	0.2857	0.2800	0.2734	0.2660	0.2574
0.0080	0.3280	0.3254	0.3226	0.3194	0.3158	0.3119	0.3075	0.3026	0.2971	0.2910	0.2840	0.2761	0.2670
0.0085	0.3409	0.3382	0.3352	0.3318	0.3280	0.3238	0.3192	0.3140	0.3082	0.3017	0.2943	0.2859	0.2761
0.0090	0.3536	0.3507	0.3475	0.3439	0.3399	0.3355	0.3305	0.3251	0.3189	0.3120	0.3043	0.2954	0.2848
0.0095	0.3659	0.3629	0.3595	0.3557	0.3515	0.3468	0.3416	0.3359	0.3294	0.3221	0.3139	0.3046	0.2930
0.0100	0.3780	0.3748	0.3712	0.3673	0.3629	0.3579	0.3525	0.3464	0.3396	0.3320	0.3233	0.3134	0.3007
0.0125	0.4352	0.4311	0.4266	0.4217	0.4161	0.4100	0.4032	0.3957	0.3872	0.3777	0.3667	0.3523	0.3305
0.0150	0.4877	0.4828	0.4774	0.4714	0.4648	0.4574	0.4493	0.4402	0.4300	0.4183	0.4029	0.3804	0.3451
0.0175	0.5366	0.5308	0.5245	0.5174	0.5097	0.5011	0.4915	0.4809	0.4687	0.4529	0.4305	0.3967	0.3425
0.0200	0.5825	0.5758	0.5685	0.5604	0.5515	0.5416	0.5307	0.5182	0.5024	0.4806	0.4487	0.4000	0.3186
0.0225	0.6258	0.6182	0.6099	0.6008	0.5907	0.5795	0.5670	0.5514	0.5305	0.5008	0.4570	0.3887	0.2698
0.0250	0.6669	0.6584	0.6491	0.6389	0.6276	0.6150	0.5999	0.5800	0.5524	0.5131	0.4545	0.3595	0.1131
0.0275	0.7060	0.6967	0.6864	0.6750	0.6625	0.6478	0.6290	0.6035	0.5681	0.5171	0.4399	0.3190	
0.0300	0.7434	0.7331	0.7218	0.7094	0.6951	0.6774	0.6539	0.6219	0.5771	0.5122	0.4124	0.2725	
0.0325	0.7792	0.7680	0.7557	0.7419	0.7251	0.7035	0.6746	0.6349	0.5793	0.4975	0.3809	0.2162	
0.0350	0.8137	0.8015	0.7880	0.7722	0.7522	0.7260	0.6908	0.6424	0.5741	0.4746	0.3465	0.1387	
0.0375	0.8468	0.8336	0.8186	0.8001	0.7763	0.7448	0.7024	0.6442	0.5610	0.4503	0.3084		
0.0400	0.8787	0.8643	0.8471	0.8254	0.7973	0.7599	0.7095	0.6400	0.5430	0.4246	0.2647		
0.0425	0.9094	0.8933	0.8735	0.8482	0.8150	0.7710	0.7117	0.6294	0.5242	0.3972	0.2123		
0.0450	0.9387	0.9205	0.8977	0.8681	0.8295	0.7783	0.7090	0.6153	0.5048	0.3678	0.1417		
0.0475	0.9666	0.9458	0.9195	0.8853	0.8407	0.7815	0.7012	0.6009	0.4846	0.3358			
0.0500	0.9928	0.9692	0.9389	0.8997	0.8485	0.7806	0.6904	0.5862	0.4635	0.3005			

Table E.9

BACKLASH IMPACT VELOCITIES	CAM LAW: CYCLOIDAL

SCCA parameters: a = .5 , b = 0 , c = .5 , Cv = 2 , Ca = 6.283185

Normalised Backlash	Normalised Natural Deceleration												
	0	.5	1	1.5	2	2.5	3	3.5	4	4.5	5	5.5	6
	Normalised Impact Velocity												
0.0000	0.0000	0.0000	0.0000	0.0000	0.0000	0.0000	0.0000	0.0000	0.0000	0.0000	0.0000	0.0000	0.0000
0.0005	0.0353	0.0351	0.0349	0.0346	0.0341	0.0336	0.0329	0.0321	0.0310	0.0295	0.0275	0.0243	0.0158
0.0010	0.0559	0.0556	0.0551	0.0546	0.0538	0.0529	0.0517	0.0503	0.0484	0.0459	0.0425	0.0367	
0.0015	0.0731	0.0727	0.0720	0.0712	0.0702	0.0689	0.0673	0.0653	0.0627	0.0593	0.0544	0.0461	
0.0020	0.0884	0.0878	0.0870	0.0860	0.0846	0.0830	0.0810	0.0785	0.0752	0.0709	0.0647	0.0538	
0.0025	0.1025	0.1017	0.1007	0.0994	0.0979	0.0959	0.0935	0.0904	0.0865	0.0813	0.0737	0.0602	
0.0030	0.1156	0.1147	0.1135	0.1120	0.1101	0.1078	0.1050	0.1015	0.0969	0.0909	0.0819	0.0657	
0.0035	0.1279	0.1269	0.1255	0.1238	0.1217	0.1191	0.1159	0.1118	0.1067	0.0997	0.0894	0.0702	
0.0040	0.1396	0.1385	0.1369	0.1350	0.1326	0.1297	0.1261	0.1216	0.1158	0.1080	0.0963	0.0741	
0.0045	0.1509	0.1495	0.1478	0.1457	0.1431	0.1398	0.1358	0.1309	0.1244	0.1157	0.1027	0.0772	
0.0050	0.1617	0.1602	0.1583	0.1560	0.1531	0.1495	0.1452	0.1397	0.1327	0.1231	0.1087	0.0797	
0.0055	0.1721	0.1705	0.1684	0.1658	0.1627	0.1589	0.1541	0.1482	0.1405	0.1301	0.1142	0.0816	
0.0060	0.1821	0.1804	0.1782	0.1754	0.1720	0.1679	0.1627	0.1564	0.1481	0.1368	0.1194	0.0828	
0.0065	0.1919	0.1900	0.1876	0.1846	0.1810	0.1766	0.1711	0.1642	0.1553	0.1431	0.1243	0.0835	
0.0070	0.2014	0.1994	0.1968	0.1936	0.1897	0.1850	0.1791	0.1718	0.1623	0.1492	0.1289	0.0834	
0.0075	0.2107	0.2085	0.2058	0.2024	0.1982	0.1932	0.1870	0.1792	0.1691	0.1551	0.1332	0.0826	
0.0080	0.2198	0.2174	0.2145	0.2109	0.2065	0.2012	0.1946	0.1863	0.1756	0.1607	0.1373	0.0810	
0.0085	0.2286	0.2261	0.2230	0.2192	0.2146	0.2089	0.2020	0.1933	0.1819	0.1661	0.1411		
0.0090	0.2373	0.2346	0.2314	0.2274	0.2225	0.2165	0.2092	0.2000	0.1880	0.1713	0.1447		
0.0095	0.2457	0.2430	0.2395	0.2353	0.2302	0.2239	0.2163	0.2066	0.1940	0.1764	0.1480		
0.0100	0.2540	0.2511	0.2475	0.2431	0.2377	0.2312	0.2231	0.2130	0.1998	0.1812	0.1512		
0.0125	0.2935	0.2899	0.2854	0.2799	0.2733	0.2653	0.2554	0.2430	0.2265	0.2032	0.1639		
0.0150	0.3301	0.3257	0.3203	0.3139	0.3060	0.2965	0.2848	0.2700	0.2503	0.2219	0.1720		
0.0175	0.3643	0.3592	0.3530	0.3455	0.3365	0.3255	0.3119	0.2946	0.2715	0.2378	0.1757		
0.0200	0.3967	0.3909	0.3838	0.3753	0.3650	0.3525	0.3370	0.3173	0.2907	0.2513	0.1746		
0.0225	0.4276	0.4210	0.4130	0.4034	0.3919	0.3779	0.3605	0.3383	0.3081	0.2626			
0.0250	0.4570	0.4497	0.4408	0.4302	0.4174	0.4019	0.3826	0.3578	0.3239	0.2718			
0.0275	0.4853	0.4772	0.4675	0.4558	0.4418	0.4247	0.4034	0.3760	0.3382	0.2790			
0.0300	0.5126	0.5037	0.4931	0.4803	0.4650	0.4464	0.4231	0.3930	0.3512	0.2844			
0.0325	0.5389	0.5292	0.5177	0.5039	0.4873	0.4671	0.4418	0.4089	0.3630	0.2878			
0.0350	0.5643	0.5539	0.5415	0.5266	0.5087	0.4869	0.4596	0.4239	0.3735	0.2893			
0.0375	0.5890	0.5778	0.5645	0.5485	0.5293	0.5059	0.4764	0.4378	0.3829	0.2887			
0.0400	0.6130	0.6010	0.5867	0.5697	0.5492	0.5241	0.4925	0.4510	0.3913	0.2860			
0.0425	0.6363	0.6235	0.6083	0.5902	0.5683	0.5416	0.5079	0.4632	0.3985				
0.0450	0.6590	0.6454	0.6293	0.6100	0.5869	0.5584	0.5225	0.4747	0.4048				
0.0475	0.6811	0.6667	0.6497	0.6293	0.6048	0.5746	0.5365	0.4855	0.4100				
0.0500	0.7027	0.6875	0.6695	0.6480	0.6221	0.5903	0.5498	0.4955	0.4142				

Table E.10

BACKLASH IMPACT VELOCITIES					CAM LAW: CYCC.50							

SCCA parameters: a = .25 , b = 0 , c = .25 , Cv = 1.333333 , Ca = 8.377581

Normalised Backlash	Normalised Natural Deceleration											
	0	.5	1	1.5	2	2.5	3	3.5	4	4.5	5	5.5	6
	Normalised Impact Velocity												
0.0000	0.0000	0.0000	0.0000	0.0000	0.0000	0.0000	0.0000	0.0000	0.0000	0.0000	0.0000	0.0000	0.0000
0.0005	0.0487	0.0485	0.0482	0.0479	0.0475	0.0470	0.0464	0.0457	0.0448	0.0439	0.0427	0.0413	0.0395
0.0010	0.0771	0.0766	0.0761	0.0754	0.0747	0.0738	0.0727	0.0715	0.0700	0.0683	0.0662	0.0638	0.0606
0.0015	0.1006	0.0999	0.0992	0.0982	0.0971	0.0959	0.0944	0.0926	0.0906	0.0881	0.0853	0.0817	0.0772
0.0020	0.1214	0.1206	0.1196	0.1183	0.1169	0.1153	0.1134	0.1111	0.1085	0.1054	0.1017	0.0971	0.0912
0.0025	0.1405	0.1394	0.1381	0.1367	0.1349	0.1329	0.1306	0.1278	0.1247	0.1209	0.1163	0.1107	0.1034
0.0030	0.1582	0.1569	0.1554	0.1536	0.1516	0.1492	0.1464	0.1432	0.1395	0.1350	0.1296	0.1229	0.1142
0.0035	0.1748	0.1733	0.1716	0.1695	0.1672	0.1644	0.1613	0.1576	0.1532	0.1481	0.1419	0.1341	0.1239
0.0040	0.1906	0.1889	0.1869	0.1846	0.1819	0.1788	0.1752	0.1710	0.1662	0.1603	0.1533	0.1444	0.1327
0.0045	0.2056	0.2037	0.2015	0.1989	0.1959	0.1924	0.1884	0.1838	0.1783	0.1718	0.1639	0.1540	0.1407
0.0050	0.2200	0.2179	0.2154	0.2126	0.2092	0.2055	0.2010	0.1959	0.1899	0.1827	0.1739	0.1628	0.1479
0.0055	0.2339	0.2316	0.2288	0.2257	0.2221	0.2179	0.2131	0.2075	0.2009	0.1930	0.1833	0.1711	0.1545
0.0060	0.2473	0.2447	0.2418	0.2383	0.2344	0.2299	0.2246	0.2185	0.2114	0.2028	0.1923	0.1788	0.1604
0.0065	0.2603	0.2575	0.2543	0.2505	0.2463	0.2414	0.2357	0.2292	0.2214	0.2121	0.2007	0.1860	0.1658
0.0070	0.2728	0.2698	0.2664	0.2624	0.2578	0.2525	0.2465	0.2394	0.2311	0.2211	0.2087	0.1928	0.1707
0.0075	0.2850	0.2818	0.2781	0.2739	0.2690	0.2633	0.2568	0.2493	0.2403	0.2296	0.2163	0.1991	0.1750
0.0080	0.2969	0.2935	0.2895	0.2850	0.2798	0.2738	0.2669	0.2588	0.2493	0.2378	0.2236	0.2051	0.1789
0.0085	0.3085	0.3049	0.3006	0.2958	0.2903	0.2839	0.2766	0.2680	0.2579	0.2457	0.2305	0.2106	0.1823
0.0090	0.3198	0.3160	0.3115	0.3064	0.3005	0.2938	0.2860	0.2770	0.2662	0.2533	0.2371	0.2158	0.1852
0.0095	0.3309	0.3268	0.3221	0.3167	0.3105	0.3034	0.2952	0.2856	0.2743	0.2606	0.2434	0.2207	0.1876
0.0100	0.3417	0.3374	0.3324	0.3267	0.3202	0.3128	0.3041	0.2940	0.2821	0.2676	0.2494	0.2252	0.1896
0.0125	0.3927	0.3872	0.3810	0.3738	0.3657	0.3563	0.3455	0.3328	0.3176	0.2991	0.2755	0.2432	0.1925
0.0150	0.4393	0.4327	0.4251	0.4165	0.4067	0.3954	0.3823	0.3668	0.3483	0.3254	0.2958	0.2540	
0.0175	0.4825	0.4747	0.4658	0.4557	0.4441	0.4309	0.4154	0.3971	0.3750	0.3474	0.3110	0.2577	
0.0200	0.5229	0.5139	0.5036	0.4920	0.4786	0.4633	0.4454	0.4241	0.3983	0.3656	0.3215	0.2538	
0.0225	0.5609	0.5506	0.5390	0.5257	0.5106	0.4931	0.4726	0.4483	0.4184	0.3802	0.3274		
0.0250	0.5968	0.5852	0.5722	0.5573	0.5403	0.5206	0.4975	0.4699	0.4357	0.3914	0.3286		
0.0275	0.6308	0.6179	0.6034	0.5869	0.5680	0.5461	0.5202	0.4891	0.4504	0.3994	0.3248		
0.0300	0.6631	0.6490	0.6330	0.6148	0.5939	0.5696	0.5410	0.5062	0.4626	0.4042			
0.0325	0.6940	0.6785	0.6610	0.6411	0.6182	0.5915	0.5599	0.5213	0.4724	0.4058			
0.0350	0.7235	0.7067	0.6876	0.6659	0.6409	0.6118	0.5770	0.5345	0.4799	0.4039			
0.0375	0.7517	0.7335	0.7129	0.6894	0.6623	0.6305	0.5926	0.5458	0.4850	0.3985			
0.0400	0.7788	0.7592	0.7370	0.7116	0.6823	0.6479	0.6066	0.5553	0.4879	0.3891			
0.0425	0.8047	0.7837	0.7599	0.7326	0.7011	0.6639	0.6191	0.5630	0.4883				
0.0450	0.8297	0.8072	0.7818	0.7525	0.7186	0.6786	0.6301	0.5690	0.4863				
0.0475	0.8537	0.8298	0.8026	0.7714	0.7351	0.6921	0.6398	0.5732	0.4817				
0.0500	0.8768	0.8514	0.8224	0.7892	0.7505	0.7045	0.6481	0.5757	0.4744				

Table E.11

BACKLASH IMPACT VELOCITIES					CAM LAW: POLYNOMIAL 3-4-5						
Polynomial parameters: A3 = 10 , A4 = -15 , A5 = 6 Cv = 1.875 , Ca = 5.773503 , Cd = 5.773503											

Normalised Backlash	Normalised Natural Deceleration											
	0	.5	1	1.5	2	2.5	3	3.5	4	4.5	5	5.5
	Normalised Impact Velocity											
0.0000	0.0000	0.0000	0.0000	0.0000	0.0000	0.0000	0.0000	0.0000	0.0000	0.0000	0.0000	0.0000
0.0005	0.0326	0.0324	0.0322	0.0319	0.0315	0.0310	0.0304	0.0293	0.0281	0.0264	0.0236	
0.0010	0.0509	0.0506	0.0502	0.0497	0.0490	0.0481	0.0470	0.0455	0.0435	0.0405	0.0355	
0.0015	0.0667	0.0663	0.0657	0.0650	0.0640	0.0628	0.0612	0.0591	0.0563	0.0522	0.0450	
0.0020	0.0807	0.0802	0.0795	0.0785	0.0773	0.0758	0.0738	0.0712	0.0676	0.0623	0.0529	
0.0025	0.0936	0.0930	0.0921	0.0910	0.0895	0.0876	0.0852	0.0821	0.0778	0.0713	0.0596	
0.0030	0.1056	0.1049	0.1039	0.1025	0.1008	0.0986	0.0958	0.0922	0.0871	0.0795	0.0654	
0.0035	0.1170	0.1161	0.1149	0.1134	0.1114	0.1090	0.1058	0.1016	0.0958	0.0871	0.0704	
0.0040	0.1278	0.1268	0.1254	0.1237	0.1215	0.1187	0.1152	0.1105	0.1040	0.0941	0.0748	
0.0045	0.1381	0.1370	0.1355	0.1336	0.1312	0.1281	0.1241	0.1189	0.1117	0.1006	0.0785	
0.0050	0.1480	0.1468	0.1452	0.1431	0.1404	0.1370	0.1327	0.1270	0.1190	0.1068	0.0817	
0.0055	0.1576	0.1563	0.1545	0.1522	0.1493	0.1456	0.1409	0.1347	0.1260	0.1125	0.0843	
0.0060	0.1669	0.1655	0.1635	0.1610	0.1579	0.1540	0.1489	0.1421	0.1327	0.1180	0.0863	
0.0065	0.1759	0.1744	0.1723	0.1696	0.1663	0.1620	0.1565	0.1493	0.1392	0.1231	0.0879	
0.0070	0.1847	0.1830	0.1808	0.1779	0.1744	0.1698	0.1640	0.1562	0.1453	0.1280	0.0889	
0.0075	0.1933	0.1915	0.1891	0.1860	0.1822	0.1774	0.1712	0.1629	0.1513	0.1326	0.0893	
0.0080	0.2016	0.1997	0.1972	0.1939	0.1899	0.1848	0.1782	0.1694	0.1570	0.1370	0.0891	
0.0085	0.2098	0.2077	0.2051	0.2017	0.1974	0.1920	0.1850	0.1757	0.1626	0.1412	0.0883	
0.0090	0.2178	0.2156	0.2128	0.2092	0.2047	0.1990	0.1917	0.1819	0.1679	0.1452	0.0867	
0.0095	0.2256	0.2233	0.2204	0.2166	0.2119	0.2059	0.1981	0.1878	0.1731	0.1489	0.0843	
0.0100	0.2333	0.2309	0.2278	0.2238	0.2189	0.2126	0.2045	0.1937	0.1782	0.1525	0.0809	
0.0150	0.3039	0.3002	0.2956	0.2898	0.2825	0.2733	0.2613	0.2452	0.2213	0.1787		
0.0200	0.3660	0.3611	0.3549	0.3472	0.3376	0.3254	0.3095	0.2875	0.2542	0.1890		
0.0250	0.4224	0.4162	0.4085	0.3989	0.3868	0.3715	0.3514	0.3233	0.2791	0.1814		
0.0300	0.4745	0.4670	0.4577	0.4461	0.4316	0.4131	0.3885	0.3537	0.2969	0.1446		
0.0350	0.5233	0.5144	0.5035	0.4899	0.4728	0.4510	0.4217	0.3795	0.3078			
0.0400	0.5693	0.5590	0.5464	0.5307	0.5111	0.4857	0.4515	0.4013	0.3117			
0.0450	0.6129	0.6012	0.5869	0.5691	0.5467	0.5177	0.4783	0.4192	0.3075			
0.0500	0.6544	0.6414	0.6253	0.6053	0.5801	0.5474	0.5023	0.4334	0.2936			

Table E.12

BACKLASH IMPACT VELOCITIES				CAM LAW: POLYNOMIAL 3-4-5-6-R					

Polynomial parameters: A3 = 13.33333 , A4 = -40 , A5 = 64 , A6 = -42.66667
Cv = 2 , Ca = 5 , Cd = 5

Normalised Backlash	Normalised Natural Deceleration									
	0	.5	1	1.5	2	2.5	3	3.5	4	4.5
	Normalised Impact Velocity									
0.0000	0.0000	0.0000	0.0000	0.0000	0.0000	0.0000	0.0000	0.0000	0.0000	0.0000
0.0005	0.0422	0.0410	0.0397	0.0381	0.0363	0.0344	0.0322	0.0293	0.0258	0.0201
0.0010	0.0654	0.0634	0.0612	0.0587	0.0560	0.0529	0.0492	0.0448	0.0389	0.0297
0.0015	0.0845	0.0818	0.0789	0.0757	0.0720	0.0679	0.0630	0.0570	0.0492	0.0371
0.0020	0.1012	0.0979	0.0943	0.0904	0.0859	0.0809	0.0749	0.0677	0.0580	0.0433
0.0025	0.1162	0.1124	0.1083	0.1036	0.0985	0.0925	0.0857	0.0771	0.0658	0.0487
0.0030	0.1302	0.1258	0.1211	0.1159	0.1100	0.1032	0.0954	0.0856	0.0726	0.0535
0.0035	0.1428	0.1380	0.1328	0.1269	0.1204	0.1129	0.1041	0.0933	0.0791	0.0580
0.0040	0.1551	0.1498	0.1440	0.1377	0.1305	0.1223	0.1126	0.1007	0.0852	0.0621
0.0045	0.1668	0.1610	0.1548	0.1478	0.1400	0.1311	0.1206	0.1077	0.0908	0.0660
0.0050	0.1779	0.1717	0.1650	0.1575	0.1491	0.1395	0.1281	0.1143	0.0961	0.0697
0.0055	0.1886	0.1820	0.1747	0.1667	0.1577	0.1475	0.1353	0.1205	0.1012	0.0732
0.0060	0.1988	0.1918	0.1841	0.1756	0.1661	0.1551	0.1422	0.1265	0.1060	0.0765
0.0065	0.2087	0.2013	0.1932	0.1842	0.1741	0.1625	0.1489	0.1322	0.1106	0.0797
0.0070	0.2183	0.2105	0.2019	0.1924	0.1818	0.1696	0.1552	0.1377	0.1151	0.0827
0.0075	0.2276	0.2194	0.2104	0.2004	0.1892	0.1764	0.1614	0.1430	0.1193	0.0856
0.0080	0.2366	0.2280	0.2186	0.2082	0.1965	0.1831	0.1673	0.1481	0.1234	0.0884
0.0085	0.2454	0.2364	0.2266	0.2157	0.2035	0.1895	0.1731	0.1531	0.1274	0.0910
0.0090	0.2539	0.2446	0.2344	0.2230	0.2103	0.1958	0.1787	0.1579	0.1313	0.0935
0.0095	0.2623	0.2526	0.2419	0.2302	0.2169	0.2018	0.1841	0.1626	0.1351	0.0959
0.0100	0.2704	0.2604	0.2493	0.2371	0.2234	0.2078	0.1894	0.1671	0.1387	0.0982
0.0150	0.3435	0.3301	0.3154	0.2992	0.2810	0.2603	0.2361	0.2072	0.1710	0.1121
0.0200	0.4058	0.3894	0.3714	0.3516	0.3295	0.3043	0.2752	0.2407	0.1978	0.0985
0.0250	0.4610	0.4418	0.4208	0.3978	0.3720	0.3429	0.3094	0.2700	0.2206	
0.0300	0.5109	0.4891	0.4654	0.4393	0.4103	0.3776	0.3402	0.2964	0.2396	
0.0350	0.5568	0.5326	0.5063	0.4774	0.4453	0.4094	0.3684	0.3205	0.2543	
0.0400	0.5995	0.5730	0.5442	0.5127	0.4778	0.4388	0.3946	0.3426	0.2641	
0.0450	0.6395	0.6109	0.5798	0.5458	0.5083	0.4665	0.4191	0.3627	0.2675	
0.0500	0.6772	0.6465	0.6133	0.5769	0.5370	0.4925	0.4422	0.3810	0.2622	

Cams for Industry

Table E.13

BACKLASH IMPACT VELOCITIES				CAM LAW: LOW IMPACT POLYNOMIAL 3-4-5-6-7-R ˋ									

Polynomial parameters: A3 = 27 , A4 = -113 , A5 = 224.4 , A6 = -228.8 A7 = 96
Cv = 1.475 , Ca = 6.48192 , Cd = 6.48192

Normalised Backlash	Normalised Natural Deceleration												
	0	.5	1	1.5	2	2.5	3	3.5	4	4.5	5	5.5	6
	Normalised Impact Velocity												
0.0000	0.0000	0.0000	0.0000	0.0000	0.0000	0.0000	0.0000	0.0000	0.0000	0.0000	0.0000	0.0000	0.0000
0.0005	0.0201	0.0231	0.0254	0.0277	0.0295	0.0309	0.0317	0.0321	0.0323	0.0318	0.0306	0.0281	
0.0010	0.0325	0.0367	0.0405	0.0440	0.0468	0.0487	0.0499	0.0505	0.0503	0.0493	0.0469	0.0417	
0.0015	0.0433	0.0488	0.0538	0.0584	0.0619	0.0642	0.0656	0.0661	0.0657	0.0639	0.0602	0.0518	
0.0020	0.0531	0.0598	0.0660	0.0715	0.0755	0.0782	0.0797	0.0801	0.0792	0.0767	0.0714	0.0592	
0.0025	0.0623	0.0700	0.0773	0.0836	0.0881	0.0910	0.0926	0.0928	0.0915	0.0881	0.0811	0.0643	
0.0030	0.0709	0.0797	0.0880	0.0950	0.0999	0.1031	0.1046	0.1046	0.1028	0.0984	0.0896	0.0672	
0.0035	0.0792	0.0891	0.0983	0.1059	0.1112	0.1145	0.1160	0.1157	0.1133	0.1080	0.0970	0.0678	
0.0040	0.0872	0.0980	0.1082	0.1163	0.1219	0.1253	0.1267	0.1261	0.1232	0.1168	0.1036	0.0656	
0.0045	0.0949	0.1068	0.1177	0.1264	0.1323	0.1357	0.1370	0.1361	0.1325	0.1250	0.1093	0.0592	
0.0050	0.1024	0.1152	0.1269	0.1361	0.1422	0.1457	0.1469	0.1456	0.1414	0.1326	0.1143	0.0452	
0.0055	0.1097	0.1235	0.1359	0.1456	0.1519	0.1554	0.1564	0.1548	0.1498	0.1397	0.1185		
0.0060	0.1168	0.1316	0.1447	0.1548	0.1613	0.1648	0.1656	0.1635	0.1578	0.1464	0.1219		
0.0065	0.1238	0.1395	0.1533	0.1637	0.1704	0.1739	0.1745	0.1720	0.1655	0.1526	0.1245		
0.0070	0.1307	0.1473	0.1617	0.1725	0.1793	0.1827	0.1831	0.1801	0.1728	0.1584	0.1264		
0.0075	0.1375	0.1549	0.1700	0.1810	0.1880	0.1913	0.1914	0.1880	0.1799	0.1638	0.1274		
0.0080	0.1442	0.1624	0.1780	0.1894	0.1964	0.1997	0.1996	0.1956	0.1866	0.1688	0.1274		
0.0085	0.1508	0.1698	0.1860	0.1977	0.2048	0.2079	0.2075	0.2030	0.1930	0.1734	0.1265		
0.0090	0.1573	0.1771	0.1938	0.2058	0.2129	0.2160	0.2152	0.2102	0.1992	0.1777	0.1244		
0.0095	0.1637	0.1843	0.2015	0.2137	0.2209	0.2238	0.2228	0.2171	0.2052	0.1816	0.1209		
0.0100	0.1700	0.1915	0.2091	0.2215	0.2287	0.2315	0.2301	0.2239	0.2109	0.1851	0.1158		
0.0150	0.2304	0.2586	0.2800	0.2937	0.3005	0.3011	0.2955	0.2820	0.2558	0.1992			
0.0200	0.2868	0.3202	0.3440	0.3579	0.3631	0.3604	0.3491	0.3257	0.2801	0.1528			
0.0250	0.3404	0.3779	0.4029	0.4162	0.4190	0.4119	0.3932	0.3569	0.2812				
0.0300	0.3918	0.4324	0.4579	0.4696	0.4692	0.4567	0.4290	0.3754	0.2472				
0.0350	0.4414	0.4841	0.5094	0.5190	0.5147	0.4956	0.4566	0.3794	0.1054				
0.0400	0.4893	0.5335	0.5579	0.5648	0.5557	0.5288	0.4759	0.3649					
0.0450	0.5357	0.5807	0.6037	0.6073	0.5926	0.5566	0.4862	0.3220					
0.0500	0.5808	0.6260	0.6470	0.6467	0.6256	0.5787	0.4862	0.2151					

APPENDIX F

D–R–R–D Tables

A dwell–rise–return–dwell cam, usually just called D–R–R–D, is a reciprocating or oscillating cam which only has a dwell period at one end of its stroke. At the other end there is an immediate return with no dwell. In an asymmetrical motion both the rise and return phases are parts of a single motion specification. A full description of this type of motion and how the following tables can be used to analyse it are given in Chapter 6.

The values in the tables are used in Equations 6.25 and 6.26, which are repeated here for convenience.

$$\frac{T1}{X_f} = \frac{(C_v2 - w') \cdot (2 \cdot u \cdot X_{fr}/X_f - 1) + (1 - w') \cdot C_v3}{(C_v2 - w') \cdot (2 \cdot u - 1) + (1 - w') \cdot C_v3 + 2 \cdot u \cdot w' \cdot C_v3/C_v1}$$

Eq. (F.1)

$$\frac{T1}{X_f} = \frac{2 \cdot u \cdot (2 - w') \cdot X_{fr}/X_f - w'}{4 \cdot u - w'}$$

Eq. (F.2)

where:

Xf = Period of the forward motion.

Xr = Period of the return motion.

Xfr = Total period of motion = $Xf + Xr = T1 + T2 + T3$.

$T1$ = Period of phase 1, the forward acceleration loop.

$T2$ = Period of phase 2, the central loop of forward deceleration and return acceleration.

$T3$ = Period of phase 3, the return deceleration loop.

H = Total output stroke.

$h1$ = Phase 1 part of the output stroke.

$h3$ = Phase 3 part of the return stroke.

C_v1 = Coefficient of velocity of the phase 1 cam law.

C_v2 = Coefficient of velocity of the phase 2 cam law.

C_v3 = Coefficient of velocity of the phase 3 cam law.

u = Normalised input displacement of the cam law part of phase 2.

w' = Normalised output velocity of the cam law part of phase 2.

Cams for Industry

Table F.1   Asymmetrical D–R–R–D motion.
Zone 1: Modified trapezoid
Zone 2: Modified trapezoid
Zone 3: Modified trapezoid

Xf/Xfr = Forward period/total period = 0.10

ZONE 1		ZONE 2			ZONE 3	
T1/Xf	h1/H	T2/Xfr	Y/2/H	U	T3/Xr	h3/H
0.00000	0.00000	0.12150	1.23712	0.41152	0.97611	0.96906
0.05000	0.04561	0.11537	1.17969	0.41171	0.97736	0.97067
0.10000	0.09165	0.10925	1.12182	0.41190	0.97861	0.97227
0.15000	0.13811	0.10313	1.06354	0.41209	0.97985	0.97387
0.20000	0.18504	0.09702	1.00478	0.41228	0.98109	0.97546
0.25000	0.23239	0.09092	0.94560	0.41247	0.98232	0.97704
0.30000	0.28020	0.08482	0.88595	0.41266	0.98354	0.97862
0.35000	0.32848	0.07872	0.82583	0.41286	0.98476	0.98019
0.40000	0.37720	0.07263	0.76527	0.41305	0.98597	0.98175
0.45000	0.42640	0.06655	0.70421	0.41324	0.98717	0.98331
0.50000	0.47607	0.06047	0.64269	0.41343	0.98837	0.98486
0.55000	0.52621	0.05440	0.58069	0.41363	0.98956	0.98641
0.60000	0.57684	0.04833	0.51820	0.41382	0.99074	0.98794
0.65000	0.62796	0.04227	0.45522	0.41401	0.99192	0.98948
0.70000	0.67958	0.03621	0.39173	0.41421	0.99310	0.99100
0.75000	0.73169	0.03016	0.32774	0.41440	0.99426	0.99252
0.80000	0.78432	0.02412	0.26323	0.41459	0.99542	0.99403
0.85000	0.83746	0.01808	0.19821	0.41479	0.99658	0.99553
0.90000	0.89111	0.01205	0.13267	0.41498	0.99772	0.99703
0.95000	0.94529	0.00602	0.06660	0.41518	0.99887	0.99852
1.00000	1.00000	0.00000	0.00000	0.41537	1.00000	1.00000
OPTIMUM POSITION						
0.52002	0.49608	0.05804	0.61794	0.41351	0.98884	0.98548

Xf/Xfr = Forward period/total period = 0.15

ZONE 1		ZONE 2			ZONE 3	
T1/Xf	h1/H	T2/Xfr	Y/2/H	U	T3/Xr	h3/H
0.00000	0.00000	0.19309	1.38355	0.38842	0.94931	0.93572
0.05000	0.04527	0.18332	1.31927	0.38866	0.95198	0.93905
0.10000	0.09100	0.17356	1.25451	0.38891	0.95463	0.94237
0.15000	0.13719	0.16382	1.18927	0.38915	0.95727	0.94567
0.20000	0.18386	0.15409	1.12354	0.38939	0.95990	0.94897
0.25000	0.23100	0.14437	1.05732	0.38964	0.96251	0.95225
0.30000	0.27862	0.13466	0.99059	0.38988	0.96511	0.95552
0.35000	0.32674	0.12496	0.92335	0.39013	0.96769	0.95878
0.40000	0.37535	0.11528	0.85561	0.39037	0.97026	0.96202
0.45000	0.42447	0.10560	0.78733	0.39062	0.97282	0.96526
0.50000	0.47409	0.09594	0.71854	0.39086	0.97536	0.96848
0.55000	0.52423	0.08629	0.64921	0.39111	0.97789	0.97169
0.60000	0.57490	0.07666	0.57933	0.39136	0.98040	0.97488
0.65000	0.62610	0.06703	0.50891	0.39161	0.98290	0.97807
0.70000	0.67783	0.05742	0.43794	0.39186	0.98539	0.98124
0.75000	0.73012	0.04782	0.36640	0.39211	0.98786	0.98440
0.80000	0.78296	0.03823	0.29429	0.39236	0.99032	0.98755
0.85000	0.83636	0.02865	0.22160	0.39261	0.99276	0.99068
0.90000	0.89032	0.01909	0.14833	0.39286	0.99519	0.99380
0.95000	0.94487	0.00954	0.07447	0.39311	0.99760	0.99691
1.00000	1.00000	0.00000	0.00000	0.39337	1.00000	1.00000
OPTIMUM POSITION						
0.51917	0.49325	0.09224	0.69202	0.39096	0.97633	0.96971

Xf/Xfr = Forward period/total period = 0.20

ZONE 1		ZONE 2			ZONE 3	
T1/Xf	h1/H	T2/Xfr	Y/2/H	U	T3/Xr	h3/H
0.00000	0.00000	0.27251	1.55340	0.36696	0.90936	0.88807
0.05000	0.04503	0.25869	1.48114	0.36724	0.91414	0.89382
0.10000	0.09055	0.24489	1.40833	0.36752	0.91889	0.89955
0.15000	0.13655	0.23110	1.33500	0.36780	0.92362	0.90527
0.20000	0.18304	0.21734	1.26113	0.36808	0.92832	0.91097
0.25000	0.23003	0.20360	1.18672	0.36837	0.93300	0.91666
0.30000	0.27753	0.18988	1.11175	0.36865	0.93765	0.92233
0.35000	0.32554	0.17618	1.03622	0.36894	0.94227	0.92799
0.40000	0.37407	0.16250	0.96013	0.36923	0.94687	0.93363
0.45000	0.42313	0.14885	0.88347	0.36951	0.95144	0.93925
0.50000	0.47273	0.13521	0.80621	0.36980	0.95599	0.94486
0.55000	0.52288	0.12159	0.72837	0.37009	0.96051	0.95045
0.60000	0.57357	0.10800	0.64993	0.37038	0.96501	0.95603
0.65000	0.62483	0.09442	0.57090	0.37068	0.96947	0.96159
0.70000	0.67666	0.08087	0.49123	0.37097	0.97391	0.96713
0.75000	0.72905	0.06734	0.41097	0.37126	0.97833	0.97265
0.80000	0.78204	0.05383	0.33006	0.37156	0.98272	0.97816
0.85000	0.83562	0.04034	0.24853	0.37185	0.98708	0.98365
0.90000	0.88980	0.02687	0.16634	0.37215	0.99141	0.98912
0.95000	0.94459	0.01342	0.08350	0.37245	0.99572	0.99457
1.00000	1.00000	0.00000	0.00000	0.37275	1.00000	1.00000
OPTIMUM POSITION						
0.51737	0.49009	0.13047	0.77924	0.36990	0.95756	0.94681

Xf/Xfr = Forward period/total period = 0.25

ZONE 1		ZONE 2			ZONE 3	
T1/Xf	h1/H	T2/Xfr	Y/2/H	U	T3/Xr	h3/H
0.00000	0.00000	0.36128	1.75230	0.34599	0.85162	0.82218
0.05000	0.04478	0.34292	1.67084	0.34629	0.85944	0.83121
0.10000	0.09008	0.32459	1.58876	0.34659	0.86722	0.84022
0.15000	0.13586	0.30629	1.50611	0.34690	0.87495	0.84922
0.20000	0.18219	0.28801	1.42280	0.34720	0.88265	0.85821
0.25000	0.22901	0.26978	1.33890	0.34751	0.89030	0.86718
0.30000	0.27638	0.25157	1.25435	0.34782	0.89791	0.87615
0.35000	0.32429	0.23339	1.16918	0.34813	0.90548	0.88510
0.40000	0.37274	0.21525	1.08335	0.34844	0.91301	0.89403
0.45000	0.42176	0.19713	0.99686	0.34875	0.92049	0.90295
0.50000	0.47133	0.17905	0.90973	0.34907	0.92793	0.91186
0.55000	0.52148	0.16100	0.82190	0.34938	0.93534	0.92075
0.60000	0.57220	0.14298	0.73342	0.34970	0.94269	0.92962
0.65000	0.62352	0.12499	0.64423	0.35002	0.95001	0.93848
0.70000	0.67543	0.10704	0.55436	0.35034	0.95728	0.94732
0.75000	0.72795	0.08912	0.46379	0.35066	0.96451	0.95615
0.80000	0.78109	0.07123	0.37249	0.35098	0.97170	0.96496
0.85000	0.83484	0.05337	0.28048	0.35131	0.97884	0.97374
0.90000	0.88926	0.03555	0.18772	0.35163	0.98594	0.98252
0.95000	0.94430	0.01776	0.09424	0.35196	0.99299	0.99127
1.00000	1.00000	0.00000	0.00000	0.35229	1.00000	1.00000
OPTIMUM POSITION						
0.51529	0.48660	0.17353	0.88295	0.34916	0.93020	0.91458

Table F.1 (continued)

Xf/Xfr = Forward period/total period = 0.30

ZONE 1		ZONE 2			ZONE 3	
T1/Xf	h1/H	T2/Xfr	Y/2/H	U	T3/Xr	h3/H
0.00000	0.00000	0.46081	1.98625	0.32551	0.77026	0.73312
0.05000	0.04451	0.43738	1.89428	0.32581	0.78232	0.74644
0.10000	0.08957	0.41397	1.80154	0.32611	0.79432	0.75978
0.15000	0.13513	0.39062	1.70813	0.32641	0.80626	0.77311
0.20000	0.18126	0.36729	1.61394	0.32671	0.81815	0.78645
0.25000	0.22791	0.34402	1.51906	0.32702	0.82997	0.79979
0.30000	0.27515	0.32078	1.42336	0.32732	0.84174	0.81314
0.35000	0.32294	0.29759	1.32695	0.32763	0.85345	0.82649
0.40000	0.37130	0.27444	1.22976	0.32794	0.86509	0.83984
0.45000	0.42024	0.25133	1.13181	0.32826	0.87667	0.85319
0.50000	0.46979	0.22826	1.03304	0.32857	0.88820	0.86655
0.55000	0.51994	0.20524	0.93348	0.32889	0.89966	0.87991
0.60000	0.57070	0.18226	0.83312	0.32921	0.91106	0.89326
0.65000	0.62207	0.15932	0.73196	0.32953	0.92240	0.90661
0.70000	0.67409	0.13642	0.62994	0.32985	0.93368	0.91997
0.75000	0.72673	0.11358	0.52711	0.33018	0.94489	0.93332
0.80000	0.78005	0.09077	0.42341	0.33050	0.95604	0.94667
0.85000	0.83401	0.06801	0.31889	0.33083	0.96713	0.96001
0.90000	0.88866	0.04529	0.21346	0.33116	0.97815	0.97335
0.95000	0.94397	0.02263	0.10719	0.33150	0.98911	0.98667
1.00000	1.00000	0.00000	0.00000	0.33183	1.00000	1.00000
			OPTIMUM POSITION			
0.51294	0.48271	0.22230	1.00735	0.32865	0.89117	0.87001

Xf/Xfr = Forward period/total period = 0.35

ZONE 1		ZONE 2			ZONE 3	
T1/Xf	h1/H	T2/Xfr	Y/2/H	U	T3/Xr	h3/H
0.00000	0.00000	0.57266	2.26240	0.30559	0.65744	0.61476
0.05000	0.04424	0.54355	2.15842	0.30586	0.67530	0.63356
0.10000	0.08900	0.51450	2.05357	0.30613	0.69308	0.65241
0.15000	0.13435	0.48548	1.94779	0.30640	0.71080	0.67132
0.20000	0.18026	0.45651	1.84108	0.30667	0.72844	0.69028
0.25000	0.22673	0.42760	1.73349	0.30695	0.74600	0.70928
0.30000	0.27380	0.39873	1.62491	0.30723	0.76350	0.72835
0.35000	0.32146	0.36991	1.51542	0.30751	0.78091	0.74745
0.40000	0.36972	0.34114	1.40495	0.30779	0.79824	0.76661
0.45000	0.41861	0.31242	1.29351	0.30808	0.81550	0.78582
0.50000	0.46811	0.28376	1.18110	0.30836	0.83268	0.80507
0.55000	0.51825	0.25514	1.06767	0.30865	0.84979	0.82437
0.60000	0.56904	0.22658	0.95325	0.30895	0.86681	0.84371
0.65000	0.62049	0.19806	0.83779	0.30924	0.88375	0.86310
0.70000	0.67260	0.16961	0.72134	0.30954	0.90060	0.88253
0.75000	0.72540	0.14120	0.60378	0.30984	0.91738	0.90201
0.80000	0.77888	0.11286	0.48523	0.31014	0.93407	0.92152
0.85000	0.83309	0.08455	0.36553	0.31044	0.95069	0.94109
0.90000	0.88798	0.05632	0.24483	0.31075	0.96720	0.96068
0.95000	0.94363	0.02813	0.12295	0.31106	0.98365	0.98032
1.00000	1.00000	0.00000	0.00000	0.31137	1.00000	1.00000
			OPTIMUM POSITION			
0.51030	0.47839	0.27786	1.15781	0.30842	0.83621	0.80904

Xf/Xfr = Forward period/total period = 0.40

ZONE 1		ZONE 2			ZONE 3	
T1/Xf	h1/H	T2/Xfr	Y/2/H	U	T3/Xr	h3/H
0.00000	0.00000	0.69855	2.58920	0.28631	0.50242	0.45952
0.05000	0.04393	0.66314	2.47171	0.28652	0.52810	0.48512
0.10000	0.08842	0.62778	2.35311	0.28673	0.55370	0.51086
0.15000	0.13351	0.59245	2.23327	0.28694	0.57924	0.53676
0.20000	0.17917	0.55720	2.11230	0.28715	0.60468	0.56279
0.25000	0.22545	0.52197	1.99008	0.28737	0.63004	0.58899
0.30000	0.27235	0.48680	1.86664	0.28759	0.65533	0.61534
0.35000	0.31986	0.45168	1.74197	0.28781	0.68053	0.64183
0.40000	0.36801	0.41662	1.61604	0.28804	0.70564	0.66847
0.45000	0.41682	0.38159	1.48880	0.28826	0.73068	0.69527
0.50000	0.46628	0.34663	1.36030	0.28849	0.75562	0.72222
0.55000	0.51641	0.31172	1.23048	0.28872	0.78047	0.74931
0.60000	0.56722	0.27686	1.09934	0.28896	0.80523	0.77656
0.65000	0.61874	0.24205	0.96683	0.28919	0.82991	0.80396
0.70000	0.67096	0.20731	0.83300	0.28943	0.85449	0.83151
0.75000	0.72392	0.17261	0.69772	0.28967	0.87899	0.85922
0.80000	0.77760	0.13797	0.56108	0.28991	0.90338	0.88707
0.85000	0.83203	0.10339	0.42301	0.29016	0.92768	0.91507
0.90000	0.88724	0.06887	0.28346	0.29041	0.95189	0.94323
0.95000	0.94322	0.03440	0.14247	0.29066	0.97599	0.97154
1.00000	1.00000	0.00000	0.00000	0.29092	1.00000	1.00000
			OPTIMUM POSITION			
0.50730	0.47355	0.34153	1.34143	0.28852	0.75925	0.72616

Xf/Xfr = Forward period/total period = 0.45

ZONE 1		ZONE 2			ZONE 3	
T1/Xf	h1/H	T2/Xfr	Y/2/H	U	T3/Xr	h3/H
0.00000	0.00000	0.84034	2.97657	0.26775	0.29029	0.25823
0.05000	0.04360	0.79797	2.84416	0.26787	0.32642	0.29200
0.10000	0.08779	0.75564	2.71020	0.26799	0.36248	0.32609
0.15000	0.13260	0.71332	2.57461	0.26811	0.39850	0.36052
0.20000	0.17802	0.67106	2.43744	0.26823	0.43445	0.39527
0.25000	0.22407	0.62883	2.29865	0.26836	0.47032	0.43038
0.30000	0.27077	0.58662	2.15815	0.26849	0.50615	0.46580
0.35000	0.31812	0.54446	2.01690	0.26862	0.54189	0.50158
0.40000	0.36614	0.50234	1.87213	0.26875	0.57757	0.53770
0.45000	0.41486	0.46024	1.72645	0.26888	0.61320	0.57419
0.50000	0.46426	0.41820	1.57906	0.26901	0.64873	0.61102
0.55000	0.51437	0.37619	1.42996	0.26914	0.68420	0.64821
0.60000	0.56522	0.33422	1.27878	0.26929	0.71960	0.68578
0.65000	0.61681	0.29229	1.12583	0.26943	0.75493	0.72371
0.70000	0.66915	0.25041	0.97099	0.26957	0.79018	0.76202
0.75000	0.72228	0.20855	0.81417	0.26971	0.82536	0.80072
0.80000	0.77616	0.16676	0.65546	0.26986	0.86044	0.83978
0.85000	0.83089	0.12499	0.49465	0.27000	0.89546	0.87925
0.90000	0.88640	0.08329	0.33189	0.27015	0.93039	0.91910
0.95000	0.94277	0.04162	0.16700	0.27030	0.96524	0.95935
1.00000	1.00000	0.00000	0.00000	0.27046	1.00000	1.00000
			OPTIMUM POSITION			
0.50390	0.46815	0.41492	1.56747	0.26902	0.65151	0.61391

Table F.1 (continued)

Xf/Xfr = Forward period/total period = 0.50

ZONE 1		ZONE 2			ZONE 3	
T1/Xf	h1/H	T2/Xfr	Y/2/H	U	T3/Xr	h3/H
0.00000	0.00000	1.00000	3.43596	0.25000	0.00000	0.00000
0.05000	0.04325	0.95000	3.28734	0.25000	0.05000	0.04325
0.10000	0.08713	0.90000	3.13659	0.25000	0.10000	0.08713
0.15000	0.13163	0.85000	2.98367	0.25000	0.15000	0.13163
0.20000	0.17678	0.80000	2.82854	0.25000	0.20000	0.17678
0.25000	0.22259	0.75000	2.67113	0.25000	0.25000	0.22259
0.30000	0.26908	0.70000	2.51141	0.25000	0.30000	0.26908
0.35000	0.31625	0.65000	2.34932	0.25000	0.35000	0.31625
0.40000	0.36413	0.60000	2.18481	0.25000	0.40000	0.36413
0.45000	0.41274	0.55000	2.01782	0.25000	0.45000	0.41274
0.50000	0.46207	0.50000	1.84829	0.25000	0.50000	0.46207
0.55000	0.51217	0.45000	1.67618	0.25000	0.55000	0.51217
0.60000	0.56303	0.40000	1.50141	0.25000	0.60000	0.56303
0.65000	0.61468	0.35000	1.32393	0.25000	0.65000	0.61468
0.70000	0.66714	0.30000	1.14368	0.25000	0.70000	0.66714
0.75000	0.72043	0.25000	0.96058	0.25000	0.75000	0.72043
0.80000	0.77457	0.20000	0.77457	0.25000	0.80000	0.77457
0.85000	0.82957	0.15000	0.58558	0.25000	0.85000	0.82957
0.90000	0.88546	0.10000	0.39354	0.25000	0.90000	0.88546
0.95000	0.94227	0.05000	0.19837	0.25000	0.95000	0.94227
1.00000	1.00000	0.00000	0.00000	0.25000	1.00000	1.00000
OPTIMUM POSITION						
0.50000	0.46207	0.50000	1.84829	0.25000	0.50000	0.46207

Xf/Xfr = Forward period/total period = 0.55

ZONE 1		ZONE 2			ZONE 3	
T1/Xf	h1/H	T2/Xfr	Y/2/H	U	T3/Xr	h3/H
0.29029	0.25824	0.84034	2.97656	0.23225	0.00000	0.00000
0.32642	0.29200	0.79797	2.84414	0.23213	0.05000	0.04360
0.36250	0.32610	0.75562	2.71015	0.23201	0.10000	0.08780
0.39850	0.36052	0.71332	2.57461	0.23189	0.15000	0.13260
0.43445	0.39528	0.67105	2.43742	0.23177	0.20000	0.17803
0.47033	0.43037	0.62882	2.29862	0.23164	0.25000	0.22408
0.50614	0.46580	0.58662	2.15816	0.23151	0.30000	0.27077
0.54189	0.50157	0.54446	2.01601	0.23138	0.35000	0.31812
0.57758	0.53770	0.50234	1.87212	0.23125	0.40000	0.36615
0.61319	0.57418	0.46025	1.72649	0.23112	0.45000	0.41485
0.64874	0.61102	0.41819	1.57905	0.23099	0.50000	0.46427
0.68421	0.64822	0.37618	1.42982	0.23085	0.55000	0.51439
0.71961	0.68578	0.33422	1.27876	0.23071	0.60000	0.56523
0.75493	0.72372	0.29229	1.12582	0.23057	0.65000	0.61681
0.79018	0.76202	0.25040	0.97098	0.23043	0.70000	0.66915
0.82535	0.80071	0.20856	0.81421	0.23029	0.75000	0.72226
0.86044	0.83978	0.16676	0.65547	0.23014	0.80000	0.77616
0.89545	0.87924	0.12501	0.49471	0.23000	0.85000	0.83087
0.93039	0.91910	0.08329	0.33188	0.22985	0.90000	0.88640
0.96524	0.95935	0.04162	0.16700	0.22970	0.95000	0.94277
1.00000	1.00000	0.00000	0.00000	0.22954	1.00000	1.00000
OPTIMUM POSITION						
0.65150	0.61390	0.41492	1.56750	0.23098	0.50389	0.46814

Xf/Xfr = Forward period/total period = 0.60

ZONE 1		ZONE 2			ZONE 3	
T1/Xf	h1/H	T2/Xfr	Y/2/H	U	T3/Xr	h3/H
0.50242	0.45952	0.69855	2.58918	0.21369	0.00000	0.00000
0.52810	0.48512	0.66314	2.47171	0.21348	0.05000	0.04393
0.55371	0.51086	0.62778	2.35309	0.21327	0.10000	0.08842
0.57923	0.53675	0.59246	2.23328	0.21306	0.15000	0.13351
0.60468	0.56280	0.55719	2.11228	0.21285	0.20000	0.17918
0.63004	0.58899	0.52197	1.99008	0.21263	0.25000	0.22545
0.65533	0.61533	0.48681	1.86666	0.21241	0.30000	0.27234
0.68052	0.64183	0.45169	1.74199	0.21219	0.35000	0.31985
0.70564	0.66848	0.41661	1.61602	0.21196	0.40000	0.36801
0.73067	0.69527	0.38160	1.48881	0.21174	0.45000	0.41681
0.75561	0.72222	0.34663	1.36031	0.21151	0.50000	0.46627
0.78047	0.74931	0.31172	1.23050	0.21128	0.55000	0.51640
0.80523	0.77656	0.27686	1.09936	0.21104	0.60000	0.56722
0.82991	0.80396	0.24205	0.96684	0.21081	0.65000	0.61874
0.85449	0.83151	0.20730	0.83298	0.21057	0.70000	0.67097
0.87898	0.85922	0.17261	0.69773	0.21033	0.75000	0.72392
0.90338	0.88707	0.13797	0.56107	0.21009	0.80000	0.77760
0.92768	0.91507	0.10339	0.42300	0.20984	0.85000	0.83203
0.95188	0.94323	0.06887	0.28348	0.20959	0.90000	0.88724
0.97599	0.97154	0.03441	0.14249	0.20934	0.95000	0.94322
1.00000	1.00000	0.00000	0.00000	0.20908	1.00000	1.00000
OPTIMUM POSITION						
0.75925	0.72617	0.34153	1.34142	0.21148	0.50730	0.47356

Xf/Xfr = Forward period/total period = 0.65

ZONE 1		ZONE 2			ZONE 3	
T1/Xf	h1/H	T2/Xfr	Y/2/H	U	T3/Xr	h3/H
0.65744	0.61476	0.57266	2.26240	0.19441	0.00000	0.00000
0.67530	0.63356	0.54356	2.15843	0.19414	0.05000	0.04423
0.69309	0.65241	0.51449	2.05356	0.19387	0.10000	0.08901
0.71080	0.67132	0.48548	1.94779	0.19360	0.15000	0.13435
0.72844	0.69028	0.45652	1.84110	0.19333	0.20000	0.18025
0.74600	0.70928	0.42760	1.73348	0.19305	0.25000	0.22673
0.76349	0.72834	0.39873	1.62493	0.19277	0.30000	0.27379
0.78091	0.74745	0.36991	1.51543	0.19249	0.35000	0.32145
0.79824	0.76661	0.34114	1.40496	0.19221	0.40000	0.36972
0.81550	0.78581	0.31243	1.29352	0.19192	0.45000	0.41860
0.83268	0.80506	0.28376	1.18111	0.19164	0.50000	0.46810
0.84978	0.82436	0.25514	1.06769	0.19135	0.55000	0.51825
0.86680	0.84371	0.22658	0.95327	0.19105	0.60000	0.56903
0.88374	0.86310	0.19807	0.83782	0.19076	0.65000	0.62048
0.90060	0.88253	0.16961	0.72134	0.19046	0.70000	0.67260
0.91738	0.90200	0.14121	0.60382	0.19016	0.75000	0.72539
0.93407	0.92152	0.11286	0.48523	0.18986	0.80000	0.77888
0.95068	0.94108	0.08456	0.36557	0.18956	0.85000	0.83307
0.96721	0.96068	0.05632	0.24481	0.18925	0.90000	0.88799
0.98365	0.98032	0.02813	0.12296	0.18894	0.95000	0.94362
1.00000	1.00000	0.00000	0.00000	0.18863	1.00000	1.00000
OPTIMUM POSITION						
0.83621	0.80904	0.27786	1.15782	0.19158	0.51030	0.47839

Table F.1 (continued)

Xf/Xfr = Forward period/total period = 0.70						
ZONE 1		ZONE 2			ZONE 3	
T1/Xf	h1/H	T2/Xfr	Y/2/H	U	T3/Xr	h3/H
0.77026	0.73312	0.46082	1.98626	0.17449	0.00000	0.00000
0.78232	0.74645	0.43738	1.89427	0.17419	0.05000	0.04452
0.79432	0.75977	0.41398	1.80156	0.17389	0.10000	0.08956
0.80626	0.77311	0.39062	1.70812	0.17359	0.15000	0.13514
0.81815	0.78645	0.36730	1.61395	0.17329	0.20000	0.18126
0.82997	0.79979	0.34402	1.51905	0.17298	0.25000	0.22792
0.84174	0.81314	0.32078	1.42338	0.17268	0.30000	0.27514
0.85344	0.82649	0.29759	1.32696	0.17237	0.35000	0.32293
0.86509	0.83984	0.27444	1.22977	0.17206	0.40000	0.37129
0.87667	0.85319	0.25133	1.13180	0.17174	0.45000	0.42024
0.88820	0.86655	0.22826	1.03304	0.17143	0.50000	0.46979
0.89966	0.87991	0.20524	0.93349	0.17111	0.55000	0.51993
0.91106	0.89326	0.18226	0.83313	0.17079	0.60000	0.57069
0.92240	0.90662	0.15932	0.73196	0.17047	0.65000	0.62207
0.93368	0.91997	0.13643	0.62996	0.17015	0.70000	0.67408
0.94489	0.93332	0.11358	0.52712	0.16982	0.75000	0.72673
0.95604	0.94666	0.09077	0.42343	0.16950	0.80000	0.78004
0.96713	0.96000	0.06801	0.31889	0.16917	0.85000	0.83400
0.97815	0.97334	0.04530	0.21348	0.16884	0.90000	0.88865
0.98911	0.98667	0.02263	0.10719	0.16850	0.95000	0.94397
1.00000	1.00000	0.00000	0.00000	0.16817	1.00000	1.00000
OPTIMUM POSITION						
0.89117	0.87001	0.22230	1.00735	0.17135	0.51294	0.48271

Xf/Xfr = Forward period/total period = 0.75						
ZONE 1		ZONE 2			ZONE 3	
T1/Xf	h1/H	T2/Xfr	Y/2/H	U	T3/Xr	h3/H
0.85162	0.82218	0.36128	1.75230	0.15401	0.00000	0.00000
0.85944	0.83121	0.34292	1.67084	0.15371	0.05000	0.04479
0.86722	0.84022	0.32459	1.58877	0.15341	0.10000	0.09007
0.87495	0.84922	0.30629	1.50610	0.15310	0.15000	0.13587
0.88264	0.85821	0.28802	1.42281	0.15280	0.20000	0.18218
0.89030	0.86718	0.26978	1.33890	0.15249	0.25000	0.22902
0.89791	0.87615	0.25157	1.25435	0.15218	0.30000	0.27638
0.90548	0.88510	0.23339	1.16918	0.15187	0.35000	0.32429
0.91301	0.89403	0.21525	1.08335	0.15156	0.40000	0.37274
0.92049	0.90295	0.19713	0.99687	0.15125	0.45000	0.42175
0.92793	0.91186	0.17905	0.90972	0.15093	0.50000	0.47133
0.93533	0.92075	0.16100	0.82191	0.15062	0.55000	0.52147
0.94269	0.92962	0.14298	0.73342	0.15030	0.60000	0.57220
0.95001	0.93848	0.12499	0.64424	0.14998	0.65000	0.62351
0.95728	0.94732	0.10704	0.55437	0.14966	0.70000	0.67543
0.96451	0.95615	0.08912	0.46379	0.14934	0.75000	0.72795
0.97169	0.96496	0.07123	0.37249	0.14902	0.80000	0.78109
0.97884	0.97375	0.05337	0.28048	0.14869	0.85000	0.83485
0.98594	0.98252	0.03555	0.18773	0.14837	0.90000	0.88925
0.99299	0.99127	0.01776	0.09426	0.14804	0.95000	0.94429
1.00000	1.00000	0.00000	0.00000	0.14771	1.00000	1.00000
OPTIMUM POSITION						
0.93020	0.91458	0.17353	0.88296	0.15084	0.51528	0.48659

Xf/Xfr = Forward period/total period = 0.80						
ZONE 1		ZONE 2			ZONE 3	
T1/Xf	h1/H	T2/Xfr	Y/2/H	U	T3/Xr	h3/H
0.90936	0.88807	0.27251	1.55340	0.13304	0.00000	0.00000
0.91414	0.89382	0.25869	1.48113	0.13276	0.05000	0.04503
0.91889	0.89955	0.24489	1.40833	0.13248	0.10000	0.09054
0.92362	0.90527	0.23110	1.33500	0.13220	0.15000	0.13654
0.92832	0.91097	0.21734	1.26113	0.13192	0.20000	0.18303
0.93300	0.91666	0.20360	1.18672	0.13163	0.25000	0.23003
0.93765	0.92233	0.18988	1.11175	0.13135	0.30000	0.27753
0.94227	0.92799	0.17618	1.03622	0.13106	0.35000	0.32554
0.94687	0.93363	0.16250	0.96013	0.13077	0.40000	0.37407
0.95144	0.93925	0.14884	0.88346	0.13049	0.45000	0.42313
0.95599	0.94486	0.13521	0.80621	0.13020	0.50000	0.47274
0.96051	0.95045	0.12159	0.72837	0.12991	0.55000	0.52288
0.96501	0.95603	0.10800	0.64993	0.12962	0.60000	0.57357
0.96947	0.96159	0.09442	0.57090	0.12932	0.65000	0.62483
0.97391	0.96713	0.08087	0.49124	0.12903	0.70000	0.67665
0.97833	0.97265	0.06734	0.41097	0.12874	0.75000	0.72905
0.98272	0.97816	0.05383	0.33006	0.12844	0.80000	0.78204
0.98708	0.98365	0.04034	0.24853	0.12815	0.85000	0.83562
0.99141	0.98912	0.02687	0.16634	0.12785	0.90000	0.88979
0.99572	0.99457	0.01342	0.08350	0.12755	0.95000	0.94459
1.00000	1.00000	0.00000	0.00000	0.12725	1.00000	1.00000
OPTIMUM POSITION						
0.95756	0.94681	0.13047	0.77924	0.13010	0.51737	0.49009

Xf/Xfr = Forward period/total period = 0.85						
ZONE 1		ZONE 2			ZONE 3	
T1/Xf	h1/H	T2/Xfr	Y/2/H	U	T3/Xr	h3/H
0.94931	0.93572	0.19309	1.38355	0.11158	0.00000	0.00000
0.95198	0.93905	0.18332	1.31926	0.11133	0.05000	0.04527
0.95463	0.94237	0.17356	1.25451	0.11109	0.10000	0.09100
0.95727	0.94567	0.16382	1.18927	0.11085	0.15000	0.13719
0.95990	0.94897	0.15409	1.12354	0.11061	0.20000	0.18386
0.96251	0.95225	0.14437	1.05732	0.11036	0.25000	0.23100
0.96511	0.95552	0.13466	0.99059	0.11012	0.30000	0.27862
0.96769	0.95878	0.12496	0.92335	0.10987	0.35000	0.32674
0.97026	0.96202	0.11528	0.85561	0.10963	0.40000	0.37535
0.97282	0.96526	0.10560	0.78734	0.10938	0.45000	0.42446
0.97536	0.96848	0.09594	0.71854	0.10914	0.50000	0.47409
0.97789	0.97169	0.08629	0.64921	0.10889	0.55000	0.52423
0.98040	0.97488	0.07666	0.57934	0.10864	0.60000	0.57490
0.98290	0.97807	0.06703	0.50891	0.10839	0.65000	0.62610
0.98539	0.98124	0.05742	0.43794	0.10814	0.70000	0.67784
0.98786	0.98440	0.04782	0.36640	0.10789	0.75000	0.73012
0.99032	0.98755	0.03823	0.29429	0.10764	0.80000	0.78296
0.99276	0.99068	0.02865	0.22160	0.10739	0.85000	0.83636
0.99519	0.99380	0.01909	0.14833	0.10714	0.90000	0.89033
0.99760	0.99691	0.00954	0.07447	0.10689	0.95000	0.94487
1.00000	1.00000	0.00000	0.00000	0.10663	1.00000	1.00000
OPTIMUM POSITION						
0.97633	0.96971	0.09224	0.69203	0.10904	0.51916	0.49325

Table F.2   Asymmetrical D–R–R–D motion.
Zone 1: Cycloidal
Zone 2: Cycloidal
Zone 3: Cycloidal

Xf/Xfr = Forward period/total period = 0.10

ZONE 1		ZONE 2			ZONE 3	
T1/Xf	h1/H	T2/Xfr	Y/2/H	U	T3/Xr	h3/H
0.00000	0.00000	0.12800	1.24088	0.39064	0.96889	0.95926
0.05000	0.04329	0.12149	1.18558	0.39098	0.97057	0.96143
0.10000	0.08722	0.11499	1.12964	0.39133	0.97223	0.96359
0.15000	0.13181	0.10851	1.07304	0.39167	0.97388	0.96573
0.20000	0.17705	0.10204	1.01578	0.39202	0.97551	0.96786
0.25000	0.22298	0.09557	0.95783	0.39236	0.97714	0.96998
0.30000	0.26957	0.08912	0.89921	0.39271	0.97875	0.97208
0.35000	0.31687	0.08269	0.83988	0.39305	0.98035	0.97416
0.40000	0.36487	0.07626	0.77984	0.39340	0.98194	0.97624
0.45000	0.41358	0.06984	0.71909	0.39375	0.98351	0.97829
0.50000	0.46302	0.06344	0.65761	0.39410	0.98507	0.98034
0.55000	0.51321	0.05704	0.59538	0.39444	0.98662	0.98237
0.60000	0.56414	0.05066	0.53239	0.39479	0.98816	0.98439
0.65000	0.61583	0.04429	0.46864	0.39514	0.98968	0.98639
0.70000	0.66830	0.03793	0.40412	0.39549	0.99119	0.98837
0.75000	0.72154	0.03158	0.33882	0.39584	0.99269	0.99035
0.80000	0.77559	0.02524	0.27270	0.39619	0.99418	0.99231
0.85000	0.83044	0.01891	0.20578	0.39654	0.99565	0.99425
0.90000	0.88612	0.01260	0.13803	0.39688	0.99711	0.99618
0.95000	0.94264	0.00629	0.06943	0.39723	0.99856	0.99810
1.00000	1.00000	0.00000	0.00000	0.39758	1.00000	1.00000
OPTIMUM POSITION						
0.53124	0.49430	0.05944	0.61881	0.39431	0.98604	0.98161

Xf/Xfr = Forward period/total period = 0.15

ZONE 1		ZONE 2			ZONE 3	
T1/Xf	h1/H	T2/Xfr	Y/2/H	U	T3/Xr	h3/H
0.00000	0.00000	0.20580	1.38968	0.36444	0.93436	0.91532
0.05000	0.04235	0.19527	1.32807	0.36487	0.93791	0.91982
0.10000	0.08543	0.18477	1.26570	0.36531	0.94144	0.92428
0.15000	0.12922	0.17430	1.20259	0.36575	0.94494	0.92872
0.20000	0.17377	0.16385	1.13870	0.36619	0.94841	0.93314
0.25000	0.21906	0.15343	1.07404	0.36663	0.95185	0.93753
0.30000	0.26513	0.14302	1.00857	0.36707	0.95526	0.94189
0.35000	0.31198	0.13265	0.94231	0.36751	0.95865	0.94622
0.40000	0.35964	0.12230	0.87521	0.36796	0.96200	0.95053
0.45000	0.40812	0.11197	0.80728	0.36840	0.96533	0.95481
0.50000	0.45742	0.10167	0.73849	0.36885	0.96863	0.95906
0.55000	0.50758	0.09139	0.66883	0.36930	0.97190	0.96329
0.60000	0.55860	0.08114	0.59829	0.36975	0.97513	0.96748
0.65000	0.61051	0.07091	0.52684	0.37021	0.97835	0.97165
0.70000	0.66331	0.06070	0.45447	0.37066	0.98153	0.97579
0.75000	0.71703	0.05052	0.38117	0.37111	0.98468	0.97990
0.80000	0.77168	0.04037	0.30692	0.37157	0.98780	0.98398
0.85000	0.82729	0.03024	0.23168	0.37203	0.99089	0.98803
0.90000	0.88386	0.02013	0.15547	0.37249	0.99396	0.99205
0.95000	0.94143	0.01006	0.07825	0.37295	0.99699	0.99604
1.00000	1.00000	0.00000	0.00000	0.37341	1.00000	1.00000
OPTIMUM POSITION						
0.53231	0.48974	0.09502	0.69357	0.36914	0.97074	0.96180

Xf/Xfr = Forward period/total period = 0.20

ZONE 1		ZONE 2			ZONE 3	
T1/Xf	h1/H	T2/Xfr	Y/2/H	U	T3/Xr	h3/H
0.00000	0.00000	0.29217	1.55939	0.34226	0.88478	0.85420
0.05000	0.04168	0.27718	1.49039	0.34274	0.89103	0.86184
0.10000	0.08412	0.26222	1.42056	0.34322	0.89723	0.86941
0.15000	0.12735	0.24730	1.34986	0.34371	0.90337	0.87701
0.20000	0.17136	0.23243	1.27832	0.34420	0.90947	0.88454
0.25000	0.21621	0.21759	1.20586	0.34469	0.91552	0.89205
0.30000	0.26187	0.20279	1.13252	0.34518	0.92151	0.89951
0.35000	0.30839	0.18803	1.05825	0.34568	0.92746	0.90695
0.40000	0.35578	0.17332	0.98305	0.34618	0.93335	0.91434
0.45000	0.40407	0.15865	0.90687	0.34668	0.93919	0.92171
0.50000	0.45325	0.14401	0.82973	0.34719	0.94498	0.92903
0.55000	0.50339	0.12942	0.75157	0.34770	0.95072	0.93631
0.60000	0.55445	0.11487	0.67243	0.34821	0.95641	0.94356
0.65000	0.60651	0.10036	0.59222	0.34873	0.96204	0.95076
0.70000	0.65954	0.08590	0.51099	0.34924	0.96762	0.95793
0.75000	0.71362	0.07147	0.42863	0.34977	0.97316	0.96505
0.80000	0.76871	0.05710	0.34521	0.35029	0.97863	0.97213
0.85000	0.82488	0.04276	0.26065	0.35082	0.98405	0.97916
0.90000	0.88213	0.02846	0.17494	0.35135	0.98942	0.98615
0.95000	0.94049	0.01421	0.08808	0.35188	0.99474	0.99310
1.00000	1.00000	0.00000	0.00000	0.35242	1.00000	1.00000
OPTIMUM POSITION						
0.53120	0.48442	0.13490	0.78109	0.34751	0.94857	0.93358

Xf/Xfr = Forward period/total period = 0.25

ZONE 1		ZONE 2			ZONE 3	
T1/Xf	h1/H	T2/Xfr	Y/2/H	U	T3/Xr	h3/H
0.00000	0.00000	0.38721	1.75201	0.32282	0.81705	0.77393
0.05000	0.04116	0.36731	1.67476	0.32330	0.82692	0.78550
0.10000	0.08312	0.34745	1.59655	0.32378	0.83673	0.79706
0.15000	0.12590	0.32765	1.51735	0.32427	0.84646	0.80859
0.20000	0.16951	0.30791	1.43715	0.32477	0.85612	0.82011
0.25000	0.21398	0.28822	1.35594	0.32527	0.86570	0.83160
0.30000	0.25933	0.26859	1.27367	0.32577	0.87521	0.84307
0.35000	0.30558	0.24902	1.19036	0.32628	0.88464	0.85451
0.40000	0.35275	0.22950	1.10596	0.32679	0.89400	0.86592
0.45000	0.40087	0.21005	1.02046	0.32731	0.90327	0.87731
0.50000	0.44996	0.19064	0.93381	0.32784	0.91247	0.88866
0.55000	0.50004	0.17131	0.84604	0.32838	0.92159	0.89997
0.60000	0.55115	0.15202	0.75706	0.32890	0.93064	0.91126
0.65000	0.60330	0.13281	0.66691	0.32943	0.93959	0.92250
0.70000	0.65654	0.11364	0.57550	0.32998	0.94848	0.93371
0.75000	0.71085	0.09455	0.48288	0.33052	0.95727	0.94487
0.80000	0.76631	0.07551	0.38897	0.33108	0.96598	0.95599
0.85000	0.82293	0.05654	0.29374	0.33163	0.97462	0.96707
0.90000	0.88071	0.03763	0.19722	0.33219	0.98316	0.97809
0.95000	0.93974	0.01878	0.09930	0.33276	0.99162	0.98907
1.00000	1.00000	0.00000	0.00000	0.33333	1.00000	1.00000
OPTIMUM POSITION						
0.52854	0.47842	0.17960	0.88385	0.32814	0.91769	0.89512

Table F.2 (continued)

Xf/Xfr = Forward period/total period = 0.30					
— ZONE 1 —		— ZONE 2 —		— ZONE 3 —	
T1/Xf	h1/H	T2/Xfr	Y/2/H	U	T3/Xr	h3/H
0.00000	0.00000	0.49111	1.96977	0.30543	0.72699	0.67225
0.05000	0.04075	0.46587	1.88354	0.30588	0.74161	0.68851
0.10000	0.08232	0.44071	1.79618	0.30633	0.75613	0.70481
0.15000	0.12473	0.41561	1.70767	0.30678	0.77056	0.72113
0.20000	0.16802	0.39057	1.61795	0.30724	0.78490	0.73748
0.25000	0.21219	0.36560	1.52704	0.30771	0.79914	0.75385
0.30000	0.25727	0.34070	1.43488	0.30818	0.81328	0.77023
0.35000	0.30328	0.31588	1.34149	0.30866	0.82732	0.78664
0.40000	0.35028	0.29112	1.24678	0.30915	0.84126	0.80306
0.45000	0.39826	0.26644	1.15077	0.30964	0.85509	0.81949
0.50000	0.44725	0.24183	1.05344	0.31014	0.86882	0.83592
0.55000	0.49730	0.21729	0.95473	0.31065	0.88244	0.85236
0.60000	0.54842	0.19283	0.85463	0.31116	0.89596	0.86881
0.65000	0.60065	0.16844	0.75310	0.31168	0.90937	0.88525
0.70000	0.65402	0.14414	0.65012	0.31220	0.92266	0.90169
0.75000	0.70855	0.11991	0.54567	0.31273	0.93584	0.91811
0.80000	0.76429	0.09576	0.43970	0.31327	0.94891	0.93453
0.85000	0.82128	0.07170	0.33218	0.31382	0.96186	0.95093
0.90000	0.87953	0.04771	0.22308	0.31437	0.97469	0.96731
0.95000	0.93909	0.02382	0.11237	0.31493	0.98741	0.98367
1.00000	1.00000	0.00000	0.00000	0.31549	1.00000	1.00000
— OPTIMUM POSITION —						
0.52467	0.47181	0.22971	1.00491	0.31039	0.87556	0.84403

Xf/Xfr = Forward period/total period = 0.35					
— ZONE 1 —		— ZONE 2 —		— ZONE 3 —	
T1/Xf	h1/H	T2/Xfr	Y/2/H	U	T3/Xr	h3/H
0.00000	0.00000	0.60409	2.21503	0.28969	0.60909	0.54670
0.05000	0.04042	0.57314	2.11930	0.29007	0.62978	0.56832
0.10000	0.08167	0.54227	2.02221	0.29045	0.65035	0.59005
0.15000	0.12378	0.51146	1.92368	0.29083	0.67083	0.61192
0.20000	0.16678	0.48073	1.82372	0.29123	0.69119	0.63391
0.25000	0.21070	0.45007	1.72225	0.29162	0.71144	0.65602
0.30000	0.25556	0.41948	1.61929	0.29203	0.73157	0.67824
0.35000	0.30139	0.38897	1.51478	0.29244	0.75159	0.70058
0.40000	0.34823	0.35852	1.40868	0.29286	0.77150	0.72304
0.45000	0.39607	0.32817	1.30101	0.29329	0.79127	0.74560
0.50000	0.44498	0.29790	1.19170	0.29372	0.81092	0.76827
0.55000	0.49499	0.26770	1.08067	0.29417	0.83045	0.79105
0.60000	0.54611	0.23760	0.96797	0.29461	0.84985	0.81392
0.65000	0.59841	0.20757	0.85348	0.29507	0.86912	0.83689
0.70000	0.65188	0.17764	0.73725	0.29554	0.88824	0.85995
0.75000	0.70659	0.14780	0.61919	0.29601	0.90723	0.88310
0.80000	0.76257	0.11805	0.49924	0.29649	0.92608	0.90633
0.85000	0.81985	0.08840	0.37742	0.29698	0.94478	0.92964
0.90000	0.87850	0.05883	0.25361	0.29747	0.96334	0.95303
0.95000	0.93852	0.02937	0.12784	0.29798	0.98174	0.97648
1.00000	1.00000	0.00000	0.00000	0.29849	1.00000	1.00000
— OPTIMUM POSITION —						
0.51976	0.46461	0.28596	1.14802	0.29390	0.81866	0.77726

Xf/Xfr = Forward period/total period = 0.40					
— ZONE 1 —		— ZONE 2 —		— ZONE 3 —	
T1/Xf	h1/H	T2/Xfr	Y/2/H	U	T3/Xr	h3/H
0.00000	0.00000	0.72640	2.49032	0.27533	0.45600	0.39466
0.05000	0.04013	0.68940	2.38482	0.27560	0.48434	0.42213
0.10000	0.08111	0.65246	2.27759	0.27588	0.51258	0.44988
0.15000	0.12299	0.61556	2.16855	0.27617	0.54072	0.47793
0.20000	0.16575	0.57875	2.05774	0.27646	0.56875	0.50627
0.25000	0.20946	0.54200	1.94506	0.27676	0.59667	0.53490
0.30000	0.25413	0.50530	1.83047	0.27706	0.62449	0.56384
0.35000	0.29979	0.46869	1.71395	0.27737	0.65218	0.59306
0.40000	0.34647	0.43215	1.59546	0.27768	0.67975	0.62258
0.45000	0.39421	0.39568	1.47493	0.27801	0.70720	0.65240
0.50000	0.44305	0.35928	1.35229	0.27833	0.73453	0.68253
0.55000	0.49300	0.32296	1.22753	0.27867	0.76173	0.71295
0.60000	0.54413	0.28672	1.10059	0.27901	0.78880	0.74366
0.65000	0.59645	0.25057	0.97145	0.27937	0.81571	0.77467
0.70000	0.65003	0.21450	0.83997	0.27972	0.84250	0.80599
0.75000	0.70489	0.17851	0.70615	0.28009	0.86915	0.83760
0.80000	0.76106	0.14262	0.56997	0.28046	0.89563	0.86950
0.85000	0.81862	0.10682	0.43132	0.28085	0.92197	0.90169
0.90000	0.87760	0.07111	0.29013	0.28124	0.94814	0.93418
0.95000	0.93803	0.03551	0.14641	0.28164	0.97415	0.96694
1.00000	1.00000	0.00000	0.00000	0.28205	1.00000	1.00000
— OPTIMUM POSITION —						
0.51396	0.45687	0.34913	1.31768	0.27843	0.74214	0.69099

Xf/Xfr = Forward period/total period = 0.45					
— ZONE 1 —		— ZONE 2 —		— ZONE 3 —	
T1/Xf	h1/H	T2/Xfr	Y/2/H	U	T3/Xr	h3/H
0.00000	0.00000	0.85829	2.79826	0.26215	0.25765	0.21339
0.05000	0.03989	0.81492	2.68298	0.26230	0.29560	0.24698
0.10000	0.08066	0.77158	2.56550	0.26245	0.33349	0.28111
0.15000	0.12231	0.72828	2.44579	0.26260	0.37130	0.31578
0.20000	0.16487	0.68504	2.32379	0.26276	0.40902	0.35101
0.25000	0.20839	0.64182	2.19942	0.26292	0.44669	0.38680
0.30000	0.25290	0.59864	2.07259	0.26309	0.48428	0.42319
0.35000	0.29841	0.55553	1.94331	0.26326	0.52177	0.46015
0.40000	0.34496	0.51246	1.81148	0.26344	0.55917	0.49772
0.45000	0.39261	0.46943	1.67696	0.26362	0.59649	0.53591
0.50000	0.44136	0.42646	1.53977	0.26380	0.63371	0.57471
0.55000	0.49128	0.38354	1.39975	0.26399	0.67084	0.61417
0.60000	0.54240	0.34067	1.25688	0.26418	0.70787	0.65427
0.65000	0.59476	0.29787	1.11106	0.26438	0.74479	0.69504
0.70000	0.64841	0.25511	0.96217	0.26459	0.78162	0.73648
0.75000	0.70338	0.21242	0.81019	0.26480	0.81832	0.77861
0.80000	0.75973	0.16980	0.65499	0.26502	0.85491	0.82143
0.85000	0.81751	0.12725	0.49648	0.26524	0.89137	0.86497
0.90000	0.87679	0.08475	0.33451	0.26547	0.92772	0.90925
0.95000	0.93759	0.04234	0.16907	0.26570	0.96393	0.95425
1.00000	1.00000	0.00000	0.00000	0.26594	1.00000	1.00000
— OPTIMUM POSITION —						
0.50734	0.44862	0.42015	1.51937	0.26383	0.63917	0.58047

Table F.2 (continued)

Xf/Xfr = Forward period/total period = 0.50						
ZONE 1		ZONE 2			ZONE 3	
T1/Xf	h1/H	T2/Xfr	Y/2/H	U	T3/Xr	h3/H
0.00000	0.00000	1.00000	3.14159	0.25000	0.00000	0.00000
0.05000	0.03970	0.95000	3.01688	0.25000	0.05000	0.03970
0.10000	0.08026	0.90000	2.88944	0.25000	0.10000	0.08026
0.15000	0.12173	0.85000	2.75917	0.25000	0.15000	0.12173
0.20000	0.16412	0.80000	2.62598	0.25000	0.20000	0.16412
0.25000	0.20748	0.75000	2.48977	0.25000	0.25000	0.20748
0.30000	0.25183	0.70000	2.35044	0.25000	0.30000	0.25183
0.35000	0.29721	0.65000	2.20787	0.25000	0.35000	0.29721
0.40000	0.34366	0.60000	2.06196	0.25000	0.40000	0.34366
0.45000	0.39121	0.55000	1.91257	0.25000	0.45000	0.39121
0.50000	0.43990	0.50000	1.75960	0.25000	0.50000	0.43990
0.55000	0.48978	0.45000	1.60291	0.25000	0.55000	0.48978
0.60000	0.54088	0.40000	1.44236	0.25000	0.60000	0.54088
0.65000	0.59326	0.35000	1.27780	0.25000	0.65000	0.59326
0.70000	0.64697	0.30000	1.10909	0.25000	0.70000	0.64697
0.75000	0.70204	0.25000	0.93606	0.25000	0.75000	0.70204
0.80000	0.75855	0.20000	0.75855	0.25000	0.80000	0.75855
0.85000	0.81653	0.15000	0.57638	0.25000	0.85000	0.81653
0.90000	0.87606	0.10000	0.38936	0.25000	0.90000	0.87606
0.95000	0.93720	0.05000	0.19730	0.25000	0.95000	0.93720
1.00000	1.00000	0.00000	0.00000	0.25000	1.00000	1.00000
OPTIMUM POSITION						
0.50000	0.43990	0.50000	1.75960	0.25000	0.50000	0.43990

Xf/Xfr = Forward period/total period = 0.55						
ZONE 1		ZONE 2			ZONE 3	
T1/Xf	h1/H	T2/Xfr	Y/2/H	U	T3/Xr	h3/H
0.25765	0.21339	0.85829	2.79827	0.23785	0.00000	0.00000
0.29560	0.24698	0.81492	2.68297	0.23770	0.05000	0.03990
0.33349	0.28111	0.77158	2.56550	0.23755	0.10000	0.08066
0.37130	0.31578	0.72828	2.44579	0.23740	0.15000	0.12231
0.40903	0.35100	0.68504	2.32380	0.23724	0.20000	0.16487
0.44669	0.38680	0.64182	2.19942	0.23708	0.25000	0.20839
0.48427	0.42319	0.59865	2.07261	0.23691	0.30000	0.25290
0.52176	0.46015	0.55553	1.94332	0.23674	0.35000	0.29841
0.55917	0.49772	0.51246	1.81147	0.23656	0.40000	0.34496
0.59648	0.53590	0.46944	1.67698	0.23638	0.45000	0.39260
0.63372	0.57472	0.42646	1.53974	0.23620	0.50000	0.44137
0.67084	0.61417	0.38354	1.39975	0.23601	0.55000	0.49128
0.70787	0.65427	0.34068	1.25689	0.23582	0.60000	0.54240
0.74480	0.69504	0.29786	1.11104	0.23562	0.65000	0.59476
0.78161	0.73647	0.25512	0.96222	0.23541	0.70000	0.64839
0.81832	0.77860	0.21243	0.81021	0.23520	0.75000	0.70338
0.85490	0.82143	0.16980	0.65500	0.23498	0.80000	0.75973
0.89137	0.86498	0.12725	0.49647	0.23476	0.85000	0.81752
0.92772	0.90925	0.08476	0.33452	0.23453	0.90000	0.87679
0.96392	0.95424	0.04234	0.16908	0.23430	0.95000	0.93759
1.00000	1.00000	0.00000	0.00000	0.23406	1.00000	1.00000
OPTIMUM POSITION						
0.63917	0.58048	0.42015	1.51936	0.23617	0.50734	0.44863

Xf/Xfr = Forward period/total period = 0.60						
ZONE 1		ZONE 2			ZONE 3	
T1/Xf	h1/H	T2/Xfr	Y/2/H	U	T3/Xr	h3/H
0.45600	0.39466	0.72640	2.49031	0.22467	0.00000	0.00000
0.48434	0.42213	0.68940	2.38481	0.22440	0.05000	0.04013
0.51258	0.44988	0.65246	2.27760	0.22412	0.10000	0.08111
0.54072	0.47793	0.61557	2.16856	0.22383	0.15000	0.12299
0.56875	0.50627	0.57875	2.05775	0.22354	0.20000	0.16575
0.59667	0.53490	0.54200	1.94507	0.22324	0.25000	0.20946
0.62449	0.56383	0.50531	1.83049	0.22294	0.30000	0.25412
0.65218	0.59306	0.46869	1.71396	0.22263	0.35000	0.29979
0.67976	0.62258	0.43215	1.59545	0.22232	0.40000	0.34647
0.70721	0.65240	0.39568	1.47492	0.22199	0.45000	0.39421
0.73453	0.68253	0.35928	1.35229	0.22167	0.50000	0.44304
0.76173	0.71294	0.32297	1.22755	0.22133	0.55000	0.49300
0.78879	0.74366	0.28673	1.10060	0.22099	0.60000	0.54413
0.81572	0.77468	0.25057	0.97143	0.22063	0.65000	0.59646
0.84250	0.80599	0.21450	0.83997	0.22028	0.70000	0.65003
0.86914	0.83760	0.17851	0.70616	0.21991	0.75000	0.70488
0.89564	0.86950	0.14262	0.56995	0.21954	0.80000	0.76107
0.92197	0.90169	0.10682	0.43131	0.21915	0.85000	0.81862
0.94814	0.93418	0.07111	0.29013	0.21876	0.90000	0.87760
0.97415	0.96694	0.03551	0.14640	0.21836	0.95000	0.93803
1.00000	1.00000	0.00000	0.00000	0.21795	1.00000	1.00000
OPTIMUM POSITION						
0.74214	0.69099	0.34913	1.31768	0.22157	0.51396	0.45688

Xf/Xfr = Forward period/total period = 0.65						
ZONE 1		ZONE 2			ZONE 3	
T1/Xf	h1/H	T2/Xfr	Y/2/H	U	T3/Xr	h3/H
0.60909	0.54670	0.60409	2.21503	0.21031	0.00000	0.00000
0.62977	0.56831	0.57315	2.11932	0.20993	0.05000	0.04041
0.65035	0.59005	0.54227	2.02222	0.20955	0.10000	0.08166
0.67083	0.61192	0.51154	1.92368	0.20917	0.15000	0.12378
0.69119	0.63391	0.48073	1.82371	0.20877	0.20000	0.16678
0.71144	0.65602	0.45006	1.72225	0.20838	0.25000	0.21071
0.73158	0.67824	0.41948	1.61928	0.20797	0.30000	0.25557
0.75159	0.70059	0.38896	1.51478	0.20756	0.35000	0.30139
0.77149	0.72304	0.35853	1.40870	0.20714	0.40000	0.34822
0.79127	0.74560	0.32818	1.30102	0.20671	0.45000	0.39607
0.81093	0.76827	0.29790	1.19169	0.20628	0.50000	0.44498
0.83045	0.79104	0.26771	1.08070	0.20583	0.55000	0.49498
0.84985	0.81392	0.23760	0.96797	0.20539	0.60000	0.54612
0.86911	0.83689	0.20758	0.85350	0.20493	0.65000	0.59840
0.88824	0.85995	0.17764	0.73726	0.20446	0.70000	0.65188
0.90723	0.88310	0.14780	0.61919	0.20399	0.75000	0.70658
0.92608	0.90633	0.11805	0.49924	0.20351	0.80000	0.76257
0.94478	0.92964	0.08839	0.37739	0.20302	0.85000	0.81986
0.96334	0.95303	0.05883	0.25359	0.20253	0.90000	0.87850
0.98175	0.97648	0.02936	0.12782	0.20202	0.95000	0.93853
1.00000	1.00000	0.00000	0.00000	0.20151	1.00000	1.00000
OPTIMUM POSITION						
0.81866	0.77726	0.28596	1.14802	0.20610	0.51976	0.46462

Table F.2 (continued)

Xf/Xfr = Forward period/total period = 0.70

ZONE 1		ZONE 2			ZONE 3	
T1/Xf	h1/H	T2/Xfr	Y/2/H	U	T3/Xr	h3/H
0.72699	0.67225	0.49111	1.96976	0.19457	0.00000	0.00000
0.74160	0.68851	0.46588	1.88355	0.19412	0.05000	0.04075
0.75613	0.70481	0.44071	1.79619	0.19367	0.10000	0.08232
0.77056	0.72113	0.41561	1.70767	0.19322	0.15000	0.12474
0.78490	0.73748	0.39057	1.61796	0.19276	0.20000	0.16801
0.79914	0.75385	0.36560	1.52704	0.19229	0.25000	0.21219
0.81328	0.77023	0.34070	1.43488	0.19182	0.30000	0.25727
0.82732	0.78664	0.31588	1.34148	0.19134	0.35000	0.30329
0.84126	0.80305	0.29112	1.24679	0.19085	0.40000	0.35027
0.85509	0.81949	0.26644	1.15077	0.19036	0.45000	0.39826
0.86882	0.83592	0.24182	1.05343	0.18986	0.50000	0.44726
0.88244	0.85236	0.21729	0.95473	0.18935	0.55000	0.49730
0.89596	0.86880	0.19283	0.85464	0.18884	0.60000	0.54841
0.90937	0.88525	0.16844	0.75310	0.18832	0.65000	0.60065
0.92266	0.90169	0.14414	0.65013	0.18780	0.70000	0.65402
0.93584	0.91811	0.11991	0.54566	0.18727	0.75000	0.70856
0.94891	0.93453	0.09576	0.43970	0.18673	0.80000	0.76430
0.96186	0.95093	0.07170	0.33217	0.18618	0.85000	0.82128
0.97470	0.96731	0.04771	0.22306	0.18563	0.90000	0.87954
0.98741	0.98367	0.02381	0.11236	0.18507	0.95000	0.93909
1.00000	1.00000	0.00000	0.00000	0.18451	1.00000	1.00000
OPTIMUM POSITION						
0.87556	0.84403	0.22971	1.00491	0.18961	0.52466	0.47181

Xf/Xfr = Forward period/total period = 0.75

ZONE 1		ZONE 2			ZONE 3	
T1/Xf	h1/H	T2/Xfr	Y/2/H	U	T3/Xr	h3/H
0.81705	0.77393	0.38721	1.75201	0.17718	0.00000	0.00000
0.82693	0.78550	0.36731	1.67476	0.17670	0.05000	0.04116
0.83673	0.79706	0.34745	1.59654	0.17622	0.10000	0.08313
0.84646	0.80859	0.32766	1.51736	0.17573	0.15000	0.12590
0.85612	0.82011	0.30791	1.43716	0.17523	0.20000	0.16951
0.86570	0.83160	0.28823	1.35595	0.17473	0.25000	0.21397
0.87521	0.84307	0.26859	1.27368	0.17423	0.30000	0.25933
0.88464	0.85451	0.24902	1.19036	0.17372	0.35000	0.30558
0.89400	0.86592	0.22950	1.10596	0.17321	0.40000	0.35275
0.90327	0.87731	0.21004	1.02046	0.17269	0.45000	0.40087
0.91247	0.88866	0.19065	0.93382	0.17216	0.50000	0.44996
0.92159	0.89998	0.17130	0.84602	0.17164	0.55000	0.50005
0.93064	0.91126	0.15202	0.75706	0.17110	0.60000	0.55116
0.93960	0.92250	0.13280	0.66690	0.17057	0.65000	0.60330
0.94847	0.93371	0.11364	0.57552	0.17002	0.70000	0.65653
0.95727	0.94487	0.09455	0.48289	0.16948	0.75000	0.71085
0.96598	0.95599	0.07551	0.38898	0.16892	0.80000	0.76630
0.97462	0.96707	0.05654	0.29375	0.16837	0.85000	0.82292
0.98316	0.97810	0.03763	0.19720	0.16781	0.90000	0.88072
0.99162	0.98907	0.01878	0.09930	0.16724	0.95000	0.93974
1.00000	1.00000	0.00000	0.00000	0.16667	1.00000	1.00000
OPTIMUM POSITION						
0.91769	0.89512	0.17960	0.88385	0.17186	0.52854	0.47843

Xf/Xfr = Forward period/total period = 0.80

ZONE 1		ZONE 2			ZONE 3	
T1/Xf	h1/H	T2/Xfr	Y/2/H	U	T3/Xr	h3/H
0.88478	0.85421	0.29217	1.55939	0.15774	0.00000	0.00000
0.89103	0.86184	0.27718	1.49030	0.15726	0.05000	0.04168
0.89723	0.86944	0.26222	1.42055	0.15678	0.10000	0.08413
0.90337	0.87701	0.24730	1.34987	0.15629	0.15000	0.12735
0.90947	0.88454	0.23242	1.27831	0.15580	0.20000	0.17137
0.91552	0.89205	0.21759	1.20586	0.15531	0.25000	0.21621
0.92151	0.89951	0.20279	1.13251	0.15482	0.30000	0.26187
0.92746	0.90695	0.18803	1.05824	0.15432	0.35000	0.30840
0.93335	0.91434	0.17332	0.98304	0.15382	0.40000	0.35578
0.93919	0.92171	0.15865	0.90687	0.15333	0.45000	0.40407
0.94498	0.92903	0.14401	0.82973	0.15281	0.50000	0.45326
0.95072	0.93631	0.12942	0.75158	0.15229	0.55000	0.50338
0.95641	0.94356	0.11487	0.67242	0.15179	0.60000	0.55446
0.96204	0.95076	0.10036	0.59222	0.15127	0.65000	0.60651
0.96763	0.95793	0.08590	0.51097	0.15076	0.70000	0.65955
0.97316	0.96505	0.07147	0.42863	0.15023	0.75000	0.71362
0.97863	0.97213	0.05710	0.34521	0.14971	0.80000	0.76871
0.98405	0.97916	0.04276	0.26065	0.14918	0.85000	0.82488
0.98942	0.98615	0.02846	0.17494	0.14865	0.90000	0.88213
0.99474	0.99310	0.01421	0.08807	0.14812	0.95000	0.94050
1.00000	1.00000	0.00000	0.00000	0.14758	1.00000	1.00000
OPTIMUM POSITION						
0.94857	0.93358	0.13490	0.78109	0.15249	0.53120	0.48442

Xf/Xfr = Forward period/total period = 0.85

ZONE 1		ZONE 2			ZONE 3	
T1/Xf	h1/H	T2/Xfr	Y/2/H	U	T3/Xr	h3/H
0.93436	0.91532	0.20580	1.38969	0.13556	0.00000	0.00000
0.93791	0.91982	0.19527	1.32807	0.13513	0.05000	0.04235
0.94144	0.92428	0.18478	1.26570	0.13469	0.10000	0.08542
0.94494	0.92872	0.17430	1.20259	0.13425	0.15000	0.12922
0.94841	0.93314	0.16385	1.13870	0.13381	0.20000	0.17376
0.95185	0.93753	0.15343	1.07404	0.13337	0.25000	0.21906
0.95526	0.94189	0.14302	1.00858	0.13293	0.30000	0.26513
0.95865	0.94622	0.13265	0.94231	0.13249	0.35000	0.31198
0.96200	0.95053	0.12230	0.87521	0.13204	0.40000	0.35964
0.96533	0.95481	0.11197	0.80728	0.13160	0.45000	0.40812
0.96863	0.95906	0.10167	0.73849	0.13115	0.50000	0.45743
0.97190	0.96329	0.09139	0.66883	0.13070	0.55000	0.50758
0.97514	0.96748	0.08114	0.59829	0.13025	0.60000	0.55860
0.97835	0.97165	0.07091	0.52684	0.12979	0.65000	0.61051
0.98153	0.97579	0.06070	0.45447	0.12934	0.70000	0.66331
0.98468	0.97990	0.05053	0.38118	0.12889	0.75000	0.71702
0.98780	0.98398	0.04037	0.30692	0.12843	0.80000	0.77168
0.99089	0.98803	0.03024	0.23169	0.12797	0.85000	0.82729
0.99396	0.99205	0.02014	0.15547	0.12751	0.90000	0.88386
0.99699	0.99604	0.01006	0.07825	0.12705	0.95000	0.94143
1.00000	1.00000	0.00000	0.00000	0.12659	1.00000	1.00000
OPTIMUM POSITION						
0.97074	0.96180	0.09502	0.69357	0.13086	0.53231	0.48974

Table F.3 Asymmetrical D–R–R–D motion.
Zone 1: Modified sine
Zone 2: Cycloidal
Zone 3: Modified sine

Xf/Xfr = Forward period/total period = 0.10

ZONE 1		ZONE 2			ZONE 3	
T1/Xf	h1/H	T2/Xfr	Y/2/H	U	T3/Xr	h3/H
0.00000	0.00000	0.12595	1.21164	0.39698	0.97117	0.96658
0.05000	0.04916	0.11965	1.15196	0.39701	0.97262	0.96825
0.10000	0.09844	0.11334	1.09213	0.39704	0.97407	0.96993
0.15000	0.14781	0.10703	1.03221	0.39708	0.97552	0.97160
0.20000	0.19726	0.10073	0.97220	0.39711	0.97697	0.97327
0.25000	0.24679	0.09443	0.91212	0.39714	0.97842	0.97495
0.30000	0.29640	0.08813	0.85195	0.39717	0.97986	0.97662
0.35000	0.34614	0.08182	0.79163	0.39720	0.98131	0.97829
0.40000	0.39591	0.07552	0.73130	0.39723	0.98275	0.97997
0.45000	0.44580	0.06922	0.67083	0.39726	0.98420	0.98164
0.50000	0.49577	0.06292	0.61027	0.39729	0.98564	0.98331
0.55000	0.54581	0.05663	0.54965	0.39732	0.98708	0.98498
0.60000	0.59592	0.05033	0.48895	0.39735	0.98852	0.98665
0.65000	0.64616	0.04404	0.42812	0.39738	0.98996	0.98832
0.70000	0.69647	0.03774	0.36721	0.39741	0.99140	0.98999
0.75000	0.74685	0.03145	0.30623	0.39744	0.99283	0.99166
0.80000	0.79729	0.02516	0.24518	0.39747	0.99427	0.99333
0.85000	0.84780	0.01887	0.18407	0.39750	0.99570	0.99500
0.90000	0.89848	0.01258	0.12276	0.39753	0.99714	0.99667
0.95000	0.94923	0.00629	0.06139	0.39755	0.99857	0.99833
1.00000	1.00000	0.00000	0.00000	0.39758	1.00000	1.00000
		OPTIMUM POSITION				
0.53094	0.52671	0.05903	0.57279	0.39731	0.98653	0.98434

Xf/Xfr = Forward period/total period = 0.15

ZONE 1		ZONE 2			ZONE 3	
T1/Xf	h1/H	T2/Xfr	Y/2/H	U	T3/Xr	h3/H
0.00000	0.00000	0.20180	1.34370	0.37166	0.93906	0.92999
0.05000	0.04813	0.19166	1.27851	0.37175	0.94216	0.93352
0.10000	0.09644	0.18153	1.21314	0.37184	0.94526	0.93705
0.15000	0.14496	0.17140	1.14753	0.37193	0.94835	0.94057
0.20000	0.19368	0.16128	1.08171	0.37202	0.95143	0.94409
0.25000	0.24259	0.15117	1.01568	0.37211	0.95451	0.94761
0.30000	0.29171	0.14105	0.94944	0.37219	0.95758	0.95112
0.35000	0.34100	0.13095	0.88301	0.37228	0.96065	0.95463
0.40000	0.39051	0.12085	0.81635	0.37237	0.96371	0.95814
0.45000	0.44022	0.11075	0.74947	0.37246	0.96676	0.96164
0.50000	0.49011	0.10066	0.68241	0.37254	0.96981	0.96514
0.55000	0.54021	0.09057	0.61511	0.37263	0.97286	0.96864
0.60000	0.59050	0.08049	0.54762	0.37272	0.97589	0.97214
0.65000	0.64098	0.07041	0.47992	0.37281	0.97893	0.97563
0.70000	0.69170	0.06034	0.41197	0.37289	0.98195	0.97912
0.75000	0.74258	0.05027	0.34385	0.37298	0.98498	0.98261
0.80000	0.79367	0.04021	0.27550	0.37307	0.98799	0.98609
0.85000	0.84495	0.03015	0.20695	0.37315	0.99100	0.98957
0.90000	0.89644	0.02009	0.13818	0.37324	0.99401	0.99305
0.95000	0.94812	0.01004	0.06919	0.37332	0.99701	0.99653
1.00000	1.00000	0.00000	0.00000	0.37341	1.00000	1.00000
		OPTIMUM POSITION				
0.53231	0.52246	0.09414	0.63895	0.37260	0.97178	0.96741

Xf/Xfr = Forward period/total period = 0.20

ZONE 1		ZONE 2			ZONE 3	
T1/Xf	h1/H	T2/Xfr	Y/2/H	U	T3/Xr	h3/H
0.00000	0.00000	0.28589	1.49654	0.34979	0.89264	0.87802
0.05000	0.04734	0.27149	1.42468	0.34992	0.89814	0.88417
0.10000	0.09496	0.25710	1.35251	0.35005	0.90362	0.89031
0.15000	0.14284	0.24273	1.28007	0.35018	0.90909	0.89644
0.20000	0.19101	0.22836	1.20728	0.35032	0.91454	0.90257
0.25000	0.23945	0.21401	1.13419	0.35045	0.91999	0.90870
0.30000	0.28817	0.19967	1.06080	0.35058	0.92541	0.91482
0.35000	0.33717	0.18534	0.98708	0.35071	0.93083	0.92094
0.40000	0.38646	0.17101	0.91305	0.35084	0.93623	0.92705
0.45000	0.43602	0.15670	0.83871	0.35098	0.94162	0.93315
0.50000	0.48586	0.14241	0.76406	0.35111	0.94699	0.93925
0.55000	0.53599	0.12812	0.68908	0.35124	0.95235	0.94535
0.60000	0.58640	0.11384	0.61380	0.35137	0.95770	0.95144
0.65000	0.63709	0.09957	0.53820	0.35150	0.96303	0.95753
0.70000	0.68807	0.08532	0.46228	0.35163	0.96835	0.96361
0.75000	0.73933	0.07107	0.38604	0.35176	0.97366	0.96969
0.80000	0.79090	0.05683	0.30946	0.35189	0.97896	0.97576
0.85000	0.84273	0.04261	0.23259	0.35203	0.98424	0.98183
0.90000	0.89486	0.02840	0.15538	0.35216	0.98950	0.98789
0.95000	0.94728	0.01419	0.07786	0.35229	0.99476	0.99395
1.00000	1.00000	0.00000	0.00000	0.35242	1.00000	1.00000
		OPTIMUM POSITION				
0.53153	0.51744	0.13339	0.71682	0.35119	0.95037	0.94310

Xf/Xfr = Forward period/total period = 0.25

ZONE 1		ZONE 2			ZONE 3	
T1/Xf	h1/H	T2/Xfr	Y/2/H	U	T3/Xr	h3/H
0.00000	0.00000	0.37861	1.67377	0.33015	0.82852	0.80776
0.05000	0.04673	0.35951	1.59403	0.33031	0.83732	0.81739
0.10000	0.09378	0.34043	1.51391	0.33047	0.84610	0.82701
0.15000	0.14118	0.32136	1.43334	0.33063	0.85486	0.83664
0.20000	0.18891	0.30231	1.35239	0.33078	0.86359	0.84626
0.25000	0.23697	0.28328	1.27104	0.33094	0.87229	0.85589
0.30000	0.28538	0.26427	1.18925	0.33110	0.88098	0.86551
0.35000	0.33412	0.24528	1.10708	0.33126	0.88963	0.87513
0.40000	0.38322	0.22630	1.02446	0.33142	0.89827	0.88475
0.45000	0.43266	0.20734	0.94144	0.33158	0.90688	0.89436
0.50000	0.48245	0.18840	0.85800	0.33174	0.91546	0.90398
0.55000	0.53259	0.16948	0.77413	0.33190	0.92403	0.91359
0.60000	0.58308	0.15058	0.68985	0.33206	0.93256	0.92320
0.65000	0.63394	0.13169	0.60512	0.33221	0.94108	0.93281
0.70000	0.68515	0.11282	0.51997	0.33237	0.94957	0.94242
0.75000	0.73672	0.09397	0.43439	0.33253	0.95803	0.95202
0.80000	0.78863	0.07514	0.34840	0.33269	0.96647	0.96162
0.85000	0.84093	0.05633	0.26196	0.33285	0.97489	0.97122
0.90000	0.89358	0.03754	0.17508	0.33301	0.98329	0.98082
0.95000	0.94661	0.01876	0.08775	0.33317	0.99166	0.99041
1.00000	1.00000	0.00000	0.00000	0.33333	1.00000	1.00000
		OPTIMUM POSITION				
0.52918	0.51167	0.17736	0.80911	0.33183	0.92046	0.90959

Table F.3 (continued)

Xf/Xfr = Forward period/total period = 0.30

ZONE 1		ZONE 2			ZONE 3	
T1/Xf	h1/H	T2/Xfr	Y/2/H	U	T3/Xr	h3/H
0.00000	0.00000	0.48058	1.88004	0.31213	0.74204	0.71539
0.05000	0.04623	0.45630	1.79112	0.31229	0.75528	0.72949
0.10000	0.09283	0.43206	1.70169	0.31246	0.76848	0.74361
0.15000	0.13982	0.40784	1.61175	0.31262	0.78166	0.75774
0.20000	0.18718	0.38365	1.52128	0.31279	0.79479	0.77188
0.25000	0.23494	0.35948	1.43030	0.31295	0.80789	0.78604
0.30000	0.28308	0.33534	1.33878	0.31312	0.82095	0.80021
0.35000	0.33162	0.31122	1.24673	0.31329	0.83398	0.81440
0.40000	0.38056	0.28712	1.15414	0.31345	0.84697	0.82860
0.45000	0.42988	0.26306	1.06103	0.31362	0.85992	0.84281
0.50000	0.47963	0.23901	0.96735	0.31379	0.87284	0.85704
0.55000	0.52977	0.21500	0.87314	0.31396	0.88572	0.87128
0.60000	0.58033	0.19100	0.77837	0.31413	0.89856	0.88554
0.65000	0.63130	0.16704	0.68306	0.31430	0.91137	0.89980
0.70000	0.68268	0.14310	0.58720	0.31447	0.92414	0.91408
0.75000	0.73450	0.11918	0.49074	0.31464	0.93688	0.92837
0.80000	0.78674	0.09530	0.39374	0.31481	0.94958	0.94267
0.85000	0.83940	0.07143	0.29616	0.31498	0.96224	0.95699
0.90000	0.89250	0.04760	0.19802	0.31515	0.97486	0.97131
0.95000	0.94603	0.02379	0.09931	0.31532	0.98745	0.98565
1.00000	1.00000	0.00000	0.00000	0.31549	1.00000	1.00000
OPTIMUM POSITION						
0.52554	0.50519	0.22674	0.91930	0.31387	0.87942	0.86431

Xf/Xfr = Forward period/total period = 0.35

ZONE 1		ZONE 2			ZONE 3	
T1/Xf	h1/H	T2/Xfr	Y/2/H	U	T3/Xr	h3/H
0.00000	0.00000	0.59256	2.12113	0.29533	0.62683	0.59598
0.05000	0.04581	0.56265	2.02167	0.29548	0.64593	0.61568
0.10000	0.09202	0.53276	1.92154	0.29563	0.66499	0.63543
0.15000	0.13867	0.50290	1.82072	0.29578	0.68400	0.65525
0.20000	0.18573	0.47307	1.71926	0.29594	0.70296	0.67510
0.25000	0.23322	0.44327	1.61712	0.29609	0.72188	0.69502
0.30000	0.28114	0.41351	1.51430	0.29625	0.74076	0.71499
0.35000	0.32949	0.38377	1.41080	0.29640	0.75958	0.73500
0.40000	0.37828	0.35406	1.30660	0.29656	0.77837	0.75508
0.45000	0.42753	0.32438	1.20169	0.29672	0.79710	0.77521
0.50000	0.47721	0.29474	1.09610	0.29687	0.81579	0.79538
0.55000	0.52736	0.26512	0.98978	0.29703	0.83443	0.81561
0.60000	0.57797	0.23554	0.88275	0.29719	0.85302	0.83589
0.65000	0.62904	0.20598	0.77499	0.29735	0.87156	0.85623
0.70000	0.68058	0.17646	0.66652	0.29751	0.89006	0.87661
0.75000	0.73260	0.14697	0.55729	0.29767	0.90850	0.89705
0.80000	0.78510	0.11751	0.44735	0.29784	0.92690	0.91754
0.85000	0.83808	0.08809	0.33664	0.29800	0.94525	0.93808
0.90000	0.89155	0.05869	0.22520	0.29816	0.96355	0.95867
0.95000	0.94552	0.02933	0.11299	0.29833	0.98180	0.97931
1.00000	1.00000	0.00000	0.00000	0.29849	1.00000	1.00000
OPTIMUM POSITION						
0.52077	0.49799	0.28244	1.05202	0.29694	0.82353	0.80378

Xf/Xfr = Forward period/total period = 0.40

ZONE 1		ZONE 2			ZONE 3	
T1/Xf	h1/H	T2/Xfr	Y/2/H	U	T3/Xr	h3/H
0.00000	0.00000	0.71558	2.40465	0.27950	0.47404	0.44309
0.05000	0.04546	0.67950	2.29306	0.27961	0.50082	0.46968
0.10000	0.09135	0.64347	2.18065	0.27973	0.52756	0.49639
0.15000	0.13769	0.60746	2.06736	0.27986	0.55424	0.52324
0.20000	0.18449	0.57147	1.95318	0.27998	0.58088	0.55021
0.25000	0.23174	0.53552	1.83814	0.28010	0.60746	0.57732
0.30000	0.27946	0.49960	1.72221	0.28022	0.63400	0.60455
0.35000	0.32766	0.46371	1.60537	0.28035	0.66049	0.63192
0.40000	0.37632	0.42785	1.48763	0.28047	0.68692	0.65942
0.45000	0.42548	0.39202	1.36896	0.28060	0.71330	0.68706
0.50000	0.47512	0.35622	1.24936	0.28073	0.73963	0.71483
0.55000	0.52527	0.32045	1.12879	0.28085	0.76592	0.74274
0.60000	0.57592	0.28471	1.00730	0.28098	0.79215	0.77078
0.65000	0.62707	0.24900	0.88485	0.28111	0.81832	0.79895
0.70000	0.67873	0.21334	0.76146	0.28124	0.84443	0.82726
0.75000	0.73094	0.17769	0.63704	0.28138	0.87051	0.85571
0.80000	0.78366	0.14209	0.51165	0.28151	0.89652	0.88429
0.85000	0.83691	0.10652	0.38528	0.28164	0.92247	0.91301
0.90000	0.89073	0.07098	0.25786	0.28178	0.94837	0.94187
0.95000	0.94508	0.03547	0.12945	0.28191	0.97421	0.97086
1.00000	1.00000	0.00000	0.00000	0.28205	1.00000	1.00000
OPTIMUM POSITION						
0.51491	0.49002	0.34555	1.21350	0.28076	0.74748	0.72314

Xf/Xfr = Forward period/total period = 0.45

ZONE 1		ZONE 2			ZONE 3	
T1/Xf	h1/H	T2/Xfr	Y/2/H	U	T3/Xr	h3/H
0.00000	0.00000	0.85087	2.74041	0.26444	0.27115	0.24824
0.05000	0.04513	0.80812	2.61508	0.26451	0.30797	0.28315
0.10000	0.09074	0.76539	2.48860	0.26458	0.34476	0.31834
0.15000	0.13684	0.72266	2.36092	0.26465	0.38152	0.35380
0.20000	0.18340	0.67998	2.23216	0.26472	0.41822	0.38950
0.25000	0.23046	0.63730	2.10216	0.26479	0.45490	0.42550
0.30000	0.27801	0.59465	1.97097	0.26486	0.49155	0.46175
0.35000	0.32606	0.55202	1.83859	0.26493	0.52814	0.49831
0.40000	0.37461	0.50943	1.70502	0.26501	0.56468	0.53513
0.45000	0.42369	0.46684	1.57015	0.26508	0.60120	0.57224
0.50000	0.47329	0.42428	1.43405	0.26515	0.63760	0.60964
0.55000	0.52342	0.38174	1.29667	0.26523	0.67410	0.64732
0.60000	0.57410	0.33923	1.15799	0.26531	0.71049	0.68530
0.65000	0.62532	0.29674	1.01802	0.26538	0.74684	0.72357
0.70000	0.67711	0.25427	0.87670	0.26546	0.78314	0.76214
0.75000	0.72946	0.21183	0.73404	0.26554	0.81940	0.80101
0.80000	0.78238	0.16941	0.59004	0.26562	0.85561	0.84019
0.85000	0.83589	0.12702	0.44463	0.26570	0.89178	0.87968
0.90000	0.88998	0.08466	0.29788	0.26578	0.92789	0.91946
0.95000	0.94468	0.04232	0.14967	0.26586	0.96397	0.95957
1.00000	1.00000	0.00000	0.00000	0.26594	1.00000	1.00000
OPTIMUM POSITION						
0.50799	0.48127	0.41748	1.41218	0.26517	0.64350	0.61564

Table F.3 (continued)

Xf/Xfr = Forward period/total period = 0.50						
ZONE 1		ZONE 2			ZONE 3	
T1/Xf	h1/H	T2/Xfr	Y/2/H	U	T3/Xr	h3/H
0.00000	0.00000	1.00000	3.14159	0.25000	0.00000	0.00000
0.05000	0.04488	0.95000	3.00061	0.25000	0.05000	0.04488
0.10000	0.09024	0.90000	2.85810	0.25000	0.10000	0.09024
0.15000	0.13610	0.85000	2.71404	0.25000	0.15000	0.13610
0.20000	0.18246	0.80000	2.56839	0.25000	0.20000	0.18246
0.25000	0.22933	0.75000	2.42114	0.25000	0.25000	0.22933
0.30000	0.27672	0.70000	2.27226	0.25000	0.30000	0.27672
0.35000	0.32464	0.65000	2.12172	0.25000	0.35000	0.32464
0.40000	0.37309	0.60000	1.96949	0.25000	0.40000	0.37309
0.45000	0.42210	0.55000	1.81554	0.25000	0.45000	0.42210
0.50000	0.47165	0.50000	1.65985	0.25000	0.50000	0.47165
0.55000	0.52178	0.45000	1.50238	0.25000	0.55000	0.52178
0.60000	0.57248	0.40000	1.34311	0.25000	0.60000	0.57248
0.65000	0.62376	0.35000	1.18200	0.25000	0.65000	0.62376
0.70000	0.67564	0.30000	1.01902	0.25000	0.70000	0.67564
0.75000	0.72812	0.25000	0.85413	0.25000	0.75000	0.72812
0.80000	0.78122	0.20000	0.68732	0.25000	0.80000	0.78122
0.85000	0.83495	0.15000	0.51853	0.25000	0.85000	0.83495
0.90000	0.88931	0.10000	0.34774	0.25000	0.90000	0.88931
0.95000	0.94432	0.05000	0.17491	0.25000	0.95000	0.94432
1.00000	1.00000	0.00000	0.00000	0.25000	1.00000	1.00000
		OPTIMUM POSITION				
0.50000	0.47165	0.50000	1.65985	0.25000	0.50000	0.47165

Xf/Xfr = Forward period/total period = 0.55						
ZONE 1		ZONE 2			ZONE 3	
T1/Xf	h1/H	T2/Xfr	Y/2/H	U	T3/Xr	h3/H
0.27115	0.24824	0.85087	2.74041	0.23556	0.00000	0.00000
0.30798	0.28316	0.80811	2.61505	0.23549	0.05000	0.04514
0.34476	0.31834	0.76538	2.48859	0.23542	0.10000	0.09075
0.38152	0.35380	0.72266	2.36092	0.23535	0.15000	0.13684
0.41823	0.38951	0.67997	2.23214	0.23528	0.20000	0.18340
0.45490	0.42550	0.63730	2.10216	0.23521	0.25000	0.23046
0.49154	0.46176	0.59466	1.97100	0.23514	0.30000	0.27800
0.52814	0.49831	0.55202	1.83860	0.23507	0.35000	0.32606
0.56469	0.53514	0.50942	1.70499	0.23499	0.40000	0.37462
0.60120	0.57224	0.46684	1.57017	0.23492	0.45000	0.42368
0.63767	0.60963	0.42428	1.43407	0.23484	0.50000	0.47328
0.67410	0.64731	0.38175	1.29669	0.23477	0.55000	0.52342
0.71049	0.68530	0.33923	1.15800	0.23469	0.60000	0.57410
0.74684	0.72357	0.29674	1.01802	0.23462	0.65000	0.62532
0.78314	0.76214	0.25427	0.87670	0.23454	0.70000	0.67711
0.81939	0.80101	0.21183	0.73406	0.23446	0.75000	0.72945
0.85560	0.84018	0.16942	0.59006	0.23438	0.80000	0.78237
0.89178	0.87967	0.12702	0.44463	0.23430	0.85000	0.83589
0.92790	0.91946	0.08466	0.29786	0.23422	0.90000	0.88998
0.96398	0.95958	0.04231	0.14962	0.23414	0.95000	0.94470
1.00000	1.00000	0.00000	0.00000	0.23406	1.00000	1.00000
		OPTIMUM POSITION				
0.64350	0.61564	0.41748	1.41217	0.23483	0.50800	0.48127

Xf/Xfr = Forward period/total period = 0.60						
ZONE 1		ZONE 2			ZONE 3	
T1/Xf	h1/H	T2/Xfr	Y/2/H	U	T3/Xr	h3/H
0.47404	0.44310	0.71557	2.40463	0.22050	0.00000	0.00000
0.50082	0.46968	0.67951	2.29308	0.22039	0.05000	0.04545
0.52756	0.49639	0.64346	2.18064	0.22027	0.10000	0.09135
0.55424	0.52324	0.60745	2.06734	0.22014	0.15000	0.13770
0.58087	0.55021	0.57148	1.95320	0.22002	0.20000	0.18448
0.60746	0.57732	0.53552	1.83814	0.21990	0.25000	0.23174
0.63400	0.60456	0.49960	1.72220	0.21978	0.30000	0.27947
0.66049	0.63192	0.46371	1.60537	0.21965	0.35000	0.32765
0.68692	0.65943	0.42785	1.48761	0.21953	0.40000	0.37633
0.71330	0.68706	0.39202	1.36895	0.21940	0.45000	0.42548
0.73964	0.71483	0.35622	1.24934	0.21927	0.50000	0.47513
0.76592	0.74274	0.32045	1.12879	0.21915	0.55000	0.52527
0.79215	0.77078	0.28471	1.00729	0.21902	0.60000	0.57592
0.81832	0.79895	0.24901	0.88485	0.21889	0.65000	0.62707
0.84444	0.82726	0.21334	0.76144	0.21876	0.70000	0.67874
0.87050	0.85571	0.17770	0.63705	0.21862	0.75000	0.73093
0.89651	0.88429	0.14209	0.51166	0.21849	0.80000	0.78366
0.92247	0.91301	0.10652	0.38529	0.21836	0.85000	0.83691
0.94837	0.94186	0.07098	0.25789	0.21822	0.90000	0.89072
0.97421	0.97087	0.03547	0.12945	0.21809	0.95000	0.94509
1.00000	1.00000	0.00000	0.00000	0.21795	1.00000	1.00000
		OPTIMUM POSITION				
0.74748	0.72314	0.34555	1.21349	0.21923	0.51491	0.49003

Xf/Xfr = Forward period/total period = 0.65						
ZONE 1		ZONE 2			ZONE 3	
T1/Xf	h1/H	T2/Xfr	Y/2/H	U	T3/Xr	h3/H
0.62683	0.59598	0.59256	2.12115	0.20467	0.00000	0.00000
0.64593	0.61568	0.56264	2.02166	0.20452	0.05000	0.04581
0.66499	0.63543	0.53276	1.92154	0.20437	0.10000	0.09203
0.68400	0.65524	0.50290	1.82074	0.20422	0.15000	0.13867
0.70297	0.67511	0.47307	1.71926	0.20406	0.20000	0.18573
0.72189	0.69502	0.44327	1.61712	0.20391	0.25000	0.23322
0.74076	0.71499	0.41351	1.51431	0.20375	0.30000	0.28113
0.75959	0.73501	0.38377	1.41078	0.20360	0.35000	0.32950
0.77837	0.75508	0.35406	1.30659	0.20344	0.40000	0.37829
0.79710	0.77521	0.32438	1.20168	0.20328	0.45000	0.42753
0.81579	0.79538	0.29474	1.09608	0.20313	0.50000	0.47722
0.83443	0.81561	0.26512	0.98977	0.20297	0.55000	0.52736
0.85302	0.83589	0.23554	0.88275	0.20281	0.60000	0.57797
0.87156	0.85622	0.20599	0.77502	0.20265	0.65000	0.62903
0.89006	0.87661	0.17646	0.66652	0.20249	0.70000	0.68058
0.90850	0.89705	0.14697	0.55731	0.20233	0.75000	0.73260
0.92690	0.91754	0.11751	0.44734	0.20216	0.80000	0.78510
0.94525	0.93808	0.08809	0.33665	0.20200	0.85000	0.83808
0.96355	0.95867	0.05869	0.22519	0.20184	0.90000	0.89156
0.98180	0.97931	0.02933	0.11298	0.20167	0.95000	0.94553
1.00000	1.00000	0.00000	0.00000	0.20151	1.00000	1.00000
		OPTIMUM POSITION				
0.82353	0.80377	0.28244	1.05204	0.20306	0.52076	0.49798

Table F.3 (continued)

Xf/Xfr = Forward period/total period = 0.70

ZONE 1		ZONE 2			ZONE 3	
T1/Xf	h1/H	T2/Xfr	Y/2/H	U	T3/Xr	h3/H
0.74204	0.71539	0.48058	1.88004	0.18787	0.00000	0.00000
0.75528	0.72949	0.45630	1.79112	0.18771	0.05000	0.04623
0.76849	0.74361	0.43206	1.70169	0.18754	0.10000	0.09283
0.78165	0.75774	0.40784	1.61176	0.18738	0.15000	0.13981
0.79479	0.77188	0.38365	1.52128	0.18721	0.20000	0.18718
0.80789	0.78604	0.35948	1.43030	0.18705	0.25000	0.23493
0.82095	0.80021	0.33533	1.33878	0.18688	0.30000	0.28308
0.83397	0.81440	0.31122	1.24674	0.18671	0.35000	0.33162
0.84697	0.82860	0.28712	1.15414	0.18655	0.40000	0.38056
0.85992	0.84281	0.26306	1.06103	0.18638	0.45000	0.42988
0.87284	0.85704	0.23901	0.96736	0.18621	0.50000	0.47962
0.88572	0.87128	0.21500	0.87314	0.18604	0.55000	0.52977
0.89856	0.88553	0.19101	0.77838	0.18587	0.60000	0.58033
0.91137	0.89980	0.16704	0.68306	0.18570	0.65000	0.63130
0.92414	0.91408	0.14310	0.58718	0.18553	0.70000	0.68269
0.93688	0.92837	0.11918	0.49074	0.18536	0.75000	0.73450
0.94958	0.94267	0.09529	0.39373	0.18519	0.80000	0.78674
0.96224	0.95699	0.07143	0.29617	0.18502	0.85000	0.83940
0.97486	0.97131	0.04759	0.19802	0.18485	0.90000	0.89250
0.98745	0.98565	0.02378	0.09930	0.18468	0.95000	0.94603
1.00000	1.00000	0.00000	0.00000	0.18451	1.00000	1.00000
OPTIMUM POSITION						
0.87942	0.86431	0.22675	0.91932	0.18613	0.52553	0.50518

Xf/Xfr = Forward period/total period = 0.75

ZONE 1		ZONE 2			ZONE 3	
T1/Xf	h1/H	T2/Xfr	Y/2/H	U	T3/Xr	h3/H
0.82852	0.80776	0.37861	1.67378	0.16985	0.00000	0.00000
0.83732	0.81739	0.35951	1.59404	0.16969	0.05000	0.04673
0.84610	0.82701	0.34043	1.51390	0.16953	0.10000	0.09379
0.85485	0.83664	0.32136	1.43336	0.16937	0.15000	0.14117
0.86358	0.84626	0.30231	1.35240	0.16922	0.20000	0.18890
0.87229	0.85589	0.28328	1.27103	0.16906	0.25000	0.23698
0.88097	0.86551	0.26427	1.18926	0.16890	0.30000	0.28537
0.88963	0.87513	0.24527	1.10706	0.16874	0.35000	0.33413
0.89827	0.88475	0.22630	1.02447	0.16858	0.40000	0.38322
0.90688	0.89436	0.20734	0.94144	0.16842	0.45000	0.43267
0.91546	0.90398	0.18840	0.85799	0.16826	0.50000	0.48246
0.92403	0.91359	0.16948	0.77412	0.16810	0.55000	0.53260
0.93256	0.92320	0.15058	0.68984	0.16794	0.60000	0.58309
0.94108	0.93281	0.13169	0.60513	0.16779	0.65000	0.63394
0.94957	0.94241	0.11282	0.51998	0.16763	0.70000	0.68514
0.95803	0.95202	0.09398	0.43440	0.16747	0.75000	0.73671
0.96647	0.96162	0.07514	0.34839	0.16731	0.80000	0.78864
0.97489	0.97122	0.05633	0.26194	0.16715	0.85000	0.84093
0.98329	0.98082	0.03753	0.17507	0.16699	0.90000	0.89359
0.99166	0.99041	0.01876	0.08775	0.16683	0.95000	0.94661
1.00000	1.00000	0.00000	0.00000	0.16667	1.00000	1.00000
OPTIMUM POSITION						
0.92046	0.90958	0.17736	0.80912	0.16817	0.52917	0.51166

Xf/Xfr = Forward period/total period = 0.80

ZONE 1		ZONE 2			ZONE 3	
T1/Xf	h1/H	T2/Xfr	Y/2/H	U	T3/Xr	h3/H
0.89264	0.87802	0.28589	1.49653	0.15021	0.00000	0.00000
0.89814	0.88417	0.27149	1.42469	0.15008	0.05000	0.04734
0.90362	0.89031	0.25710	1.35252	0.14995	0.10000	0.09496
0.90909	0.89644	0.24273	1.28006	0.14982	0.15000	0.14284
0.91454	0.90257	0.22836	1.20727	0.14968	0.20000	0.19101
0.91999	0.90870	0.21401	1.13419	0.14955	0.25000	0.23945
0.92541	0.91482	0.19967	1.06079	0.14942	0.30000	0.28817
0.93083	0.92094	0.18534	0.98708	0.14929	0.35000	0.33717
0.93623	0.92705	0.17102	0.91306	0.14916	0.40000	0.38645
0.94162	0.93315	0.15671	0.83872	0.14902	0.45000	0.43601
0.94699	0.93925	0.14241	0.76406	0.14889	0.50000	0.48586
0.95235	0.94535	0.12812	0.68909	0.14876	0.55000	0.53599
0.95770	0.95144	0.11384	0.61380	0.14863	0.60000	0.58640
0.96303	0.95753	0.09957	0.53820	0.14850	0.65000	0.63709
0.96835	0.96361	0.08532	0.46228	0.14837	0.70000	0.68807
0.97366	0.96969	0.07107	0.38603	0.14824	0.75000	0.73934
0.97896	0.97576	0.05684	0.30948	0.14811	0.80000	0.79089
0.98424	0.98183	0.04261	0.23259	0.14797	0.85000	0.84273
0.98950	0.98789	0.02840	0.15538	0.14784	0.90000	0.89487
0.99476	0.99395	0.01419	0.07786	0.14771	0.95000	0.94728
1.00000	1.00000	0.00000	0.00000	0.14758	1.00000	1.00000
OPTIMUM POSITION						
0.95037	0.94310	0.13340	0.71683	0.14881	0.53152	0.51743

Xf/Xfr = Forward period/total period = 0.85

ZONE 1		ZONE 2			ZONE 3	
T1/Xf	h1/H	T2/Xfr	Y/2/H	U	T3/Xr	h3/H
0.93906	0.92999	0.20180	1.34370	0.12834	0.00000	0.00000
0.94216	0.93352	0.19166	1.27852	0.12825	0.05000	0.04812
0.94526	0.93705	0.18153	1.21312	0.12816	0.10000	0.09645
0.94835	0.94057	0.17140	1.14752	0.12807	0.15000	0.14496
0.95143	0.94409	0.16128	1.08172	0.12798	0.20000	0.19367
0.95451	0.94761	0.15117	1.01569	0.12789	0.25000	0.24259
0.95758	0.95112	0.14106	0.94946	0.12781	0.30000	0.29169
0.96065	0.95463	0.13095	0.88302	0.12772	0.35000	0.34100
0.96371	0.95814	0.12085	0.81637	0.12763	0.40000	0.39050
0.96676	0.96164	0.11075	0.74948	0.12754	0.45000	0.44021
0.96981	0.96514	0.10066	0.68241	0.12746	0.50000	0.49011
0.97286	0.96864	0.09057	0.61512	0.12737	0.55000	0.54021
0.97589	0.97214	0.08049	0.54761	0.12728	0.60000	0.59051
0.97893	0.97563	0.07041	0.47991	0.12719	0.65000	0.64099
0.98195	0.97912	0.06034	0.41199	0.12711	0.70000	0.69168
0.98498	0.98261	0.05027	0.34385	0.12702	0.75000	0.74258
0.98799	0.98609	0.04021	0.27551	0.12693	0.80000	0.79366
0.99100	0.98957	0.03015	0.20694	0.12685	0.85000	0.84495
0.99401	0.99305	0.02009	0.13817	0.12676	0.90000	0.89644
0.99701	0.99653	0.01004	0.06919	0.12668	0.95000	0.94812
1.00000	1.00000	0.00000	0.00000	0.12659	1.00000	1.00000
OPTIMUM POSITION						
0.97178	0.96741	0.09414	0.63894	0.12740	0.53232	0.52247

Table F.4 Asymmetrical D–R–R–D motion.
Zone 1: Modified sine
Zone 2: Modified trapezoid
Zone 3: Modified sine

Xf/Xfr = Forward period/total period = 0.10

ZONE 1		ZONE 2			ZONE 3	
T1/Xf	h1/H	T2/Xfr	Y/2/H	U	T3/Xr	h3/H
0.00000	0.00000	0.11993	1.20831	0.41692	0.97786	0.97460
0.05000	0.05162	0.11395	1.14631	0.41684	0.97894	0.97584
0.10000	0.10309	0.10797	1.08446	0.41677	0.98003	0.97709
0.15000	0.15435	0.10200	1.02283	0.41669	0.98112	0.97833
0.20000	0.20544	0.09601	0.96136	0.41661	0.98221	0.97958
0.25000	0.25636	0.09003	0.90005	0.41653	0.98330	0.98083
0.30000	0.30711	0.08404	0.83891	0.41645	0.98440	0.98209
0.35000	0.35768	0.07806	0.77795	0.41637	0.98549	0.98335
0.40000	0.40809	0.07206	0.71712	0.41629	0.98660	0.98461
0.45000	0.45831	0.06607	0.65650	0.41622	0.98770	0.98587
0.50000	0.50838	0.06008	0.59602	0.41614	0.98880	0.98714
0.55000	0.55826	0.05408	0.53572	0.41606	0.98991	0.98841
0.60000	0.60800	0.04808	0.47556	0.41598	0.99102	0.98969
0.65000	0.65756	0.04208	0.41558	0.41591	0.99214	0.99097
0.70000	0.70695	0.03607	0.35575	0.41583	0.99325	0.99225
0.75000	0.75620	0.03007	0.29606	0.41575	0.99437	0.99353
0.80000	0.80527	0.02406	0.23655	0.41568	0.99549	0.99482
0.85000	0.85420	0.01804	0.17717	0.41560	0.99662	0.99611
0.90000	0.90295	0.01203	0.11797	0.41552	0.99774	0.99740
0.95000	0.95156	0.00602	0.05890	0.41545	0.99887	0.99870
1.00000	1.00000	0.00000	0.00000	0.41537	1.00000	1.00000
OPTIMUM POSITION						
0.58203	0.59014	0.05024	0.49717	0.41601	0.99062	0.98923

Xf/Xfr = Forward period/total period = 0.15

ZONE 1		ZONE 2			ZONE 3	
T1/Xf	h1/H	T2/Xfr	Y/2/H	U	T3/Xr	h3/H
0.00000	0.00000	0.18984	1.33754	0.39506	0.95313	0.94705
0.05000	0.05123	0.18039	1.26957	0.39498	0.95542	0.94963
0.10000	0.10231	0.17093	1.20173	0.39489	0.95773	0.95222
0.15000	0.15328	0.16147	1.13399	0.39481	0.96003	0.95482
0.20000	0.20410	0.15201	1.06639	0.39472	0.96235	0.95742
0.25000	0.25479	0.14254	0.99890	0.39463	0.96466	0.96003
0.30000	0.30537	0.13306	0.93150	0.39455	0.96698	0.96265
0.35000	0.35578	0.12359	0.86426	0.39446	0.96931	0.96527
0.40000	0.40609	0.11410	0.79711	0.39438	0.97164	0.96790
0.45000	0.45627	0.10462	0.73008	0.39429	0.97398	0.97054
0.50000	0.50632	0.09513	0.66316	0.39421	0.97632	0.97318
0.55000	0.55624	0.08563	0.59635	0.39412	0.97867	0.97583
0.60000	0.60604	0.07613	0.52965	0.39404	0.98102	0.97849
0.65000	0.65571	0.06663	0.46307	0.39396	0.98337	0.98116
0.70000	0.70526	0.05713	0.39660	0.39387	0.98573	0.98383
0.75000	0.75470	0.04761	0.33022	0.39379	0.98810	0.98651
0.80000	0.80399	0.03810	0.26397	0.39370	0.99047	0.98919
0.85000	0.85317	0.02858	0.19783	0.39362	0.99285	0.99188
0.90000	0.90223	0.01906	0.13178	0.39353	0.99523	0.99458
0.95000	0.95117	0.00953	0.06584	0.39345	0.99761	0.99729
1.00000	1.00000	0.00000	0.00000	0.39337	1.00000	1.00000
OPTIMUM POSITION						
0.58142	0.58755	0.07966	0.55442	0.39407	0.98014	0.97750

Xf/Xfr = Forward period/total period = 0.20

ZONE 1		ZONE 2			ZONE 3	
T1/Xf	h1/H	T2/Xfr	Y/2/H	U	T3/Xr	h3/H
0.00000	0.00000	0.26688	1.48848	0.37470	0.91640	0.90730
0.05000	0.05097	0.25360	1.41338	0.37460	0.92050	0.91181
0.10000	0.10183	0.24032	1.33835	0.37451	0.92460	0.91633
0.15000	0.15258	0.22703	1.26341	0.37440	0.92872	0.92086
0.20000	0.20321	0.21373	1.18855	0.37431	0.93284	0.92541
0.25000	0.25376	0.20043	1.11375	0.37421	0.93697	0.92997
0.30000	0.30422	0.18711	1.03899	0.37411	0.94111	0.93454
0.35000	0.35456	0.17379	0.96435	0.37401	0.94526	0.93913
0.40000	0.40480	0.16046	0.88976	0.37391	0.94942	0.94373
0.45000	0.45494	0.14713	0.81524	0.37382	0.95359	0.94834
0.50000	0.50498	0.13379	0.74079	0.37372	0.95776	0.95297
0.55000	0.55491	0.12044	0.66642	0.37362	0.96195	0.95761
0.60000	0.60476	0.10709	0.59210	0.37352	0.96614	0.96227
0.65000	0.65450	0.09372	0.51786	0.37343	0.97034	0.96694
0.70000	0.70414	0.08036	0.44370	0.37333	0.97455	0.97162
0.75000	0.75369	0.06698	0.36958	0.37323	0.97877	0.97631
0.80000	0.80314	0.05360	0.29555	0.37313	0.98300	0.98102
0.85000	0.85250	0.04021	0.22155	0.37304	0.98724	0.98575
0.90000	0.90176	0.02681	0.14764	0.37294	0.99148	0.99049
0.95000	0.95093	0.01341	0.07378	0.37284	0.99574	0.99524
1.00000	1.00000	0.00000	0.00000	0.37275	1.00000	1.00000
OPTIMUM POSITION						
0.57979	0.58461	0.11249	0.62215	0.37356	0.96444	0.96038

Xf/Xfr = Forward period/total period = 0.25

ZONE 1		ZONE 2			ZONE 3	
T1/Xf	h1/H	T2/Xfr	Y/2/H	U	T3/Xr	h3/H
0.00000	0.00000	0.35266	1.66764	0.35445	0.86312	0.85144
0.05000	0.05070	0.33513	1.58409	0.35434	0.86983	0.85864
0.10000	0.10133	0.31759	1.50054	0.35423	0.87655	0.86586
0.15000	0.15187	0.30004	1.41705	0.35412	0.88328	0.87310
0.20000	0.20235	0.28247	1.33355	0.35401	0.89003	0.88037
0.25000	0.25276	0.26490	1.25007	0.35390	0.89680	0.88766
0.30000	0.30308	0.24732	1.16662	0.35380	0.90358	0.89498
0.35000	0.35332	0.22972	1.08320	0.35369	0.91037	0.90232
0.40000	0.40349	0.21212	0.99979	0.35358	0.91718	0.90969
0.45000	0.45359	0.19450	0.91640	0.35347	0.92400	0.91708
0.50000	0.50361	0.17688	0.83304	0.35336	0.93084	0.92449
0.55000	0.55357	0.15923	0.74966	0.35325	0.93769	0.93193
0.60000	0.60345	0.14158	0.66631	0.35315	0.94455	0.93940
0.65000	0.65325	0.12393	0.58300	0.35304	0.95143	0.94689
0.70000	0.70301	0.10625	0.49965	0.35293	0.95833	0.95440
0.75000	0.75267	0.08857	0.41637	0.35282	0.96524	0.96194
0.80000	0.80227	0.07088	0.33307	0.35272	0.97216	0.96950
0.85000	0.85180	0.05318	0.24979	0.35261	0.97910	0.97709
0.90000	0.90126	0.03546	0.16653	0.35250	0.98605	0.98470
0.95000	0.95067	0.01774	0.08325	0.35240	0.99302	0.99234
1.00000	1.00000	0.00000	0.00000	0.35229	1.00000	1.00000
OPTIMUM POSITION						
0.57784	0.58135	0.14941	0.70325	0.35319	0.94151	0.93609

Table F.4 (continued)

Xf/Xfr = Forward period/total period = 0.30

ZONE 1		ZONE 2			ZONE 3	
T1/Xf	h1/H	T2/Xfr	Y/2/H	U	T3/Xr	h3/H
0.00000	0.00000	0.44904	1.88330	0.33405	0.78709	0.77372
0.05000	0.05043	0.42673	1.78957	0.33393	0.79753	0.78465
0.10000	0.10080	0.40440	1.69582	0.33382	0.80799	0.79562
0.15000	0.15111	0.38207	1.60204	0.33371	0.81847	0.80663
0.20000	0.20139	0.35971	1.50819	0.33360	0.82898	0.81768
0.25000	0.25162	0.33735	1.41431	0.33348	0.83950	0.82877
0.30000	0.30181	0.31496	1.32036	0.33337	0.85005	0.83991
0.35000	0.35194	0.29256	1.22640	0.33326	0.86062	0.85108
0.40000	0.40204	0.27015	1.13237	0.33315	0.87122	0.86229
0.45000	0.45210	0.24772	1.03829	0.33304	0.88183	0.87354
0.50000	0.50211	0.22527	0.94416	0.33293	0.89247	0.88484
0.55000	0.55209	0.20281	0.84997	0.33282	0.90312	0.89617
0.60000	0.60201	0.18034	0.75576	0.33271	0.91380	0.90754
0.65000	0.65189	0.15785	0.66149	0.33260	0.92450	0.91896
0.70000	0.70174	0.13534	0.56714	0.33249	0.93522	0.93041
0.75000	0.75155	0.11282	0.47275	0.33238	0.94597	0.94191
0.80000	0.80131	0.09029	0.37832	0.33227	0.95673	0.95345
0.85000	0.85105	0.06774	0.28381	0.33216	0.96752	0.96502
0.90000	0.90073	0.04518	0.18928	0.33205	0.97832	0.97664
0.95000	0.95039	0.02259	0.09466	0.33194	0.98915	0.98830
1.00000	1.00000	0.00000	0.00000	0.33183	1.00000	1.00000
OPTIMUM POSITION						
0.57561	0.57766	0.19131	0.80173	0.33276	0.90859	0.90199

Xf/Xfr = Forward period/total period = 0.35

ZONE 1		ZONE 2			ZONE 3	
T1/Xf	h1/H	T2/Xfr	Y/2/H	U	T3/Xr	h3/H
0.00000	0.00000	0.55830	2.14683	0.31345	0.67954	0.66587
0.05000	0.05010	0.53056	2.04071	0.31335	0.69529	0.68199
0.10000	0.10018	0.50281	1.93447	0.31324	0.71106	0.69817
0.15000	0.15025	0.47504	1.82811	0.31313	0.72686	0.71441
0.20000	0.20031	0.44725	1.72161	0.31303	0.74270	0.73072
0.25000	0.25035	0.41944	1.61500	0.31292	0.75856	0.74708
0.30000	0.30039	0.39160	1.50824	0.31282	0.77446	0.76351
0.35000	0.35041	0.36376	1.40137	0.31271	0.79038	0.78000
0.40000	0.40043	0.33589	1.29436	0.31261	0.80633	0.79655
0.45000	0.45044	0.30799	1.18721	0.31250	0.82231	0.81316
0.50000	0.50043	0.28009	1.07997	0.31240	0.83832	0.82983
0.55000	0.55043	0.25216	0.97254	0.31230	0.85436	0.84658
0.60000	0.60039	0.22423	0.86506	0.31219	0.87043	0.86337
0.65000	0.65037	0.19626	0.75737	0.31209	0.88653	0.88023
0.70000	0.70033	0.16828	0.64961	0.31199	0.90265	0.89715
0.75000	0.75030	0.14027	0.54165	0.31188	0.91881	0.91414
0.80000	0.80024	0.11226	0.43360	0.31178	0.93499	0.93119
0.85000	0.85020	0.08422	0.32539	0.31168	0.95120	0.94830
0.90000	0.90013	0.05616	0.21707	0.31158	0.96744	0.96567
0.95000	0.95007	0.02809	0.10860	0.31147	0.98370	0.98270
1.00000	1.00000	0.00000	0.00000	0.31137	1.00000	1.00000
OPTIMUM POSITION						
0.57307	0.57348	0.23927	0.92297	0.31225	0.86177	0.85432

Xf/Xfr = Forward period/total period = 0.40

ZONE 1		ZONE 2			ZONE 3	
T1/Xf	h1/H	T2/Xfr	Y/2/H	U	T3/Xr	h3/H
0.00000	0.00000	0.68346	2.47462	0.29263	0.52756	0.51562
0.05000	0.04975	0.64948	2.35298	0.29254	0.55086	0.53900
0.10000	0.09950	0.61549	2.23118	0.29245	0.57418	0.56247
0.15000	0.14930	0.58147	2.10910	0.29236	0.59755	0.58603
0.20000	0.19911	0.54743	1.98686	0.29227	0.62095	0.60967
0.25000	0.24895	0.51337	1.86436	0.29219	0.64438	0.63341
0.30000	0.29882	0.47928	1.74163	0.29210	0.66786	0.65723
0.35000	0.34872	0.44518	1.61869	0.29201	0.69136	0.68114
0.40000	0.39863	0.41106	1.49555	0.29193	0.71490	0.70514
0.45000	0.44856	0.37693	1.37221	0.29184	0.73846	0.72922
0.50000	0.49854	0.34276	1.24860	0.29176	0.76207	0.75339
0.55000	0.54855	0.30857	1.12476	0.29167	0.78572	0.77766
0.60000	0.59860	0.27436	1.00067	0.29159	0.80940	0.80201
0.65000	0.64865	0.24015	0.87643	0.29150	0.83310	0.82644
0.70000	0.69876	0.20589	0.75189	0.29142	0.85685	0.85098
0.75000	0.74889	0.17162	0.62715	0.29133	0.88062	0.87559
0.80000	0.79903	0.13735	0.50222	0.29125	0.90443	0.90029
0.85000	0.84923	0.10304	0.37699	0.29116	0.92828	0.92509
0.90000	0.89945	0.06871	0.25156	0.29108	0.95215	0.94997
0.95000	0.94972	0.03436	0.12588	0.29100	0.97606	0.97494
1.00000	1.00000	0.00000	0.00000	0.29092	1.00000	1.00000
OPTIMUM POSITION						
0.57009	0.56866	0.29483	1.07492	0.29164	0.79523	0.78743

Xf/Xfr = Forward period/total period = 0.45

ZONE 1		ZONE 2			ZONE 3	
T1/Xf	h1/H	T2/Xfr	Y/2/H	U	T3/Xr	h3/H
0.00000	0.00000	0.82874	2.89144	0.27150	0.31141	0.30391
0.05000	0.04936	0.78742	2.74980	0.27145	0.34558	0.33767
0.10000	0.09874	0.74614	2.60804	0.27139	0.37974	0.37150
0.15000	0.14822	0.70483	2.46590	0.27134	0.41395	0.40545
0.20000	0.19773	0.66352	2.32354	0.27129	0.44816	0.43950
0.25000	0.24736	0.62215	2.18069	0.27123	0.48244	0.47370
0.30000	0.29704	0.58079	2.03758	0.27118	0.51674	0.50799
0.35000	0.34676	0.53943	1.89424	0.27113	0.55105	0.54238
0.40000	0.39661	0.49800	1.75039	0.27107	0.58543	0.57692
0.45000	0.44648	0.45660	1.60638	0.27102	0.61981	0.61154
0.50000	0.49643	0.41518	1.46199	0.27097	0.65422	0.64629
0.55000	0.54646	0.37373	1.31727	0.27092	0.68867	0.68115
0.60000	0.59656	0.33227	1.17224	0.27086	0.72315	0.71612
0.65000	0.64674	0.29078	1.02684	0.27081	0.75766	0.75121
0.70000	0.69699	0.24929	0.88114	0.27076	0.79220	0.78641
0.75000	0.74730	0.20778	0.73512	0.27071	0.82676	0.82172
0.80000	0.79770	0.16625	0.58873	0.27066	0.86135	0.85715
0.85000	0.84816	0.12471	0.44206	0.27061	0.89597	0.89269
0.90000	0.89868	0.08317	0.29509	0.27056	0.93061	0.92833
0.95000	0.94931	0.04159	0.14769	0.27051	0.96530	0.96412
1.00000	1.00000	0.00000	0.00000	0.27046	1.00000	1.00000
OPTIMUM POSITION						
0.56658	0.56306	0.35999	1.26923	0.27090	0.70010	0.69273

Table F.4 (continued)

Xf/Xfr = Forward period/total period = 0.50

ZONE 1		ZONE 2			ZONE 3	
T1/Xf	h1/H	T2/Xfr	Y/2/H	U	T3/Xr	h3/H
0.00000	0.00000	1.00000	3.43596	0.25000	0.00000	0.00000
0.05000	0.04888	0.95000	3.26803	0.25000	0.05000	0.04888
0.10000	0.09787	0.90000	3.09970	0.25000	0.10000	0.09787
0.15000	0.14697	0.85000	2.93096	0.25000	0.15000	0.14697
0.20000	0.19620	0.80000	2.76183	0.25000	0.20000	0.19620
0.25000	0.24554	0.75000	2.59230	0.25000	0.25000	0.24554
0.30000	0.29500	0.70000	2.42236	0.25000	0.30000	0.29500
0.35000	0.34457	0.65000	2.25202	0.25000	0.35000	0.34457
0.40000	0.39427	0.60000	2.08127	0.25000	0.40000	0.39427
0.45000	0.44408	0.55000	1.91011	0.25000	0.45000	0.44408
0.50000	0.49402	0.50000	1.73854	0.25000	0.50000	0.49402
0.55000	0.54407	0.45000	1.56656	0.25000	0.55000	0.54407
0.60000	0.59424	0.40000	1.39417	0.25000	0.60000	0.59424
0.65000	0.64453	0.35000	1.22136	0.25000	0.65000	0.64453
0.70000	0.69495	0.30000	1.04814	0.25000	0.70000	0.69495
0.75000	0.74548	0.25000	0.87450	0.25000	0.75000	0.74548
0.80000	0.79614	0.20000	0.70045	0.25000	0.80000	0.79614
0.85000	0.84692	0.15000	0.52597	0.25000	0.85000	0.84692
0.90000	0.89782	0.10000	0.35107	0.25000	0.90000	0.89782
0.95000	0.94885	0.05000	0.17575	0.25000	0.95000	0.94885
1.00000	1.00000	0.00000	0.00000	0.25000	1.00000	1.00000
OPTIMUM POSITION						
0.56244	0.55654	0.43756	1.52372	0.25000	0.56244	0.55654

Xf/Xfr = Forward period/total period = 0.55

ZONE 1		ZONE 2			ZONE 3	
T1/Xf	h1/H	T2/Xfr	Y/2/H	U	T3/Xr	h3/H
0.31142	0.30392	0.82872	2.89139	0.22850	0.00000	0.00000
0.34556	0.33765	0.78744	2.74988	0.22855	0.05000	0.04933
0.37974	0.37150	0.74614	2.60804	0.22861	0.10000	0.09874
0.41393	0.40544	0.70484	2.46595	0.22866	0.15000	0.14820
0.44817	0.43951	0.66351	2.32350	0.22871	0.20000	0.19774
0.48244	0.47369	0.62216	2.18070	0.22877	0.25000	0.24736
0.51673	0.50798	0.58080	2.03763	0.22882	0.30000	0.29702
0.55106	0.54238	0.53942	1.89422	0.22887	0.35000	0.34677
0.58542	0.57691	0.49802	1.75046	0.22893	0.40000	0.39659
0.61981	0.61154	0.45660	1.60637	0.22898	0.45000	0.44648
0.65422	0.64628	0.41518	1.46201	0.22903	0.50000	0.49643
0.68867	0.68114	0.37373	1.31728	0.22908	0.55000	0.54646
0.72314	0.71577	0.33227	1.17226	0.22914	0.60000	0.59655
0.75766	0.75121	0.29078	1.02684	0.22919	0.65000	0.64674
0.79219	0.78640	0.24930	0.88116	0.22924	0.70000	0.69698
0.82675	0.82171	0.20779	0.73515	0.22929	0.75000	0.74729
0.86134	0.85714	0.16627	0.58879	0.22934	0.80000	0.79768
0.89596	0.89268	0.12473	0.44211	0.22939	0.85000	0.84814
0.93061	0.92834	0.08316	0.29507	0.22944	0.90000	0.89869
0.96529	0.96411	0.04160	0.14773	0.22949	0.95000	0.94930
1.00000	1.00000	0.00000	0.00000	0.22954	1.00000	1.00000
OPTIMUM POSITION						
0.70012	0.69275	0.35996	1.26913	0.22910	0.56661	0.56310

Xf/Xfr = Forward period/total period = 0.60

ZONE 1		ZONE 2			ZONE 3	
T1/Xf	h1/H	T2/Xfr	Y/2/H	U	T3/Xr	h3/H
0.52755	0.51561	0.68347	2.47465	0.20737	0.00000	0.00000
0.55085	0.53900	0.64949	2.35302	0.20746	0.05000	0.04974
0.57418	0.56247	0.61549	2.23117	0.20755	0.10000	0.09951
0.59755	0.58603	0.58147	2.10911	0.20764	0.15000	0.14930
0.62095	0.60967	0.54743	1.98685	0.20773	0.20000	0.19911
0.64439	0.63341	0.51336	1.86434	0.20781	0.25000	0.24896
0.66786	0.65723	0.47928	1.74164	0.20790	0.30000	0.29882
0.69136	0.68114	0.44519	1.61872	0.20799	0.35000	0.34871
0.71489	0.70513	0.41107	1.49558	0.20807	0.40000	0.39862
0.73847	0.72922	0.37691	1.37216	0.20816	0.45000	0.44858
0.76208	0.75340	0.34275	1.24857	0.20824	0.50000	0.49856
0.78572	0.77766	0.30857	1.12473	0.20833	0.55000	0.54856
0.80940	0.80201	0.27436	1.00067	0.20841	0.60000	0.59860
0.83311	0.82645	0.24013	0.87639	0.20850	0.65000	0.64866
0.85685	0.85097	0.20589	0.75190	0.20858	0.70000	0.69875
0.88063	0.87559	0.17162	0.62714	0.20867	0.75000	0.74889
0.90444	0.90030	0.13734	0.50217	0.20875	0.80000	0.79905
0.92827	0.92509	0.10304	0.37699	0.20884	0.85000	0.84923
0.95215	0.94997	0.06871	0.25155	0.20892	0.90000	0.89946
0.97606	0.97494	0.03437	0.12590	0.20900	0.95000	0.94971
1.00000	1.00000	0.00000	0.00000	0.20908	1.00000	1.00000
OPTIMUM POSITION						
0.79524	0.78744	0.29482	1.07489	0.20836	0.57010	0.56868

Xf/Xfr = Forward period/total period = 0.65

ZONE 1		ZONE 2			ZONE 3	
T1/Xf	h1/H	T2/Xfr	Y/2/H	U	T3/Xr	h3/H
0.67954	0.66587	0.55830	2.14682	0.18655	0.00000	0.00000
0.69529	0.68199	0.53057	2.04072	0.18665	0.05000	0.05009
0.71106	0.69817	0.50281	1.93447	0.18676	0.10000	0.10018
0.72687	0.71442	0.47504	1.82809	0.18687	0.15000	0.15025
0.74270	0.73072	0.44724	1.72160	0.18697	0.20000	0.20031
0.75857	0.74709	0.41943	1.61497	0.18708	0.25000	0.25036
0.77446	0.76351	0.39160	1.50822	0.18718	0.30000	0.30040
0.79038	0.78000	0.36376	1.40137	0.18729	0.35000	0.35041
0.80633	0.79655	0.33588	1.29435	0.18739	0.40000	0.40044
0.82231	0.81316	0.30800	1.18722	0.18750	0.45000	0.45044
0.83832	0.82984	0.28009	1.07995	0.18760	0.50000	0.50044
0.85436	0.84657	0.25216	0.97256	0.18770	0.55000	0.55042
0.87043	0.86337	0.22422	0.86503	0.18781	0.60000	0.60040
0.88653	0.88023	0.19626	0.75738	0.18791	0.65000	0.65037
0.90265	0.89715	0.16828	0.64960	0.18801	0.70000	0.70033
0.91880	0.91414	0.14028	0.54167	0.18812	0.75000	0.75029
0.93499	0.93119	0.11226	0.43361	0.18822	0.80000	0.80024
0.95120	0.94830	0.08422	0.32539	0.18832	0.85000	0.85019
0.96744	0.96547	0.05617	0.21708	0.18842	0.90000	0.90013
0.98370	0.98270	0.02809	0.10861	0.18853	0.95000	0.95007
1.00000	1.00000	0.00000	0.00000	0.18863	1.00000	1.00000
OPTIMUM POSITION						
0.86177	0.85432	0.23927	0.92296	0.18775	0.57307	0.57348

Table F.4 (continued)

Xf/Xfr = Forward period/total period = 0.70

ZONE 1		ZONE 2			ZONE 3	
T1/Xf	h1/H	T2/Xfr	Y/2/H	U	T3/Xr	h3/H
0.78709	0.77372	0.44904	1.88330	0.16595	0.00000	0.00000
0.79753	0.78465	0.42673	1.78960	0.16607	0.05000	0.05041
0.80799	0.79562	0.40441	1.69584	0.16618	0.10000	0.10078
0.81847	0.80663	0.38207	1.60204	0.16629	0.15000	0.15111
0.82898	0.81768	0.35971	1.50819	0.16640	0.20000	0.20139
0.83950	0.82877	0.33735	1.41432	0.16652	0.25000	0.25161
0.85005	0.83990	0.31496	1.32038	0.16663	0.30000	0.30180
0.86062	0.85108	0.29256	1.22639	0.16674	0.35000	0.35195
0.87121	0.86229	0.27015	1.13238	0.16685	0.40000	0.40204
0.88183	0.87354	0.24772	1.03829	0.16696	0.45000	0.45210
0.89246	0.88484	0.22528	0.94417	0.16707	0.50000	0.50211
0.90312	0.89617	0.20282	0.84999	0.16718	0.55000	0.55208
0.91380	0.90754	0.18034	0.75575	0.16729	0.60000	0.60201
0.92450	0.91896	0.15785	0.66147	0.16740	0.65000	0.65190
0.93522	0.93041	0.13534	0.56714	0.16751	0.70000	0.70174
0.94597	0.94191	0.11282	0.47275	0.16762	0.75000	0.75155
0.95673	0.95345	0.09029	0.37832	0.16773	0.80000	0.80131
0.96751	0.96502	0.06774	0.28383	0.16784	0.85000	0.85104
0.97832	0.97664	0.04518	0.18928	0.16795	0.90000	0.90073
0.98915	0.98830	0.02260	0.09467	0.16806	0.95000	0.95038
1.00000	1.00000	0.00000	0.00000	0.16817	1.00000	1.00000
OPTIMUM POSITION						
0.90859	0.90199	0.19130	0.80170	0.16724	0.57562	0.57767

Xf/Xfr = Forward period/total period = 0.75

ZONE 1		ZONE 2			ZONE 3	
T1/Xf	h1/H	T2/Xfr	Y/2/H	U	T3/Xr	h3/H
0.86312	0.85144	0.35266	1.66762	0.14555	0.00000	0.00000
0.86983	0.85864	0.33513	1.58407	0.14566	0.05000	0.05070
0.87655	0.86586	0.31759	1.50056	0.14577	0.10000	0.10132
0.88328	0.87310	0.30004	1.41703	0.14588	0.15000	0.15188
0.89003	0.88037	0.28248	1.33355	0.14599	0.20000	0.20235
0.89680	0.88766	0.26490	1.25008	0.14610	0.25000	0.25275
0.90358	0.89498	0.24732	1.16663	0.14620	0.30000	0.30307
0.91037	0.90232	0.22972	1.08321	0.14631	0.35000	0.35332
0.91718	0.90969	0.21212	0.99979	0.14642	0.40000	0.40349
0.92400	0.91708	0.19450	0.91639	0.14653	0.45000	0.45359
0.93084	0.92450	0.17687	0.83302	0.14664	0.50000	0.50362
0.93769	0.93193	0.15924	0.74966	0.14675	0.55000	0.55357
0.94455	0.93940	0.14159	0.66632	0.14685	0.60000	0.60345
0.95143	0.94689	0.12393	0.58299	0.14696	0.65000	0.65326
0.95833	0.95440	0.10625	0.49966	0.14707	0.70000	0.70300
0.96524	0.96194	0.08857	0.41635	0.14718	0.75000	0.75268
0.97216	0.96950	0.07088	0.33306	0.14728	0.80000	0.80228
0.97910	0.97709	0.05318	0.24978	0.14739	0.85000	0.85181
0.98605	0.98470	0.03546	0.16652	0.14750	0.90000	0.90127
0.99302	0.99234	0.01773	0.08324	0.14760	0.95000	0.95067
1.00000	1.00000	0.00000	0.00000	0.14771	1.00000	1.00000
OPTIMUM POSITION						
0.94151	0.93609	0.14941	0.70325	0.14681	0.57784	0.58135

Xf/Xfr = Forward period/total period = 0.80

ZONE 1		ZONE 2			ZONE 3	
T1/Xf	h1/H	T2/Xfr	Y/2/H	U	T3/Xr	h3/H
0.91640	0.90730	0.26688	1.48848	0.12530	0.00000	0.00000
0.92050	0.91181	0.25360	1.41337	0.12540	0.05000	0.05097
0.92460	0.91633	0.24032	1.33836	0.12550	0.10000	0.10182
0.92872	0.92086	0.22703	1.26341	0.12560	0.15000	0.15257
0.93284	0.92541	0.21373	1.18854	0.12569	0.20000	0.20322
0.93697	0.92997	0.20042	1.11373	0.12579	0.25000	0.25377
0.94111	0.93454	0.18711	1.03900	0.12589	0.30000	0.30422
0.94526	0.93913	0.17379	0.96435	0.12599	0.35000	0.35455
0.94942	0.94373	0.16047	0.88977	0.12609	0.40000	0.40479
0.95359	0.94834	0.14713	0.81525	0.12618	0.45000	0.45493
0.95776	0.95297	0.13379	0.74080	0.12628	0.50000	0.50497
0.96195	0.95761	0.12044	0.66643	0.12638	0.55000	0.55491
0.96614	0.96227	0.10709	0.59211	0.12648	0.60000	0.60475
0.97034	0.96694	0.09373	0.51787	0.12657	0.65000	0.65449
0.97455	0.97162	0.08036	0.44369	0.12667	0.70000	0.70414
0.97877	0.97631	0.06698	0.36958	0.12677	0.75000	0.75369
0.98300	0.98102	0.05360	0.29554	0.12687	0.80000	0.80314
0.98724	0.98575	0.04021	0.22157	0.12696	0.85000	0.85249
0.99148	0.99049	0.02681	0.14764	0.12706	0.90000	0.90176
0.99574	0.99524	0.01341	0.07379	0.12716	0.95000	0.95093
1.00000	1.00000	0.00000	0.00000	0.12725	1.00000	1.00000
OPTIMUM POSITION						
0.96444	0.96038	0.11249	0.62214	0.12644	0.57979	0.58462

Xf/Xfr = Forward period/total period = 0.85

ZONE 1		ZONE 2			ZONE 3	
T1/Xf	h1/H	T2/Xfr	Y/2/H	U	T3/Xr	h3/H
0.95313	0.94705	0.18984	1.33755	0.10494	0.00000	0.00000
0.95542	0.94963	0.18039	1.26958	0.10502	0.05000	0.05122
0.95773	0.95222	0.17093	1.20172	0.10511	0.10000	0.10232
0.96003	0.95482	0.16147	1.13399	0.10519	0.15000	0.15328
0.96235	0.95742	0.15201	1.06638	0.10528	0.20000	0.20410
0.96466	0.96003	0.14254	0.99889	0.10537	0.25000	0.25480
0.96698	0.96265	0.13306	0.93151	0.10545	0.30000	0.30536
0.96931	0.96527	0.12358	0.86424	0.10554	0.35000	0.35580
0.97164	0.96790	0.11410	0.79709	0.10562	0.40000	0.40610
0.97398	0.97054	0.10462	0.73008	0.10571	0.45000	0.45627
0.97632	0.97318	0.09513	0.66315	0.10579	0.50000	0.50633
0.97867	0.97583	0.08563	0.59635	0.10588	0.55000	0.55624
0.98102	0.97849	0.07613	0.52965	0.10596	0.60000	0.60603
0.98337	0.98116	0.06663	0.46307	0.10604	0.65000	0.65571
0.98573	0.98383	0.05713	0.39659	0.10613	0.70000	0.70526
0.98810	0.98651	0.04761	0.33022	0.10621	0.75000	0.75469
0.99047	0.98919	0.03810	0.26397	0.10630	0.80000	0.80399
0.99285	0.99188	0.02858	0.19782	0.10638	0.85000	0.85317
0.99523	0.99458	0.01906	0.13177	0.10647	0.90000	0.90224
0.99761	0.99729	0.00953	0.06584	0.10655	0.95000	0.95117
1.00000	1.00000	0.00000	0.00000	0.10663	1.00000	1.00000
OPTIMUM POSITION						
0.98014	0.97750	0.07967	0.55443	0.10593	0.58142	0.58755

APPENDIX G

Software

The software available with this book consists of three executable computer programs which run under Microsoft MSDOS on a personal computer, to perform some of the cam geometry calculations described in the book.

RECOMMENDED MINIMUM COMPUTER CONFIGURATION

Intel 80286 or equivalent processor.
VGA graphics monitor screen.
MSDOS version 2.1 or later operating system.

CAM-DISK

This program calculates the profile geometry of disc cams with either roller or flat-faced followers, with either translating or swinging motions.

CAM-CYL

This program calculates the profile geometry of cylindrical cams with roller followers, with either translating or swinging motions.

CAMMERGE

This program merges the DXF files that are produced by either of the above programs to make a single DXF file for several segments of the *same cam profile*.

MOTION LAWS

The programs process any SCCA cam law (including Parabolic, Simple Harmonic, Mod. Sine, Mod. Trap. and Cycloidal) with or without a central constant velocity zone. The motions may be either symmetrical or asymmetrical.

They also deal with polynomials, non-reflective and reflective with or without a central constant velocity zone, symmetrical or asymmetrical.

The cam laws may be combined with constant velocity to facilitate cam law blending as described in the book.

Program input for CAM-DISK and CAM-CYL programs is by entering the following data in boxes on a screen menu:

Follower type: roller or flat-faced.
Follower motion: swinging or translating.
Cam law (motion law) parameters.
Input stroke (period).
Output stroke.
Number of calculation steps in the period.
Dimensions of the mechanism (in millimetres or inches).

Program output consists of a drawing of the cam profile segment on screen and two ASCII file(s):

(1)   A file of the cam profile drawing in DXF format for importing into an AutoCad drawing or a compatible CAD drawing system.
(2)   A data file containing:
Pitch curve co-ordinates.
Profile co-ordinates.
Cutter curve co-ordinates.
All pressure angles and indication of the maximum value.
All radii of curvature and indication of the minimum value.

Complete cam profile. The profiles of *all* the motion segments of a cam can be shown on screen simultaneously and the separate DXF files can be merged into a single DXF file so that the complete profile can be imported into a CAD drawing where it can be modified, moved, scaled up or down, etc. as required, using the CAD editing facilities.

The programs enable the rapid optimisation of a cam design by repetitive trials. The on-screen drawing, and the critical values of pressure angle and radius of curvature (also on screen), allow the operator immediately to re-run the calculations with revised input data until satisfied with the design. The data files produced can be printed out to give complete hard-copy geometric data of the cam profile. The DXF file(s) can be used to facilitate the production of a CAD drawing of the cam.

The cutter curve co-ordinates in the ASCII data files are useful as input to Numerical Control profile machining programs. The data can be extracted from the files with standard text processing techniques and further manipulated as necessary. The number and variety of N.C. systems available are too great for particular methods of data processing to be included here.

Typical examples of the input menus, on-screen drawings and output data files are shown. The data in the input screen menu boxes are user input. The text to the right of those boxes is either confirmation of the user's choice or information that is helpful in context.

CAM-DISK PROGRAM OPERATION

The program is activated by typing CAM-DISK and pressing [Return]. The first two screens give brief information about the program and about the data input procedure on the following screens. These are best explained by way of examples.

The first input screen, Fig. G.1, is for the first motion segment of a disc cam with a swinging roller follower and Mod. Sine cam law. A menu of SCCA cam law options has appeared when SCCA cam law type (1) was selected from which Mod Sine (5) has been chosen.

The second screen, Fig. G.2, shows the cam law name and parameters, which have been automatically filled in because option 5 was chosen, but these can be overridden from the keyboard if required. Also on this screen is the menu for the cam geometry data, which is entered from the keyboard. A different set of data would have been requested if different follower motion and type options had been chosen.

The last option on the screen is to continue the processing with the existing data, to modify the data already entered or to exit the program forthwith. Any or all of the input data may be changed at this stage or it

```
JOB/SEGMENT REFERENCE  [DEMONSTRATION                                    ]
FOLLOWER MOTION   (1 = SWINGING  or  2 = TRANSLATING)   [1]  SWINGING MOTION
FOLLOWER TYPE     (1 = ROLLER or  2 = FLAT-FACED)       [1]  ROLLER FOLLOWER
UNITS OF MEASURE  (1 = millimetres or  2 = inches)      [1]  millimetres
CAM LAW TYPE      (1 = SCCA or  2 = POLYNOMIAL)         [1]  SCCA cam law

                                  SCCA STANDARD CAM LAW MENU
                                 ENTER A NUMBER TO CHOOSE A LAW  [5]
                                 1   CONSTANT VELOCITY or DWELL
                                 2   PARABOLIC
                                 3   SIMPLE HARMONIC
                                 4   MODIFIED TRAPEZOID
                                 5   MODIFIED SINE
                                 6   CYCLOIDAL
                                 7   MSC.50
                                 8   CYCC.50
                                 9   OTHER SCCA CAM LAW
```

Fig. G.1 CAM-DISK input screen 1

```
JOB/SEGMENT REFERENCE  [DEMONSTRATION                                 ]
FOLLOWER MOTION   (1 = SWINGING  or  2 = TRANSLATING)   [1]  SWINGING MOTION
FOLLOWER TYPE     (1 = ROLLER or  2 = FLAT-FACED)       [1]  ROLLER FOLLOWER
UNITS OF MEASURE  (1 = millimetres or  2 = inches)      [1]  millimetres
CAM LAW TYPE      (1 = SCCA or  2 = POLYNOMIAL)         [1]  SCCA cam law
CAM LAW NAME    [MODIFIED SINE                      ]   0 ≤ a+b+c ≤ 1
SCCA PARAMETERS     a [.25  ]  b [     ]  c [.75  ]     CV zone = 1-a-b-c
ASYMMETRY FACTOR         AF           [.5   ]  0=no acceleration, 1=no deceleration
CAM START ANGLE          G (degrees)  [30   ]  Motion start after cam datum
INPUT STROKE (PERIOD)    A (degrees)  [120  ]  Positive leading, negative trailing
FOLLOWER START ANGLE     D (degrees)  [50   ]  Common centre-line to staret posn.
TOTAL OUTPUT STROKE      B (degrees)  [30   ]  Positive rising, negative falling
CV PART OF OUTPUT      Bcv (degrees)  [     ]  Only if cam law is combined with CV
CENTRE DISTANCE          C (mm)       [100  ]  Centre of cam to follower pivot
FOLLOWER ARM LENGTH      F (mm)       [105  ]  Roller centre to follower pivot
DIAMETER OF ROLLER *     d (mm)       [25   ]  DIAMETER OF CUTTER (mm)  38.1
NUMBER OF CALCULATION POINTS          [30   ]  Cutter may be different from roller
PATH for output files, if not current directory [                              ]
FILE NAME** [DEMO1  ]  (extensions .DAT and >DXF are added automatically)
* To plot an internal cam track use negative roller and cutter diameters
** Files with extension .DAT and .DXF (from which a cam drawing can be
produced) are written with this base name.
CONTINUE, EXIT OR MODIFY DATA? (C/X/M)   [C]
```

Fig. G.2 CAM-DISK input screen 2

may be modified at the end of processing for a re-run: as many re-runs as necessary may be made until the user is satisfied with the design.

The table, Fig. G.7, is a print-out of DEMO1.DAT file produced from the input data shown on the first two screens. There are only 30 calculation points for the purpose of the demonstration (every 4° of cam rotation) which is adequate for making design decisions based on the on-screen drawing and the calculated critical values of pressure angle and curvatures. For accurate profile cutting, however, it may be necessary to use steps of 1° or even 0.25°. Such small steps may also be necessary to obtain accurate critical values if there is a risk of interference. In either case, once the operator is satisfied with a cam segment design, the program can be re-run immediately with a new number of steps before moving on.

The third screen, Fig. G.3, appears after processing the first profile segment is complete and when option 'A' – add a segment – has been selected. In this case a polynomial cam law has been selected (option 2) and a menu of such laws has appeared on screen. The standard option 3 is chosen.

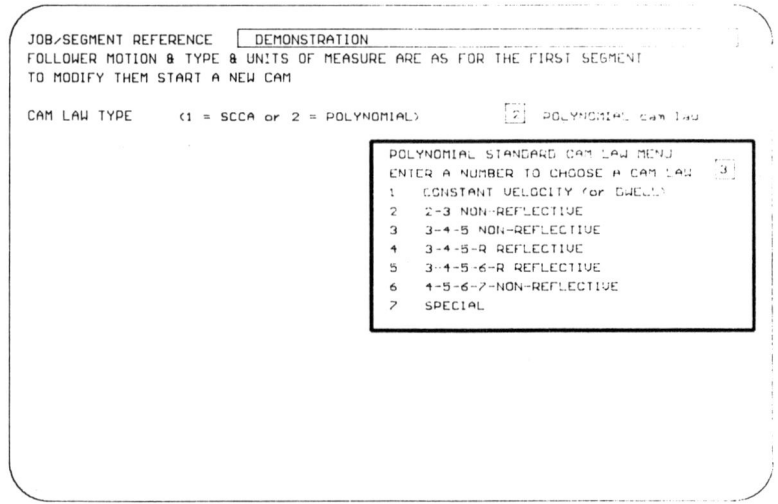

Fig. G.3   CAM-DISK input screen 3

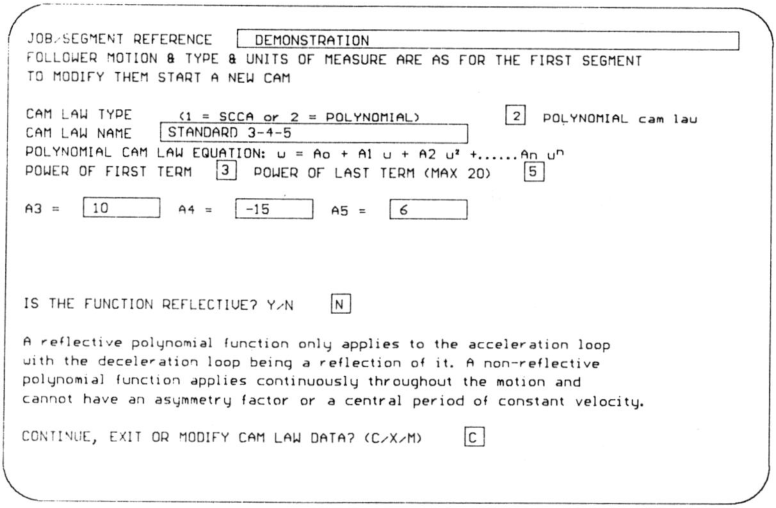

```
JOB/SEGMENT REFERENCE    DEMONSTRATION
FOLLOWER MOTION & TYPE & UNITS OF MEASURE ARE AS FOR THE FIRST SEGMENT
TO MODIFY THEM START A NEW CAM

CAM LAW TYPE      (1 = SCCA or 2 = POLYNOMIAL)        2   POLYNOMIAL cam law
CAM LAW NAME     STANDARD 3-4-5
POLYNOMIAL CAM LAW EQUATION: u = Ao + A1 u + A2 u² +......An uⁿ
POWER OF FIRST TERM   3   POWER OF LAST TERM (MAX 20)   5

A3 =   10       A4 =   -15      A5 =   6



IS THE FUNCTION REFLECTIVE? Y/N     N

A reflective polynomial function only applies to the acceleration loop
with the deceleration loop being a reflection of it. A non-reflective
polynomial function applies continuously throughout the motion and
cannot have an asymmetry factor or a central period of constant velocity.

CONTINUE, EXIT OR MODIFY CAM LAW DATA? (C/X/M)    C
```

Fig. G.4   CAM-DISK input screen 4

The next screen, Fig. G.4, automatically displays the cam law name and polynomial coefficients, but these can be changed if required. Once the cam law data is accepted with option 'C' – continue – the cam geometry data can be entered on the next screen, Fig. G.5, for the new segment, but this time previous segment data is displayed for convenience. For example, the new cam start angle is the sum of the old cam start angle and old input stroke, and the new follower start angle is the sum of the old follower start angle and the old output stroke, both of which are displayed on screen. In most cases, when a sequence of consecutive cam segments are being processed, this is convenient. The new strokes are re-set to zero. All input data displayed, including the pre-set values, can be changed if required. That geometry data which *must* remain unchanged throughout the whole of the cam profile is not available for input to any but the first segment.

In this case the new strokes have been entered manually, and because the second segment is a falling motion a negative output stroke has been used. Positive input strokes (periods) have been used for all segments because this mechanism has a leading follower arm: negative values

```
JOB/SEGMENT REFERENCE    [ DEMONSTRATION                                    ]
FOLLOWER MOTION & TYPE & UNITS OF MEASURE ARE AS FOR THE FIRST SEGMENT
TO MODIFY THEM START A NEW CAM

CAM LAW TYPE         (1 = SCCA or  2 = POLYNOMIAL)        [2]  POLYNOMIAL cam law
CAM LAW NAME      [ STANDARD 3-4-5                    ]
POLYNOMIAL CAM LAW EQUATION: u = Ao + A1 u + A2 u² +......An uⁿ

CAM START ANGLE         G (degrees)  [ 150 ]   Motion start after cam datum
INPUT STROKE (PERIOD)   A (degrees)  [  90 ]   Positive leading, negative trailing
FOLLOWER START ANGLE    D (degrees)  [  80 ]   Common centre-line to staret posn.
TOTAL OUTPUT STROKE     B (degrees)  [ -30 ]   Positive rising, negative falling
CV PART OF OUTPUT     Bcv (degrees)  [     ]   Only if cam law is combined with CV

MECHANISM GEOMETRY, CUTTER (& ROLLER) ARE AS FOR THE FIRST SEGMENT

NUMBER OF CALCULATION POINTS       [  30 ]
PATH for output files, if not current directory  [                         ]
FILE NAME**  [ DEMO2 ]   (extensions .DAT and .DXF are added automatically)
* To plot an internal cam track use negative roller and cutter diameters
** Files with extension .DAT and .DXF (from which a cam drawing can be
produced) are written with this base name.
CONTINUE, EXIT OR MODIFY DATA? (C/X/M)         [C]
```

Fig. G.5   CAM-DISK input screen 5

would have been used for a trailing follower arm. For conventional cams all the input strokes must have the same sign.

The output data for the second segment is shown in the table, Fig. G.8. Note that a slightly different file name has been used, which avoids overwriting the first segment files.

The third segment (with yet another slightly different file name) is a dwell period, which is processed with the 'A' – add a segment – option as for segment 2. It is usually unnecessary to process dwell segments of a cam, as these are very simple to draw and require very little processing for numerical control machining. The pressure angles and radii of curvature are constant throughout a dwell period and are usually the same as at the end points of adjacent segments. The input screen and output data table for this segment are shown in Figs G.6 and G.9.

The various segments of the same cam do not have to be processed in any particular order, but a consecutive sequence is a little easier to deal with. All that is necessary is that their file names be different. The program prevents the *inadvertent* overwriting of existing files, but does allow it when required, e.g. during the modification of a particular segment.

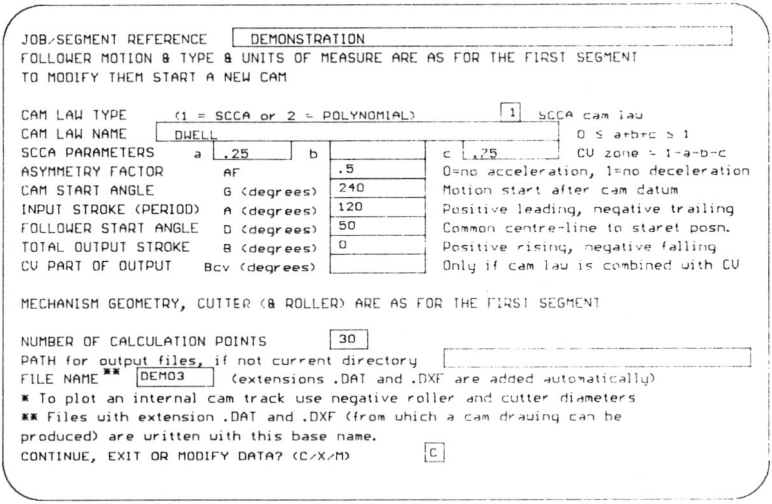

```
JOB/SEGMENT REFERENCE    [ DEMONSTRATION                                        ]
FOLLOWER MOTION & TYPE & UNITS OF MEASURE ARE AS FOR THE FIRST SEGMENT
TO MODIFY THEM START A NEW CAM

CAM LAW TYPE        (1 = SCCA or 2 = POLYNOMIAL)       [ 1 ]  SCCA cam law
CAM LAW NAME        [ DUELL                              ]  0 ≤ a+b+c ≤ 1
SCCA PARAMETERS     a [ .25    ]  b           c [ .25    ]  CV zone = 1-a-b-c
ASYMMETRY FACTOR        AF         [ .5   ]      0=no acceleration, 1=no deceleration
CAM START ANGLE     G (degrees)    [ 240  ]      Motion start after cam datum
INPUT STROKE (PERIOD)  A (degrees) [ 120  ]      Positive leading, negative trailing
FOLLOWER START ANGLE   D (degrees) [ 50   ]      Common centre-line to staret posn.
TOTAL OUTPUT STROKE    B (degrees) [ 0    ]      Positive rising, negative falling
CV PART OF OUTPUT   Bcv (degrees)  [      ]      Only if cam law is combined with CV

MECHANISM GEOMETRY, CUTTER (& ROLLER) ARE AS FOR THE FIRST SEGMENT

NUMBER OF CALCULATION POINTS       [ 30 ]
PATH for output files, if not current directory  [                              ]
FILE NAME ** [ DEMO3 ]  (extensions .DAT and .DXF are added automatically)
* To plot an internal cam track use negative roller and cutter diameters
** Files with extension .DAT and .DXF (from which a cam drawing can be
produced) are written with this base name.
CONTINUE, EXIT OR MODIFY DATA? (C/X/M)       [ C ]
```

Fig. G.6   CAM-DISK input screen 6

Should it be necessary to modify the basic mechanism geometry, such as the centre distance in this case, then the first segment of a sequence must be modified using the 'S' – start a new cam – options and all subsequent segments must be re-done with the 'A' option.

It is interesting to note that, although the curvature is acceptable in all three profile segments, the worst pressure angle in the second profile segment is $-54.88°$, which is quite high but may be considered acceptable for a falling motion in many applications. The drawing screen display has not only shown this value, but has also highlighted the fact that it is high, and has briefly recommended possible modifications to the input data to improve the situation. At that stage it is simple to opt for 'M' – modify input data – for a re-run of that segment only.

Another interesting fact that is shown by the on-screen drawing is that part of the cam profile overlaps the pivot shaft centre. This means that the designer must decide on either a cantilevered follower roller on an arm that is mounted on the end of the pivot shaft, or if he prefers a pivot shaft that can pass the cam he must use a very much larger follower roller or a smaller follower start angle to decrease the cam size.

Alterations to any segment or to the basic mechanism geometry can be

```
========================== PROGRAM CAM-DISK =================================
JOB/REFERENCE: DEMONSTRATION
FILE: DEMO1.DAT (DRAWING FILE DEMO1.DXF), RUN AT 17:39 on 11-02-95
EXTERNAL DISC CAM WITH SWINGING ROLLER FOLLOWER (LEADING)
============================================================================
MODIFIED SINE, RISING MOTION
SCCA CAMLAW PARAMETERS   a =  .25   b = 0    c =  .75
ASYMMETRY FACTOR   AF =  .5 : COEFF. OF VELOCITY  Cv =  1.759603
COEFF. OF ACCELN. Ca =  5.527958 : COEFF. OF DECELN.  Cd =  5.527958
CAM START ANGLE        G (degrees)  30
INPUT STROKE           A (degrees) 120     ----------------------------
FOLLOWER START ANGLE   D (degrees)  50     ¦ CRITICAL RESULTS MARKED ¦
TOTAL OUTPUT STROKE    B (degrees)  30     ¦ THUS ** ->,  ** <-      ¦
CV PART OF OUTPUT    Bcv (degrees)   0     ¦                         ¦
CENTRE DISTANCE        C (mm)      100     ¦ UNITS OF MEASURE ARE     ¦
FOLLOWER ARM LENGTH    F (mm)      105     ¦ DEGREES AND MILLIMETRES  ¦
ROLLER DIAMETER        d (mm)       25     ----------------------------
CUTTER DIAMETER          (mm)     38.1
NUMBER OF CALCULATION POINTS ON PROFILE  30
============================================================================
 CAM       FOLLOWER¦ PRESSR. RADIUS OF¦         POLAR CO-ORDINATES
ROTATION  POSITION¦ ANGLE  CURVATURE¦ CAM PROFILE  ¦ CUTTER PITCH CURVE
 ALPHA      BETA  ¦  PHI      Rc     ¦   R     THETA ¦   Rcut    THETAcut

  0.000   50.000   -27.99    74.255    74.255   97.994     93.305    97.994
  4.000   50.013   -27.41   109.946    74.277  102.085     93.326   102.038
  8.000   50.099   -25.77   182.481    74.427  106.319     93.461   106.137
 12.000   50.321   -23.32   332.179    74.826  110.616     93.803   110.233
 16.000   50.715   -20.49   480.631    75.546  114.877     94.404   114.264
 20.000   51.293   -17.70   525.818    76.602  119.023     95.283   118.184
 24.000   52.052   -15.08   482.172    77.980  123.025     96.435   121.982
 28.000   52.986   -12.75   388.058    79.654  126.870     97.854   125.651
 32.000   54.085   -10.78   297.330    81.592  130.552     99.526   129.190
 36.000   55.337    -9.25   228.805    83.753  134.072    101.434   132.607
 40.000   56.725    -8.20   180.719    86.093  137.443    103.556   135.910
 44.000   58.232    -7.65   147.234    88.568  140.678    105.863   139.113
 48.000   59.836    -7.61   123.547    91.134  143.799    108.321   142.232
 52.000   61.515    -8.07   106.437    93.750  146.824    110.895   145.284
 56.000   63.245    -8.99    93.859    96.375  149.776    113.544   148.286
 60.000   65.000   -10.35    84.523    98.973  152.677    116.226   151.256
 64.000   66.755   -12.09    77.613   101.509  155.547    118.899   154.211
 68.000   68.485   -14.16    72.602   103.955  158.406    121.519   157.170
 72.000   70.164   -16.50    69.137   106.282  161.275    124.048   160.147
 76.000   71.768   -19.05    66.975   108.468  164.172    126.448   163.160
 80.000   73.275   -21.71    65.934   110.492  167.115    128.684   166.223
**84.000   74.663    24.44    65.864<- 112.337  170.121    130.727   169.349
 88.000   75.915   -27.15    66.631   113.986  173.204    132.553   172.551
 92.000   77.014   -29.80    68.102   115.426  176.376    134.142   175.839
 96.000   77.948   -32.32    70.148   116.645  179.618    135.481   179.222
100.000   78.707   -34.68    72.635   117.634  183.026    136.560   182.704
104.000   79.285   -36.85    75.434   118.388  186.516    137.376   186.291
108.000   79.679   -38.80    79.634   118.903  190.120    137.930   189.984
112.000   79.901   -40.35    88.011   119.194  193.850    138.240   193.787
116.000   79.987   -41.33   100.914   119.310  197.706    138.360   197.690
**120.000   80.000 ->-41.66  119.327   119.327  201.665    138.377   201.665
============================================================================
```

Fig. G.7   CAM-DISK data print-out: first segment

carried out while the program is running, and their effects assessed immediately: repeated trials enable a satisfactory mechanism design to be reached quite soon.

Similar procedures to the one described above can be used for a swinging flat-faced follower, a translating roller follower or a translating

```
======================== PROGRAM CAM-DISK ============================
JOB/REFERENCE: DEMONSTRATION
FILE: DEMO2.DAT (DRAWING FILE DEMO2.DXF), RUN AT 17:41 on 11-02-95
EXTERNAL DISC CAM WITH SWINGING ROLLER FOLLOWER (LEADING)
======================================================================
STANDARD 3-4-5 POLYNOMIAL (NON-REFLECTIVE), FALLING MOTION
w =  +10.0000 u^3  -15.0000 u^4  +6.0000 u^5
ASYMMETRY FACTOR  AF =  .5 : COEFF. OF VELOCITY  Cv =  1.875
COEFF. OF ACCELN. Ca =  5.773503 : COEFF. OF DECELN.  Cd =  5.773503
CAM START ANGLE         G (degrees)  150
INPUT STROKE            A (degrees)   90        -----------------------
FOLLOWER START ANGLE    D (degrees)   80       | CRITICAL RESULTS MARKED |
TOTAL OUTPUT STROKE     B (degrees)  -30       | THUS ** ->,  ** <-      |
CV PART OF OUTPUT     Bcv (degrees)    0        |                         |
CENTRE DISTANCE         C (mm)       100       | UNITS OF MEASURE ARE     |
FOLLOWER ARM LENGTH     F (mm)       105       | DEGREES AND MILLIMETRES  |
ROLLER DIAMETER         d (mm)        25        -----------------------
CUTTER DIAMETER           (mm)      38.1
NUMBER OF CALCULATION POINTS ON PROFILE  30
======================================================================
CAM        FOLLOWER| PRESSR. RADIUS OF|       POLAR CO-ORDINATES
ROTATION POSITION| ANGLE  CURVATURE|  CAM PROFILE  | CUTTER PITCH CURVE
 ALPHA     BETA  |  PHI      Rc     |   R     THETA |  Rcut    THETAcut
------------------------------------------------------------------------
    0.000   80.000  -41.66  119.327   119.327  201.665   138.377  201.665
    3.000   79.989  -42.01   95.173   119.313  204.634   138.363  204.651
    6.000   79.920  -42.92   82.584   119.221  207.572   138.266  207.636
    9.000   79.743  -44.20   76.090   118.990  210.522   138.019  210.655
   12.000   79.424  -45.68   73.118   118.575  213.519   137.569  213.734
** 15.000   78.935  -47.22   72.276<-  117.941  216.588   136.879  216.893
   18.000   78.262  -48.72   72.725   117.064  219.748   135.924  220.147
   21.000   77.398  -50.12   73.922   115.928  223.007   134.693  223.501
   24.000   76.344  -51.36   75.507   114.530  226.368   133.183  226.957
   27.000   75.108  -52.43   77.250   112.870  229.830   131.401  230.512
   30.000   73.704  -53.30   79.019   110.961  233.386   129.363  234.160
   33.000   72.152  -53.99   80.762   108.822  237.027   127.093  237.891
   36.000   70.477  -54.48   82.493   106.480  240.742   124.621  241.693
   39.000   68.706  -54.78   84.282   103.966  244.517   121.986  245.552
** 42.000   66.869  ->-54.78  86.252   101.319  248.340   119.230  249.455
   45.000   65.000  -54.78   88.590    98.581  252.197   116.401  253.385
   48.000   63.131  -54.45   91.562    95.797  256.075   113.550  257.329
   51.000   61.294  -53.90   95.560    93.016  259.963   110.731  261.271
   54.000   59.523  -53.09  101.173    90.285  263.851   107.998  265.198
   57.000   57.848  -52.01  109.341    87.651  267.733   105.402  269.100
   60.000   56.296  -50.62  121.651    85.160  271.603   102.994  272.964
   63.000   54.892  -48.90  140.974    82.854  275.460   100.816  276.785
   66.000   53.656  -46.84  172.904    80.774  279.303    98.901  280.555
   69.000   52.602  -44.43  229.206    78.955  283.130    97.276  284.270
   72.000   51.738  -41.70  335.357    77.426  286.937    95.952  287.924
   75.000   51.065  -38.74  531.055    76.208  290.712    94.926  291.511
   78.000   50.576   35.70  716.935    75.310  294.432    94.185  295.020
   81.000   50.257  -32.81  525.229    74.720  298.062    93.700  298.436
   84.000   50.080   30.34  261.882    74.397  301.560    93.430  301.745
   87.000   50.011  -28.63  132.557    74.273  304.882    93.322  304.932
   90.000   50.000  -27.99   74.255    74.255  307.994    93.305  307.994
========================================================================
```

Fig. G.8 CAM-DISK data print-out: second segment

flat-faced follower by choosing the appropriate options on the input data screens. The units of measure can be either metric or imperial, but must of course be the same for all segments of the same cam.

Internal cam profiles can also be generated by the same program. If required, both internal and external profiles can be generated for the

```
======================= PROGRAM CAM-DISK ==============================
JOB/REFERENCE: DEMONSTRATION
FILE: DEMO3.DAT (DRAWING FILE DEMO3.DXF), RUN AT 17:44 on 11-02-95
EXTERNAL DISC CAM WITH SWINGING ROLLER FOLLOWER (LEADING)
-----------------------------------------------------------------------
DWELL
CCCA CAMLAW PARAMETERS    a =   0    b =  0    c =  0
ASYMMETRY FACTOR    AF =  .5 : COEFF. OF VELOCITY   Cv =  1
COEFF. OF ACCELN.  Ca =   0 : COEFF. OF DECELN.  Cd =  0
CAM START ANGLE          G (degrees)    240
INPUT STROKE             A (degrees)    120
FOLLOWER START ANGLE     D (degrees)    50      | CRITICAL RESULTS MARKED |
TOTAL OUTPUT STROKE      B (degrees)    0       | THUS ** >,   ** <       |
CV PART OF OUTPUT        Bcv (degrees)  0       |                         |
CENTRE DISTANCE          C (mm)         100     | UNITS OF MEASURE ARE     |
FOLLOWER ARM LENGTH      F (mm)         105     | DEGREES AND MILLIMETRES  |
ROLLER DIAMETER          d (mm)         25      ...........................
CUTTER DIAMETER          (mm)           38.1
NUMBER OF CALCULATION POINTS ON PROFILE   30
=======================================================================
     CAM       FOLLOWER| PRESSR. RADIUS OF|       POLAR CO-ORDINATES
     ROTATION POSITION| ANGLE  CURVATURE| CAM PROFILE   | CUTTER PITCH CURVE
     ALPHA    BETA    | PHI    Rc       | R      THETA  | Rcut   THETAcut
**   0.000   50.000  > -27.99  74.255<-  74.255  307.994  93.305  307.994
     4.000   50.000    -27.99  74.255    74.255  311.994  93.305  311.994
     8.000   50.000    -27.99  74.255    74.255  315.994  93.305  315.994
    12.000   50.000    -27.99  74.255    74.255  319.994  93.305  319.994
    16.000   50.000    -27.99  74.255    74.255  323.994  93.305  323.994
    20.000   50.000    -27.99  74.255    74.255  327.994  93.305  327.994
    24.000   50.000    -27.99  74.255    74.255  331.994  93.305  331.994
    28.000   50.000    -27.99  74.255    74.255  335.994  93.305  335.994
    32.000   50.000    -27.99  74.255    74.255  339.994  93.305  339.994
    36.000   50.000    -27.99  74.255    74.255  343.994  93.305  343.994
    40.000   50.000    -27.99  74.255    74.255  347.994  93.305  347.994
    44.000   50.000    -27.99  74.255    74.255  351.994  93.305  351.994
    48.000   50.000    -27.99  74.255    74.255  355.994  93.305  355.994
    52.000   50.000    -27.99  74.255    74.255  359.994  93.305  359.994
    56.000   50.000    -27.99  74.255    74.255  363.994  93.305  363.994
    60.000   50.000    -27.99  71.255    74.255  367.994  93.305  367.994
    64.000   50.000    -27.99  74.255    74.255  371.994  93.305  371.994
    68.000   50.000    -27.99  74.255    74.255  375.994  93.305  375.994
    72.000   50.000    -27.99  74.255    74.255  379.994  93.305  379.994
    76.000   50.000    -27.99  74.255    71.255  383.994  93.305  383.994
    80.000   50.000    -27.99  74.255    74.255  387.994  93.305  387.994
    84.000   50.000    -27.99  74.255    74.255  391.994  93.305  391.994
    88.000   50.000    -27.99  74.255    74.255  395.994  93.305  395.994
    92.000   50.000    -27.99  74.255    74.255  399.994  93.305  399.994
    96.000   50.000    -27.99  74.255    74.255  403.994  93.305  403.994
   100.000   50.000    -27.99  74.255    74.255  407.994  93.305  407.994
   104.000   50.000     27.99  74.255    74.255  411.994  93.305  411.994
   108.000   50.000    -27.99  74.255    74.255  415.994  93.305  415.994
   112.000   50.000    -27.99  74.255    74.255  419.994  93.305  419.994
   116.000   50.000    -27.99  74.255    74.255  423.994  93.305  423.994
   120.000   50.000    -27.99  74.255    74.255  427.994  93.305  427.994
=======================================================================
```

Fig. G.9   CAM-DISK data print-out: third segment

same segment to analyse the geometry of an *enclosed track* cam, the data being output to separate files, but both profiles appearing together on the on-screen drawing display.

The cutter co-ordinates output to the .DAT files relate to the cutter *pitch* curve, i.e. the path of the cutter centre. For an open track cam the

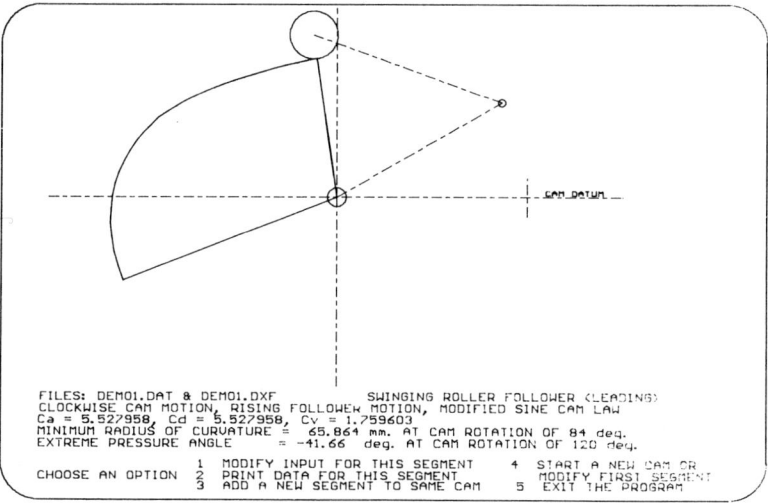

Fig. G.10   CAM-DISK drawing screen 1

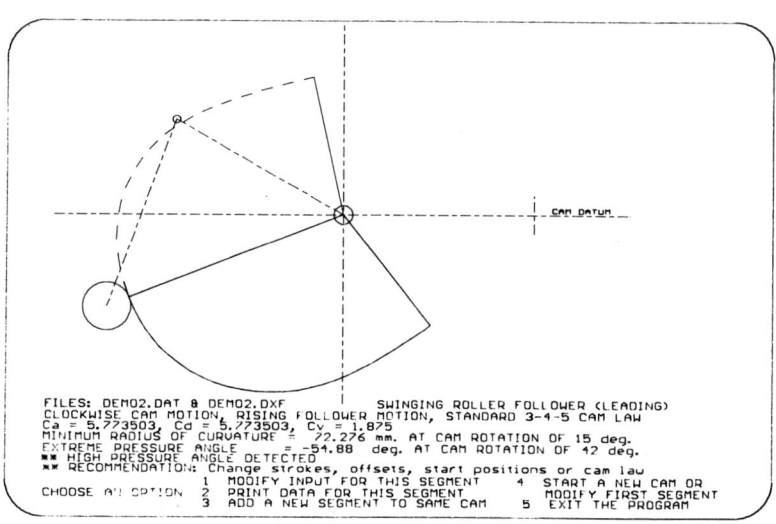

Fig. G.11   CAM-DISK drawing screen 2

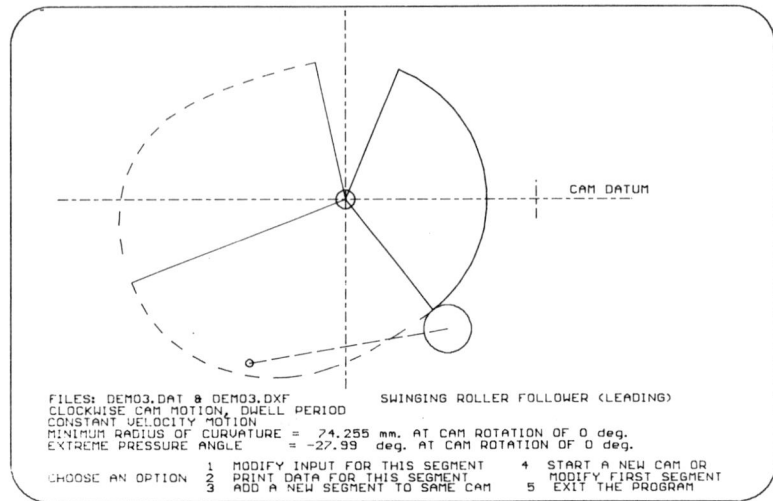

Fig. G.12   CAM-DISK drawing screen 3

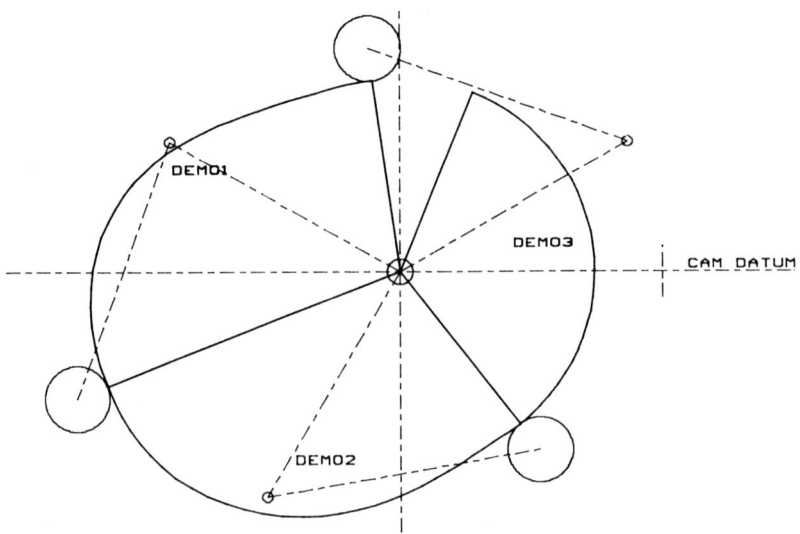

Fig. G.13   CAM-DISK CAD drawing of whole profile

cutter may be a different size from the follower roller, subject to certain limitations (see Chapter 15) but for an enclosed track cam it must be the same size or smaller. The cutter diameter is therefore input separately from the roller size.

Any number of cams can be processed in one run. The run is terminated by pressing 'X' at the appropriate prompt, or pressing 'Ctrl+X' or 'Ctrl+C' or 'Ctrl+Esc' at any other time.

The individual output screen displays of the three motion segments are shown in Figs G.10, G.11 and G.12. The relevant DXF files for these are similar, but without showing the critical values. By running the CAM-MERGE program a DXF file is produced (DEMO.DXF in this example – different again from the other file names) which combines these segments into a complete profile. That DXF file was imported into a CAD system to produce the drawing shown in Fig. G.13.

CAM-CYL PROGRAM OPERATION

The CAM-CYL program operates for cylindrical cams in a similar way that CAM-DISK does for disc cams, but without the option of flat-faced followers. It is started by typing CAM-CYL at the DOS prompt and pressing [Return]. However, this program produces on-screen drawings of profile *developments* for both sides of an enclosed track simultaneously. The .DAT and .DXF files generated also have data for both track walls. If the cam is an open track one then the unwanted side can be ignored, both in the on-screen drawings and in the data files. In the case of a CAD drawing produced from the .DXF file the unwanted track can be deleted using the CAD editing facility.

With cylindrical cams the pressure angle and curvature at the top of the track (at the outside diameter of the cam) are different from those at the bottom of the track. It is essential therefore to use the program to analyse the properties at the bottom of the track where they have the worst values. Both top and bottom track profiles (at outside and root diameters of the cam) can be processed in the same session, of course, using the common mechanism geometry. The cam cutting co-ordinates in the .DAT files are identical for both top and bottom of the track, so either set may be used. It is absolutely essential that the cutter be virtually the same size and *shape* as the follower roller (see Chapter 15).

It is convenient to use CAM-CYL to design a cylindrical *indexing cam* because the on-screen drawing gives a good visual indication of the tooth thickness between tracks – an important design criterion.

CAMMERGE PROGRAM OPERATION

The CAMMERGE program, which is started by typing 'CAMMERGE' [Return] at the DOS prompt, simply reads the individual .DXF files produced by the other two programs and merges them into a single new .DXF file. It is used to combine several profile segments of a cam so that a drawing of the whole profile can be easily imported from that file into a CAD drawing system.

Only source file names and a destination file name are entered as input data. All source files must be in the current directory when running the program. The basic source file names, without the .DXF extension, are entered in any order, terminating with a blank entry. The destination file name, without extension, is then entered. The files are processed and the destination file, with extension .DXF added, is written to the current directory.

As an aid to file recognition it is convenient when using CAM-DISK or CAM-CYL to use similar (although unique) file names for the various segments of a cam, such as CAM254A, CAM254B, CAM254C, etc. and CAM254 (.DXF) for the merged file.

Fig. G.14 shows a cylindrical indexing cam where the track developments were produced by CAM-CYL and CAMMERGE and then edited in a CAD drawing system. It is immediately obvious from the track layouts, which are true to scale, that there is a more than adequate tooth (or blade) thickness at both the outside diameter and at the root diameter of the cam. This indicates that there is scope for a substantial improvement in the load capacity of the mechanism by increasing the size of the rollers by at least 50%, leaving all the other dimensions unchanged. The screens displayed during the run of CAM-CYL revealed that neither the pressure angle nor the curvature were excessive: facts that could be verified from the data files.

The maximum pressure angle is not affected by a change in the roller diameter, and the minimum radius of curvature is simply altered by the difference between the two roller radii. However, the width of the cam may have to be altered slightly in order to ensure that there will still be a reasonable overlap at the control transfer positions.

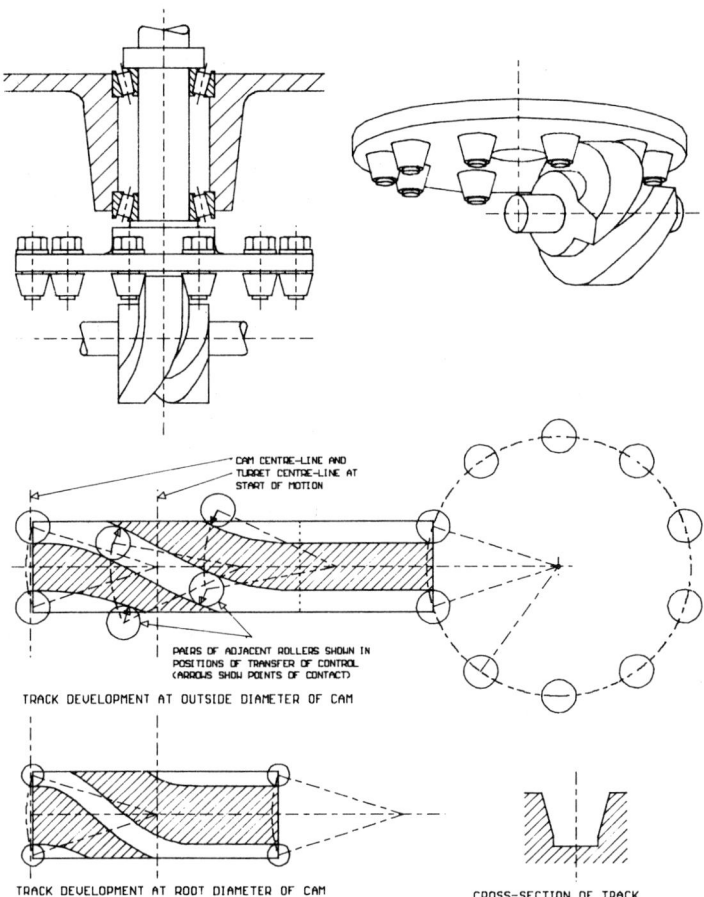

Fig. G.14   Cylinder cam indexing mechanism with track development

APPENDIX H

Recommended Further Reading

Of the many books already published in the fields of cam design and tribology the following few are recommended for those who require a more thorough treatment of some aspects of the subject than is given here.

CHEN, F. Y.	Mechanics & Design of Cam Mechanisms. Pergamon, 1982.
ESDU International	Mechanisms, Volume 3, Cams (16 'items') ESDU International, 1979 onwards.
JENSEN, P.	Cam Design & Manufacture. Dekker, 1987.
STOLARSKY, T. A.	Tribology in Machine Design. Heinemann Newnes, 1990.

INDEX